OSHA

STALLCUP'S®
CONSTRUCTION
REGULATIONS
SIMPLIFIED

Written by Mike Bahr and James Stallcup, Sr.
Edited by James Stallcup, Jr.
Design, graphics and layout by Billy G. Stallcup

Notice Concerning Liability: Publication of this work is for the purpose of circulating information and opinion among those concerned for fire and electrical safety and related subjects. While every effort has been made to achieve a work of high quality, neither the NFPA nor the authors and contributors to this work guarantee the accuracy or completeness of or assume any liability in connection with the information and opinions contained in this work. The NFPA and the authors and contributors shall in no event be liable for any personal injury, property, or other damages of any nature whatsoever, whether special, indirect, consequential, or compensatory, directly or indirectly resulting from the publication, use of or reliance upon this work.

This work is published with the understanding that the NFPA and the authors and contributors to this work are supplying information and opinion but are not attempting to render engineering or other professional services. If such services are required, the assistance of an appropriate professional should be sought.

NFPA No.: SOC02
ISBN: 0-87765-511-1
Library of Congress Card Catalog No.: 2003116671

Printed in the United States of America
04   05   06   07   08   5   4   3   2   1

Table of Contents

Introduction

The OSHA Standards present necessary and constant considerations by an employer for their implementation. Stallcup has authored a series of publications to make these regulations easier to understand, by correlating these rules and regulations with other codes and Standards, such as the NEC, NFPA 70E, and the NESC. And by illustrating their application, the adherence by employers and workers is promoted.

This book is an excerpt from OSHA - Stallcup's® Construction Regulations and represents only the portion that covers the electrical standards for construction.

Other OSHA publications by Stallcup:

OSHA - Stallcup's® Electrical Regulations Simplified
OSHA - Stallcup's® High-Voltage and Telecommunications Regulations Simplified
OSHA - Stallcup's® Electrical Construction Regulations Simplified
OSHA - Stallcup's® Construction Regulations Simplified

General

Protecting construction workers from injury and disease is among the greatest challenges in occupational safety and health. The following construction concerns are the reason this book was developed to help assist in promoting worker safety and a safe work environment:

- There are more than 7 million persons employed in construction, representing 5 percent of the labor force. About 1.5 million are self-employed.

- Of 636,000 construction companies, 90 percent employ fewer than 50 workers. Few of these construction companies have formal safety and health programs.

- There are 1000 construction workers killed on the job each year. These construction fatalities are more than in any other industry.

- 15 percent of workers' compensation costs are spent on construction injuries.

COMPLEX TASKS

The following are complex tasks for assuring safety and health in construction:

- Involving short term worksites,
- Changing hazards and
- Multiple operations and crews working in close proximity.

SAFE WORKSITES

The keys to safe worksites are knowledge, understanding, application and communications. The following are keys to achieving construction safety goals:

- Knowing what is to be done,

- Understanding the reason why certain actions must be taken and the consequences of not taking them and

- Making sure all personnel are aware of the necessity for safety procedures in their work assignments. No job is so important or so urgent that time cannot be taken to perform it safety.

STANDARDS

Environmental Protection, Safety, and Health Protection Standards require compliance with the Occupational Safety and Health Administration (OSHA) regulations Title 29 Code of Federal Regulations (CFR) 1926, "Safety and Health Regulations for Construction," and related parts of 29 CFR 1910, "Occupational Safety and Health Standards." This guide is based on these OSHA regulations and, where appropriate, incorporates additional standards, codes and work practices that are recognized and accepted by the construction industry.

In most cases, the regulation, standard or code has been reworded to make it more easily understood and user friendly. Illustrations have been incorporated into the guide to draw attention to a subject or to clarify the text. The Construction Safety Guide is not intended to be an all-encompassing comprehensive safety document. It is intended to be a reference guide for the most common safety concerns encountered during construction activities.

SUMMARY

It is recommended that this safety guide for construction be used to the fullest as a tool for promoting and encouraging construction site safety, worker safety, the control or elimination of property damage and protection of the environment.

2

Advanced planning and quality site preparations are important for a safe, well-organized construction site and will ensure the efficient layout for the site, movement of material, storage and servicing. A well-arranged and well-controlled construction site will also positively influence attitudes towards the attitudes of employees that will improve work quality and safety of site.

The chapter will be helpful to those conducting the initial site preparation and layout.

REVIEW OF OVERALL CONSTRUCTION SITE

The employer should visit the site location and review upcoming plans and procedures with the site supervisor. Using a plot plan, the employer and site supervisor should evaluate the site for the following items:

- Access
- Neighbors
- Site security
- Facilities for worker parking
- Storage for heavy equipment
- Site superintendent's office
- Location for material storage
- On-site roadways
- Utilities
- Safety and health program and inspections
- Other items

ACCESS

The following must be considered by the employer and site supervisor when determining access to the construction site:

- How many points of access to the site are needed and available?

- Are access points near a highway, busy street or business area?

- Can access to the site be controlled or does access need to be controlled?

- How will the number and location of access points influence the accessibility of the site to workers and equipment?

- What access actions or controls will be needed?

NEIGHBORS

The following must be considered by the employer and site supervisor when determining operating procedures close to neighbors:

- How far from the site are the closest neighbors and how will the construction operations affect them?

- Will any hazardous materials, which may affect the neighbors, be used on the construction site?

- Are appropriate control measures in place?

- Are there any nuisance operations that may need control, such as dust, noise or odor?

- Are there other potential problems that may affect the neighbors?

- A scheduled visit to the neighbors may be necessary to work out or avoid problems.

SITE SECURITY

The following must be considered by the employer and site supervisor for site security:

- Will it be necessary to provide a security fence around the site?

- Will it be necessary to have security patrol after hours?

- What other security precautions must be taken?

FACILITIES FOR WORKER PARKING

The following must be considered by the employer and site supervisor for facilities when determining worker parking:

- Is the site large enough to allow worker parking inside the site, or will workers have to park in adjacent areas?

- Will outside parking be a problem to neighbor's or cause congestion?

- Parking problems should be discussed with workers.

STORAGE OF HEAVY EQUIPMENT

The following must be considered by the employer and site supervisor for storage of heavy equipment:

- Where will heavy equipment, accessories and parts be stored?

- Where and how will the equipment be fueled and serviced?

- Will there be an on-site service or fueling station?

SITE SUPERINTENDENT'S OFFICE

The answers to the following may help in determining the location of the superintendent's office:

- What will be the ideal location from which to supervise the site?

- How will the office be used?

- Will deliveries be made through contact with the superintendent's office?

- Will employees report and leave via the superintendent's office?

- Will visitors report through the superintendent's office?

- Will the office be used as a communications center?

LOCATION FOR MATERIAL STORAGE

The following factors should be taken into account when material storage areas are identified:

- Type and quantity of materials,

- Type and quantity of generated waste and

- Requirements to meet fire and other safety needs and regulations and restrictions by the owner

ON-SITE ROADWAYS

Roadways within the site must be adequate to accommodate delivery of materials, movement of large equipment and emergency fire and other rescue vehicles. Roadways should be built so as not to interfere with other site operations, such as excavations or electrical service installations. If there is a major street or highway adjacent to the site, a connection to these roadways may be necessary; and all local traffic control ordinances must be observed. All appropriate traffic control and warning signs shall be posted where necessary.

UTILITIES

The following should be considered with reference to utilities on the construction site:

- Where are utility connections located, especially potable water, fire water mains or hydrants and electrical? Where are the best locations to tie in branch circuits for serving the site and eliminating damage?

- What location requires the most use of water?

- If utilities come into the site over an active work area or roadway on which equipment over 16 ft. high is moved, all equipment operators shall be warned and signs posted as necessary.

SAFETY AND HEALTH PROGRAM AND INSPECTIONS

The following shall be considered by the employer and site supervisor for safety and health program and inspections:

- Is there an established and written safety and health program ready for site use?

- Have arrangements been made for available competent personnel to conduct frequent and regular site safety inspections of all areas and operations?

OTHER ITEMS

Construction activities usually require some special coordination. Activities such as security and control present special problems and should be carefully planned. Other items of concern are:

- Supply, servicing and placement of toilets and washing facilities;

- Supply and issue of general-use safety items, glasses, hard hats, etc.;

- Site-posting information (warning signs, site name, safety and other postings, receiving, etc.);

- Drainage control;

- Site history, use or misuse;

- Information about hazards, irritants, and toxic plants and instruction in first aid or methods of medical treatment for personnel clearing the site.

SUMMARY

Understanding and taking the above actions should help in the prearrangement of a construction site and help in planning toward avoidance of major and minor items that could be a conflict to the site operations. It is understood that not every item of site-preparation concern is included in the above list; however, the list is a key to the start of and continuance to a safe construction site.

3

Main General Safety and Health Provisions

Under the requirements of the Occupational Safety Health Act (OSHA) all employers have the responsibility to initiate and maintain an effective health and safety program. OSHA's recommended "Safety and Health Program Management Guidelines" issued in 1989 can provide a blueprint for employers who are seeking guidance on how to effectively manage and protect worker safety and health. The four main elements of an effective occupational safety and health program are:

- Management commitment and employee involvement,

- Worksite analysis,

- Hazard prevention and control and

- Safety and health training.

Those elements encompass principles such as establishing and communicating clear goals of a safety and health management program; conducting worksite examinations to identify existing hazards and the conditions under which changes might occur; effectively designing the job site or job to prevent hazards; and providing essential training to address the safety and health responsibilities of both management and employees.

OCCUPATIONAL SAFETY AND HEALTH PROGRAM

The following elements of an effective occupational safety and health program are:

- Management commitment and leadership
- Assignment of responsibility
- Identification and control of hazards
- Training and education
- Recordkeeping and hazard analysis
- First aid and medical assistance

MANAGEMENT COMMITMENT AND LEADERSHIP

The following must be considered for management and leadership:

- Policy statement: goals established, issued and communicated to employees.
- Program revised annually.
- Participation in safety meetings, inspections; agenda items in meetings.
- Commitment of resources is adequate.
- Safety rules and procedures incorporated into site operations.
- Management observes safety rules.

ASSIGNMENT OF RESPONSIBILITY

The following must be considered for assignment of responsibilities:

- Safety designee on site, knowledgeable and accountable.
- Supervisors (including foremen), safety and health responsibilities understood.
- Employees adhere to safety rules.

IDENTIFICATION AND CONTROL OF HAZARDS

The following must be considered for identification and control of hazards:

- Periodic site safety inspection program involves supervisors.
- Preventative controls in place (personal protective equipment, maintenance and engineering controls).
- Action taken to address hazards.
- Safety Committee, where appropriate.
- Technical references available.
- Enforcement procedures by management.

TRAINING AND EDUCATION
OSHA 1926.21

OSHA Required Training
1926.20(b)(2) and (4)
1926.21(b)(1) thru (b)(6)(i) and (ii)

The following must be considered for training and education:

- Supervisors receive basic training.
- Specialized training taken when needed.
- Employee training program exists, is ongoing and is effective.

RECORDKEEPING AND HAZARD ANALYSIS

The following must be considered for recordkeeping and hazard analysis:

- Records maintained of employee illnesses/injuries, and posted.
- Supervisors perform accident investigations, determine causes and propose corrective action.
- Injuries, near misses and illnesses are evaluated for trends, similar causes; corrective action initiated.

FIRST AID AND MEDICAL ASSISTANCE
OSHA 1926.23

OSHA Required Training
1926.50(c)

The following must be considered for first aid and medical assistance:

- First aid supplies and medical service available.
- Employees informed of medical results.
- Emergency procedures and training, where necessary.

CONFINED SPACES
OSHA 1926.21(b)(6)(ii)

The following must be considered for confined spaces:

- Confined space requirements
- Training requirements for confined spaces
- Ten rules for confined-spaced entry

CONFINED SPACE REQUIREMENTS
OSHA 1910.146

OSHA Required Training
1910.146(g)(1) and (g)(2)(i) thru (iv)(3); (4) and (k)(1)(i) thru (iv)

Because of the nature of confined spaces, the possibility of dangerous atmospheres and the type of operations performed in them, confined-space workers and supervisors shall take every possible precaution. **(See Figure 3-1)**

Pits more than 4 ft. deep, tanks, vessels, silos, utility vaults, pipelines, ducts, sewers, digesters, tunnels, process tanks and vats are examples of confined spaces. Confined spaces are areas with one or more of the following characteristics:

- Atmosphere that contains or has the potential to contain less than 19.5 percent oxygen;

- Atmosphere that is or can become flammable, combustible, explosive or toxic;

- Area unprotected from entry of water, gas, sand, earth, ore, grain, gravel, coal, biological waste, radiation, corrosive chemicals or other hazardous substances present; ventilation inadequate (natural or mechanical); entry or exit restricted; or

- Design not suitable for normal occupancy.

Confined spaces found on construction sites typically include pits or excavations more than 4 ft. deep with a potential for accumulating hazardous materials or vapors or for displacing oxygen. **(See Figure 3-2)**

All confined spaces shall be identified by a sign stating "DANGER - CONFINED SPACE," with other appropriate warning or information. **(See Figure 3-3)**

Figure 3-1. This illustration shows examples of confined spaces.

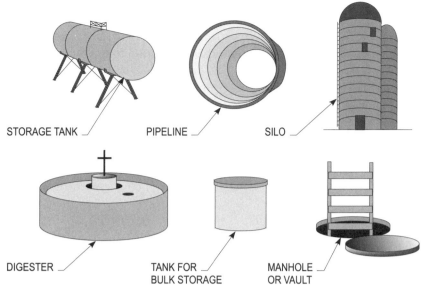

STORAGE TANK   PIPELINE   SILO

DIGESTER   TANK FOR BULK STORAGE   MANHOLE OR VAULT

CONFINED SPACE REQUIREMENTS
OSHA 1910.146

Figure 3-2. This illustration shows that confined spaces found on construction sites typically include pits or excavations.

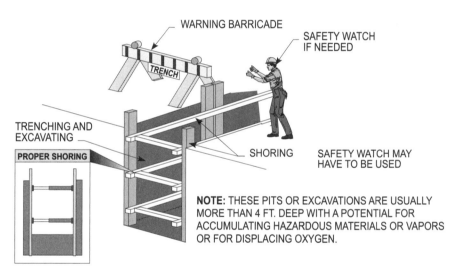

WARNING BARRICADE

SAFETY WATCH IF NEEDED

TRENCH

TRENCHING AND EXCAVATING

PROPER SHORING

SHORING

SAFETY WATCH MAY HAVE TO BE USED

NOTE: THESE PITS OR EXCAVATIONS ARE USUALLY MORE THAN 4 FT. DEEP WITH A POTENTIAL FOR ACCUMULATING HAZARDOUS MATERIALS OR VAPORS OR FOR DISPLACING OXYGEN.

CONFINED SPACE REQUIREMENTS
OSHA 1910.146

Figure 3-3. This illustration shows that confined spaces shall be identified by a sign stating "DANGER – CONFINED SPACES," with other appropriate warning or information.

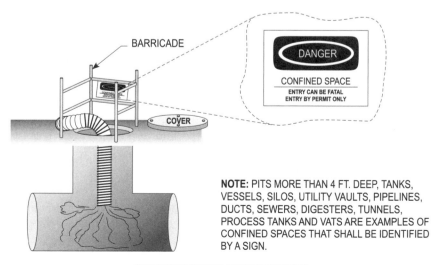

BARRICADE

DANGER

CONFINED SPACE
ENTRY CAN BE FATAL
ENTRY BY PERMIT ONLY

COVER

NOTE: PITS MORE THAN 4 FT. DEEP, TANKS, VESSELS, SILOS, UTILITY VAULTS, PIPELINES, DUCTS, SEWERS, DIGESTERS, TUNNELS, PROCESS TANKS AND VATS ARE EXAMPLES OF CONFINED SPACES THAT SHALL BE IDENTIFIED BY A SIGN.

CONFINED SPACE REQUIREMENTS
OSHA 1910.146

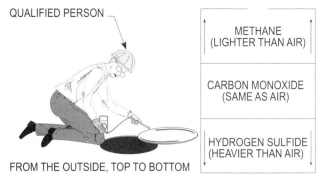

QUALIFIED PERSON	METHANE (LIGHTER THAN AIR)	**TESTING ATMOSPHERE**

FROM THE OUTSIDE, TOP TO BOTTOM

METHANE (LIGHTER THAN AIR)

CARBON MONOXIDE (SAME AS AIR)

HYDROGEN SULFIDE (HEAVIER THAN AIR)

TESTING ATMOSPHERE

• NEVER TRUST YOUR SENSES TO DETERMINE IF THE AIR IN A CONFINED SPACE IS SAFE! YOU CANNOT SEE OR SMELL MANY TOXIC GASES AND VAPORS, NOR CAN YOU DETERMINE THE LEVEL OF OXYGEN PRESENT.
• NORMAL SUPPLY OF AIR IS APPROXIMATELY 20% OXYGEN.
• ABNORMAL SUPPLY OF AIR IS LESS THAN 19.5% OXYGEN.

CONFINED SPACE REQUIREMENTS
OSHA 1910.146

Figure 3-4(a). This illustration shows that confined space atmospheres shall be monitored by a qualified person.

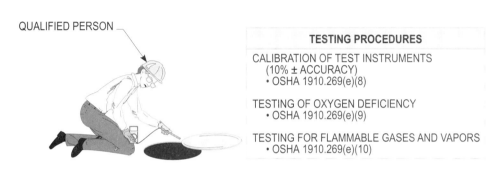

QUALIFIED PERSON

TESTING PROCEDURES

CALIBRATION OF TEST INSTRUMENTS (10% ± ACCURACY)
• OSHA 1910.269(e)(8)

TESTING OF OXYGEN DEFICIENCY
• OSHA 1910.269(e)(9)

TESTING FOR FLAMMABLE GASES AND VAPORS
• OSHA 1910.269(e)(10)

CONFINED SPACE REQUIREMENTS
OSHA 1910.146

Figure 3-4(b). This illustration shows that confined spaces shall be monitored by a qualified person using approved meters and sampling methods.

OSHA Required Training
1910.146

To determine if the atmosphere is hazardous, confined space atmospheres shall be monitored by a qualified person using approved meters and sampling methods. To determine safe levels of a contaminant, see Subpart Z of "Toxic and Hazardous Substances," 29 CFR 1910.1000, "Air Contaminants." **(See Figures 3-4(a) and (b))**

SafetyTip: Confined space atmospheres must be monitored by a qualified person.

All personal protective equipment and other equipment used in a confined space shall be inspected before it is used or brought into a confined space. Defective or dangerous equipment shall not be used. Personal protective equipment or other safety equipment or devices likely to be used in a confined space entry include, but are not limited to, respiratory protection (air purifying, air supplied, self-contained breathing apparatus [SCBA]), fresh-air supply system, powered exhaust system, safety harness and life line, atmospheric sampler or monitor and regular personal protective equipment.

Safety Tip: An inspection of all PPE that is to be used in a confined space must be inspected before it is used.

A supervisor shall complete the following actions before allowing workers to enter confined spaces:

- Conduct a planning session with confined-space workers to determine work objectives, date and time the work will begin and who will be assigned as confined space workers, safety watch and other associated workers,

- Perform an activity hazard analysis of the proposed operations to identify (as applicable),

- Chemicals or materials presently or previously stored or used in the confined space,

- Unexplained leaks or spills of hazardous chemicals or materials,

- Hazardous properties or incompatible combinations of materials or by-products, hazardous or toxic gases, vapors, dust, mist or fumes,

- Existing and probable physical hazards such as slipping, tripping, falling or potential contact with moving equipment; electrical hazards, rodents, snakes or insects,

- Methods of operation for necessary clearing, purging, ventilation or guarding,

- Safety procedures including first aid, cardiopulmonary resuscitation (CPR), decontamination, communications methods and other safety and emergency rescue methods,

- Date and time of pre-entry work session,

- Make initial test for hazardous atmospheres and oxygen level and provide continuous monitoring method as necessary,

- Determine need for specific personal protective equipment identified in the activity hazard analysis, provide required equipment and make certain all workers required to use personal protective equipment are knowledgeable and qualified for its operation,

- Provide adequate ventilation where necessary (ignitionproof or explosionproof as may be required),

- Lockout and tagout all electrical and mechanical equipment and entry pipes or valves in the confined space that may pose a hazard and

- Select and complete an entry permit.

See Figure 3-5 for a detailed illustration pertaining to a sample entry permit for confined spaces.

TRAINING REQUIREMENTS FOR CONFINED SPACES

The following shall be considered for training requirements in confined spaces:

- General training
- Specific training

GENERAL TRAINING

Confined-space workers, standby workers and supervisors shall be trained in the following:

- Emergency and routine entry/exit procedures; use of respirators; first aid and CPR;

- Lockout and tagout procedures and isolation of entry sources (electricity, gas and water);

See Figure 3-6 for a detailed illustration pertaining to lockout and tagout procedures and isolation of energy sources.

- Rescue methods; and

- Use of safety equipment.

Confined Space Entry Permit

Note: To Be Completed On Initial Entry and Each Entry Thereafter

Assessment Date: _____ Time: _____

Entry Date: _____ Time: _____

Workers Assigned: _____ _____

_____ _____

Location: _____

Description of Work: _____

Precautions/Actions	Yes	No	N/A	Notes
Qualified Person				
Safety Watch				
Space Clean				
Atmosphere Safe				
Periodic Monitor For Atmosphere				
Continuous Monitor For Atmosphere				
Lines Shut/Capped				
Lockout Complete				
Safety Lights				
Communications System Established				
Safety Belt/Life Line				
Respirator (Air Line/Filter Type)				
Filter Respirator				
Warning Signs				
Protective Gear				
Rescue Gear				
Fire Protection Equipment				
Tool/Equipment Inspected				
Other/Miscellaneous (List				
Test Instruments Used				

Name of Safety Watch: _____

Permit Issued By: _____ Date: _____ Time: _____

Use Reverse Side for Additional Comments:

Figure 3-5. This illustration shows a sample entry permit for confined spaces.

NOTE: LOCKOUT AND TAGOUT PROCEDURES ARE USED FOR THE ISOLATION OF ENTRY SOURCES SUCH AS ELECTRICITY, GAS AND WATER.

Figure 3-6. This illustration shows the different types of lockout and tagout control procedures used to control hazardous electrical energy.

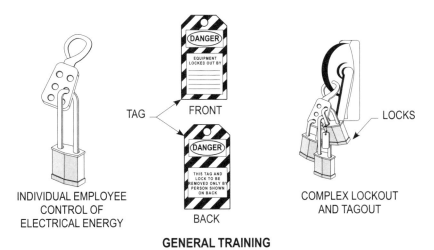

INDIVIDUAL EMPLOYEE CONTROL OF ELECTRICAL ENERGY

TAG → FRONT

BACK

COMPLEX LOCKOUT AND TAGOUT

LOCKS

GENERAL TRAINING

Figure 3-7. This illustration shows confined space workers performing a rescue and using safety equipment.

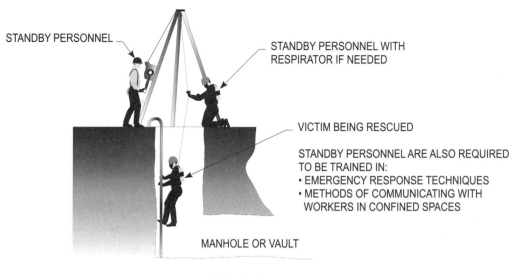

STANDBY PERSONNEL

STANDBY PERSONNEL WITH RESPIRATOR IF NEEDED

VICTIM BEING RESCUED

STANDBY PERSONNEL ARE ALSO REQUIRED TO BE TRAINED IN:
• EMERGENCY RESPONSE TECHNIQUES
• METHODS OF COMMUNICATING WITH WORKERS IN CONFINED SPACES

MANHOLE OR VAULT

GENERAL TRAINING

See Figure 3-7 for a detailed illustration pertaining to rescue methods and use of safety equipment.

Standby workers shall also be trained in:

- Emergency response techniques and
- Methods of communicating with workers in confined spaces.

SPECIFIC TRAINING

Before beginning a confined-space entry operation, the supervisor shall provide pre-phase training to all workers involved in the operation. This training shall at a minimum include a thorough review of the activity hazard analysis for the operation.

TEN RULES FOR CONFINED-SPACE ENTRY

- Planning sessions shall be conducted by a qualified person,
- Atmospheres shall be tested,
- Ventilation shall be adequate,
- Workers shall be trained,
- Lockout/Tagout procedures shall be used,
- Standby workers shall use proper communications methods,
- Safe tools and equipment shall be used,
- Entry permits shall be up-to-date,
- Atmospheres shall be monitored continuously and
- Record-keeping requirements shall be followed.

HOUSEKEEPING
OSHA 1926.25

To ensure the safety of construction-site workers and the public, each work site shall be kept broom clean and orderly.

Scrap, waste material and debris shall be kept away from work areas, passageways, stairs, ladders, elevator openings and from open sides of floors and landings. This material shall also be kept off roads inside the worksite.

Safety Tip: Worksites must be cleared of scrap and waste on a daily basis.

Worksites shall be cleared of all scrap and waste material daily or at the end of the workshift. Debris that may be a fire hazard shall be cleared during or immediately following an operation. Combustible scrap from welding and burning operations shall be removed at regular intervals throughout the day.

Flammable, toxic and other hazardous waste (for example, containers, applicators, oily rags and used tubes of caulk) shall be stored in closed, approved containers and removed from the worksite on a regular basis throughout the day.

Garbage and other waste material shall be kept in closed containers and disposed of on a regular basis.

WORKSITE LIGHTING
OSHA 1926.26

Safety Tip: Artificial lighting must be provided when a specific level of light is necessary for the safety of workers.

If the level of natural light is inadequate, artificial lighting shall be provided for ramps, runways, corridors, workshops, storage areas, offices and other areas where a specific level of light is necessary for the safe movement of workers and for the performance of operations.

See Figure 3-8 for a detailed illustration pertaining to the minimum lighting requirements for specific activities. As the visual demands increase, the lighting levels should also increase.

OPERATIONS/LOCATIONS	MINIMUM FT. CANDLES
GENERAL CONSTRUCTION AREA	5
GENERAL CONSTRUCTION AREAS, CONCRETE PLACEMENT EXCAVATION AND WASTE AREAS, ACCESSWAYS, ACTIVE STORAGE, LOADING PLATFORMS, REFUELING, FIELD MAINTENANCE	3
WAREHOUSE, CORRIDORS, HALLWAYS, EXITWAYS	
TUNNELS, SHAFTS, GENERAL UNDERGROUND WORKAREAS. (EXCEPTION - MINIMUM 10 FT. CANDLES ARE REQUIRED AT TUNNEL AND SHAFT HEADINGS DURING DRILLING, MUCKING AND SCALING. BUREAU-OF-MINES APPROVED CAP LIGHT SHALL BE ACCEPTABLE FOR USE IN TUNNEL HEADING)	5
GENERAL CONSTRUCTION PLANT AND SHOPS (e.g., BATCH PLANTS, SCREENING PLANTS, MECHANICAL AND ELECTRICAL EQUIPMENT ROOMS, CARPENTER SHOPS, RIGGING LOFTS AND ACTIVE STOREROOMS, BAR RACKS OR LIVING QUARTERS, LOCKER OR DRESSING ROOMS, MESS HALLS AND INDOOR TOILETS AND WORKROOMS)	10
FIRST AID STATIONS, INFIRMARIES AND OFFICES 29 CFR 1926.56(a)	30

Figure 3-8. This illustration shows the minimum lighting requirements for specific activities.

Emergency lighting shall be provided for emergency exiting from work areas, stairways and other passageways.

Safety Note: See National Fire Protection Association (NFPA) 101.

SafetyTip: Use only approved electrical devices in work areas that contain or may contain a flammable atmosphere.

If a work area contains or may contain a flammable atmosphere, only approved electrical devices and lighting for such an atmosphere shall be used.

Safety Note: For further information, see OSHA 1926.407, Chapter 5 and Article 500 of the NEC.

PERSONNEL PROTECTIVE EQUIPMENT
OSHA 1926.28

Each employer is responsible for requiring the wearing of the appropriate personnel protective equipment where there is an exposure to hazardous conditions. For more information about the selection and use of personnel protective equipment, refer to Chapter 5.

General Safety and Health Provisions

Section	Answer	
_____	T	F
_____	T	F
_____	T	F
_____	T	F
_____	T	F
_____	T	F
_____	T	F
_____	T	F
_____	T	F
_____	T	F

General Safety and Health Provisions

1. Under the requirements of the Occupational Safety Health Act (OSHA) only employers that have more than 10 employees have the responsibility to initiate and maintain an effective health and safety program.

2. OSHA's Construction Safety and Health Regulations Part 1926 do not contain a permit-required confined space regulation, these requirements can be found in OSHA's General Industry Regulation, 1910.146.

3. All employees required to enter into confined or enclosed spaces shall be instructed as to the nature of the hazards involved, the necessary precautions to be taken and in the use of protective and emergency equipment required.

4. If an area has an atmosphere that contains or has the potential to contain less than 20.5 percent oxygen, it is considered a confined space.

5. Confined spaces found on construction sites typically include pits or excavations more than 4 ft. deep with a potential for accumulating hazardous materials or vapors or for displacing oxygen.

6. All personal protective equipment (PPE) and other equipment used in a confined space shall be inspected at least annually before it is used or brought into a confined space.

7. To determine if the atmosphere is hazardous, confined space atmospheres shall be monitored by an unqualified person using approved meters and sampling methods.

8. The four main elements of an effective occupational safety and health program are:

   • Management commitment and employee involvement,
   • Worksite analysis,
   • Hazard prevention and control and
   • Safety and health training.

9. OSHA's General Industry Regulation, 1910.146 *Permit-required confined spaces,* contains requirements for practices and procedures to protect employees in general industry from the hazards of entry into permit-required confined spaces. This regulation applies to construction.

10. Because of the nature of confined spaces, the possibility of dangerous atmospheres, and the type of operations performed in them, confined-space workers and supervisors shall take every possible precaution.

Protection for Worker Safety, Health and the Environment

The employer shall ensure that medical personnel are available to advise and consult on matters of occupational health. Before the start of a project, provisions shall be made for prompt medical attention in the event of serious injury. Supervisors are responsible for arranging transportation of an injured worker to a source of medical attention, either by company vehicle or ambulance. A means of communication, along with telephone numbers and numbers of the local ambulance service, hospital and designated health-care provider, shall be provided in a location accessible to all employees.

If a health-care provider or facility is not reasonably accessible to the construction site, an individual certified in first aid (by US Bureau of Mines, American Red Cross or an equivalent organization) shall be available on site to give first aid.

First aid supplies approved by a consulting physician shall be wrapped in individually sealed packages and kept in a weatherproof container. The contents of this container shall be examined before the job begins and weekly thereafter to ensure that all used items are replaced.

Where corrosives are used, a method for drenching or flushing shall be provided, such as eyewash or safety shower.

Safety Tip: Drinking water must be provided at the worksite.

SANITATION
OSHA 1926.51

Drinking (potable) water shall be provided at the construction site. If no permanent source is available, a sealed container of water (with a dispensing tap) marked "DRINKING WATER" shall be available. Disposable cups (kept in a sanitary container) shall also be available.

Nonpotable water outlets shall be identified with a sign bearing the words "UNSAFE FOR DRINKING, COOKING OR WASHING." There shall be no cross connection between potable and nonpotable water supply systems.

Anti-siphon or backflow prevention methods or devices should be used at any potable water service location where there is a possibility of siphoning or backflow of impurities.

Washing facilities shall be available near areas where workers apply paints, coatings, herbicides, insecticides or other contaminants.

Food service facilities at the construction site shall conform to local sanitation ordinances.

All temporary sleeping facilities shall have heat, ventilation and light.

Safety Tip: Toilet facilities must be provided for all workers.

Toilet facilities shall be provided for all construction workers, except mobile crews who have access to nearby facilities. For worksites with no sanitary sewer system, a chemical, recirculating or combustion toilet (and toilet supplies) shall be provided, unless they are prohibited by local code.

Toilets shall be provided for workers according to the following requirements.

Number of workers	Number of facilities
20 or less	1 toilet seat
20 or more	1 toilet seat and 1 urinal per 40 workers
200 or more	1 toilet seat and 1 urinal per 50 workers

HEARING PROTECTION PROGRAM
OSHA 1926.52

Worker exposure to noise shall not exceed the limits provided in Figure 4-1, "Noise Exposure Levels."

Safety Note: F value greater than 1.0 indicates the noise level exposure exceeds permissible limits.

If a worker's noise exposure includes two or more periods at different noise levels, the combined noise exposure level is determined as follows:

$$F = (T1/L1) + (T2/L2) + (Tn/Ln)$$

Where:
F = equivalent noise exposure factor
T = actual time period of noise exposure at a given noise level
L = permissible duration of exposure at a given noise level

See Figure 4-1 for a detailed illustration pertaining to noise exposure levels.

DURATION - HRS. PER DAY AT EXPOSED LEVEL (T)	PERMISSIBLE LEVEL DATA SLOW RESPONSE (L)
8	90
6	92
4	95
3	97
2	100
1.5	102
1	105
0.5	110
0.25	115

Figure 4-1. This illustration shows the employee noise exposure levels.

NOTE: IF A WORKER'S NOISE EXPOSURE INCLUDES TWO OR MORE PERIODS AT DIFFERENT NOISE LEVELS, THE COMBINED NOISE EXPOSURE LEVEL IS REQUIRED TO BE DETERMINED.

For example, a worker was exposed to noise levels as follows: (Note: dBA indicates noise level measured on an A-weighted sound level.)

110 dBA for a time period (T1) of 1/4 or 0.25 hours
100 dBA for a time period (T2) of 1/2 or 0.5 hours
90 dBA for a time period (T3) of 1 1/2 or 1.5 hours

The values of T and L or the example are as follows: **(L values from Figure 4-1)**

for 110 dBA T1 = 0.25   L1 = 0.5
for 100 dBA T2 = 0.5   L2 = 2
for 90 dBA T3 = 1.5   L3 = 8

Substituting the values for T and L in the above equation yields
F = (0.25/0.5) + (0.5/2) + (1.5/8) F = 0.5 + 0.25 + 0.19 F = 0.94

Because F does not exceed 1, the exposure is within permissible limits. However, F is greater than the noise exposure action level of 0.5; therefore, the employer shall implement a hearing conservation program.

Safety Tip: In areas where noise levels exceed the permissible limits, exposure levels must be controlled.

When noise levels exceed the permissible limits, worker exposure levels shall be controlled through the use of engineering controls, administrative controls, personnel protective equipment, or any combination thereof. The following shall be considered by the employer when implementing a hearing conservation program:

• Engineering controls may consist of isolating, enclosing, or insulating noise producing equipment or operations, dampening equipment vibration or substituting quieter alternatives for noisy equipment or operations. Engineering controls eliminate or lessen the actual physical hazard and are therefore the preferred method of limiting noise exposure.

• Administrative controls typically involve altering worker assignments or work shifts to reduce the amount of time the worker spends in a high-noise-level area. Administrative controls are preferred over the use of personal protective equipment because they are more reliable and more easily implemented.

• Personal protective equipment, ear muffs or plugs shall be used by workers to reduce the noise exposure to within acceptable limits when engineering or administrative controls fail or are not feasible. See Chapter 5 in this book for more information on hearing protection devices.

A noise exposure of 85 dBA averaged over an 8-hour workday (time-weight-average or TWA) or an equivalent noise exposure factor (F) of 0.5 is referred to as the noise exposure action level. Where exposures are at or above the action level, regardless of the use of personal protective equipment, the employer shall implement a hearing conservation program that at a minimum includes the following elements:

- Monitoring
- Employee notification
- Observation of monitoring
- Audiometric testing (initial and annual)
- Audiogram evaluation
- Noise-training program and
- Record keeping.

See Figure 4-2 for a chart that is used as a guide for common information concerning produced noise factors, levels, program and protection.

Figure 4-2. This illustration shows the noise protection guidelines for common produced noise factors, levels, program and protection.

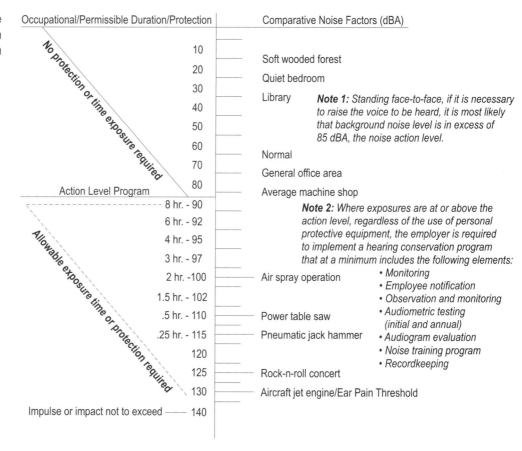

IONIZING RADIATION
OSHA 1926.53

See OSHA 10 CFR Part 20 and 29 CFR 1910.96 "Ionizing Radiation" that applies to construction and related activities that use ionizing radiation. See OSHA 29 CFR 1926.53(a) for further information.

Any activity that involves the use of radioactive materials or x-rays, whether or not under license, shall be performed by a competent person specially trained in the proper use and safe operation of such equipment. In the case of materials used under license, only persons actually licensed or a competent person under direction of the licensee shall perform such work.

The contractor shall ensure that all workers who could be exposed to radiation sources or materials are aware of the operations and are instructed in necessary safeguards. For complete discussion on requirements relating to ionizing radiation, refer to OSHA1910.96.

OSHA Required Training
1926.53(b)

NONIONIZING RADIATION (LASERS)
OSHA 1926.54

Only qualified and trained workers shall install, adjust and operate laser equipment. Proof of the laser equipment operator's qualifications shall be in his or her possession at all times.

OSHA Required Training
1926.54(a) and (b)

Workers who may be exposed to direct or reflected laser light shall be provided with appropriate anti-laser eye protection. Refer to paragraph III, A, 2, b, (2), (h) for more information.

Areas in which lasers are used shall be posted with warning placards.

Safety Tip: Warning placards must be posted in areas where lasers are used.

A laser that is to be unattended for 1 hour or more shall be turned off.

Only mechanical or electronic detectors shall be used to internally align the laser.

The laser beam shall not be directed at workers.

If practical, lasers shall not be operated during rain, snow or in dusty or foggy conditions.

Laser equipment shall be labeled with its maximum output.

The following worker exposure limits for laser light shall be observed:

- Direct staring - 1 microwatt per square centimeter
- Incidental observing - 1 milliwatt per square centimeter
- Diffused reflected light- 2 1/2 watts per square centimeter.

When possible, laser units should be set up above the heads of employees and not at eye level.

Workers shall not be exposed to microwave power densities in excess of 10 milliwatts per square centimeter.

VENTILATION
OSHA 1926.57

Exhaust systems operate in one of two ways:

- They draw inside air out to the open and replace it with clean air; or they remove, scrub or separate hazardous substances in the air and return the cleaned air to the work area.

- On the construction site, if harmful quantities of dust fumes, mist, vapors or gases are produced, they shall be removed by the appropriate mechanical exhaust system; only clean (or cleaned) air shall be returned to the work area.

GENERAL REQUIREMENTS FOR MECHANICAL-EXHAUST SYSTEMS
OSHA 1926.57(a)

Hazardous concentrations of contaminants in the air shall be determined as specified in the "Threshold Limit Values of Toxic Chemicals of the American Conference of Governmental Industrial Hygienists" and in OSHA 1910.1000, "Limits of Air Contaminants," Tables B-1, 2 and 3.

Where local exhaust ventilation systems are used, they shall be designed to prevent the dispersion of harmful concentrations of hazardous materials from the source or operation (i.e., welding, sanding and grinding) into the worker's breathing zone.

Exhaust systems shall operate for the time duration they are designed to serve.

Exhaust systems shall operate for as long as it takes to clear an occupied work area of hazardous substances.

Workers wearing respiratory protective equipment (RPE) in an area where an exhaust system is used to remove dust shall not remove the respirator until the atmosphere in the work area is clear.

Exhaust systems shall be designed, built and maintained so that hazardous substances are not drawn through the work area. **(See Figure 4-3)**

Figure 4-3. This illustration shows that exhaust systems shall be designed, built and maintained so that hazardous substances are not drawn through the work area.

NOTE 1: EXHAUST SYSTEMS SHALL OPERATE FOR THE DURATION THEY ARE DESIGNED TO SERVE.

NOTE 2: EXHAUST SYSTEMS SHALL OPERATE FOR AS LONG AS IT TAKES TO CLEAR AN OCCUPIED WORK AREA OF HAZARDOUS SUBSTANCES.

NOTE 3: WORKERS WEARING RESPIRATORY PROTECTIVE EQUIPMENT IN AN AREA WHERE AN EXHAUST SYSTEM IS USED TO REMOVE DUST SHALL NOT REMOVE THE RESPIRATOR UNTIL THE ATMOSPHERE IN THE WORK AREA IS CLEAR.

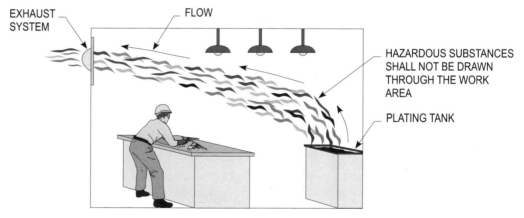

GENERAL REQUIREMENTS FOR MECHANICAL-EXHAUST SYSTEMS
OSHA 1926.57(a)

Exhaust systems shall be designed, built, maintained and operated in a manner that ensures the volume and velocity of exhaust air is sufficient to capture harmful substances from the work place and carry them to areas of safe disposal. **(See Figure 4-4)**

Equipment and technical measures used to keep employee exposure to air contaminants within the prescribed limits shall be approved by a competent person, that is, an industrial hygienist or other technically qualified person.

For assistance in identifying hazardous substances and atmospheres, refer to the appropriate Material Safety Data Sheets (MSDS).

For further information about ventilation for welding operations, see Chapter 10 in this book.

NOTE 1: EQUIPMENT AND TECHNICAL MEASURES USED TO KEEP EMPLOYEE EXPOSURE TO AIR CONTAMINANTS WITHIN THE PRESCRIBED LIMITS SHALL BE APPROVED BY AN INDUSTRIAL HYGIENIST OR OTHER TECHNICALLY QUALIFIED PERSON.

NOTE 2: FOR ASSISTANCE IN IDENTIFYING HAZARDOUS SUBSTANCES AND ATMOSPHERES, REFER TO THE APPROPRIATE MATERIAL SAFETY DATA SHEETS.

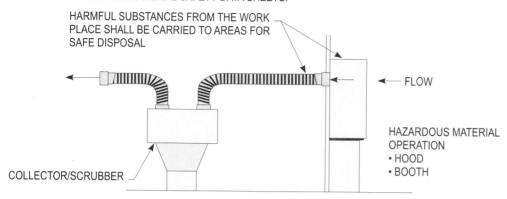

GENERAL REQUIREMENTS FOR MECHANICAL-EXHAUST SYSTEMS
OSHA 1926.57(a)

Figure 4-4. This illustration shows that exhaust systems shall be designed, built, maintained and operated in a manner that ensures the volume and velocity of exhaust air is sufficient to capture harmful substances from the work place and carry them to areas of safe disposal.

HAZARD COMMUNICATION
OSHA 1910.59

OSHA Required Training
1926.59

The Hazard Communication Program (HAZ COM) requirements applicable to Construction work are identical to those that are set forth in the General Industry standards (OSHA 1910.1200). The following shall be considered for hazard communication:

- Written hazard communication program
- Labels and other forms of warning
- Material safety data sheets
- Employee information and training

WRITTEN HAZARD COMMUNICATION PROGRAM
OSHA 1910.1200(e)

OSHA's Haz Com standard requires that employers develop, implement and maintain at the workplace a written, comprehensive hazard communication program that includes provisions for container labeling, collection and availability of material safety data sheets and an employee training program. It also shall contain a list of the hazardous chemicals in each work area, the means the employer will use to inform employees of the hazards of non-routine tasks (for example, the cleaning of reactor vessels) and the hazards associated with chemicals in unlabeled pipes. If the workplace has multiple employers on-site (for example, a construction site), the rule requires these employers to ensure that information regarding hazards and protective measures be made available to the other employers on-site, where appropriate. **(See Figure 4-5)**

Safety Tip: Employers must develop, implement and maintain a written Haz Com program.

The written program does not have to be lengthy or complicated, and some employers may be able to rely on existing hazard communication programs to comply with the above requirements. The written program shall be available to employees, their designated representatives, the Assistant Secretary of Labor for Occupational Safety and Health and the Director of the National Institute for Occupational Safety and Health (NIOSH).

OSHA Required Training
1910.1096(h)(1)(2)(i) thru (iii) and (3)(i) thru (iv).

NOTE 1: EMPLOYERS SHALL ENSURE THAT INFORMATION REGARDING HAZARDOUS AND PROTECTIVE MEASURES BE MADE AVAILABLE TO OTHER EMPLOYEES ON SITE, WHERE APPROPRIATE.

NOTE 2: THE WRITTEN PROGRAM SHALL BE AVAILABLE TO EMPLOYEES, THEIR DESIGNATED REPRESENTATIVES, THE ASSISTANT SECRETARY OF LABOR FOR OCCUPATIONAL SAFETY AND HEALTH AND THE DIRECTOR OF NATIONAL INSTITUTE FOR OCCUPATIONAL SAFETY AND HEALTH.

WRITTEN HAZARD COMMUNICATION PROGRAM
OSHA 1926.1200(e)

LABELS AND OTHER FORMS OF WARNING
OSHA 1910.1200(f)

Chemical manufacturers, importers and distributors shall be sure that containers of hazardous chemicals leaving the workplace are labeled, tagged or marked with the identity of the chemicals, appropriate hazard warnings and the name and address of the manufacturer or other responsible party.

Safety Tip: All containers in the workplace must be properly labeled.

In the workplace, each container shall be labeled, tagged or marked with the identity of hazardous chemicals contained therein, and shall show hazard warnings appropriate for employee protection. The hazard warning can be any type of message, words, pictures or symbols that convey the hazards of the chemical(s) in the container. Labels shall be legible, in English (plus other languages, if desired) and prominently displayed. **(See Figure 4-6)**

EXEMPTION TO THE REQUIREMENT FOR IN-PLANT INDIVIDUAL CONTAINER LABELS

The following are exemptions to the requirements for in-plant individual container labels:

- Employers can post signs or placards that convey the hazard information if there are a number of stationary containers within a work area that have similar contents and hazards.

- Employers can substitute various types of standard operating procedures, process sheets, batch tickets, blend tickets and similar written materials for container labels on stationary process equipment if they contain the same information and are readily available to employees in the work area.

- Employers are not required to label portable containers into which hazardous chemicals are transferred from labeled containers and that are intended only for the immediate use of the employee who makes the transfer.

- Employers are not required to label pipes or piping systems.

NOTE 1: THE HAZARD WARNING CAN BE ANY TYPE OF MESSAGE, WORDS, PICTURES OR SYMBOLS THAT CONVEY THE HAZARDOUS OF THE CHEMICAL(S) IN THE CONTAINER.

NOTE 2: LABELS SHALL BE LEGIBLE, IN ENGLISH (PLUS OTHER LANGUAGES, IF DESIRED) AND PROMINENTLY DISPLAYED.

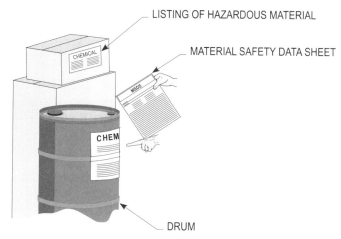

LISTING OF HAZARDOUS MATERIAL

MATERIAL SAFETY DATA SHEET

DRUM

LABELS AND OTHER FORMS OF WARNING
OSHA 1926.1200(f)

Figure 4-6. This illustration shows each container shall be labeled, tagged or marked with the identity of hazardous chemicals contained therein, and shall show hazard warning appropriate for employee protection.

MATERIAL SAFETY DATA SHEETS
OSHA 1910.1200(g)

Chemical manufacturers and importers shall develop an material safety data sheet for each hazardous chemical they produce or import, and shall provide the material safety data sheet automatically at the time of the initial shipment of a hazardous chemical to a downstream distributor or user. Distributors shall also ensure that downstream employers are similarly provided an material safety data sheet.

Each material safety data sheet shall be in English and include information regarding the specific chemical identity of the hazardous chemical(s) involved and the common names. In addition, information shall be provided on the physical and chemical characteristics of the hazardous chemical; known acute and chronic health effects and related health information; exposure limits; whether the chemical is considered to be a carcinogen by NTP, IARC or OSHA; precautionary measures; emergency and first-aid procedures; and the identification of the organization responsible for preparing the sheet.

Safety Tip: MSDS's must be readily accessible to all employees in the work area.

Copies of the material safety data sheet for hazardous chemicals in a given work site are to be readily accessible to employees in that area. As a source of detailed information on hazards, they shall be located close to workers, and readily available to them during each workshift.

Employers shall prepare a list of all hazardous chemicals in the workplace. When the list is complete, it should be checked against the collected material safety data sheets that the employer has been sent. If there are hazardous chemicals used for which no material safety data sheet has been received, the employer shall write to the supplier, manufacturer or importer to obtain the missing material safety data sheet. If employers do not receive the material safety data sheet within a reasonable period of time, they should contact the nearest OSHA office.

OSHA Required Training
1910.1200(h)

EMPLOYEE INFORMATION AND TRAINING
OSHA 1910.1200(h)

Safety Tip: A training program must be established for employees exposed to hazardous chemicals

Employers shall establish a training and information program for employees exposed to hazardous chemicals in their work area at the time of initial assignment and whenever a new hazard is introduced into their work area.

INFORMATION

At a minimum, the discussion topics shall include the following:

- The existence of the hazard communication standard and the requirements of the standard.

- The components of the hazard communication program in the employees' workplaces.

- Operations in work areas where hazardous chemicals are present.

- Where the employer will keep the written hazard evaluation procedures, communications program, lists of hazardous chemicals and the required material safety data sheet forms.

TRAINING

The employee training plan shall consist of the following elements:

- How the hazard communication program is implemented in that workplace, how to read and interpret information on labels and the material safety data sheet and how employees can obtain and use the available hazard information.

- The hazards of the chemicals in the work area. (The hazards may be discussed by individual chemical or by hazard categories such as flammability.)

- Measures employees can take to protect themselves from the hazards.

- Specific procedures put into effect by the employer to provide protection such as engineering controls, work practices and the use of personal protective equipment.

Safety Tip: Employees must implement a lockout/tagout program that includes procedures for enforcement.

- Methods and observations (such as visual appearance or smell) workers can use to detect the presence of a hazardous chemical to which they may be exposed.

OSHA Required Training
1910.147(A)(3)(ii); (4)(i)(1); (7)(i)(A) thru (C); (ii)(A) thru (F); (iii)(A) thru (C)(iv) and (8)

LOCKOUT-TAGOUT PROGRAM
OSHA 1910.147

It is the responsibility of the employer to implement a lockout and tagout program with procedures of enforcing such a program. Employees shall be trained on how to use the program safely and the steps to be taken before deenergizing the electrical system. The employee has the responsibility to learn his or her part in the lockout and tagout procedures concerning the lockout and tagout program.

LOCKOUT-TAGOUT PROCEDURE
OSHA 1910.147(c)

OSHA requires that all employers establish a written lockout-tagout procedure, and also establish procedures for enforcement of such a program. The lockout-tagout procedure is intended to protect personnel from injury during servicing and/or maintenance of machines and equipment. Lockout-tagout is not intended to cover normal production operations. In order to have a safe and reliable lockout-tagout program, the procedure shall describe the scope, purpose, responsibilities, authorization, rules and techniques needed to control all hazardous energy sources.

Work on cord-and-plug connected electrical equipment does not require lockout-tagout where the equipment is unplugged and the plug is under the exclusive control of the employee who is performing the servicing and maintenance on the equipment. **(See Figure 4-7)**

In general, the kinds of activities that are covered under lockout-tagout are activities such as lubrication, cleaning or unjamming, servicing of machines or equipment and making adjustments or tool changes. However, activities that are normal production operations are not covered, such as minor tool changes and adjustments if they are routine, repetitive and integral to the use of the equipment for production, and the work is performed using methods that provide effective employee protection.

Figure 4-7. This illustration shows that a cord-and-plug shall be permitted to serve as a disconnecting means as long as it is in control of the user.

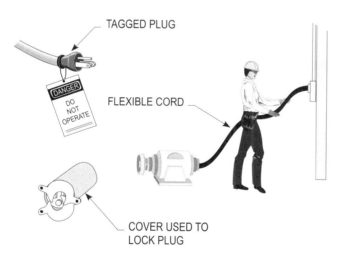

TAGGED PLUG

DANGER
DO
NOT
OPERATE

FLEXIBLE CORD

COVER USED TO
LOCK PLUG

LOCKOUT-TAGOUT PROCEDURE
OSHA 1910.147(c)

LOCKOUT-TAGOUT
OSHA 1910.147(c)(2)

OSHA has determined that lockout is by far the most effective means of providing employee protection and is preferred over tagout. However, if the energy isolating device is not capable of being locked out, a tagout program can be used, provided that the tagout program will provide the same level of safety equivalent to that obtained when using a lockout program. Additional means beyond those necessary for lockout are required, these means include; removal of an isolating circuit element, blocking of a controlling switch, opening of an extra disconnecting device or the removal of a valve handle to reduce the likelihood of inadvertent energization. If the energy isolating equipment is capable of being locked out, the employer shall utilize a lockout program, unless it can be demonstrated that the tagout program will provide the same level of safety. **(See Figure 4-8)**

Safety Tip: OSHA has determined that lockout is preferred over tagout.

Figure 4-8. This illustration shows the different means of lockout-tagout used to provide employee protection.

NOTE: IF THE ENERGY ISOLATING EQUIPMENT IS CAPABLE OF BEING LOCKED OUT, THE EMPLOYER SHALL UTILIZE A LOCKOUT PROGRAM, UNLESS IT CAN BE DEMONSTRATED THAT THE TAGOUT PROGRAM WILL PROVIDE THE SAME LEVEL OF SAFETY.

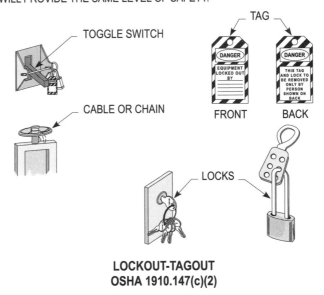

**LOCKOUT-TAGOUT
OSHA 1910.147(c)(2)**

On all equipment that is replaced, repaired, renovated or modified after October 31, 1989, the energy isolating devices for this equipment shall be designed to accept a lockout device.

ENERGY CONTROL PROCEDURE
OSHA 1910.147(c)(4)

Safety Tip: Employers must develop a written lockout/tagout procedure.

A written documented procedure shall be developed and enforced for all employees who may be injured by the unexpected startup or reenergization of machines and equipment while performing servicing and maintenance. However, if all of the following elements are met for a particular machine or piece of equipment, the employer does not have to document the procedure.

- The machine or equipment has no potential for stored or residual energy.

- The machine or equipment has a single source of energy that can be readily identified and isolated.

- The isolation and locking out of the energy source will completely deenergize and deactivate the machine or equipment.

- The machine or equipment is isolated from the energy source and locked out during servicing or maintenance.

- A single lockout device will achieve a locked out condition.

- The lockout device is under the exclusive control of the authorized employee performing the servicing and maintenance of the equipment.

- The servicing and maintenance of the equipment does not create a hazard for other employees.

See Figure 4-9 for a detailed illustration pertaining to a written documented energy control procedure.

THE EMPLOYER DOES NOT HAVE TO DOCUMENT THE PROCEDURE FOR A PARTICULAR MACHINE OR PIECE OF EQUIPMENT IF ALL OF THE FOLLOWING ELEMENTS ARE COMPLIED WITH:
• THE MACHINE OR EQUIPMENT HAS NO POTENTIAL FOR STORED OR RESIDUAL ENERGY
• THE MACHINE OR EQUIPMENT HAS A SINGLE SOURCE OF ENERGY THAT CAN BE READILY IDENTIFIED AND ISOLATED
• THIS ISOLATION AND LOCKING OUT OF THE ENERGY SOURCE WILL COMPLETELY DEENERGIZE AND DEACTIVATE THE MACHINE OR EQUIPMENT
• A SINGLE LOCKOUT DEVICE WILL ACHIEVE A LOCKED OUT CONDITION
• THE LOCKOUT DEVICE IS UNDER THE EXCLUSIVE CONTROL OF THE AUTHORIZED EMPLOYEE PERFORMING THE SERVICING AND MAINTENANCE OF THE EQUIPMENT
• THE SERVICING AND MAINTENANCE OF THE EQUIPMENT DOES NOT CREATE A HAZARD FOR OTHER EMPLOYEES

AFFECTED EMPLOYEES

NOTE: IF THE PROCEDURE IS NOT DOCUMENTED, THE EMPLOYEE SHALL HAVE NOT HAD ANY ACCIDENTS INVOLVING THE UNEXPECTED ACTIVATION OR REENERGIZATION OF THE MACHINE OR EQUIPMENT.

ENERGY CONTROL PROCEDURE
OSHA 1910.147(c)(4)

Figure 4-9. This illustration shows that a written documented procedure shall be developed and enforced for all employees who may be injured by the unexpected startup or reenergization of machines and equipment.

If this exception is used, the employer shall have not had any accidents involving the unexpected activation or reenergization of the machine or equipment.

The employer in this written procedure shall clearly and specifically outline the scope, purpose, authorization, rules and techniques that will be utilized for the control of hazardous energy. Such items to be included are the administrative responsibilities for implementing, training, compliance and the following:

Safety Tip: The written procedure must clearly outline the scope, purpose, authorization, rules and techniques that will be used for the control of hazardous energy.

• Intended use of procedure.

• Steps for shutting down, isolating, blocking and securing machines or equipment.

• Steps for placement, removal and transfer of lockout-tagout devices and the responsibility for them.

• Requirements for testing a machine or piece of equipment to determine the effectiveness of the lockout-tagout.

LOCKOUT-TAGOUT DEVICES
OSHA 1910.147(c)(5)

Locks, tags, chains and other hardware that are used for the isolation of hazardous energy sources shall be furnished by the employer. These devices shall be uniquely identified and used for no other purpose and shall meet the following requirements:

• Standardized, using one or more of the following:

(a) Color
(b) Shape
(c) Size
(d) Type
(e) Format

• Clearly visible, distinctive where seen and easy to recognize.

• Designed so that all needed information is conveyed that is necessary for the application.

- Designed in such a manner as to deter accidental or unauthorized removal, and substantial enough to prevent removal without the use of excessive force or unusual techniques.

- Designed for the conditions of the environment in which they are installed and they shall be capable of withstanding the environment to which they will be exposed for the maximum time that exposure is expected.

See Figure 4-10 for a detailed illustration pertaining to lockout devices

Figure 4-10. This illustration shows that locks, tags, chains and other hardware shall be furnished by the employer.

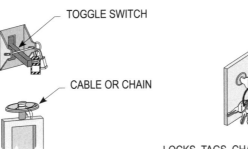

TOGGLE SWITCH

CABLE OR CHAIN

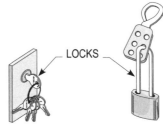

LOCKS

TAG

DANGER
EQUIPMENT
LOCKED
OUT BY

DANGER
THIS TAG AND
LOCK TO BE
REMOVED
ONLY BY
PERSON
SHOWN ON
BACK

FRONT   BACK

STANDARDIZED, USING ONE OR MORE OF THE FOLLOWING:
- COLOR
- SIZE
- FORMAT
- SHAPE
- TYPE

LOCKS, TAGS, CHAINS AND OTHER HARDWARE SHALL BE UNIQUELY IDENTIFIED AND USED FOR NO OTHER PURPOSE AND SHALL MEET THE FOLLOWING REQUIREMENTS:

- STANDARDIZED
- CLEARLY VISIBLE, DISTINCTIVE WHERE SEEN AND EASY TO RECOGNIZE
- DESIGNED SO THAT ALL NEEDED INFORMATION IS CONVEYED THAT IS NECESSARY FOR THE APPLICATION
- DESIGNED IN SUCH A MANNER AS TO DETER ACCIDENTAL OR UNAUTHORIZED REMOVAL, AND SUBSTANTIAL ENOUGH TO PREVENT REMOVAL WITHOUT THE USE OF EXCESSIVE FORCE OR UNUSUAL TECHNIQUES
- DESIGNED FOR THE CONDITIONS OF THE ENVIRONMENT IN WHICH THEY ARE INSTALLED AND CAPABLE OF WITHSTANDING THE ENVIRONMENT TO WHICH THEY WILL BE EXPOSED FOR THE MAXIMUM TIME THAT EXPOSURE IS EXPECTED

LOCKOUT-TAGOUT DEVICES
OSHA 1910.147(c)(5)

Tagout devices and their attachment means shall be substantial and durable enough to prevent accidental or inadvertent removal and have the following characteristics:

Safety Tip: A means of identifying the person applying the lockout/tagout device must be implemented.

- Attachment means must be non-reusable
- Attachable by hand
- Self-locking
- Have an unlocking strength of no less than 50 pounds
- Be at least equivalent in general design to a one-piece, all environment-tolerant nylon cable tie

See Figure 4-11 for a detailed illustration pertaining to tagout devices and their attachment means.

IDENTIFICATION
OSHA 1910.147(c)(5)(d)

All lockout-tagout devices shall have a means of identifying the person applying the lockout-tagout devices, and the tagout device shall warn against hazardous conditions if the equipment is reenergized and shall include at least one of the following warnings:

- Do not start
- Do not open
- Do not close
- Do not energize
- Do not operate

See Figure 4-12 for a detailed illustration pertaining to the identification of lockout-tagout devices.

TAGOUT DEVICES AND THEIR ATTACHMENT MEANS SHALL BE SUBSTANTIAL AND DURABLE ENOUGH TO PREVENT ACCIDENTAL OR INADVERTENT REMOVAL AND HAVE THE FOLLOWING CHARACTERISTICS:

- ATTACHMENT MEANS SHALL BE NON-REUSABLE
- ATTACHABLE BY HAND
- SELF-LOCKING
- HAVE AN UNLOCKING STRENGTH NO LESS THAN 50 POUNDS
- BE AT LEAST EQUIVALENT IN GENERAL DESIGN TO A ONE-PIECE ALL ENVIRONMENT-TOLERANT NYLON CABLE TIE

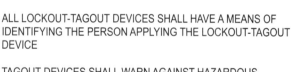

LOCKOUT-TAGOUT DEVICES
OSHA 1910.147(c)(5)

Figure 4-11. This illustration shows tagout devices and their attachment means.

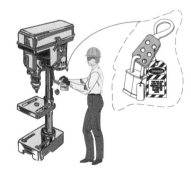

ALL LOCKOUT-TAGOUT DEVICES SHALL HAVE A MEANS OF IDENTIFYING THE PERSON APPLYING THE LOCKOUT-TAGOUT DEVICE

TAGOUT DEVICES SHALL WARN AGAINST HAZARDOUS CONDITIONS IF THE EQUIPMENT IS REENERGIZED AND IS REQUIRED TO INCLUDE AT LEAST ONE OF THE FOLLOWING WARNINGS:

- DO NOT START
- DO NOT OPEN
- DO NOT CLOSE
- DO NOT ENERGIZE
- DO NOT OPERATE

IDENTIFICATION
OSHA 1910.147(c)(5)(d)

Figure 4-12. This illustration shows the identification requirements for lockout-tagout devices.

PERIODIC INSPECTIONS
OSHA 1910.147(c)(6)

At least annually, an authorized employee other than the one(s) utilizing the lockout-tagout procedure shall inspect and verify the effectiveness of the lockout-tagout procedure. These inspections shall provide for a demonstration of the procedure and be implemented through random audits and planned visual observations. These inspections are intended to ensure that the lockout-tagout procedures are being properly implemented and to correct any deviations or inadequacies observed.

Safety Tip: An annual audit of the lockout/tagout program must be done by an authorized employee.

When locks are used without tags, the inspection shall include a review of the responsibilities of each authorized employee implementing the procedure with that employee. Group meetings of all authorized employees who implement the procedure would constitute compliance with this requirement. If tags are used without locks, the employer shall conduct this review with each affected and authorized employee.

If the lockout-tagout procedures are used less than once a year, they only need to be inspected when used. The periodic inspection shall provide for and ensure effective correction of identified deficiencies. These periodic inspections shall be documented, identify the machine or equipment on which the lockout-tagout was applied, the date of inspection, the employees included in the inspection and the person performing the inspection.

OSHA Required Training
1910.147(c)(7)

Safety Tip: Employees must be properly trained before lockout/tagout can take place.

TRAINING AND COMMUNICATION
OSHA 1910.147(c)(7)

Before lockout-tagout can take place, the employer shall ensure that all employees are trained in the purpose and function of the lockout-tagout procedure. The employer shall ensure that all employees understand and have the knowledge and skills required for the safe application, usage and removal other energy controls. OSHA 1910.147 recognizes the following three types of employees:

- Authorized
- Affected
- Other

Different levels of training are required for each type of employee based upon the roles of each employee in the role of lockout-tagout and the level of knowledge they shall have to accomplish tasks and to ensure the safety of their fellow workers.

AUTHORIZED EMPLOYEE

Any employee who is allowed to lock or implement a tagout procedure on a machine or piece of equipment to perform servicing or maintenance is considered an authorized employee. These employees shall receive training on the recognition of hazardous energy sources and the methods and means necessary of the control of the hazardous energy.

AFFECTED EMPLOYEE

An affected employee is one whose job requires them to operate or use a machine or piece of equipment that is being serviced or maintained under a lockout-tagout, or those who are working in an area in which this work is being performed. These employees shall receive training in the purpose and use of the lockout-tagout procedure.

All other employees who work around or are in the area where lockout-tagout is utilized shall receive training about the procedure and about the serious consequences relating to the attempts to restart or reenergize the equipment.

MINIMUM TRAINING REQUIREMENTS

Each training program shall cover at a minimum the following three areas:

- The energy control program
- The elements of the energy control procedures relevant to the employees duties
- The pertinent requirements of the lockout-tagout standard OSHA 1910.147

RETRAINING

Retraining for authorized and affected employees shall be done under the following conditions:

- Whenever there is a change in job assignments.

- Whenever a new hazard is introduced due to a change in machines, equipment or process.

- Whenever there is a change in the lockout-tagout procedure.

- Whenever the required or other periodic inspections by the employer reveals inadequacies in the company procedures or a lack of knowledge of the employees.

See Figure 4-13 for a detailed illustration pertaining to the training and retraining of employees.

EACH TRAINING PROGRAM SHALL COVER AT A MINIMUM THE FOLLOWING THREE AREAS:
- THE ENERGY CONTROL PROGRAM
- THE ELEMENTS OF THE ENERGY CONTROL PROCEDURES RELEVANT TO THE EMPLOYEE DUTIES
- THE PERTINENT REQUIREMENTS OF THE LOCKOUT-TAGOUT STANDARD OSHA 1910.147

RETRAINING FOR AUTHORIZED AND AFFECTED EMPLOYEES SHALL BE DONE UNDER THE FOLLOWING CONDITIONS:
- WHENEVER THERE IS A CHANGE IN JOB ASSIGNMENTS
- WHENEVER A NEW HAZARD IS INTRODUCED DUE TO CHANGE IN MACHINES, EQUIPMENT OR PROCESS
- WHENEVER THERE IS A CHANGE IN THE LOCKOUT-TAGOUT PROCEDURE
- WHENEVER THE REQUIRED OR OTHER PERIODIC INSPECTIONS BY THE EMPLOYER REVEALS INADEQUACIES IN THE COMPANY PROCEDURES OF A LACK OF KNOWLEDGE OF THE EMPLOYEES

Figure 4-13. This illustration shows that employees shall be trained and retrained as necessary to assure that the lockout and tagout procedure is maintained and a high level of safety is always provided.

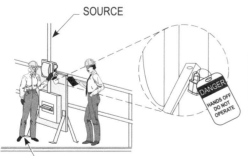

SOURCE

WORKERS BEING TRAINED OR RETRAINED

NOTE: ALL TRAINING THAT HAS BEEN COMPLETED SHALL BE DOCUMENTED, KEPT ON FILE AND UP TO DATE. THE DOCUMENTATION SHALL INCLUDE THE EMPLOYEE'S NAME, DATES THE TRAINING WAS COMPLETED AND NAME OF INSTRUCTOR.

TRAINING AND COMMUNICATION
OSHA 1910.147(c)(7)

CERTIFICATION OF TRAINING

All training that has been completed shall be documented, kept on file and up to date. The documentation shall include the employee's name, dates the training was completed and name of instructor.

ISOLATION OF CIRCUITS AND EQUIPMENT
OSHA 1910.147(c)(8)

Lockout or tagout shall only be performed by the authorized employees who are performing the servicing or maintenance of circuits and equipment.

PREPLANNING

Preplanning of the safe manner in which equipment or circuits are going to be deenergized shall be done. Before the machine, equipment or circuits are deenergized, the authorized employee shall have knowledge of the type of hazardous energy and the methods to control it. The preplanning stages would include the following:

- Types and amount of energy involved

- Individual machine involved

- Processing machine with stored energy involved

- Verification of energy isolation devices

Safety Tip: All affected employees must be notified before machines and equipment are shutdown and deenergized.

Before machines and equipment are shut down and deenergized, all affected employees shall be notified. These notifications shall take place before the controls are applied and after they are removed from the equipment. The implementation of the lockout-tagout controls can only be applied by an authorized employee. The methods for disconnecting electric circuits and equipment shall include the following procedures:

- The circuits and equipment shall be disconnected from all energy sources.

- Disconnecting is done only by personnel authorized by the employer.

- The sole disconnecting means shall not be control circuit devices such as push buttons, selector switches or electrical interlocks that deenergize electric power circuits indirectly through contactors or controllers.

- The sole disconnecting means is not allowed to be an electrically operated disconnecting device such as a panic button operating a shunt trip on a large power circuit breaker.

- The authorized employee who turns off a machine or equipment shall have knowledge of the type and the magnitude of the energy, the hazards involved and the methods or means to control the energy.

- An orderly shutdown shall be utilized to avoid any additional hazards.

LOCKOUT-TAGOUT DEVICE APPLICATION
OSHA 1910.147(d)(4)

The lockout-tagout procedure for applying locks or tags or both to disconnected circuits and equipment from all sources of energy shall be included in the written procedure. Where these requirements in the procedure are applied, they will prevent reenergizing the circuits and equipment. These lockout-tagout devices shall only be applied by an authorized employee. **(See Figure 4-15)**

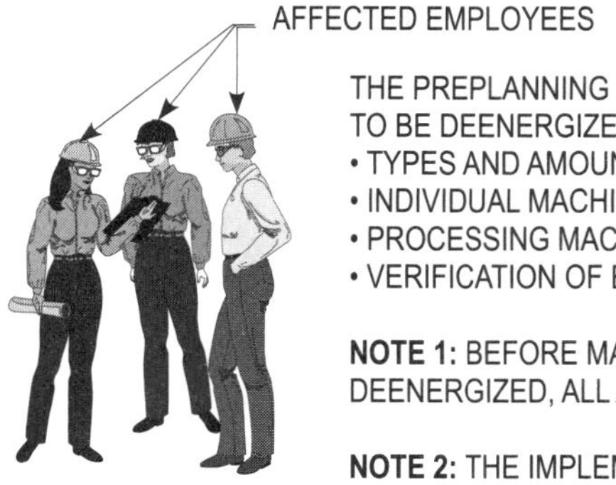

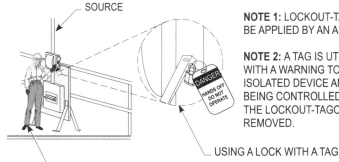

AFFECTED EMPLOYEES

THE PREPLANNING STAGES IN WHICH EQUIPMENT OR CIRCUITS ARE TO BE DEENERGIZED INCLUDE THE FOLLOWING:
• TYPES AND AMOUNT OF ENERGY INVOLVED
• INDIVIDUAL MACHINE INVOLVED
• PROCESSING MACHINE WITH STORED ENERGY INVOLVED
• VERIFICATION OF ENERGY ISOLATION DEVICES

NOTE 1: BEFORE MACHINES AND EQUIPMENT ARE SHUT DOWN AND DEENERGIZED, ALL AFFECTED EMPLOYEES SHALL BE NOTIFIED.

NOTE 2: THE IMPLEMENTATION OF THE LOCKOUT-TAGOUT CONTROLS CAN ONLY BE APPLIED BY AN AUTHORIZED EMPLOYEE.

PREPLANNING
OSHA 1910.147(c)(8)

Figure 4-14. This illustration shows the requirements for preplanning of equipment or circuits that are to be energized.

SOURCE

NOTE 1: LOCKOUT-TAGOUT DEVICES SHALL ONLY BE APPLIED BY AN AUTHORIZED EMPLOYEE.

NOTE 2: A TAG IS UTILIZED TO SUPPLEMENT A LOCK WITH A WARNING TO INDICATE THAT THE ENERGY ISOLATED DEVICE AND THE ELECTRICAL EQUIPMENT BEING CONTROLLED CANNOT BE OPERATED UNTIL THE LOCKOUT-TAGOUT DEVICES HAVE BEEN REMOVED.

USING A LOCK WITH A TAG

WORKER APPLYING LOCK AND TAG

Figure 4-15. This illustration shows that the lockout and tagout procedure for applying locks or tags or both to disconnected circuits and equipment from all sources of energy shall be included in the written procedure.

LOCKOUT-TAGOUT DEVICE APPLICATION
OSHA 1910.147(d)(4)

Tags are utilized to supplement the locks. A tag is a warning to indicate that the energy isolated device and the electrical equipment being controlled cannot be operated until the lockout-tagout devices have been removed. The basic rule is "Only the authorized employee placing the lockout-tagout device is allowed to remove it." This requirement prohibits an unauthorized person from removing the lockout-tagout device and energizing the circuit or equipment that could cause serious injury to an employee working on the equipment. Many electrical shocks and other injuries are due to an unauthorized person removing a lockout-tagout device and reenergizing the circuit or equipment.

Control circuit devices such as push buttons, selector switches and interlocks shall not be used as the sole disconnecting means for disconnecting energized circuits and parts. To serve as a disconnect, they shall disconnect all ungrounded (phase) conductors. **(See Figure 4-16)**

Safety Tip: Only the authorized employee who paced the lockout/tagout device is allowed to remove it.

NOTE: CONTROL DEVICES SUCH AS PUSHBUTTONS OR SELECTOR SWITCHES ARE NOT BE USED AS PRIMARY ISOLATING DEVICES.

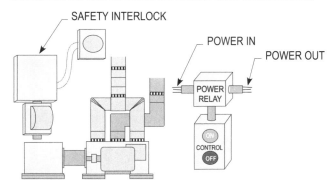

SAFETY INTERLOCK
POWER IN
POWER OUT
POWER RELAY
ON
CONTROL
OFF

LOCKOUT-TAGOUT DEVICE APPLICATION
OSHA 1910.147(d)(4)

Figure 4-16. This illustration shows that control circuit devices such as push buttons, selector switches and interlocks are not permitted to be used as the sole disconnecting means for disconnecting energized circuits or parts. Note: To serve as a disconnect, such devices shall disconnect all ungrounded (phase) conductors.

APPLYING TAGS ONLY
OSHA 1910.147(c)(3)

Safety Tip: Tags can be used without locks, only because of design limitations or if it can be demonstrated that tagging is as safe as applying a lock.

Tags are allowed to be used without locks where locks cannot be applied to an energy isolating device because of design limitations, or if the employer demonstrates that tagging alone will provide safety that is equivalent to applying a lock. However, because a person could operate the disconnecting means before reading or seeing the tag, one additional safety measure should be provided. These safety measures include the following:

• By removal of a fuse or fuses for a circuit.

• By removal of a draw-out circuit breaker from a switchboard.

• By placement of a blocking mechanism over the operating handle of a disconnecting means. The handle shall be blocked from being placed in the closed position.

• The opening of a switch and not the disconnecting means that opens the circuit between the source of power and the exposed circuits and parts.

• The opening of a switch for a control circuit that operates a disconnect and disables the system.

• Grounding the circuit where the work is to be done.

See Figure 4-17 for a detailed illustration pertaining to tags that are allowed to be used without locks.

Figure 4-17. This illustration shows that tags can be used without locks where locks cannot be used due to a given installation or the employer can demonstrate that tagging provides safety equivalent to a lock.

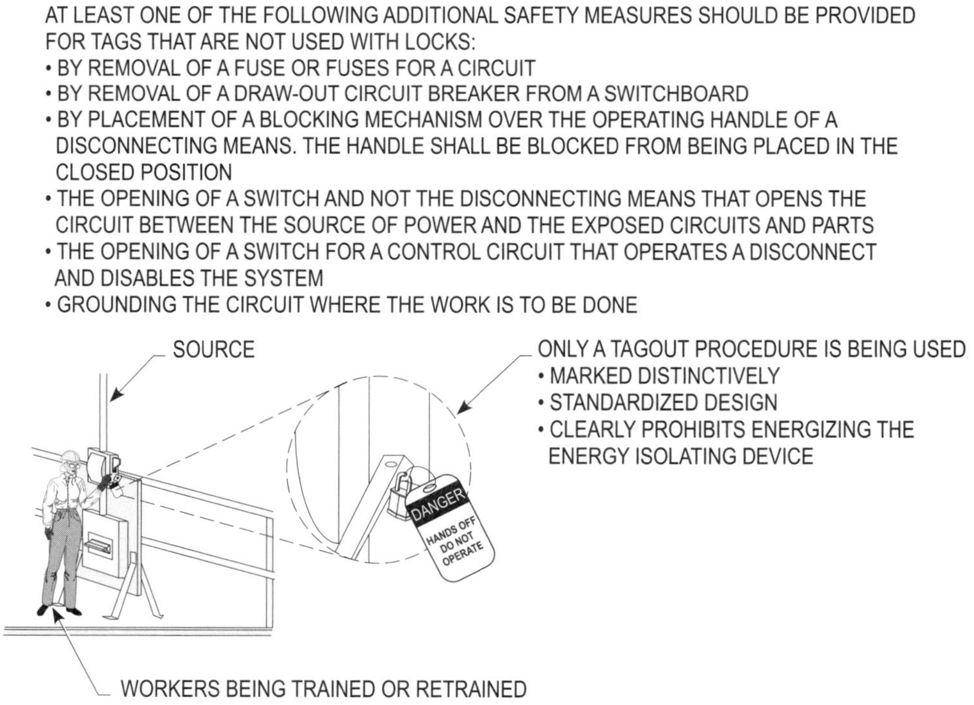

AT LEAST ONE OF THE FOLLOWING ADDITIONAL SAFETY MEASURES SHOULD BE PROVIDED FOR TAGS THAT ARE NOT USED WITH LOCKS:
• BY REMOVAL OF A FUSE OR FUSES FOR A CIRCUIT
• BY REMOVAL OF A DRAW-OUT CIRCUIT BREAKER FROM A SWITCHBOARD
• BY PLACEMENT OF A BLOCKING MECHANISM OVER THE OPERATING HANDLE OF A DISCONNECTING MEANS. THE HANDLE SHALL BE BLOCKED FROM BEING PLACED IN THE CLOSED POSITION
• THE OPENING OF A SWITCH AND NOT THE DISCONNECTING MEANS THAT OPENS THE CIRCUIT BETWEEN THE SOURCE OF POWER AND THE EXPOSED CIRCUITS AND PARTS
• THE OPENING OF A SWITCH FOR A CONTROL CIRCUIT THAT OPERATES A DISCONNECT AND DISABLES THE SYSTEM
• GROUNDING THE CIRCUIT WHERE THE WORK IS TO BE DONE

SOURCE

ONLY A TAGOUT PROCEDURE IS BEING USED
• MARKED DISTINCTIVELY
• STANDARDIZED DESIGN
• CLEARLY PROHIBITS ENERGIZING THE ENERGY ISOLATING DEVICE

DANGER
HANDS OFF DO NOT OPERATE

WORKERS BEING TRAINED OR RETRAINED

APPLYING TAGS ONLY
OSHA 1910.147(c)(3)

The additional safety measures are necessary because tagging alone is considered less safe than locking out. A disconnecting means without a lock can be closed by an employee who has failed to recognize the purpose of the tag. The disconnect is also capable of being accidentally closed by an employee who thinks it controls their equipment.

Tags are required to have the following included:

- Marked distinctively
- Standardized design
- Clearly prohibits energizing the energy isolating device

STORED ENERGY RELEASE
OSHA 1910.147(d)(5)

All stored or mechanical energy that might endanger personnel shall be released or restrained. Stored energy such as capacitors shall be discharged either through the motor windings or other effective means. High capacitance elements shall be short circuited and grounded before the electrical equipment or components are worked on. If the possibility for reaccummulation of this energy to a hazardous level exists, verification of a safe condition shall be continued until the work is complete. Mechanical energy, such as springs, shall be released or restrained so as to immobilize. This procedure of containing stored energy of all types will prevent unexpected power or energizing of devices that could cause injury to employees not expecting such. The following are methods that can be used to release or restrain mechanical energy:

Safety Tip: Stored or mechanical energy must be released or restrained.

- Slide gate
- Slip blend
- Line valve and block
- Grounding

Safety Note: The pushbuttons, selector switches and other control circuit type devices are not considered to be approved for such use. **(See Figure 4-18)**

NOTE 1: PUSHBUTTONS, SELECTOR SWITCHES, AND OTHER CONTROL CIRCUIT TYPE DEVICES ARE NOT CONSIDERED TO BE APPROVED FOR SUCH USE.

NOTE 2: MECHANICAL ENERGY, SUCH AS SPRINGS, SHALL BE RELEASED OR RESTRAINED SO AS TO IMMOBILIZE.

Figure 4-18. This illustration shows that not only does the power have to be disconnected but all stored energy or mechanical energy has to be released or restrained.

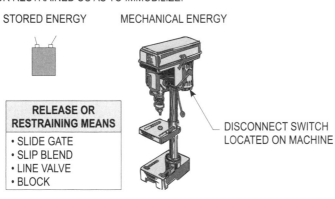

STORED ENERGY   MECHANICAL ENERGY

RELEASE OR RESTRAINING MEANS
• SLIDE GATE
• SLIP BLEND
• LINE VALVE
• BLOCK

DISCONNECT SWITCH LOCATED ON MACHINE

STORED ENERGY RELEASE
OSHA 1910.147(d)(5)

VERIFY THAT EQUIPMENT CANNOT BE RESTARTED
OSHA 1910.147(d)(6)

Only qualified personnel are permitted to operate the equipment operating controls or otherwise verify that the equipment cannot be restarted or reenergized. Where present, the following devices shall be activated to verify that it is impossible to restart the equipment by energizing the circuits and parts:

• Push buttons
• Selector switches
• Electrical interlocks
• Verify systems can't be restarted
• The opening of a switch for a control circuit that operates a disconnect and disables the system
• Grounding the circuit where the work is to be done

OSHA Required Training
1910.147(e)(3)

PROCEDURES FOR RESTORING SERVICE
OSHA 1910.147 (e)

There are certain procedures that have to be adhered to before power can be restored to circuits and equipment. The following is to be included in the lockout-tagout procedure for restoring power to circuits and equipment:

• Clear the machine or equipment of tools or materials
• Remove employees from the machine or equipment area
• Remove the lockout-tagout devices
• Energize and proceed with testing or positioning
• Deenergize all systems and reapply energy control measures
• Remove all electrical jumpers
• Remove shorts and grounds, if any have been applied
• Blocking of mechanical equipment has been cleared

Safety Tip: All affected employees in a work area must be notified before power is restored to circuits and equipment.

The work area shall be inspected to ensure that all nonessential items have been removed and to ensure that all machines or equipment components are operationally intact.

Before disconnects are closed and power restored to circuits and equipment, all affected employees in the work area that are near or working on the circuits or equipment shall be warned. A check is to be made to ensure all employees have been safely positioned or removed and are clear of circuits, parts and equipment. After a visual check to verify all employees are indeed clear, the power can restored.

A lockout device is to be removed only by the authorized employee who applied them. However, if the employee who applied the lockout-tagout devices is absent from the workplace, the locks and tags may be removed by some other authorized employee. The following rules are to be followed:

• The employer shall ensure that the employee who applied the lockout-tagout is absent from the workplace and all reasonable efforts have been made to contact the employee to inform them that the lockout-tagout has been removed.

• The employee shall be notified of the removal of the lockout-tagout before they resume work at the facility.

• There shall be unique operating conditions involving complex systems present and the employer can demonstrate that it is infeasible to do otherwise.

When circuits and equipment are ready to be reenergized, employees should be available to assist in any way necessary to ensure that circuits and equipment can be safely energized. Employees who are responsible for operating the equipment or process should be notified that the system is ready to be energized. **(See Figure 4-20)**

ONLY QUALIFIED PERSONNEL ARE ALLOWED TO OPERATE THE CONTROLS OR OTHERWISE VERIFY THAT THE EQUIPMENT CANNOT BE REENERGIZED, WHERE PRESENT, THE FOLLOWING DEVICES SHALL BE ACTIVATED TO VERIFY THAT IT IS IMPOSSIBLE TO RESTART THE EQUIPMENT BY ENERGIZING THE CIRCUITS AND PARTS:
• PUSHBUTTONS
• SELECTOR SWITCHES
• ELECTRICAL INTERLOCKS
• VERIFY SYSTEMS CANNOT BE RESTARTED

VERIFY THAT EQUIPMENT CANNOT BE RESTARTED
OSHA 1910.147(d)(6)

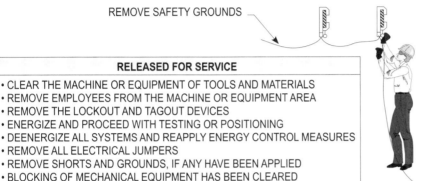

REMOVE SAFETY GROUNDS

RELEASED FOR SERVICE
• CLEAR THE MACHINE OR EQUIPMENT OF TOOLS AND MATERIALS
• REMOVE EMPLOYEES FROM THE MACHINE OR EQUIPMENT AREA
• REMOVE THE LOCKOUT AND TAGOUT DEVICES
• ENERGIZE AND PROCEED WITH TESTING OR POSITIONING
• DEENERGIZE ALL SYSTEMS AND REAPPLY ENERGY CONTROL MEASURES
• REMOVE ALL ELECTRICAL JUMPERS
• REMOVE SHORTS AND GROUNDS, IF ANY HAVE BEEN APPLIED
• BLOCKING OF MECHANICAL EQUIPMENT HAS BEEN CLEARED

PROCEDURES FOR RESTORING SERVICE
OSHA 1910.147(e)

Figure 4-20. This illustration shows that before electrical power can be reenergized to circuits and parts, certain tests and visual checks shall be made.

TESTING OR POSITIONING OF MACHINES OR EQUIPMENT OSHA 1910.147(f)

Before circuits and equipment can be reenergized for testing or repositioning, certain tests and visual check shall be made. The following procedures shall be applied before testing or temporary operation of circuits or equipment is allowed:

- Clear the machine or equipment of tools and materials.
- Remove employees from the machine or equipment area.
- Remove the lockout-tagout devices.
- Energize and proceed with testing or positioning.
- Deenergize all systems and reapply energy control measures.

See Figure 4-21 for a detailed illustration of the proper procedures for testing and temporary operation of circuits and equipment.

THE FOLLOWING STEPS SHALL BE APPLIED BEFORE RESTORING POWER
• CLEAR THE MACHINE OR EQUIPMENT OF TOOLS AND MATERIALS
• REMOVE EMPLOYEES FROM THE MACHINE OR EQUIPMENT AREA
• REMOVE THE LOCKOUT AND TAGOUT DEVICES
• ENERGIZE AND PROCEED WITH TESTING OR POSITIONING
• DEENERGIZE ALL SYSTEMS AND REAPPLY ENERGY CONTROL MEASURES

TESTING OR POSITIONING OF MACHINES OF EQUIPMENT
OSHA 1910.147(f)

Figure 4-21. This illustration shows that sometimes it is necessary to check the continuity of control circuits, etc. to reposition the equipment for servicing.

CONTRACTOR LOCKOUT-TAGOUT
OSHA 1910.147(f)(2)

Safety Tip: Contractors must use the lockout/tagout procedure that is in place at the worksite.

Whenever contractors and their employees are utilized, they shall follow the lockout-tagout procedure that is in place at the work site. If the contractor personnel have not been properly trained in the lockout-tagout procedure, the employer shall provide an authorized employee to implement a lockout-tagout of the energy isolating devices.

OSHA Required Training
1910.147(f)(2)(i)

GROUP LOCKOUT-TAGOUT
OSHA 1910.147(f)(3)

If more than one individual, craft, crew or department is required to lockout-tagout equipment or processes, the lockout-tagout procedure shall afford the same level of protection for the group as that provided by personal locks and tags. The group lockout-tagout shall be used in but not necessarily limited to the following:

- Primary responsibility for the group lockout-tagout is delegated to one authorized employee for a set number of employees working under the protection of a group lockout-tagout.

- Provisions have been made for the authorized employee to ascertain the exposure status of individual group members with regard to the lockout-tagout.

- When more than one crew, craft or department is involved, an authorized employee has been assigned overall responsibility for the group lockout-tagout to coordinate affected work forces and ensure continuity of protection.

- Each authorized employee shall place their own personal lockout-tagout devices to the group lockout-tagout device, group lockbox or comparable mechanism when they begin work. These devices shall be removed by each employee when they are done working on the machine or equipment.

Release from the group lockout-tagout shall be accomplished by the following:

Safety Tip: When group lockout/tagout is used, each employee must be protected by their personal lockout/tagout device.

- The machine or equipment area shall be cleared of nonessential items to prevent malfunctions that could result in employee injuries.

- All authorized employees shall remove their respective locks or tags from the energy isolating devices or from the group lockbox, following procedures established by the company.

- In all cases, the lockout-tagout procedure shall provide a system that identifies each authorized employee involved in the servicing and maintenance operation.

- Before reenergization, all employees in the area shall be safely positioned or removed from the area, and all affected employees shall be notified that the lockout-tagout devices have been removed.

During all group lockout-tagout operations where the release of hazardous energy is possible, each authorized employee shall be protected by their personal lockout-tagout device and by the company procedure. A master danger tag used for group lockout-tagout can be used as a personnel tagout device if each employee personally signs on and signs off on it and the tag clearly identifies each authorized employee who is being protected by it. **(See Figure 4-22)**

NOTE: THE PERSON-IN-CHARGE WILL BE HELD ACCOUNTABLE FOR SAFE EXECUTION OF THE COMPLEX LOCKOUT AND TAGOUT SCHEME.

PERSON-IN-CHARGE WITH ELECTRICAL SINGLE-LINE DIAGRAM

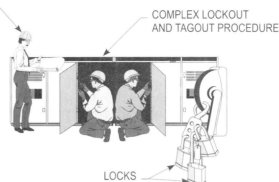

COMPLEX LOCKOUT AND TAGOUT PROCEDURE

COMPLEX PROCEDURES APPLY WHEN THERE ARE:
- MULTIPLE ENERGY SOURCES
- MULTIPLE CREWS
- MULTIPLE CRAFTS
- MULTIPLE LOCATIONS
- DIFFERENT DISCONNECTING MEANS
- PARTICULAR SEQUENCES
- CONTINUES FOR MORE THAN ONE WORK PERIOD

LOCKS

GROUP LOCKOUT-TAGOUT
OSHA 1910.147(f)(3)

Figure 4-22. This illustration shows that a complex lockout and tagout procedure shall identify and account for all persons involved in such procedures.

PROCEDURE FOR SHIFT CHANGES
OSHA 1910.147(f)(4)

The lockout-tagout procedure shall ensure continuity of protection if the lockout-tagout is going to extend beyond the shift change or personnel changes. The procedure shall provide provisions for the orderly transfer of the lockout-tagout devices between offgoing and oncoming employees. **(See Figure 4-23)**

SOURCE

SHIFT CHANGE
- EMPLOYER ENSURES THAT THE EMPLOYEE WHO APPLIED THE LOCK OR TAG IS NOT AVAILABLE AT THE WORKPLACE
- EMPLOYER ENSURES THAT THE EMPLOYEE IS AWARE THAT THE LOCK OR TAG HAS BEEN REMOVED BEFORE THEY RESUME WORK

DANGER HANDS OFF DO NOT OPERATE

CONTROLLER

PROCEURES FOR SHIFT CHANGES
OSHA 1910.147(f)(4)

Figure 4-23. This illustration shows that the lockout and tagout procedure (main rule) require that each lock and tag be removed by the employee who applied it or under his or her supervision. Note that the supervisor can remove the lock and tag if the employee is absent from the workplace due to illness, etc.

LEAD EXPOSURE IN CONSTRUCTION WORKER PROTECTION PROGRAMS
OSHA 1926.63

OSHA Required Training
1926.62(l)(1)(i) through (iiv); 2(i) thru (viii) and (3)(i) and (ii)

Lead has been poisoning workers for thousands of years. In the construction industry, traditionally most over exposures to lead have been found in the trades, such as plumbing, welding and painting.

In building construction, lead is frequently used for roofs, cornices, tank linings and electrical conduits. In plumbing, soft solder, used chiefly for soldering tinplate and copper pipe joints, is an alloy of lead and tin. Soft solder, in fact, has been banned for many uses in the United States. The use of lead-based paint in residential application has also been banned by the Consumer Product Safety Commission. However, since lead-based paint inhibits the rusting and corrosion of iron and steel, it is still used on bridges, railways, ships, lighthouses and other steel structures, although substitute coatings are available.

Significant lead exposures can also arise from removing paint from surfaces previously coated with lead-based paint, such as in bridge repair, residential renovation and demolition. With the increase in highway work, including bridge repair, residential lead abatement and residential remodeling, the potential for exposure to lead-based paint has become more common. The trades potentially exposed to lead include iron work, demolition work, painting, lead-based paint abatement work, plumbing, heating/air-conditioning, electrical work and carpentry/renovation/remodeling.

Operations that generate lead dust and fume include the following:

- Flame-torch cutting, welding, the use of heat guns, sanding, scraping and grinding of lead painted surfaces in repair, reconstruction, dismantling, and demolition work

- Abrasive blasting of bridges and other structures containing lead-based paints

- Use of torches and heat guns, and sanding, scraping, and grinding lead-based paint surfaces during remodeling or abating lead-based paint

- Maintaining process equipment or exhaust duct work

Safety Tip: Employers must implement a program that minimizes worker's risk of lead exposure.

The employer of construction workers is responsible for the development and implementation of a worker protection program in accordance with OSHA 1926.20 and OSHA 1926.62(e). This program is essential in minimizing worker risk of lead exposure. Construction projects vary in their scope and potential for exposing workers to lead and other hazards. Many projects may involve limited exposure, such as the removal of paint from a few interior residential doors. Others may involve the removal, or stripping off, of substantial quantities of lead-based paints on large bridges. The employer should, as needed, consult a qualified safety and health professional to develop and implement an effective worker protection program.

The most effective way to protect workers is to minimize exposure through the use of engineering controls and good work practices. It is OSHA policy that respirators are not to be used in lieu of engineering and work practices to reduce employee exposures to below the PEL. Respirators can only be used in combination with engineering controls and work practices to control employee exposures.

OSHA's standard for lead in construction limits worker exposures to 50 micrograms of lead per cubic meter of air averaged over an eight-hour workday.

At the minimum, the following elements should be included in the employer's worker protection program for employees exposed to lead:

- Hazard determination, including exposure assessment
- Engineering and work practice controls:

    Respiratory protection
    Protective clothing and equipment
    Housekeeping
    Hygiene facilities and practices
    Medical surveillance and provisions for medical removal
    Training
    Signs
    Recordkeeping

To implement the worker protection program properly, the employer needs to designate a competent person, i.e., one who is capable of identifying existing and predictable hazards or working conditions that are hazardous or dangerous to employees, in accordance with the general safety and health provisions of OSHA's construction standards. The competent person must have the authorization to take prompt corrective measures to eliminate such problems. Qualified medical personnel shall be available to advise the employer and employees on the health effects of employee lead exposure and supervise the medical surveillance program.

Protection for Worker Safety, Health and the Environment

Section	Answer	
_____	T	F
_____	T	F
_____	T	F
_____	T	F
_____	T	F
_____	T	F
_____	T	F
_____	T	F
_____	T	F
_____	T	F
_____	T	F
_____	T	F
_____	T	F
_____	T	F

Protection for Worker Safety, Health and the Environment

1. The employer shall ensure that medical personnel are available to advise and consult on matters of occupational health.

2. If a health-care provider or facility is not reasonably accessible to the construction site, an individual certified in first aid (by the National First Aid Association or an equivalent organization) shall be available on site to give first aid.

3. Where corrosives are used, a method for drenching or flushing shall be provided, such as eyewash or safety shower.

4. Drinking (potable) water shall be provided at the construction site. If no permanent source is available, a sealed container of water (with a dispensing tap) marked "Cold Drinks" shall be available.

5. Toilet facilities shall be provided for all construction workers, except mobile crews who have access to nearby facilities.

6. A noise exposure of 95 dBA averaged over an 8-hour workday (time-weight-average or TWA) or an equivalent noise exposure factor (F) of 0.5 is referred to as the noise exposure action level.

7. Any activity that involves the use of radioactive materials or x-rays, whether or not under license, shall be performed by a supervisor specially trained in the proper use and safe operation of such equipment.

8. Areas in which lasers are used shall be posted with warning placards.

9. When possible, laser units should be set at eye level.

10. Exhaust systems shall operate for at least 50 percent of the operation it is designed to serve.

11. Exhaust systems shall be designed, built and maintained so that hazardous substances are not drawn through the work area.

12. The Hazard Communication Program (HAZ COM) requirements applicable to Construction work are identical to those that are set forth in the General Industry standards (OSHA 1910.1200).

13. OSHA's Haz Com standard requires that employers develop, implement, and maintain at the workplace a written, comprehensive hazard communication program that includes provisions for container labeling, collection and availability of material safety data sheets, and an employee training program.

14. In the workplace, it is not necessary for the employer to label, tag or mark each container with the identity of hazardous chemicals contained therein, or show hazard warnings appropriate for employee protection.

_____ T F **15.** Copies of the Material Safety Data Sheets for hazardous chemicals in a given work site are to be readily accessible to employees in that area.

_____ T F **16.** Employers shall establish a training and information program for employees exposed to hazardous chemicals in their work area within 90 days of initial assignment and whenever a new hazard is introduced into their work area.

_____ T F **17.** OSHA requires that all employers establish a written Lockout/Tagout procedure, and also establish procedures for enforcement of such a program.

_____ T F **18.** Locks, tags, chains and other hardware that are used for the isolation of hazardous energy sources are required to be furnished by the employee.

_____ T F **19.** All lockout/tagout devices shall have a means of identifying the person applying the lockout/tagout devices.

_____ T F **20.** At least semi-annually, an authorized employee other than the one(s) utilizing the lockout/tagout procedure shall inspect and verify the effectiveness of the lockout/tagout procedure.

Personal Protective Equipment

The need for and use of personal protective equipment is as essential to the job as is any tool used for the job. The minimum personal protection for any worker or visitor to a construction site shall be safety-approved eye, head, and foot protection generally referred to as safety glasses, hard hat, and safety shoes or boots. Other safety protective devices or equipment may be required for specific jobs or operations and shall be worn or used as prescribed.

CRITERIA FOR PERSONAL PROTECTIVE EQUIPMENT
OSHA 1926.95

The following is required to be considered when applying the criteria for personal protective equipment:

- Application
- Employee-owned equipment
- Design

APPLICATION
OSHA 1926.95(a)

Protective equipment, including personal protective equipment for eyes, face, head and extremities, protective clothing, respiratory devices, and protective shields and barriers, shall be provided, used and maintained in a sanitary and reliable condition wherever it is necessary by reason of hazards of processes or environment, chemical hazards, radiological hazards or mechanical irritants encountered in a manner capable of causing injury or impairment in the function of any part of the body through absorption, inhalation or physical contact.

EMPLOYEE-OWNED EQUIPMENT
OSHA 1926.95(b)

Where employees provide their own protective equipment, the employer shall be responsible to assure its adequacy, including proper maintenance and sanitation of such equipment.

DESIGN
OSHA 1926.95(c)

All personal protective equipment shall be of safe design and construction for the work to be performed.

FOOT PROTECTION
OSHA 1926.96

Safety Tip: Workers and visitors must wear proper footwear when exposed to hazardous construction sites.

To prevent injury to feet and toes, construction workers and visitors exposed to foot hazards shall wear safety shoes or safety boots with protective toes. Safety shoes and safety boots shall meet the foot and toe protection standards of ANSI Z4 1.1 at a minimum. **(See Figure 5-1)**

Figure 5-1. This illustration shows the different types of protective footwear to protect the feet while performing various job functions.

SAFETY SHOE OR BOOT
• BUILT-IN TOE GUARD

SAFETY SHOE OR BOOT
• METATARSAL GUARD

SAFETY SHOE OR BOOT
• METAL OR FIBER GUARD

ADDITIONAL FOOT PROTECTION
• NONCONDUCTIVE AND STATIC-RESISTANT SAFETY SHOES ARE ALSO AVAILABLE TO MEET SPECIFIC ELECTRICAL AND OTHER SAFETY REQUIREMENTS
• SPECIAL PROTECTIVE FOOTWEAR SHALL BE USED UNDER CONDITIONS
• RUBBER OVERSHOES SHOULD BE USED WHEN THE GROUND IS WET AND USED IN EXTREMELY WET WEATHER
• RUBBER OVERSHOES OR BOOT SHOULD BE USED TO PROTECT THE WORKER FROM GROUND CONTAMINATIONS
• DISPOSABLE PLASTIC SHOE COVERS MAY BE NECESSARY TO PROTECT FROM TOXIC OR HIGHLY CONTAMINATED MATERIAL

NOTE: SAFETY SHOES AND SAFETY BOOTS SHALL MEET THE FOOT AND TOE PROTECTION STANDARDS OF ANSI Z4 1.1 AT A MINIMUM.

FOOT PROTECTION
OSHA 1926.96

Safety shoes or boots with metatarsal guard should be used for operations that may be very hazardous to the feet. They are used in heavy metal fabrication and heavy demolition where additional foot hazards may be introduced.

Attachable metal or fiber shoe guards to fit over regular shoes may be provided when necessary.

Nonconductive and static-resistant safety shoes are also available to meet specific electrical and other safety requirements.

Special protective footwear shall be used under certain conditions. Rubber overshoes should be used when the ground is wet, and rubber boots should be used in extremely wet weather. Rubber overshoes or boots should also be used to protect the worker from ground contamination such as a chemical spill. Disposable plastic shoe covers may be necessary to protect from toxic or highly contaminated material. Contaminated footwear shall be handled and disposed of in accordance with applicable instructions and regulations.

HEAD PROTECTION
OSHA 1926.100

Workers and visitors on a construction site shall wear a nonconductive safety-approved hat or cap (hard hat) meeting the requirements of American National Standards Institute (ANSI) Z89.1-1969 (standard use) or ANSI Z89.2-1971 (electrical use) for job being performed. Hard hats shall bear a safety approval label or marking. **(See Figure 5-2)**

HEAD PROTECTION REQUIREMENTS

• HARD HATS SHALL BEAR A SAFETY APPROVAL LABEL OR MARKING

• HARD HATS SHALL NOT BE MODIFIED OR CHANGED IN ANY WAY

• FOR OPERATIONS THAT REQUIRE BLASTING HOODS, WELDING HOODS, RESPIRATORS, FACE SHIELDS AND OTHER HEAD-WORN PROTECTIVE EQUIPMENT, APPROPRIATE HEAD PROTECTION SHALL BE PROVIDED AS PART OF THE SPECIFIC EQUIPMENT, OR HARD HATS SHALL BE PROVIDED THAT ACCOMMODATE THE REQUIRED EQUIPMENT WITHOUT INTERFERING WITH ITS FUNCTION

Figure 5-2. This illustration shows the requirements for a nonconductive safety-approved hard hat or cap to protect the head while performing various job functions.

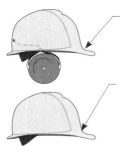

SAFETY APPROVED HARD HAT
WITH ATTACHED EAR MUFFS

SAFETY APPROVED HARD HAT

NOTE: A NONCONDUCTIVE SAFETY-APPROVED HAT OR CAP (HARD HAT) MEETING THE REQUIREMENTS OF AMERICAN NATIONAL STANDARDS INSTITUTE (ANSI) Z89.1 - 1969 (STANDARD USE) OR ANSI Z89.2 - 1971 (ELECTRICAL USE) SHALL BE USED.

HEAD PROTECTION
OSHA 1926.100

Hard hats shall not be modified or changed in any way. Hard hats that have been altered or damaged or are otherwise unsafe shall be removed from service immediately.

Safety Tip: Hard hats must not be altered in any way, unless specifically designed for that purpose.

A hard hat shall not be altered so that additional appliances such as hearing protective devices, face shields and goggles can be added. Any appliance or device added to a hard hat shall be specifically designed for that purpose and shall meet the hard hat manufacturer's specifications. Any badges, logos, decals and symbols affixed to a hard hat shall be of nonconductive material.

Where no overhead hazard exists, workers may not be required to wear head protection in roofed offices, living quarters or inside the cab of over-the-road vehicles.

For operations that require blasting hoods, welding hoods, respirators, face shields and other head-worn protective equipment, appropriate head protection shall be provided as part of the specific equipment, or hard hats shall be provided that accommodate the required equipment without interfering with its function.

OSHA Required Training
1926.101(b)

HEARING PROTECTION DEVICES
OSHA 1926.101

Hearing protective devices are generally used to control worker noise exposure on a construction site. Such devices are commonly found as ear muffs or ear plugs.

Workers required to use hearing protective devices shall have the opportunity to select the type of device to be worn from a variety of suitable hearing protectors.

Safety Tip: Workers required to wear hearing protection devices must be fitted and properly instructed in their use.

Ear muffs cover the user's ears, are usually cup shaped, and are attached to a tension band that is worn about the user's head or attached to a hard hat. Ear muffs have a rated noise attenuation factor that varies with type and manufacturer.

Workers using hearing protective devices shall be fitted with the protective device and instructed in its proper use and care. (See Figure 5-3)

Ear plugs are devices that fit into the ear canal. Such devices are commonly made of fibrous material, soft plastic or sponge-like material. Ear plugs may be found as individual plugs or in pairs attached with a cord or tension head band. Ear plugs also have a rated noise attenuation factor that differs by type and manufacturer.

Workers using hearing protective devices shall be fitted with the protective device and instructed in its proper use and care. **(See Figure 5-3)**

Hearing protection devices shall provide the necessary noise exposure attenuation to limit a worker's 8-hour time-weight-average noise exposure to no more than 90 decibel (dB), 85 dB for workers who have experienced a standard threshold shift.

Figure 5-3. This illustration shows that workers using hearing protective devices shall be fitted with the protective device and instructed in its proper use and care.

NOTE 1: EAR MUFFS AND EAR PLUGS HAVE A RATED NOISE ATTENUATION FACTOR THAT VARIES WITH TYPE AND MANUFACTURER.

NOTE 2: HEARING PROTECTION DEVICES SHALL PROVIDE THE NECESSARY NOISE EXPOSURE ATTENUATION TO LIMIT A WORKER'S 8-HOUR TIME-WEIGHT-AVERAGE NOISE EXPOSURE TO NO MORE THAN 90 DECIBEL (dB), 85 dB FOR WORKERS WHO HAVE EXPERIENCED A STANDARD THRESHOLD SHIFT.

EAR MUFFS

EAR PLUGS

HEARING PROTECTION DEVICES
OSHA 1926.101

EYE AND FACE PROTECTION
OSHA 1926.102

Eye and face protective equipment shall meet ANSI Z97.1-1968 standards for occupational use. All eye protective equipment shall be appropriately marked as meeting this standard. Manufacturers of eye and face protection equipment produce a variety of styles of safety-approved occupational eye and face wear and devices that meet ANSI Z87.1. One should be selective in choosing the best applicable protection.

Safety Note: Street-wear glasses approved by the Food and Drug Administration (FDA), often referred to as safety glasses, do not meet occupational safety standards of ANSI Z87.1 and should not be considered as approved eye protection for construction-site safety.

Safety-approved eye protection shall be worn by all workers and visitors on a construction site, with the minimum protection being safety glasses or spectacles. This is important because varied construction operations may produce hazards to the eyes in the form of chips and fines, dust and flying particles, possible chemical splash or contact, radiant energy light rays and other work-associated eye hazards. Additional safety-approved eye and face protection shall be worn as required for protection from specific operations and their associated hazards.

The recommended and most-often-required standard eye protection for construction site workers is safety glasses with side shields.

See Figure 5-4 for a detailed illustration pertaining to basic eye protection operations recommendations.

Figure 5-4. This illustration shows a chart pertaining to basic eye protection operations recommendations.

BASIC EYE PROTECTION OPERATIONS RECOMMENDATIONS		
OPERATION	**HAZARD**	**PROTECTION TYPE DESCRIPTION**
ACETYLENE (BURNING, CUTTING AND WELDING)	SPARKS, HARMFUL RAYS, MOLTEN METAL AND FLYING PARTICLES	5-BURNING GOGGLES
CHEMICAL HANDLING	SPLASH, CORROSIVE BURNS, FUMES	4-CHEMICAL GOGGLES, 8-PLASTIC FACE SHIELD
ARC WELDING (ELECTRIC)	FLYING PARTICLES	2,3-COVER GOGGLES, 8-PLASTIC FACE SHIELD
FURNACE VIEWING	GLARE, HEAT, MOLTEN METAL	9-WELDING HELMET
GRINDING (LIGHT)	FLYING PARTICLES	2,3-COVER GOGGLES, 8-PLASTIC FACE SHIELD
GRINDING (HEAVY)	FLYING PARTICLES	1A,2,3,8-PLASTIC FACE SHIELD
LABORATORY	CHEMICAL SPLASH, BROKEN GLASS	4-CHEMICAL GOGGLE, 8-PLASTIC FACE SHIELD
MACHINING	FLYING PARTICLES	1A-SAFETY GLASSES, 2,3-COVER GOGGLES
MOLTEN METAL	HEAT, GLARE, SPARKS, SPLASH	5-FURNACE GOGGLES, 8-PLASTIC OR WIRE SHIELD, 9-WELDING HELMET
SPOT WELDING	FLYING PARTICLES, SPARKS	2,3-COVER GOGGLES, 8-PLASTIC FACE SHIELD, 9-WELDING HELMET
VISITOR (OCCASIONAL, NOT EXPOSED TO DIRECT HAZARD OPERATIONS)	GENERAL SITE TOURING	1,1A-SAFETY GLASSES, 2,3 COVER GOGGLES, 8-PLASTIC FACE SHIELD
BASIC WORKER REQUIREMENT	CONSTRUCTION SITE	1,1A-SAFETY GLASSES

SAFETY GLASSES, GOGGLES AND SHIELDS

The following types of safety glasses, goggles and shields are used for protection of workers:
- Safety glasses – Type 1
- Safety glasses with side shields – Type 1A
- Cover goggles – Type 2
- Cover goggles – Type 3
- Chemical goggles – Type 4
- Burning goggles – Type 5
- Furnace goggles – Type 6
- Laser glasses/goggles – Type 7
- Face shield – Type 8
- Welding helmet – Type 9

SAFETY GLASSES – TYPE 1

Type 1 safety glasses may have plain (nonprescription) or Rx (prescription) lenses mounted in a metal or plastic frame without side shields. They provide basic protection from frontal impact hazards. The lenses are usually available in clear and a variety of shades. **(See Figure 5-5)**

SAFETY GLASSES WITH SIDE SHIELDS - TYPE IA

These are the same as Type 1 except with fixed or attached side shields that offer added protection to the sides. Some safety glasses are designed with an upper frame extended to the wearer's head. This is generally referred to as a "brow guard," which offers added protection at the top of the glasses. **(See Figure 5-5)**

Figure 5-5. This illustration shows the different types of spectacles used to protect the eyes while performing various job functions.

TYPE 1 SAFETY GLASSES PROVIDE BASIC PROTECTION FROM FRONTAL IMPACT HAZARDS.

SAFETY GLASSES
• TYPE 1
• NONPRESCRIPTION OR PRESCRIPTION LENSES
• METAL OR PLASTIC FRAME

SAFETY GLASSES WITH SIDE SHIELDS
• TYPE 1A

TYPE 1A SAFETY GLASSES WITH SHIELDS PROVIDE BASIC PROTECTION FROM FRONTAL IMPACT HAZARDS WITH ADDED PROTECTION TO THE SIDES

SAFETY GLASSES

COVER GOGGLES -TYPE 2

These goggles are generally a one-piece full-view lens affixed in a rigid or flexible plastic face-fitting body held in wearing position with an adjustable head piece, band or attachment to a hard hat. Most often the body has holes or perforations on the top and sides for venting. Their general use is for frontal impact protection and protection from heavy chips and fines. They may be worn as protection over nonsafety prescription glasses or by workers and visitors for basic eye protection. They are available in clear and a variety of shades. **(See Figure 5-6)**

COVER GOGGLES - TYPE 3

These cover goggles are the same as Type 2 except with indirect or shielded venting. They provide additional protection from fines and dust. **(See Figure 5-6)**

CHEMICAL GOGGLES - TYPE 4

Chemical goggles are the same as Types 2 and 3 except with protected indirect or filtered venting on the body. They provide additional protection for chemical splashes, heavy mist and overspray. **(See Figure 5-6)**

BURNING GOGGLES - TYPE 5

These goggles have either a single full-lens or eye-cup design equipped with a filtered lens appropriate to the radiant-light ray hazard. The body of each is usually made of solid plastic with indirect venting designed to prevent entry of radiant light. Some full-body goggles are designed to hold a standard 2 in. x 4 1/2 in. filter lens. These goggles are also available with a flip-up front filtered lens with a clear safety lens behind it allowing protection for chipping and grinding. Goggles may be held in wearing position by a head piece or band or may be attached to a hard hat in head hazard areas. See Figure 5-8 for selection of recommended filter lenses. **(See Figure 5-7)**

See Figure 5-8 for a detailed illustration pertaining to the selection of filtered and tinted lens shades.

FRESH AIR VENTS

COVER GOGGLES
• TYPE 2

TYPE 2 COVER GOGGLES PROVIDE FRONTAL IMPACT PROTECTION AND PROTECTION FROM HEAVY CHIPS AND FINES

Figure 5-6. This illustration shows the different types of goggles used to protect the eyes while performing various job functions.

SHIELDED VENTS

COVER GOGGLES
• TYPE 3

TYPE 3 COVER GOGGLES PROVIDE FRONTAL IMPACT PROTECTION AND PROTECTION FROM HEAVY CHIPS, FINES AND DUST

VENTS

COVER GOGGLES
• TYPE 4

TYPE 4 COVER GOGGLES PROVIDE FRONTAL IMPACT PROTECTION AND PROTECTION FROM HEAVY CHIPS, FINES AND DUSTS AND ADDITIONAL PROTECTION FOR CHEMICAL SPLASHES, HEAVY MIST AND OVERSPRAY

COVER AND CHEMICAL GOGGLES

BURNING GOGGLES
• SINGLE FULL LENS

Figure 5-7. This illustration shows the different types of burning goggles used to protect the eyes while performing various job functions.

NOTE: THE BODY OF BURNING GOGGLES IS USUALLY MADE OF SOLID PLASTIC WITH INDIRECT VENTING DESIGNED TO PREVENT ENTRY OF RADIANT LIGHT.

BURNING GOGGLES
• EYE CUP

BURNING GOGGLES

FURNACE GOGGLES - TYPE 6

Furnace goggles are the same as Type 5 fitted with the appropriate tinted or filtering lens to protect from glare, radiant light and heat (they can be worn behind a standard face shield or metal mesh face shield to protect the face from heat.)

Figure 5-8. This illustration shows a chart pertaining to the selection of filtered and tinted lens shades.

GUIDE TO SELECTING FILTERED AND TINTED LENS SHADES	
WELDING OPERATIONS	**SHADE NUMBER**
SHIELDED METAL ARC 1/16" TO 5/32" DIAMETER ELECTRODE	10
GAS-SHIELDED ARC (NONFERROUS) 1/16" TO 5/32" DIAMETER ELECTRODE	11
GAS-SHIELDED ARC (FERROUS) 1/16" TO 5/32" DIAMETER ELECTRODE	12
SHIELDED METAL ARC 3/16" TO 1/4", 5/16" TO 3/8" DIAMETER EXCHANGE	13
ATOMIC HYDROGEN	10 TO 14
CARBON ARC	14

GAS OPERATIONS	**SHADE NUMBER**
TORCH SOLDERING	2
TORCH BRAZING	3 OR 4
CUTTING TO 1"	3 OR 4
CUTTING TO 1" TO 6" THICK	4 OR 5
CUTTING OVER 6" THICK	5 OR 6
GAS WELDING TO 1/8" THICK	4 OR 5
GAS WELDING FROM 1/8" TO 1/2" THICK	5 OR 6
GAS WELDING OVER 1/2" THICK	6 OR 8

LASER GLASSES/GOGGLES - TYPE 7

Laser glasses/goggles are worn for protection from harmful laser radiation. They are available in safety glasses or goggle design (as Type IA or 5) fitted with the appropriate laser protective lens. For lens absorption and attenuation factors, refer to the laser equipment manufacturer's recommendation in accordance with the watts per square centimeter (CW) power density, optical density and attenuation factor. **(See Figure 5-9)**

Figure 5-9. This illustration shows a chart pertaining to the selection of laser protection.

INTENSITY, CW	OPTICAL DENSITY (O.D.)	ATTENUATION FACTOR
10-2	5	105
10-1	6	106
1.0	7	107
10.0	8	108

NOTE: OUTPUT LEVELS FALLING BETWEEN LINE IN THIS TABLE SHALL REQUIRE THE HIGHER O.D.

FACE SHIELD - TYPE 8

A face shield is generally a plastic face piece or shield attached to an adjustable fitting head band or attached to a hard hat for use in head hazard areas. It provides face protection from chemicals and other face hazards and is to be worn over other eye protection. The face piece or shield is available in clear, shades and in see-through reflective plastic with a metal mesh for protection from flying debris, heat and glare. **(See Figure 5-10)**

WELDING HELMET - TYPE 9

Welding helmets are designed for protection from arc-welding radiant-light rays and molten metal and slag. They are made of a solid fabricated face piece or hood that is equipped with a filter-lens holder at the wearer's eye level and an adjustable head band. The helmets may be attached to a hard hat for wear in head hazard areas. On some welding hoods, the filter-lens holder can be a flip-up type with a backup clear safety lens providing viewing and protection while chipping and grinding without having to remove the hood. Welding hoods are not limited to use in welding as they may be used for added eye and face protection in other high energy light and heat-producing operations. See Figure 5-8 for recommended filter-lens selection. **(See Figure 5-11)**

TYPE 8 FACE SHIELDS PROVIDE FACE PROTECTION FROM CHEMICALS AND OTHER FACE HAZARDS AND IS TO BE WORN OVER OTHER EYE PROTECTION

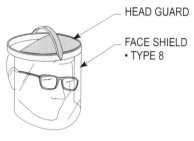

HEAD GUARD

FACE SHIELD
• TYPE 8

FACE SHIELD

Figure 5-10. This illustration shows a face shield used to protect the eyes while performing various job functions.

TYPE 9 WELDING HELMETS PROVIDE PROTECTION FROM ARC-WELDING RADIANT-LIGHT RAYS AND MOLTEN METAL AND SLAG

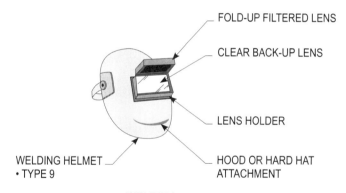

FOLD-UP FILTERED LENS

CLEAR BACK-UP LENS

LENS HOLDER

WELDING HELMET
• TYPE 9

HOOD OR HARD HAT ATTACHMENT

WELDING HELMET

Figure 5-11. This illustration shows a welding helmet used to protect the eyes while performing various job functions.

SHADED, TINTED AND FILTER LENSES

Shaded, tinted or filter lenses are available in almost all types of eye and face protection. These are provided for specific purposes such as for protection from bright sunlight reflection (sun reflected off sheet-metal roofing) to radiant-light rays (furnace operations and weld-flash). Shaded or tinted lenses come in a variety of colors and color densities from light to dark and commonly in colors of green, gray or special-purpose design. For filter shades in operations of radiant-light rays, refer to Figure 5-7 or recommendations specified in ANSI Z87.1. For glare or other purposes, refer to the manufacturer's information and recommendations.

INSPECTION AND CARE

All eye protection devices should be inspected daily. Welding helmets should be periodically examined for cracked or broken shade lenses and light leakage. Lenses should be cleaned to maintain good vision. When slag or abrasives may damage the lenses of safety glasses, especially prescription glasses, added cover protection should always be used, such as cover goggles or a face shield. Damaged eye and face wear should be replaced.

VISITORS' EYE PROTECTION

Visitors shall be furnished with and wear approved eye protection that shall at a minimum include safety glasses with or without side shields. Those wearing non-safety prescription glasses may wear any type of safety approved eye cover protection that fits securely over their prescription glasses (i.e., cover goggles, face shield or safety-approved visitor specs). Should visitors be subject to operational eye hazards, they shall be provided with appropriate eye and face protection for the specific hazard.

ARC WELDING FLASH BURN PROTECTION

Precautions should be taken to avoid flash hazards due to electric arc welders, and those working nearby are subject to flash burns to the eyes. Control methods include correctly using the following:

- Nonreflective shields,

- Light-absorbent curtains,

- Appropriate safety glasses fitted with solid or colored side shields and colored lenses and

- Welding helmets.

Safety Note: In multiwelder operations, welders shall wear safety-approved glasses with side shields and filtered lenses under their helmet. The lenses shall be of a shade number consistent with the maximum weld-flash hazard in the welding operation. Welders should affix a dark flame retardant cloth to the entire back (top and sides) of their welding helmet that extends past the collar to protect exposed skin and neck from other welding operations. To the greatest extent possible, shields or curtains should be used to separate welding operations.

HAND AND WRIST PROTECTION

Gloves are worn at construction sites to protect the hands and wrists from injuries caused by sharp instruments, rough material, friction, heat, chemicals and other hazards.

Cotton gloves are worn when workers handle material with rough surfaces or have minimal contact with material such as tar, paint and grease. Cotton gloves shall not be worn by those working near flames or handling material that splinters, snags or can be absorbed through the skin. **(See Figure 5-12)**

Leather gloves provide the best all-purpose hand protection and should be worn when workers handle rough material or material that splinters or snags. Leather gloves shall be worn when the worker is handling moderately hot material or when the worker is welding, cutting, servicing equipment or rigging. **(See Figure 5-12)**

Leather gloves with attached gauntlets should be worn to protect the wrist from impact, flying particles and radiant heat. **(See Figure 5-12)**

Cloth gloves with leather palms, with or without gauntlets, should be worn to protect the worker from impacts and flying particles. They are not recommended for heat or flame work.

Cotton gloves with synthetic palms should be worn when handling material that will splinter or has sharp edges. **(See Figure 5-13)**

Synthetic gloves, with or without gauntlets, should be worn when working with adhesives, caulks, resins or other similar material, as well as with concrete. These gloves are not recommended for handling hot material or when working near flames. **(See Figure 5-13)**

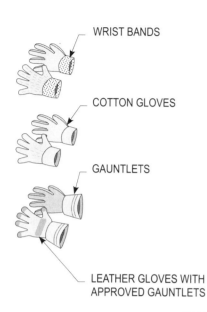

WRIST BANDS

COTTON GLOVES

GAUNTLETS

LEATHER GLOVES WITH
APPROVED GAUNTLETS

LEATHER AND COTTON GLOVE REQUIREMENTS

- COTTON GLOVES ARE WORN WHEN WORKERS HANDLE MATERIAL WITH ROUGH SURFACES OR HAVE MINIMAL CONTACT WITH MATERIAL SUCH AS TAR, PAINT AND GREASE

- COTTON GLOVES SHALL NOT BE WORN BY THOSE WORKING NEAR FLAMES OR HANDLING MATERIAL THAT SPLINTERS, SNAGS OR CAN BE ABSORBED THROUGH THE SKIN

- LEATHER GLOVES SHOULD BE WORN WHEN THE WORKER IS HANDLING MODERATELY HOT MATERIAL OR WHEN THE WORKER IS WELDING, CUTTING, SERVICING EQUIPMENT OR RIGGING

- LEATHER GLOVES WITH ATTACHED GAUNTLETS SHOULD BE WORN TO PROTECT THE WRIST FROM IMPACT, FLYING PARTICLES AND RADIANT HEAT

Figure 5-12. This illustration shows the different types of gloves used to protect the hands while performing various job functions.

HAND AND WRIST PROTECTION

COTTON GLOVES
- LEATHER PALMS
- SYNTHETIC PALMS

COTTON GLOVES WITH LEATHER AND SYNTHETIC AND SYNTHETIC GLOVE REQUIREMENTS

- COTTON GLOVES WITH LEATHER PALMS, WITH OR WITHOUT GAUNTLETS, SHOULD BE WORN TO PROTECT THE WORKER FROM IMPACTS AND FLYING PARTICLES

- COTTON GLOVES WITH SYNTHETIC PALMS SHOULD BE WORN WHEN HANDLING MATERIAL THAT WILL SPLINTER OR HAS SHARP EDGES

- SYNTHETIC GLOVES, WITH OR WITHOUT GAUNTLETS, SHOULD BE WORN WHEN WORKING WITH ADHESIVES, CAULKS, RESINS, OR OTHER SIMILAR MATERIAL, AS WELL AS WITH CONCRETE

Figure 5-13. This illustration shows the different types of gloves used to protect the hands while performing various job functions.

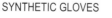

SYNTHETIC GLOVES

HAND AND WRIST PROTECTION

Chemical gloves (rubber, neoprene and latex) should be worn for protection of hand exposure to hazardous chemicals such as corrosives, solvents, epoxies and other materials. It is very important to review the glove selection for each chemical exposure as some chemicals will readily deteriorate or penetrate certain types of synthetic gloves. Refer to the product material safety data sheet or the glove manufacturer's recommendation for the appropriate glove protection against specific chemicals or materials. **(See Figure 5-14)**

Lineman's gloves, which are protected with an over glove, shall be used for protection from electric shock. They are voltage-rated and require special inspection and testing.

Special protective hand wear such as gloves and mitts are available for handling specific material (for example, material that is extremely hot or cold and certain hazardous chemicals). **(See Figure 5-14)**

Figure 5-14. This illustration shows the different types of gloves used to protect the hands while performing various job functions.

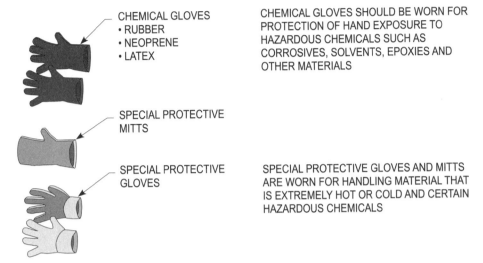

CHEMICAL GLOVES
• RUBBER
• NEOPRENE
• LATEX

CHEMICAL GLOVES SHOULD BE WORN FOR PROTECTION OF HAND EXPOSURE TO HAZARDOUS CHEMICALS SUCH AS CORROSIVES, SOLVENTS, EPOXIES AND OTHER MATERIALS

SPECIAL PROTECTIVE MITTS

SPECIAL PROTECTIVE GLOVES

SPECIAL PROTECTIVE GLOVES AND MITTS ARE WORN FOR HANDLING MATERIAL THAT IS EXTREMELY HOT OR COLD AND CERTAIN HAZARDOUS CHEMICALS

HAND AND WRIST PROTECTION

Gloves should be free of tears and holes that may cause unwanted exposure during use. Dirty gloves should never be washed in solvents because they may deteriorate the glove material, especially leather, and may leave a harmful residue. Certain gloves, cotton or other fabric, may be washed in laundry detergent and rinsed well for reuse. For specific guidance on appropriate glove cleaning methods, refer to the glove manufacturer's recommendations.

Gloves should not be worn near moving machinery (for example, chain drives, belt drives, pulleys, transmission drives, gear trains and cutting tools). Wearing gloves near such equipment could result in an amputation or other serious injury.

SPECIAL PERSONAL PROTECTIVE EQUIPMENT

The following are required to be considered for special personal protective equipment:

• Welding
• Working in traffic
• Disposable apparel
• Personal attire

WELDING

Workers need special protection from molten metal and slag during an overhead or other heavy welding operation. Leather or flame-proof aprons, leggings, chaps, leather capes and arm protection may be used for this purpose. To avoid weld-flash burns, welders and other workers in the vicinity subject to flash burn should be certain that the entire body is covered and their clothing has no holes or gaps. See heading "Welding Helmet – Type 9" in this chapter for specific eye and face protection for welding operations. **(See Figure 5-15)**

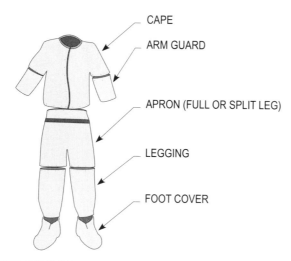

CAPE

ARM GUARD

APRON (FULL OR SPLIT LEG)

LEGGING

FOOT COVER

NOTE: TO AVOID WELD-FLASH BURNS, WELDERS AND OTHER WORKERS IN THE VICINITY SUBJECT TO FLASH BURN SHOULD BE CERTAIN THAT THE ENTIRE BODY IS COVERED AND THEIR CLOTHING HAS NO HOLES OR GAPS.

WELDING

Figure 5-15. This illustration shows a welder using special protection for a overhead weld.

WORKING IN TRAFFIC

Department of Transportation (DOT) orange vests shall be worn by workers who direct traffic or are engaged in construction adjacent to highways or secondary roads. **(See Figure 5-16)**

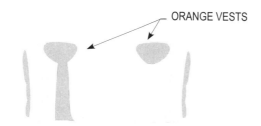

ORANGE VESTS

DEPARTMENT OF TRANSPORTATION (DOT) ORANGE VESTS SHOULD BE WORN BY WORKERS WHO DIRECT TRAFFIC OR ARE ENGAGED IN CONSTRUCTION ADJACENT TO HIGHWAYS OR SECONDARY ROADS

WORKING IN TRAFFIC

Figure 5-16. This illustration shows the types of orange vests that shall be worn by workers who direct traffic or are engaged in construction adjacent to highways or secondary roads.

DISPOSABLE APPAREL

If disposable apparel (for example, coveralls, shoe covers and head covers) is necessary for specific operations such as when working with polychlorinated biphenyls (PCB's), asbestos, hazardous waste or other toxic materials, appropriate protective clothing shall be furnished to affected personnel. Personnel required to use disposable apparel shall be trained in the appropriate use and disposal procedures.

PERSONAL ATTIRE

Workers should consider both the demands of the job assignment and current weather conditions when dressing for the worksite. Because appropriate dress contributes to each employee's personal safety, it should always be of primary concern to the worker and supervisor.

Safety Note: The contractor should establish a worksite dress code. Workers should follow it at all times.

Workers should dress in clothing suitable to their job assignment. Clothing should be durable and washable and protect workers from on-the-job hazards, inclement weather and sunburn. Clothing should fit close to the body so that it will not catch on power machinery or other objects.

Workers subject to hazards of weld-flash, sunburn or other radiant light energy should wear clothing that covers the entire body, including the neck and arms.

Because of the potential for accidents, workers should not wear jewelry such as rings, necklaces and bracelets. Wrist watches with a rip-away band are acceptable. Conductive jewelry of any type should never be worn for jobs that involve electric wiring.

RESPIRATORY PROTECTION
OSHA 1926.103

The requirements for respiratory protection in construction are Identical to those that are set forth in the General Industry Standard 1910.134. The following shall be considered for respiratory protection in construction:

- Permissible practice
- Requirements for a minimal acceptable program
- Type of respiratory protective devices
- Prefilters
- Selection of respirators

PERMISSIBLE PRACTICE
OSHA 1910.134(a)

Safety Tip: When required, suitable respirators must be provided by the employer.

In the control of those occupational diseases caused by breathing air contaminated with harmful dusts, fogs, fumes, mists, gases, smokes, sprays or vapors, the primary objective shall be to prevent atmospheric contamination. This shall be accomplished as far as feasible by accepted engineering control measures (for example, enclosure or confinement of the operation, general and local ventilation and substitution of less toxic materials). When effective engineering controls are not feasible, or while they are being instituted, appropriate respirators shall be used pursuant to the following requirements.

Respirators shall be provided by the employer when such equipment is necessary to protect the health of the employee. The employer shall provide the respirators that are applicable and suitable for the purpose intended. The employer shall be responsible for the establishment and maintenance of a respiratory protective program, which shall include the requirements outlined in the following section. The employee shall use the provided respiratory protection in accordance with instructions and training received.

OSHA Required Training
1926.103(c)(1)

REQUIREMENTS FOR A MINIMAL ACCEPTABLE PROGRAM
OSHA 1910.134(c)

The following is required to be considered for requirements pertaining to a minimal acceptable program:

- Written standard operating procedures governing the selection and use of respirators shall be established.

- Respirators shall be selected on the basis of hazards to which the worker is exposed.

• The user shall be instructed and trained in the proper use of respirators and their limitations.

• Respirators shall be regularly cleaned and disinfected. Those used by more than one worker shall be thoroughly cleaned and disinfected after each use.

• Respirators shall be stored in a convenient, clean and sanitary location.

• Respirators used routinely shall be inspected during cleaning. Worn or deteriorated parts shall be replaced. Respirators for emergency use such as self-contained devices shall be thoroughly inspected at least once a month and after each use.

• Appropriate surveillance of work area conditions and degree of employee exposure or stress shall be maintained.

• Regular inspection and evaluation is required to determine the continued effectiveness of the program.

• Persons should not be assigned to tasks requiring use of respirators unless it has been determined that they are physically able to perform the work and use the equipment. The local physician shall determine what health and physical conditions are pertinent. The respirator user's medical status should be reviewed periodically (for instance, annually).

• Respirators shall be selected from among those jointly approved by the Mine Safety and Health Administration and the National Institute for Occupational Safety and Health under the provisions of 30 CFR Part 11.

TYPES OF RESPIRATORY PROTECTIVE DEVICES

Three basic types of respiratory protective devices, commonly referred to as respirators, are used on a construction site. **(See Figure 5-17)**

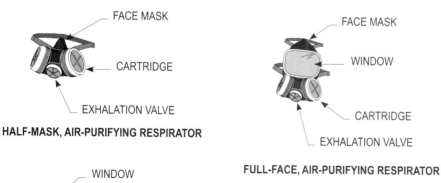

Figure 5-17. This illustration shows the three basic types of respiratory protective devices.

HALF-MASK, AIR-PURIFYING RESPIRATOR

FULL-FACE, AIR-PURIFYING RESPIRATOR

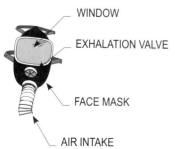

FULL-FACE, AIR-SUPPLIED RESPIRATOR

DISPOSABLE-TYPE MASK

TYPES OF RESPIRATORY PROTECTIVE DEVICES

Disposable-Type Mask. These masks are used for protection from nuisance dust, mist, vapors and fumes that are not of a toxic or hazardous nature. Disposable-type masks cover the breathing zone (nose and mouth) of the wearer with a fine mesh or filtering media and are usually discarded following their use. These masks have very limited protective values and should not be used for protection against toxic or hazardous materials.

Half-Mask Respirator. A half-mask respirator is a face piece that fits tightly (forming a seal) against the wearer's face covering only the nose and mouth. This type of respirator may be an air-purifying type that is equipped with filter cartridges or canisters or air-supplied (provided with breathing air).

Full-Face Respirator. A full-face respirator is a face piece that fits tightly (forming a seal) over the wearer's face, covering the nose, mouth, eyes and full face. They also have a sealed window for vision purposes. This type of respirator may be an air-purifying type that is equipped with filter cartridges or canisters or be air supplied (provided with breathing air).

Cartridges and Canisters. These air-purifying devices are fitted to respirators to filter limited amounts of specific types of hazardous materials from the user's breathing zone. Users should refer to the manufacturer's literature for assistance in selecting the appropriate cartridge or canister for the specific hazardous materials in the work area and to ensure that the equipment is used within its specific limitations. Generally, air-purifying cartridges or canisters are used as limited protection against the following substances:

- asbestos
- organic vapors
- acid gases
- ammonia
- hydrogen chloride
- chlorine
- pesticides
- radionuclides
- hazardous dust
- formaldehyde
- methylamine
- combinations of two or more substances

See Figure 5-18 for a detailed illustration pertaining to cartridges and canisters used for respiratory protective devices.

Figure 5-18. This illustration shows the different types of cartridges and canisters used for respiratory protective devices.

TYPES OF CARTRIDGES FOR RESPIRATORS

AIR-PURIFYING CARTRIDGES OR CANISTERS ARE USED AS LIMITED PROTECTION AGAINST THE FOLLOWING SUBSTANCES:
• ASBESTOS • ORGANIC VAPORS • ACID GASES • AMMONIA • HYDROGEN CHLORIDE • CHLORINE • PESTICIDES • RADIONUCLIDES • HAZARDOUS DUST • FORMALDEHYDE • METHYLAMINE • COMBINATIONS OF TWO OR MORE SUBSTANCES

NOTE: USERS SHOULD REFER TO MANUFACTURER'S LITERATURE FOR ASSISTANCE IN SELECTING THE APPROPRIATE CARTRIDGE OR CANISTER FOR THE SPECIFIC HAZARDOUS MATERIALS IN THE WORK AREA AND TO ENSURE THAT THE EQUIPMENT IS USED WITHIN ITS SPECIFIC LIMITATIONS.

CARTRIDGES AND CANISTERS

PREFILTERS

A prefilter is an additional filtering device that fits outside and over an air-purifying cartridge to trap or prefilter materials, such as dust and mist, preventing clogging of the primary cartridge.

SELECTION OF RESPIRATORS
OSHA 1910.134(d)

The chemical and physical properties of contaminants and the toxicity and concentration of hazardous material shall be considered when respiratory protective equipment is selected. Only National Institute of Occupational Safety and Health (NIOSH) / Mine Safety and Health Administration (MSHA) approved respiratory protective equipment shall be used.

The following shall be considered when respiratory protective equipment is selected:

- Work requirements.

- Nature and extent of the hazards.

- Limitations of each piece of equipment.

See Figure 5-19 for a detailed illustration pertaining to the types of respiratory protective equipment required for specific hazards.

GUIDE TO SELECTING RESPIRATORY PROTECTIVE EQUIPMENT	
OXYGEN DEFICIENCY; IMMEDIATELY DANGEROUS TO LIFE AND HEALTH	SELF-CONTAINED BREATHING APPARATUS (SCBA), HOSE MASK WITH BLOWER, COMBINATION AIR-LINE RESPIRATOR WITH AUXILIARY SELF CONTAINED AIR SUPPLY, AIR-STORAGE RECEIVER WITH ALARM
GAS AND VAPOR CONTAMINANTS IMMEDIATELY DANGEROUS TO LIFE AND HEALTH	SCBA, HOSE MASK WITH BLOWER, AIR-PURIFYING, FULL-FACE RESPIRATOR WITH CHEMICAL CANISTER (GAS MASK), SELF-RESCUE MOUTHPIECE RESPIRATOR (FOR ESCAPE ONLY), COMBINATION AIR-LINE RESPIRATOR WITH AUXILIARY SELF-CONTAINED AIR SUPPLY, AIR-STORAGE RECEIVER WITH ALARM
GAS AND VAPOR CONTAMINANTS IMMEDIATELY DANGEROUS TO LIFE AND HEALTH	AIR-LINE RESPIRATOR, HOSE MASK WITHOUT BLOWER, AIR-PURIFYING, HALF MASK OR MOUTHPIECE RESPIRATOR WITH CHEMICAL CARTRIDGE
PARTICULATE CONTAMINANTS IMMEDIATELY DANGEROUS TO LIFE AND HEALTH	SCBA, HOSE MASK WITH BLOWER, AIR-PURIFYING, FULL-FACE RESPIRATOR WITH APPROPRIATE FILTER, SCBA (FOR ESCAPE ONLY), COMBINATION AIR-LINE RESPIRATOR WITH AUXILIARY SELF-CONTAINED AIR SUPPLY, AIR-STORAGE RECEIVER WITH ALARM
PARTICULATE CONTAMINANTS IMMEDIATELY DANGEROUS TO LIFE AND HEALTH	AIR-PURIFYING, HALF-MASK OR MOUTHPIECE RESPIRATOR WITH FILTER PAD OR CARTRIDGE, AIR-LINE RESPIRATOR, AIR-LINE ABRASIVE-BLASTING RESPIRATOR, HOSE MASK WITHOUT BLOWER
COMBINATION GAS, VAPOR AND PARTICULATE CONTAMINANTS IMMEDIATELY DANGEROUS TO LIFE AND HEALTH	SCBA, HOSE MASK WITH BLOWER, AIR-PURIFYING, FULL-FACE RESPIRATOR WITH CHEMICAL CANISTER AND APPROPRIATE FILTER (GAS MASK WITH FILTER), SELF RESCUE MOUTHPIECE RESPIRATOR (FOR ESCAPE ONLY), COMBINATION AIR-LINE RESPIRATOR WITH AUXILIARY SELF CONTAINED AIR SUPPLY, OR AIR SUPPLY OR AIR-STORAGE RECEIVER WITH ALARM
COMBINATION GAS, VAPOR AND PARTICULATE CONTAMINANTS IMMEDIATELY DANGEROUS TO LIFE AND HEALTH	AIR-LINE RESPIRATOR, HOSE MASK WITHOUT BLOWER, AIR-PURIFYING, HALF-MASK OR MOUTHPIECE WITH CHEMICAL CARTRIDGE AND APPROPRIATE FILTER

Figure 5-19. This illustration shows a chart for selecting respiratory protective equipment.

NOTE: FOR THE PURPOSE OF THIS SECTION "IMMEDIATELY DANGEROUS TO LIFE AND HEALTH" IS DEFINED AS A CONDITION THAT EITHER EXPOSES AN IMMEDIATE THREAT TO LIFE AND HEALTH OR AN IMMEDIATE THREAT OF SEVERE EXPOSURE TO CONTAMINANTS SUCH AS RADIOACTIVE MATERIALS, WHICH MAY HAVE ADVERSE DELAYED EFFECTS ON HEALTH.

Workers shall be trained in the safe use and limitations of the respiratory equipment to be used.

Workers who issue respiratory protective equipment shall be trained in how to select the appropriate equipment for the job, including respirators and cartridges.

Medical surveillance and determination of workers' health are required if workers are required to wear respirators.

Safety Note: A clean shaven face is required to maintain a good seal.

A respirator must not be worn if an adequate seal between the face and respirator cannot be made. The following may cause an inadequate seal:

- Sideburns
- A beard
- Absence of upper and lower dentures
- Temples on glasses

Respiratory protective equipment shall be inspected daily and maintained in good condition. Canisters, cartridges and air-line filters shall be replaced as necessary. For further information, refer to the manufacturer's recommendations for each respirator.

SCBA equipment, including air cylinders, shall be thoroughly inspected at least monthly and before each use. Records of all inspections, including inspections of emergency rescue equipment, shall be maintained.

Air supplied to respiratory protective equipment from bottles, cylinders, or compressors shall be clean, respirable and free of contaminants. The air shall be at least grade "D" breathing air, as set forth in Compressed Gas Association Commodity Specification G07.1.

Hose connections used to supply air to respiratory protective equipment shall be incompatible with connections on other gas cylinders or fittings.

Safety Tip: When workers are in a contaminated area, they must not remove their respiratory protective equipment.

Workers who use air-line- or SCBA-respiratory protective equipment in high-hazard atmospheres or those immediately dangerous to life shall be equipped with a safety harness and lifeline. Also, a safety watch or stand-by worker with the same equipment shall be immediately outside the area. A method of communication between the respirator user and the safety watch shall be established before workers enter these areas.

Once used, respiratory protective equipment shall be thoroughly cleaned, disinfected, inspected and determined to be in good condition before it is reissued. New canisters or cartridges that are appropriate for the hazard shall be supplied with the respirator.

Emergency rescue equipment shall be cleaned and disinfected after each use.

Workers shall not remove respiratory protective equipment when workers are in a contaminated area.

SAFETY BELTS, LIFELINES AND LANYARDS OSHA 1926.104

Lifelines, safety belts, and lanyards shall be used only for employee safeguarding. Any lifeline, safety belt or lanyard actually subjected to in-service loading, as distinguished from static load testing, shall be immediately removed from service and shall not be used again for employee safeguarding.

Lifelines shall be secured above the point of operation to an anchorage or structural member capable of supporting a minimum dead weight of 5400 pounds.

Lifelines used on rock-scaling operations, or in areas where the lifeline may be subjected to cutting or abrasion, shall be a minimum of 7/8 in. wire core manila rope. For all other lifeline applications, a minimum of 3 or 4 in. manila or equivalent, with a minimum breaking strength of 5400 pounds, shall be used. **(See Figure 5-20)**

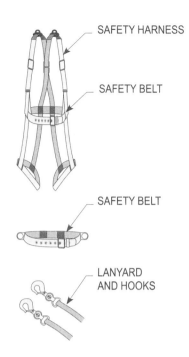

SAFETY HARNESS

SAFETY BELT

SAFETY BELT

LANYARD AND HOOKS

SAFETY BELTS, LIFELINES AND LANYARD REQUIREMENTS

- LIFELINES SHALL BE SECURED ABOVE THE POINT OF OPERATION TO AN ANCHORAGE OR STRUCTURAL MEMBER CAPABLE OF SUPPORTING A MINIMUM DEAD WEIGHT OF 5400 POUNDS

- LIFELINES USED ON ROCK-SCALING OPERATIONS, OR IN AREAS WHERE THE LIFELINE MAY BE SUBJECTED TO CUTTING OR ABRASION, SHALL BE A MINIMUM OF 7/8 IN. WIRE CORE MANILA ROPE. FOR ALL OTHER LIFELINE APPLICATIONS, A MINIMUM OF 3 OR 4 IN. MANILA OR EQUIVALENT, WITH A MINIMUM BREAKING STRENGTH OF 5400 POUNDS, SHALL BE USED

- SAFETY BELT LANYARDS SHALL BE A MINIMUM OF 1/2 IN. NYLON OR EQUIVALENT, WITH A MAXIMUM LENGTH TO PROVIDE FOR A FALL OF NO GREATER THAN 6 FT. THE ROPE SHALL HAVE A NOMINAL BREAKING STRENGTH OF 5400 POUNDS

- ALL SAFETY BELT AND LANYARD HARDWARE SHALL BE DROP FORGED OR PRESSED STEEL OR CADMIUM PLATED

- ALL SAFETY BELT AND LANYARD HARDWARE, EXCEPT RIVETS, SHALL BE CAPABLE OF WITHSTANDING A TENSILE LOADING OF 4000 POUNDS WITHOUT CRACKING, BREAKING OR TAKING A PERMANENT DEFORMATION

SAFETY BELTS, LIFELINES AND LANYARDS
OSHA 1926.104

Figure 5-20. This illustration shows that lifelines, safety belts and lanyards shall be used for employee protection.

Safety belt lanyards shall be a minimum of 1/2 in. nylon, or equivalent, with a maximum length to provide for a fall of no greater than 6 ft. The rope shall have a nominal breaking strength of 5400 pounds.

All safety belt and lanyard hardware shall be drop forged or pressed steel, cadmium plated in accordance with Type 1, Class B plating specified in Federal Specification QQ-P-416. Surface shall be smooth and free of sharp edges.

All safety belt and lanyard hardware, except rivets, shall be capable of withstanding a tensile loading of 4000 pounds without cracking, breaking or taking a permanent deformation.

SAFETY NETS
OSHA 1926.105

Safety nets shall be provided when workplaces are more than 25 ft. above the ground or water surface, or other surfaces where the use of ladders, scaffolds, catch platforms, temporary floors, safety lines or safety belts is impractical.

Where safety net protection is required by this section, operations shall not be undertaken until the net is in place and has been tested.

Nets shall extend 8 ft. beyond the edge of the work surface where employees are exposed and shall be installed as close under the work surface as practical but in no case more than 25 ft. below such work surface. Nets shall be hung with sufficient clearance to prevent user's contact with the surfaces or structures below. Such clearances shall be determined by impact load testing.

It is intended that only one level of nets be required for bridge construction.

The mesh size of nets shall not exceed 6 in. by 6 in. All new nets shall meet accepted performance standards of 17,500 foot-pounds minimum impact resistance as determined and certified by the manufacturers, and shall bear a label of proof test. Edge ropes shall provide a minimum breaking strength of 5000 pounds.

Forged steel safety hooks or shackles shall be used to fasten the net to its supports.

Connections between net panels shall develop the full strength of the net.

See Figure 5-21 for a detailed illustration pertaining to safety nets used above ground or water surfaces.

Figure 5-21. This illustration shows the requirements for safety nets used above the ground or water surfaces.

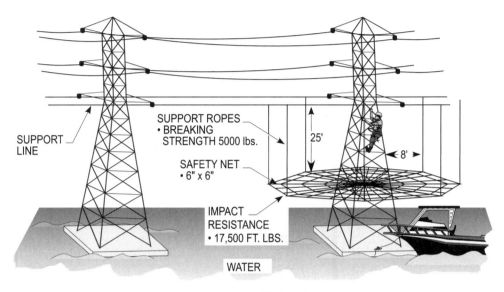

OSHA 1926.105

WORKING OVER OR NEAR WATER
OSHA 1926.106

Employees working over or near water, where the danger of drowning exists, shall be provided with U.S. Coast Guard-approved life jacket or buoyant work vests.

Prior to and after each use, the buoyant work vests or life preservers shall be inspected for defects that would alter their strength or buoyancy. Defective units shall not be used.

Ring buoys with at least 90 ft. of line shall be provided and readily available for emergency rescue operations. Distance between ring buoys shall not exceed 200 ft.

At least one lifesaving skiff shall be immediately available at locations where employees are working over or adjacent to water.

Personal Protective Equipment

Section	Answer
_____	T F
_____	T F
_____	T F
_____	T F
_____	T F
_____	T F
_____	T F
_____	T F
_____	T F
_____	T F
_____	T F
_____	T F
_____	T F

Personal Protective Equipment

1. Protective equipment, including personal protective equipment for eyes, face, head and extremities, protective clothing, respiratory devices and protective shields and barriers, shall be provided by the employee.

2. A hard hat shall not be altered so that additional appliances such as hearing protective devices, face shields and goggles can be added.

3. Hearing protection devices shall provide the necessary noise exposure attenuation to limit a worker's 8-hour time-weight-average noise exposure to no more than 100 decibel (dB), 95 dB for workers who have experienced a standard threshold shift.

4. Street-wear glasses approved by the Food and Drug Administration (FDA), often referred to as safety glasses, do not meet occupational safety standards of ANSI Z87.1 and should not be considered as approved eye protection for construction-site safety.

5. The recommended and most-often-required standard eye protection for construction site workers is safety glasses with side shields.

6. Precautions should be taken to avoid flash hazards because electric arc welders and those working nearby are subject to flash burns to the eyes.

7. Leather gloves provide the best all-purpose hand protection and should be worn when workers handle rough material or material that splinters or snags.

8. All eye protection devices should be inspected weekly.

9. Visitors shall furnish and wear their own approved eye protection.

10. Department of Transportation (DOT) red vests shall be worn by workers who direct traffic or are engaged in construction adjacent to highways or secondary roads.

11. Because appropriate dress contributes to each employee's personal safety, it should always be of primary concern to the worker and supervisor.

12. In the control of those occupational diseases caused by breathing air contaminated with harmful dusts, fogs, fumes, mists, gases, smokes, sprays or vapors, the primary objective shall be to prevent atmospheric contamination.

13. Respirators shall be regularly cleaned and disinfected. Those used by more than one worker shall be thoroughly cleaned and disinfected weekly.

14. Persons should not be assigned to tasks requiring use of respirators unless it has been determined by their supervisor that they are physically able to perform the work and use the equipment.

15. Lifelines, safety belts and lanyards can be used for purposes other than employee safeguarding.

16. Lifelines shall be secured above the point of operation to an anchorage or structural member capable of supporting a minimum dead weight of 6400 pounds.

17. Safety belt lanyard shall be a minimum of 1/2 in. nylon, or equivalent, with a maximum length to provide for a fall of no greater than 8 ft.

18. All safety belt and lanyard hardware, except rivets, shall be capable of withstanding a tensile loading of 4000 pounds without cracking, breaking or taking a permanent deformation.

19. Safety nets shall be provided when workplaces are more than 25 ft. above the ground or water surface, or other surfaces where the use of ladders, scaffolds, catch platforms, temporary floors, safety lines or safety belts is impractical.

20. Ring buoys with at least 50 ft. of line shall be provided and readily available for emergency rescue operations. Distance between ring buoys shall not exceed 200 ft.

6

Fire Protection

This chapter covers fire protection, fire prevention, flammable and combustible liquids, liquefied petroleum gas (LP-gas) and temporary heating devices. The employer shall be responsible for the development of a fire-protection program to be followed throughout all phases of the construction and demolition work. The employer shall consider the ignition hazards, temporary buildings, open yard storage and indoor storage when providing fire prevention for employees.

FIRE PROTECTION
OSHA 1926.150

The following shall be considered for fire protection:

- General requirements
- Water supply
- Portable firefighting equipment
- Fixed firefighting equipment
- Fire alarm devices
- Fire cutoffs

GENERAL REQUIREMENTS
OSHA 1926.150(a)

The employer shall be responsible for the development of a fire protection program to be followed throughout all phases of the construction and demolition work, and shall provide for the firefighting equipment as specified in this subpart. As fire hazards occur, there shall be no delay in providing the necessary equipment.

Access to all available firefighting equipment shall be maintained at all times. All firefighting equipment, provided by the employer, shall be conspicuously located.

All fire fighting equipment shall be periodically inspected and maintained in operating condition. Defective equipment shall be immediately replaced. **(See Figure 6-1)**

Figure 6-1. Fire protective equipment shall be periodically inspected and maintained in an operating condition.

NOTE 1: ACCESS TO ALL AVAILABLE FIREFIGHTING EQUIPMENT SHALL BE MAINTAINED AT ALL TIMES.

NOTE 2: ALL FIREFIGHTING EQUIPMENT, PROVIDED BY THE EMPLOYER, SHALL BE CONSPICUOUSLY LOCATED.

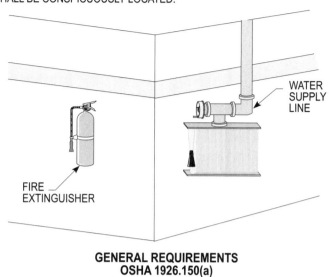

GENERAL REQUIREMENTS
OSHA 1926.150(a)

As warranted by the project, the employer shall provide a trained and equipped firefighting organization (fire brigade) to assure adequate protection to life.

WATER SUPPLY
OSHA 1926.150(b)

A temporary or permanent water supply, of sufficient volume, duration and pressure required to properly operate the fire fighting equipment, shall be made available as soon as combustible materials accumulate.

Where underground water mains are to be provided, they shall be installed, completed and made available for use as soon as practicable.

OSHA Required Training
1926.150(c)(1)(viii)

PORTABLE FIREFIGHTING EQUIPMENT
OSHA 1926.150(c)

The following shall be considered for portable firefighting equipment:

- Fire extinguishers and small hose lines
- Fire hose and connections

FIRE EXTINGUISHERS
AND SMALL HOSE LINES

A fire extinguisher, rated not less than 2A, shall be provided for each 3000 sq. ft. of the protected building area, or major fraction thereof. Travel distance from any point of the protected area to the nearest fire extinguisher shall not exceed 100 ft.

One 55-gallon open drum of water with two fire pails may be substituted for a fire extinguisher having a 2A rating.

A 1/2-inch diameter garden-type hose line, not to exceed 100 ft. in length and equipped with a nozzle, may be substituted for a 2A-rated fire extinguisher, providing it is capable of discharging a minimum of 5 gallons per minute with a minimum hose stream range of 30 ft. horizontally. The garden-type hose lines shall be mounted on conventional racks or reels. The number and location of hose racks or reels shall be such that at least one hose stream can be applied to all points in the area.

One or more fire extinguishers, rated not less than 2A, shall be provided on each floor. In multistory buildings, at least one fire extinguisher shall be located adjacent to stairway.

Extinguishers and water drums subject to freezing shall be protected from freezing.

A fire extinguisher, rated not less than 10B, shall be provided within 50 ft. of wherever more than 5 gallons of flammable or combustible liquids or 5 pounds of flammable gas are being used on the jobsite. This requirement does not apply to the integral fuel tanks of motor vehicles.

There are a variety of types and classes of fire extinguishers that may be used to meet the construction-site fire prevention requirements. The most common types and sizes used on construction sites are as follows:

- 2 1/2-gallon pressurized water Class A fire extinguisher for emergency use on combustible materials (wood, paper and trash)

- 2 1/2-to-5-gallon pump-type water Class A fire extinguisher for emergency use and protective wetting in or around hot work on combustible material (wood, paper and trash)

- 5-to-30-pound pressurized multipurpose dry chemical Class A/B/C fire extinguisher for emergency use on combustible material, flammable liquid and gas and electrical fires. **(See Figures 6-2 and 6-3)**

NOTE 1: ONE OR MORE FIRE EXTINGUISHERS, RATED NOT LESS THAN 2A, SHALL BE PROVIDED ONE EACH FLOOR. IN MULTISTORY BUILDINGS, AT LEAST ONE FIRE EXTINGUISHER SHALL BE LOCATED ADJACENT TO STAIRWAY.

NOTE 2: EXTINGUISHERS AND WATER DRUMS SUBJECT TO FREEZING SHALL BE PROTECTED FROM FREEZING.

Figure 6-2. Class A fire extinguishers can be used for construction-site fire prevention requirements.

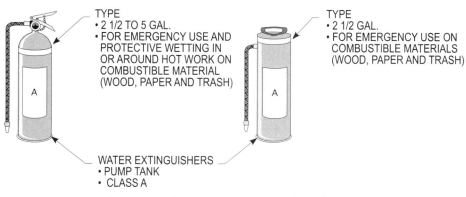

TYPE
- 2 1/2 TO 5 GAL.
- FOR EMERGENCY USE AND PROTECTIVE WETTING IN OR AROUND HOT WORK ON COMBUSTIBLE MATERIAL (WOOD, PAPER AND TRASH)

TYPE
- 2 1/2 GAL.
- FOR EMERGENCY USE ON COMBUSTIBLE MATERIALS (WOOD, PAPER AND TRASH)

WATER EXTINGUISHERS
- PUMP TANK
- CLASS A

FIRE EXTINGUISHERS AND SMALL HOSE LINES
OSHA 1926.150(c)(1)(x)

Carbon tetrachloride and other toxic vaporizing liquid fire extinguishers are prohibited.

Portable fire extinguishers shall be inspected periodically and maintained in accordance with Maintenance and Use of Portable Fire Extinguishers, NFPA No. 10-1970. Fire extinguishers that have been listed or approved by a nationally recognized testing laboratory shall be used to meet the requirements of this subpart.

Safety Tip: Employers must train their employees in the use of portable fire extinguishers.

Table 6-1 in 1926.150(c)(1)(x) may be used as a guide for selecting the appropriate portable fire extinguishers.

Where the employer has provided portable fire extinguishers for employee use in the workplace, the employer shall also provide an educational program to familiarize employees with the general principles of fire extinguisher use and the hazards involved with incipient stage fire fighting. The employer shall assure that portable fire extinguishers are maintained in a fully charged and operable condition and kept in their designated places at all times except during use.

The employer shall assure that portable fire extinguishers are subjected to an annual maintenance check. Stored pressure extinguishers do not require an internal examination. The employer shall record the annual maintenance date and retain this record for one year after the last entry or the life of the shell, whichever is less. The record shall be available to the Assistant Secretary upon request. **(See Figure 6-4)**

Figure 6-3. Class A/B/C fire extinguishers can be used for emergency use on combustible material, flammable liquid and gas and electric fires.

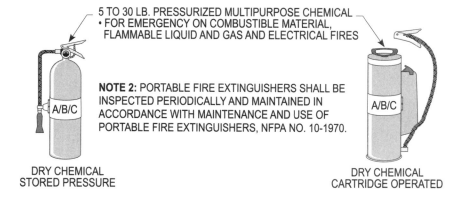

NOTE 1: A FIRE EXTINGUISHER, RATED NOT LESS THAN 10B, SHALL BE PROVIDED WITHIN 50 FT. OF WHEREVER MORE THAN 5 GALLONS OF FLAMMABLE OR COMBUSTIBLE LIQUIDS OR 5 POUNDS OF FLAMMABLE GAS ARE BEING USED ON THE JOBSITE. THIS REQUIREMENT DOES NOT APPLY TO THE INTEGRAL FUEL TANKS OF MOTOR VEHICLES.

5 TO 30 LB. PRESSURIZED MULTIPURPOSE CHEMICAL
• FOR EMERGENCY ON COMBUSTIBLE MATERIAL, FLAMMABLE LIQUID AND GAS AND ELECTRICAL FIRES

NOTE 2: PORTABLE FIRE EXTINGUISHERS SHALL BE INSPECTED PERIODICALLY AND MAINTAINED IN ACCORDANCE WITH MAINTENANCE AND USE OF PORTABLE FIRE EXTINGUISHERS, NFPA NO. 10-1970.

A/B/C

A/B/C

DRY CHEMICAL
STORED PRESSURE

DRY CHEMICAL
CARTRIDGE OPERATED

**FIRE EXTINGUISHERS AND SMALL HOSE LINES
OSHA 1926.150(c)(1)(x)**

Figure 6-4. Portable fire extinguishers shall be inspected periodically and maintained per NFPA 10.

NOTE 1: WHERE THE EMPLOYER HAS PROVIDED PORTABLE FIRE EXTINGUISHERS FOR EMPLOYEE USE IN THE WORKPLACE, THE EMPLOYER SHALL ALSO PROVIDE AN EDUCATIONAL PROGRAM TO FAMILIARIZE EMPLOYEES WITH THE GENERAL PRINCIPLES OF FIRE EXTINGUISHER USE AND THE HAZARDS INVOLVED WITH INCIPIENT STAGE FIRE FIGHTING.

NOTE 2: THE EMPLOYER SHALL ASSURE THE PORTABLE FIRE EXTINGUISHERS ARE SUBJECTED TO AN ANNUAL MAINTENANCE CHECK. STORED PRESSURE EXTINGUISHERS DO NOT REQUIRE AN INTERNAL EXAMINATION. THE EMPLOYER SHALL RECORD THE ANNUAL MAINTENANCE DATE AND RETAIN THIS RECORD FOR ONE YEAR AFTER THE LAST ENRTY OR THE LIFE OF THE SHELL, WHICHEVER IS LESS.

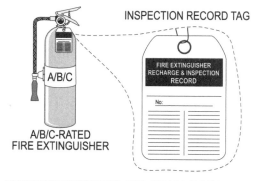

INSPECTION RECORD TAG

FIRE EXTINGUISHER
RECHARGE & INSPECTION
RECORD

No:

A/B/C

A/B/C-RATED
FIRE EXTINGUISHER

**FIRE EXTINGUISHERS AND SMALL HOSE LINES
OSHA 1926.150(c)**

FIRE HOSE AND CONNECTIONS

One hundred feet, or less, of 1 1/2-inch hose, with a nozzle capable of discharging water at 25 gallons or more per minute, may be substituted for a fire extinguisher rated not more than 2A in the designated area provided that the hose line can reach all points in the area.

If fire hose connections are not compatible with local firefighting equipment, the contractor shall provide adapters, or equivalent, to permit connections.

During demolition involving combustible materials, charged hose lines, supplied by hydrants, water tank trucks with pumps, or equivalent, shall be made available.

Safety Tip: Contractors must provide adapters for local firefighting equipment.

FIXED FIREFIGHTING EQUIPMENT
OSHA 1910.150(d)

The following shall be considered for fixed firefighting equipment:

• Sprinkler protection
• Standpipes

SPRINKLER PROTECTION

If the facility being constructed includes the installation of automatic sprinkler protection, the installation shall closely follow the construction and be placed in service as soon as applicable laws permit following completion of each story.

During demolition or alterations, existing automatic sprinkler installations shall be retained in service as long as reasonable. The operation of sprinkler control valves shall be permitted only by properly authorized persons. Modification of sprinkler systems to permit alterations or additional demolition should be expedited so that the automatic protection may be returned to service as quickly as possible. Sprinkler control valves shall be checked daily at close of work to ascertain that the protection is in service.

STANDPIPES

In all structures in which standpipes are required, or where standpipes exist in structures being altered, they shall be brought up as soon as applicable laws permit, and shall be maintained as construction progresses in such a manner that they are always ready for fire protection use. The standpipes shall be provided with siamese fire department connections on the outside of the structure, at the street level, which shall be conspicuously marked. There shall be at least one standard hose outlet at each floor. **(See Figure 6-5)**

FIRE ALARM DEVICES
OSHA 1926.150(e)

An alarm system, e.g., telephone system, siren, etc., shall be established by the employer whereby employees on the site and the local fire department call be alerted for an emergency. The alarm code and reporting instructions shall be conspicuously posted at phones and at employee entrances. **(See Figure 6-6)**

Figure 6-5. This illustration depicts a building-faced connection and freestanding connection used as siamese connections.

NOTE 1: IN ALL STRUCTURES IN WHICH STANDPIPES ARE REQUIRED, OR WHERE STANDPIPES EXIST IN STRUCTURES BEING ALTERED, THEY SHALL BE BROUGHT UP AS SOON AS APPLICABLE LAWS PERMIT, AND SHALL BE MAINTAINED AS CONSTRUCTION PROGRESSES IN SUCH A MANNER THAT THEY ARE ALWAYS READY FOR FIRE PROTECTION USE.

NOTE 2: THE STANDPIPES SHALL BE PROVIDED WITH SIAMESE FIRE DEPARTMENT CONNECTIONS ON THE OUSIDE OF THE STRUCTURE, AT THE STREET LEVEL, WHICH SHALL BE CONSPICUOUSLY MARKED.

FREE-STANDING
CONNECTION

BUILDING-FACED
CONNECTION

STANDPIPES
OSHA 1910.150(d)

Figure 6-6. This illustration depicts emergency notification devices used to sound or report a fire.

NOTE 1: AN ALARM SYSTEM, E.G., TELEPHONE SYSTEM, SIREN, ETC., SHALL BE ESTABLISHED BY THE EMPLOYER WHEREBY EMPLOYEES ON THE SITE AND THE LOCAL FIRE DEPARTMENT CALL BE ALERTED FOR AN EMERGENCY.

NOTE 2: THE ALARM CODE AND REPORTING INSTRUCTIONS SHALL BE CONSPICUOUSLY POSTED AT PHONES AND AT EMPLOYEE ENTRANCES.

FIRE ALARM DEVICES
OSHA 1926.150(e)

FIRE CUTOFFS
OSHA 1926.150(f)

Fire walls and exit stairways, required for the completed buildings, shall be given construction priority. Fire doors, with automatic closing devices, shall be hung on openings as soon as practicable. **(See Figure 6-7)**

Fire cutoffs shall be retained in buildings undergoing alterations or demolition until operations necessitate their removal.

FIRE PREVENTION
OSHA 1926.151

The following shall be considered for fire prevention:

- Ignition hazards
- Temporary buildings
- Open yard storage
- Indoor storage

NOTE 1: FIRE DOORS, WITH AUTOMATIC CLOSING DEVICES, SHALL BE HUNG AS SOON AS PRACTICABLE.

NOTE 2: FIRE CUTOFFS SHALL BE RETAINED IN BUILDINGS UNDERGOING ALTERATIONS OR DEMOLITION UNTIL OPERATIONS NECESSITATE THEIR REMOVAL.

NO STORAGE PERMITTED THAT COULD BLOCK EXIT

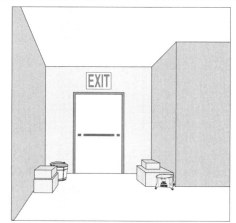

UNOBSTRUCTED EXIT WAY ACCESS

FIRE CUTOFFS
OSHA 1926.150(f)

Figure 6-7. Storage shall not block the area leading to an exit door. Note that fire doors shall be equipped with automatic closing devices.

IGNITION HAZARDS
OSHA 1926.151(a)

Electrical wiring and equipment for light, heat or power purposes shall be installed in compliance with the requirements of Subpart K, Electrical.

Internal combustion engine powered equipment shall be so located that the exhausts are well away from combustible materials. When the exhausts are piped to outside the building under construction, a clearance of at least 6 in. shall be maintained between such piping and combustible material.

Smoking shall be prohibited at or in the vicinity of operations that constitute a fire hazard, and shall be conspicuously posted: "No Smoking or Open Flame." **(See Figure 6-8)**

NOTE: INTERNAL COMBUSTION ENGINE POWERED EQUIPMENT SHALL BE SO LOCATED THAT THE EXHAUSTS ARE WELL AWAY FROM COMBUSTIBLE MATERIALS. WHEN THE EXHAUSTS ARE PIPED TO OUTSIDE THE BUILDING UNDER CONSTRUCTION, A CLEARANCE OF AT LEAST 6 IN. SHALL BE MAINTAINED BETWEEN SUCH PIPING AND COMBUSTIBLE MATERIAL.

Figure 6-8. Smoking signs shall be posted to prohibit smoking that could constitute a fire hazard.

POST SIGNS THAT PROHIBIT SMOKING

IGNITION HAZARDS
OSHA 1926.151(a)

Portable battery powered lighting equipment used in connection with the storage, handling or use of flammable gases or liquids, shall be of the type approved for the hazardous locations.

The nozzle of air, inert gas and steam lines or hoses, when used in the cleaning or ventilation of tanks and vessels that contain hazardous concentrations of flammable gases or vapors, shall be bonded to the tank or vessel shell. Bonding devices shall not be attached or detached in hazardous concentrations of flammable gases or vapors.

TEMPORARY BUILDINGS
OSHA 1926.151(b)

No temporary building shall be erected where it will adversely affect any means of exit.

Temporary buildings, when located within another building or structure, shall be of either noncombustible construction or of combustible construction having a fire resistance of not less than 1 hour.

Temporary buildings, located other than inside another building and not used for the storage, handling or use of flammable or combustible liquids, flammable gases, explosives, or blasting agents or similar hazardous occupancies, shall be located at a distance of not less than 10 ft. from another building or structure. Groups of temporary buildings, not exceeding 2000 sq. ft. in aggregate, shall, for the purposes of this part, be considered a single temporary building.

OPEN YARD STORAGE
OSHA 1926.151(c)

Safety Tip: Driveways must be kept free from accumulation of rubbish.

Combustible materials shall be piled with due regard to the stability of piles and in no case higher than 20 ft.

Driveways between and around combustible storage piles shall be at least 15 ft. wide and maintained free from accumulation of rubbish, equipment or other articles or materials. Driveways shall be so spaced that a maximum grid system unit of 50 ft. by 150 ft. is produced.

The entire storage site shall be kept free from accumulation of unnecessary combustible materials. Weeds and grass shall be kept down and a regular procedure provided for the periodic cleanup of the entire area. When there is a danger of an underground fire, that land shall not be used for combustible or flammable storage.

Method of piling shall be solid wherever possible and in orderly and regular piles. No combustible material shall be stored outdoors within 10 ft. of a building or structure.

Portable fire extinguishing equipment, suitable for the fire hazard involved, shall be provided at convenient, conspicuously accessible locations in the yard area. Portable fire extinguishers, rated not less than 2A, shall be placed so that maximum travel distance to the nearest unit shall not exceed 100 ft.

INDOOR STORAGE
OSHA 1926.151(d)

Storage shall not obstruct, or adversely affect, means of exit. All materials shall be stored, handled and piled with due regard to their fire characteristics.

Noncompatable materials, which may create a fire hazard, shall be segregated by a barrier having a fire resistance of at least 1 hour.

Material shall be piled to minimize the spread of fire internally and to permit convenient access for firefighting. Stable piling shall be maintained at all times. Aisle space shall be maintained to safely accommodate the widest vehicle that may be used within the building for firefighting purposes.

Clearance of at least 36 in. shall be maintained between the top level of the stored material and the sprinkler deflectors.

Clearance shall be maintained around lights and heating units to prevent ignition of combustible materials.

A clearance of 24 in. shall be maintained around the path of travel of fire doors unless a barricade is provided, in which case no clearance is needed. Material shall not be stored within 36 in. of a fire door opening.

FLAMMABLE AND COMBUSTIBLE LIQUIDS OSHA 1926.152

The following shall be considered for flammable and combustible liquids:

- General requirements
- Indoor storage of flammable and combustible liquids
- Storage outside buildings
- Fire control for flammable or combustible liquid storage
- Dispensing liquids
- Handling liquids at point of final use
- Service and refueling areas
- Scope
- Tank storage
- Piping, valves and fittings
- Marine service stations

GENERAL REQUIREMENTS OSHA 1926.152(a)

Only approved containers and portable tanks shall be used for storage and handling of flammable and combustible liquids. Approved metal safety cans shall be used for the handling and use of flammable liquids in quantities greater than one gallon, except that this shall not apply to those flammable liquid materials which are highly viscid (extremely hard to pour), which may be used and handled in original shipping containers. For quantities of one gallon or less, only the original container or approved metal safety cans shall be used for storage, use and handling of flammable liquids. **(See Figures 6-9(a) and (b))**

Flammable or combustible liquids shall not be stored in areas used for exits, stairways or normally used for the safe passage of people. **(See Figure 6-10)**

Safety Tip: Use only approved containers for storage of flammable and combustible liquids.

Figure 6-9(a). This illustration shows the requirements for tanks that are used to store flammables.

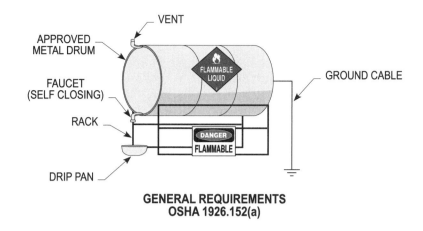

**GENERAL REQUIREMENTS
OSHA 1926.152(a)**

Figure 6-9(b). This illustration shows the requirements for approved safety cans that are used to store flammables.

NOTE: APPROVED SAFETY CANS SHALL BE USED FOR THE HANDLING AND USE OF FLAMMABLE LIQUIDS IN QUANTITIES GREATER THAN ONE GALLON, EXCEPT THAT THIS SHALL NOT APPLY TO THOSE FLAMMABLE LIQUID MATERIALS THAT ARE HIGHLY VISCID (EXTREMELY HARD TO POUR), WHICH MAY BE USED AND HANDLED IN ORIGINAL SHIPPING CONTAINERS.

**GENERAL REQUIREMENTS
OSHA 1926.152(a)**

Figure 6-10. Flammable or combustible liquids shall not be stored in areas used for exits, stairways or safe passageways for people.

NOTE: FOR QUANTITIES OF ONE GALLON OR LESS, ONLY THE ORIGINAL CONTAINER OR APPROVED METAL SAFETY CANS SHALL BE USED FOR STORAGE, USE AND HANDLING OF FLAMMABLE LIQUIDS.

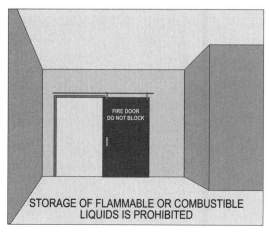

**GENERAL REQUIREMENTS
OSHA 1926.152(a)**

INDOOR STORAGE OF FLAMMABLE AND COMBUSTIBLE LIQUIDS
OSHA 1926.152(b)

No more than 25 gallons of flammable or combustible liquids shall be stored in a room outside of an approved storage cabinet. For storage of liquefied petroleum gas, see OSHA 1926.153.

Quantities of flammable and combustible liquid in excess of 25 gallons shall be stored in an acceptable or approved cabinet meeting the following requirements:

- Acceptable wooden storage cabinets shall be constructed in the following manner, or equivalent: The bottom, sides and top shall be constructed of an exterior grade of plywood at least 1 in. in thickness, which shall not break down or delaminate under standard fire test conditions. All joints shall be rabbeted and shall be fastened in two directions with flathead wood screws. When more than one door is used, there shall be a rabbeted overlap of not less than 1 in. Steel hinges shall be mounted in such a manner as to not lose their holding capacity due to loosening or burning out of the screws when subjected to fire. Such cabinets shall be painted inside and out with fire retardant paint. **(See Figure 6-11)**

Safety Tip: Wooden storage cabinets must be properly constructed.

Figure 6-11. This illustration shows the requirements for wooden storage cabinets that are used to store flammable liquids.

ACCEPTABLE WOODEN STORAGE CABINETS SHALL BE CONSTRUCTED IN THE FOLLOWING MANNER:

- THE BOTTOM, SIDES AND TOP SHALL BE CONSTRUCTED OF AN EXTERIOR GRADE OF PLYWOOD AT LEAST 1 IN. IN THICKNESS
- ALL JOINTS SHALL BE RABBETED AND SHALL BE FASTENED IN TWO DIRECTIONS WITH FLATHEAD WOOD SCREWS
- WHEN MORE THAN ONE DOOR IS USED, THERE SHALL BE A RABBETED OVERLAP OF NOT LESS THAN 1 IN.
- STEEL HINGES SHALL BE MOUNTED IN SUCH A MANNER AS TO NOT LOSE THEIR HOLDING CAPACITY
- CABINETS SHALL BE PAINTED INSIDE AND OUT WITH FIRE RETARDANT PAINT

STORAGE CABINET FOR FLAMMABLE LIQUIDS

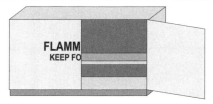

FLAMM
KEEP FO

**INDOOR STORAGE OF FLAMMABLE AND COMBUSTIBLE LIQUIDS
OSHA 1926.152(b)**

- Approved metal storage cabinets will be acceptable.

- Cabinets shall be labeled in conspicuous lettering, "Flammable-Keep Fire Away."

- Not more than 60 gallons of flammable or 120 gallons of combustible liquids shall be stored in any one storage cabinet. Not more than three such cabinets may be located in a single storage area. Quantities in excess of this shall be stored in an inside storage room.

Inside storage rooms shall be constructed to meet the required fire-resistive rating for their use. Such construction shall comply with the test specifications set forth in Standard Methods of Fire Test of Building Construction and Material, NFPA 251-1999.

Where an automatic extinguishing system is provided, the system shall be designed and installed in an approved manner. Openings to other rooms or buildings shall be provided with noncombustible liquid-tight raised sills or ramps at least 4 in. in height, or the floor in the storage area shall be at least 4 in. below the surrounding floor. Openings shall be provided with approved self-closing fire doors. The room shall be liquid-tight where the walls join the floor. A permissible alternate to the sill or ramp is an open-grated trench, inside of the room, which drains to a safe location. Where other portions of the building or other buildings are exposed, windows shall be protected as set forth in the Standard for Fire Doors and Windows, NFPA No. 80-1970, for Class E or F openings. Wood of at least 1 in. nominal thickness may be used for shelving, racks, dunnage, scuffboards, floor overlay and similar installations.

Materials which will react with water and create a fire hazard shall not be stored in the same room with flammable or combustible liquids.

Note: Fire protection system shall be sprinkler, water spray, carbon dioxide or other system approved by a nationally recognized testing laboratory for this purpose.

Safety Tip: Electrical wiring located inside a storage room must be approved.

Electrical wiring and equipment located in inside storage rooms shall be approved for Class I, Division 1, Hazardous Locations. For definition of Class I, Division I, Hazardous Locations, see 1926.449.

Every inside storage room shall be provided with either a gravity or a mechanical exhausting system. Such system shall commence not more than 12 in. above the floor and be designed to provide for a complete change of air within the room at least 6 times per hour. If a mechanical exhausting system is used, it shall be controlled by a switch located outside of the door. The ventilating equipment and any lighting fixtures shall be operated by the same switch. An electric pilot light shall be installed adjacent to the switch if flammable liquids are dispensed within the room. Where gravity ventilation is provided, the fresh air intake, as well as the exhausting outlet from the room, shall be on the exterior of the building in which the room is located.

In every inside storage room there shall be maintained one clear aisle at least 3 ft. wide. Containers over 30 gallons capacity shall not be stacked one upon the other.

Flammable and combustible liquids in excess of that permitted in inside storage rooms shall be stored outside of buildings in accordance with paragraph "Storage Outside Buildings" of this section.

The quantity of flammable or combustible liquids kept in the vicinity of spraying operations shall be the minimum required for operations and should ordinarily not exceed a supply for 1 day or one shift. Bulk storage of portable containers of flammable or combustible liquids shall be in a separate, constructed building detached from other important buildings or cut off in a standard manner.

STORAGE OUTSIDE BUILDINGS
OSHA 1926.152(c)

Storage of containers (not more than 60 gallons each) shall not exceed 1100 gallons in any one pile or area. Piles or groups of containers shall be separated by a 5 ft. clearance. Piles or groups of containers shall not be nearer than 20 ft. to a building. **(See Figure 6-12)**

Within 200 ft. of each pile of containers, there shall be a 12 ft. wide access way to permit approach of fire control apparatus.

The storage area shall be graded in a manner to divert possible spills away from buildings or other exposures, or shall be surrounded by a curb or earth dike at least 12 in. high. When curbs or dikes are used, provisions shall be made for draining off accumulations of ground or rain water, or spills of flammable or combustible liquids. Drains shall terminate at a safe location and shall be accessible to operation under fire conditions.

NOTE: THE STORAGE AREA SHALL BE GRADED IN A MANNER TO DIVERT POSSIBLE SPILLS AWAY FROM BUILDINGS OR OTHER EXPOSURES, OR SHALL BE SURROUNDED BY A CURB OR EARTH DIKE AT LEAST 12 IN. HIGH.

STORAGE CONTAINERS
• 60 GAL. OR LESS (EACH)

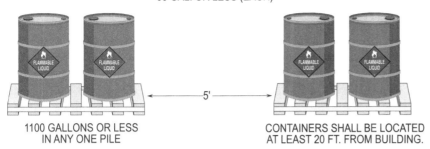

←———5'———→

1100 GALLONS OR LESS
IN ANY ONE PILE

CONTAINERS SHALL BE LOCATED
AT LEAST 20 FT. FROM BUILDING.

**INDOOR STORAGE OF FLAMMABLE AND COMBUSTIBLE LIQUIDS
OSHA 1926.152(e)**

Outdoor portable tank storage:

Portable tanks shall not be nearer than 20 ft. from any building. Two or more portable tanks, grouped together, having a combined capacity in excess of 2200 gallons, shall be separated by a 5 ft. clear area. Individual portable tanks exceeding 1100 gallons shall be separated by a 5 ft. clear area.

Safety Tip: Portable tanks must not be closer than 20 ft. from any building.

Within 200 ft. of each portable tank, there shall be a 12 ft. wide access way to permit approach of fire control apparatus.

Storage areas shall be kept free of weeds, debris and other combustible material not necessary to the storage.

Portable tanks, not exceeding 660 gallons, shall be provided with emergency venting and other devices, as required by chapters III and IV of NFPA 30-1996, The Flammable and Combustible Liquids Code.

Portable tanks, in excess of 660 gallons, shall have emergency venting and other devices, as required by chapters II and III of The Flammable and Combustible Liquids Code, NFPA 30-1996.

FIRE CONTROL FOR FLAMMABLE OR COMBUSTIBLE LIQUID STORAGE OSHA 1926.152(d)

At least one portable fire extinguisher, having a rating of not less than 20-B units, shall be located outside of, but not more than 10 feet from, the door opening into any room used for storage of more than 60 gallons of flammable or combustible liquids.

At least one portable fire extinguisher having a rating of not less than 20-B units shall be located not less than 25 feet, nor more than 75 feet, from any flammable liquid storage area located outside.

When sprinklers are provided, they shall be installed in accordance with the Standard for the Installation of Sprinkler Systems, NFPA 13-1969.

At least one portable fire extinguisher having a rating of not less than 20-B:C units shall be provided on all tank trucks or other vehicles used for transporting and/or dispensing flammable or combustible liquids.

DISPENSING LIQUIDS
OSHA 1926.152(e)

Safety Tip: When transferring flammable liquids from one container to another, they must be electrically interconnected.

Areas in which flammable or combustible liquids are transferred at one time, in quantities greater than 5 gallons from one tank or container to another tank or container, shall be separated from other operations by 25 ft. distance or by construction having a fire resistance of at least 1 hour. Drainage or other means shall be provided to control spills. Adequate natural or mechanical ventilation shall be provided to maintain the concentration of flammable vapor at or below 10 percent of the lower flammable limit.

Transfer of flammable liquids from one container to another shall be done only when containers are electrically interconnected (so that they are bonded together). **(See Figure 6-13)**

Flammable or combustible liquids shall be drawn from or transferred into vessels, containers or tanks within a building or outside only through a closed piping system, from safety cans, by means of a device drawing through the top, or from a container, or portable tanks, by gravity or pump, through an approved self-closing valve. Transferring by means of air pressure on the container or portable tanks is prohibited. **(See Figure 6-14)**

Figure 6-13. Containers shall be bonded together and connected to an earth ground.

NOTE 1: AREAS IN WHICH FLAMMABLE OR COMBUSTIBLE LIQUIDS ARE TRANSFERRED AT ONE TIME, IN QUANTITIES GREATER THAN 5 GALLONS FROM ONE TANK OR CONTAINER TO ANOTHER TANK OR CONTAINER, SHALL BE SEPARATED FROM OTHER OPERATIONS BY 25 FT. DISTANCE OR BY CONSTRUCTION HAVING A FIRE RESISTANCE OF AT LEAST 1 HOUR.

NOTE 2: ADEQUATE NATURAL OR MECHANICAL VENTILATION SHALL BE PROVIDED TO MAINTAIN THE CONCENTRATION OF FLAMMABLE VAPOR AT OR BELOW 10 PERCENT OF THE LOWER FLAMMABLE LIMIT.

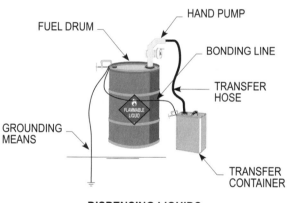

HAND PUMP
FUEL DRUM
BONDING LINE
TRANSFER HOSE
GROUNDING MEANS
TRANSFER CONTAINER

DISPENSING LIQUIDS
OSHA 1910.152(e)

Figure 6-14. This illustration depicts a self-closing fuel faucet.

NOTE: FLAMMABLE OR COMBUSTIBLE LIQUIDS SHALL BE DRAWN FROM OR TRANSFERRED INTO VESSELS, CONTAINERS OR TANKS WITHIN A BUILDING OR OUTSIDE ONLY THROUGH A CLOSED PIPING SYSTEM, FROM SAFETY CANS, BY MEANS OF A DEVICE DRAWING THROUGH THE TOP, OR FROM A CONTAINER, OR PORTABLE TANKS, BY GRAVITY OR PUMP, THROUGH AN APPROVED SELF-CLOSING VALVE.

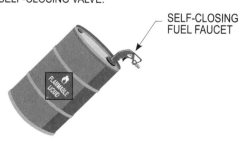

SELF-CLOSING FUEL FAUCET

DISPENSING LIQUIDS
OSHA 1926.152(e)

The dispensing units shall be protected against collision damage. Dispensing devices and nozzles for flammable liquids shall be of an approved type.

HANDLING LIQUIDS AT
POINT OF FINAL OF USE
OSHA 1926.152(f)

Flammable liquids shall be kept in closed containers when not actually in use. Leakage or spillage of flammable or combustible liquids shall be disposed of promptly and safely.

Flammable liquids may be used only where there are no open flames or other sources of ignition within 50 ft. of the operation, unless conditions warrant greater clearance.

SERVICE AND
REFUELING AREAS
OSHA 1926.152(g)

Flammable or combustible liquids shall be stored in approved closed containers, in tanks located underground or in aboveground portable tanks.

The tank trucks shall comply with the requirements covered in the Standard for Tank Vehicles for Flammable and Combustible Liquids, NFPA 385-2000.

The dispensing hose shall be an approved type, and the dispensing nozzle shall be an approved automatic-closing type without a latch-open device. **(See Figure 6-15)**

NOTE 1: FLAMMABLE OR COMBUSTIBLE LIQUIDS SHALL BE STORED IN APPROVED CLOSED CONTAINERS, IN TANKS LOCATED UNDERGROUND OR IN ABOVEGROUND PORTABLE TANKS.

NOTE 2: THE DISPENSING HOSE SHALL BE AN APPROVED TYPE, AND THE DISPENSING NOZZLE SHALL BE AN APPROVED AUTOMATIC-CLOSING TYPE WITHOUT A LATCH-OPEN DEVICE.

Figure 6-15. This illustration depicts a self-closing fuel nozzle.

SELF-CLOSING FUEL NOZZLE

**SERVICE AND REFUELING AREAS
OSHA 1926.152(g)**

Underground tanks shall not be abandoned.

Clearly identified and easily accessible switches shall be provided at a location remote from dispensing devices to shut off the power to all dispensing devices in the event of an emergency.

Heating equipment of an approved type may be installed in the lubrication or service area where there is no dispensing or transferring of flammable liquids, provided the bottom of the heating unit is at least 18 in. above the floor and is protected from physical damage.

Heating equipment installed in lubrication or service areas, where flammable liquids are dispensed, shall be of an approved type for garages, and shall be installed at least 8 ft. above the floor.

There shall be no smoking or open flames in the areas used for fueling, servicing fuel systems for internal combustion engines, receiving or dispensing of flammable or combustible liquids. Conspicuous and legible signs prohibiting smoking shall be posted. **(See Figure 6-16)**

Figure 6-16. Signs prohibiting smoking or open flames in areas used for fueling or servicing purposes shall be posted.

NOTE 1: CLEARLY IDENTIFIED AND EASILY ACCESSIBLE SWITCHES SHALL BE PROVIDED AT A LOCATION REMOTE FROM DISPENSING DEVICES TO SHUT OFF THE POWER TO ALL DISPENSING DEVICES IN THE EVENT OF AN EMERGENCY.

NOTE 2: EACH SERVICE OR FUELING AREA SHALL BE PROVIDED WITH AT LEAST ONE FIRE EXTINGUISHER HAVING A RATING OF NOT LESS THAN 20-B:C LOCATED SO THAT AN EXTINGUISHER WILL BE WITHIN 75 FT. OF EACH PUMP, DISPENSER, UNDERGROUND FILL PIPE OPENING AND LUBRICATION OR SERVICE AREA.

SERVICE AND REFUELING AREAS
OSHA 1926.152(g)

The motors of all equipment being fueled shall be shut off during the fueling operation.

Each service or fueling area shall be provided with at least one fire extinguisher having a rating of not less than 20-B:C located so that an extinguisher will be within 75 ft. of each pump, dispenser, underground fill pipe opening and lubrication or service area.

SCOPE
OSHA 1926.152(h)

This section applies to the handling, storage and use of flammable and combustible liquids with a flash point below 200°F (93.33°C). This section does not apply to: (1) Bulk transportation of flammable and combustible liquids; and (2) Storage, handling and use of fuel oil tanks and containers connected with oil burning equipment.

TANK STORAGE
OSHA 1926.152(i)

Refer to 1926.152(i) for design, construction and installation requirements for flammable or combustible liquid storage tanks.

PIPING, VALVES AND FITTINGS
OSHA 1926.152(j)

Refer to 1926.152(j) for design, fabrication, assembly, test and inspection requirements for piping systems containing flammable or combustible liquids.

MARINE SERVICE STATIONS
OSHA 1926.152(k)

Refer to 1926.152(k) for dispensing, tanks and pumps and piping service stations where flammable or combustible liquids used as fuels are stored and dispensed.

LIQUEFIED PETROLEUM GAS (LP-GAS)
OSHA 1926.153

The following shall be considered for liquefied petroleum gas (LP-GAS):

- Approval of equipment and systems
- Welding on LP-GAS containers
- Container valves and container accessories
- Safety devices
- Dispensing
- Requirements for appliances
- Containers and regulating equipment installed outside of buildings or structures
- Containers and equipment used inside of buildings or structures
- Multiple container systems
- Storage LPG containers
- Storage outside of buildings
- Fire protection
- Systems utilizing containers other than DOT containers
- Marking of gas cylinders
- Damage from vehicles

APPROVAL OF EQUIPMENT AND SYSTEMS
OSHA 1926.153(a)

Each system shall have containers, valves, connectors, manifold valve assemblies and regulators of an approved type. **(See Figure 6-17)**

NOTE: ALL CYLINDERS SHALL MEET THE DEPARTMENT OF TRANSPORTATION SPECIFICATION IDENTIFICATION REQUIREMENTS PUBLISHED IN 49 CFR PART 178, SHIPPING CONTAINER SPECIFICATIONS.

Figure 6-17. Components of each system shall be of an approved type.

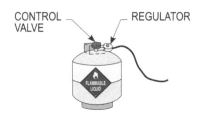

CONTROL VALVE — REGULATOR

FLAMMABLE LIQUID

APPROVAL OF EQUIPMENT AND SYSTEMS OSHA 1926.153(a)

All cylinders shall meet the Department of Transportation specification identification requirements published in 49 CFR Part 178, Shipping Container Specifications.

As used in this section, "containers" are defined as all vessels, such as tanks, cylinders or drums, used for transportation or storing liquefied petroleum gases.

WELDING ON LP-GAS CONTAINERS
OSHA 1926.153(b)

Welding is prohibited on containers.

CONTAINER VALVES AND
CONTAINER ACCESSORIES
OSHA 1926.153(c)

Valves, fittings and accessories connected directly to the container, including primary shutoff valves, shall have a rated working pressure of at least 250 p.s.i.g. and shall be of material and design suitable for LP-GAS service.

Connections to containers, except safety relief connections, liquid level gauging devices and plugged openings, shall have shutoff valves located as close to the container as practicable.

SAFETY DEVICES
OSHA 1926.153(d)

Every container and every vaporizer shall be provided with one or more approved safety relief valves or devices. These valves shall be arranged to afford free vent to the outer air with discharge not less than 5 ft. horizontally away from any opening into a building which is below such discharge.

Shutoff valves shall not be installed between the safety relief device and the container, or the equipment or piping to which the safety relief device is connected, except that a shutoff valve may be used where the arrangement of this valve is such that full required capacity flow through the safety relief device is always afforded.

Container safety relief devices and regulator relief vents shall be located not less than 5 ft. in any direction from air openings into sealed combustion system appliances or mechanical ventilation air intakes.

DISPENSING
OSHA 1926.153(e)

Filling of fuel containers for trucks or motor vehicles from bulk storage containers shall be performed not less than 10 ft. from the nearest masonry-walled building, or not less than 25 ft. from the nearest building or other construction and, in any event, not less than 25 ft. from any building opening.

Filling of portable containers or containers mounted on skids from storage containers shall be performed not less than 50 ft. from the nearest building.

REQUIREMENTS FOR APPLIANCES
OSHA 1926.153(f)

Any appliance that was originally manufactured for operation with a gaseous fuel other than LP-GAS, and is in good condition, may be used with LP-GAS only after it is properly converted, adapted and tested for performance with LP-GAS before the appliance is placed in use.

CONTAINERS AND REGULATING EQUIPMENT INSTALLED OUTSIDE OF BUILDINGS OR STRUCTURES
OSHA 1926.153(g)

Containers shall be upright upon firm foundations or otherwise firmly secured. The possible effect on the outlet piping of settling shall be guarded against by a flexible connection or special fitting.

CONTAINERS AND EQUIPMENT USED INSIDE OF BUILDINGS OR STRUCTURES
OSHA 1926.153(h)

When operational requirements make portable use of containers necessary, and their location outside of buildings or structures is impracticable, containers and equipment shall be permitted to be used inside of buildings or structures in accordance with paragraphs (h)(2) through (h)(11) of 1926.153(h)

"Containers in use" means connected for use.

Systems utilizing containers having a water capacity greater than 2 to 5 pounds (nominal 1 pound LP-GAS capacity) shall be equipped with excess flow valves. Such excess flow valves shall be either integral with the container valves or in the connections to the container valve outlets.

Regulators shall be either directly connected to the container valves or to manifolds connected to the container valves. The regulator shall be suitable for use with LP-GAS. Manifolds and fittings connecting containers to pressure regulator inlets shall be designed for at least 250 p.s.i.g, service pressure.

Valves on containers having water capacity greater than 50 pounds (nominal 20 pounds LP-GAS capacity) shall be protected from damage while in use or storage.

Aluminum piping or tubing shall not be used.

Hose shall be designed for a working pressure of at least 250 p.s.i.g. Design, construction and performance of hose, and hose connections shall have their suitability determined by listing by a nationally recognized testing agency. The hose length shall be as short as practicable. Hoses shall be long enough to permit compliance with spacing provisions of paragraphs (h)(1) through (h)(13) of 1926.153, without kinking or straining, or causing hose to be so close to a burner as to be damaged by heat.

Safety Tip: An approved automatic device to shut off the gas must be provided on all portable gas heaters.

Portable heaters, including salamanders, shall be equipped with an approved automatic device to shut off the flow of gas to the main burner, and pilot if used, in the event of flame failure. Such heaters, having inputs above 50,000 B.t.u. per hour, shall be equipped with either a pilot, which must be lighted and proved before the main burner can be turned on, or an electrical ignition system.

Note: The provisions of this subparagraph do not apply to portable heaters under 7500 B.t.u. per hour input when used with containers having a maximum water capacity of 2 pounds.

Container valves, connectors, regulators, manifolds, piping and tubing shall not be used as structural supports for heaters.

Containers, regulating equipment, manifolds, pipe, tubing and hose shall be located to minimize exposure to high temperatures or physical damage.

Containers having a water capacity greater than 2 pounds (nominal 1 pound LP-GAS capacity) connected for use shall stand on a firm and substantially level surface and, when necessary, shall be secured in an upright position.

The maximum water capacity of individual containers shall be 245 pounds (nominal 100 pounds LP-GAS capacity).

For temporary heating, heaters (other than integral heater-container units) shall be located at least 6 ft. from any LP-Gas container. This shall not prohibit the use of heaters specifically designed for attachment to the container or to a supporting standard, provided they are designed and installed so as to prevent direct or radiant heat application from the heater onto the containers. Blower and radiant type heaters shall not be directed toward any LP-GAS container within 20 ft. **(See Figure 6-18)**

Figure 6-18. For temporary heating, heaters other than integral type shall be located at least 6 ft. away from any LP-Gas container.

NOTE 1: BLOWER AND RADIANT TYPE HEATERS SHALL NOT BE DIRECTED TOWARD ANY LP-GAS CONTAINER WITHIN 20 FT.

NOTE 2: IF TWO OR MORE HEATER-CONTAINER UNITS, OF EITHER THE INTEGRAL OR NONINTEGRAL TYPE, ARE LOCATED IN AN UNPARTITIONED AREA ON THE SAME FLOOR, THE CONTAINER OR CONTAINERS OF EACH UNIT SHALL BE SEPARATED FROM THE CONTAINER OR CONTAINERS OF ANY OTHER UNIT AT LEAST 20 FT.

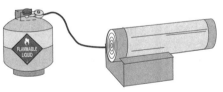

LP-GAS HEATER

**CONTAINER AND EQUIPMENT USED
INSIDE OF BUILDINGS OR STRUCTURES
OSHA 1926.153(h)**

If two or more heater-container units, of either the integral or nonintegral type, are located in an unpartitioned area on the same floor, the container or containers of each unit shall be separated from the container or containers of any other unit by at least 20 ft.

When heaters are connected to containers for use in an unpartitioned area on the same floor, the total water capacity of containers, manifolded together for connection to a heater or heaters, shall not be greater than 735 pounds (nominal 300 pounds LP-GAS capacity). Such manifolds shall be separated by at least 20 ft.

Storage of containers awaiting use shall be in accordance with paragraphs (j) and (k) of this section.

MULTIPLE CONTAINER SYSTEMS
OSHA 1926.153(i)

Valves in the assembly of multiple container systems shall be arranged so that replacement of containers can be made without shutting off the flow of gas in the system. This provision is not to be construed as requiring an automatic changeover device.

Heaters shall be equipped with an approved regulator in the supply line between the fuel cylinder and the heater unit. Cylinder connectors shall be provided with an excess flow valve to minimize the flow of gas in the event the fuel line becomes ruptured.

Regulators and low-pressure relief devices shall be rigidly attached to the cylinder valves, cylinders, supporting standards, the building walls or otherwise rigidly secured, and shall be so installed or protected from the elements.

STORAGE LPG CONTAINERS
OSHA1926.153(j)

Storage of LPG within buildings is prohibited.

STORAGE OUTSIDE OF BUILDINGS
OSHA 1926.153(k)

Storage outside of buildings, for containers awaiting use, shall be located from the nearest building or group of buildings, in accordance with the following:

Containers shall be in a suitable ventilated enclosure or otherwise protected against tampering.

FIRE PROTECTION
OSHA 1926.153(l)

Storage locations shall be provided with at least one approved portable fire extinguisher having a rating of not less than 20-B:C.

SYSTEMS UTILIZING CONTAINERS
OTHER THAN DOT CONTAINERS
OSHA 1926.153(m)

This paragraph applies specifically to systems utilizing storage containers other than those constructed in accordance with DOT specifications. Paragraph (b) of this section applies to this paragraph unless otherwise noted in paragraph (b) of this section.

Storage containers shall be designed and classified in accordance with Table F-31 of 1926.153(m)(2).

Containers with foundations attached (portable or semiportable containers with suitable steel "runners" or "skids" and popularly known in the industry as "skid tanks") shall be designed, installed and used in accordance with these rules subject to the following provisions:

If they are to be used at a given general location for a temporary period not to exceed 6 months, they need not have fire-resisting foundations or saddles but shall have adequate ferrous metal supports.

They shall not be located with the outside bottom of the container shell more than 5 ft. (1.52 m) above the surface of the ground unless fire-resisting supports are provided.

The bottom of the skids shall not be less than 2 in. (5.08 cm) or more than 12 in. (30.48 cm) below the outside bottom of the container shell.

Flanges, nozzles, valves, fittings and the like, having communication with the interior of the container, shall be protected against physical damage.

When not permanently located on fire-resisting foundations, piping connections shall be sufficiently flexible to minimize the possibility of breakage or leakage of connections if the container settles, moves or is otherwise displaced.

Skids, or lugs for attachment of skids, shall be secured to the container in accordance with the code or rules under which the container is designed and built (with a minimum factor of safety of four) to withstand loading in any direction equal to four times the weight of the container and attachments when filled to the maximum permissible loaded weight.

Field welding, where necessary, shall be made only on saddle plates or brackets that were applied by the manufacturer of the tank.

MARKING OF GAS CYLINDERS
OSHA 1926.153(n)

When LP-GAS and one or more other gases are stored or used in the same area, the containers shall be marked to identify their contents. Marking shall be in compliance with American National Standard Z48.1-1954, Method of Marking Portable Compressed Gas Containers To Identify the Material Contained.

DAMAGE FROM VEHICLES
OSHA 1926.153(o)

When damage to LP-GAS systems from vehicular traffic is a possibility, precautions against such damage shall be taken.

TEMPORARY HEATING DEVICES
OSHA 1926.154

The following shall be considered for temporary heating devices:

- Ventilation
- Clearance and mounting
- Stability
- Solid fuel salamanders
- Oil-fire heaters

VENTILATION
OSHA 1926.154(a)

Fresh air shall be supplied in sufficient quantities to maintain the health and safety of workers. Where natural means of fresh air supply is inadequate, mechanical ventilation shall be provided.

When heaters are used in confined spaces, special care shall be taken to provide sufficient ventilation in order to ensure proper combustion, maintain the health and safety of workers and limit temperature rise in the area.

CLEARANCE AND MOUNTING
OSHA 1926.154(b)

Temporary heating devices shall be installed to provide clearance to combustible material not less than the amount shown in Table F-4 in 1926.154(b)(1).

Temporary heating devices, which are listed for installation with lesser clearances than specified in Table F-4, may be installed in accordance with their approval.

Heaters that are not suitable for use on wood floors shall not be set directly upon them or other combustible materials. When such heaters are used, they shall rest on suitable heat insulating material or at least 1 in. concrete, or equivalent. The insulating material shall extend beyond the heater 2 ft. or more in all directions.

Heaters used in the vicinity of combustible tarpaulins, canvas or similar coverings shall be located at least 10 ft. from the coverings. The coverings shall be securely fastened to prevent ignition or upsetting of the heater due to wind action on the covering or other material.

STABILITY
OSHA 1926.154(c)

Heaters, when in use, shall be set horizontally level, unless otherwise permitted by the manufacturer's markings.

SOLID FUEL SALAMANDERS
OSHA 1926.154(d)

Solid fuel salamanders are prohibited in buildings and on scaffolds. **(See Figure 6-19)**

OIL-FIRED HEATERS
OSHA 1926.154(e)

Flammable liquid-fired heaters shall be equipped with a primary safety control to stop the flow of fuel in the event of flame failure. Barometric or gravity oil feed shall not be considered a primary safety control.

Heaters designed for barometric or gravity oil feed shall be used only with the integral tanks.

Heaters specifically designed and approved for use with separate supply tanks may be directly connected for gravity feed, or an automatic pump, from a supply tank.

Figure 6-19. Solid fuel salamanders are prohibited in buildings and on scaffolds.

NOTE 1: HEATERS THAT ARE NOT SUITABLE FOR USE ON WOOD FLOORS SHALL NOT BE SET DIRECTLY UPON THEM OR OTHER COMBUSTIBLE MATERIALS. WHERE SUCH HEATERS ARE USED, THEY SHALL REST ON SUITABLE HEAT INSULATING MATERIAL OR AT LEAST 1 IN. CONCRETE, OR EQUIVALENT. THE INSULATING MATERIAL SHALL EXTEND BEYOND THE HEATER 2 FT. OR MORE IN ALL DIRECTIONS.

NOTE 2: HEATER USED IN THE VICINITY OF COMBUSTIBLE TARPAULINS, CANVAS OR SIMILAR COVERINGS SHALL BE LOCATED AT LEAST 10 FT. FROM THE COVERINGS. THE COVERINGS SHALL BE SECURELY FASTENED TO PREVENT IGNITION OR UPSETTING OF THE HEATER DUE TO WIND ACTION ON THE COVERING OR OTHER MATERIAL.

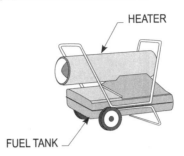

HEATER

FUEL TANK

SOLID FUEL SALAMANDERS
OSHA 1926.154(d)

Fire Protection

Section	Answer
_____	T F
_____	T F
_____	T F
_____	T F
_____	T F
_____	T F
_____	T F
_____	T F
_____	T F
_____	T F
_____	T F

Fire Protection

1. The employer shall be responsible for the development of a fire protection program to be followed throughout all phases of the construction and demolition work.

2. All fire fighting equipment shall be periodically inspected and maintained in operating condition. Defective equipment shall be replaced by the end of the shift.

3. A fire extinguisher, rated not less than 2A, shall be provided for each 2000 sq. ft. of the protected building area, or major fraction thereof.

4. A 1/2 in. diameter garden-type hose line, not to exceed 100 ft. in length and equipped with a nozzle, may be substituted for a 2A-rated fire extinguisher.

5. A fire extinguisher, rated not less than 10B, shall be provided within 50 ft. of wherever more than 5 gallons of flammable or combustible liquids or 5 pounds of flammable gas are being used on the jobsite.

6. Where the employer has provided portable fire extinguishers for employee use in the workplace, the employer shall also provide an educational program to familiarize employees with the general principles of fire extinguisher use and the hazards involved with incipient stage fire fighting.

7. The employer shall assure that portable fire extinguishers are subjected to an semi-annual maintenance check.

8. One hundred feet, or less, of 1 1/2 in. hose, with a nozzle capable of discharging water at 15 gallons or more per minute, may be substituted for a fire extinguisher rated not more than 2A in the designated area provided that the hose line can reach all points in the area.

9. An alarm system, e.g., telephone system, siren, etc., shall be established by the employer whereby employees on the site and the local fire department call be alerted for an emergency.

10. Internal combustion engine powered equipment shall be so located that the exhausts are well away from combustible materials. When the exhausts are piped to outside the building under construction, a clearance of at least 8 in. shall be maintained between such piping and combustible material.

11. Temporary buildings, when located within another building or structure, shall be of either noncombustible construction or of combustible construction having a fire resistance of not less than 1 hour.

T   F

12. Driveways between and around combustible storage piles shall be at least 10 ft. wide and maintained free from accumulation of rubbish, equipment or other articles or materials.

T   F

13. Storage shall not obstruct, or adversely affect, means of exit. All materials shall be stored, handled and piled with due regard to their fire characteristics.

T   F

14. Clearance of at least 24 in. shall be maintained between the top level of the stored material and the sprinkler deflectors.

T   F

15. Approved metal safety cans shall be used for the handling and use of flammable liquids in quantities greater than five gallons.

T   F

16. No more than 25 gallons of flammable or combustible liquids shall be stored in a room outside of an approved storage cabinet.

T   F

17. Storage of containers outside of buildings (not more than 60 gallons each) shall not exceed 1100 gallons in any one pile or area.

T   F

18. Areas in which flammable or combustible liquids are transferred at one time, in quantities greater than 5 gallons from one tank or container to another tank or container, shall be separated from other operations by 25 ft. distance or by construction having a fire resistance of at least 1 hour.

T   F

19. Flammable liquids may be used only where there are no open flames or other sources of ignition within 25 ft. of the operation, unless conditions warrant greater clearance.

T   F

20. Heating equipment of an approved type may be installed in the lubrication or service area where there is no dispensing or transferring of flammable liquids, provided the bottom of the heating unit is at least 12 in. above the floor and is protected from physical damage.

Signs, Signals and Barricades

This chapter covers the different types of signs, signals and barricades to protect employees from accidents. These appropriate signs, signals and barricades shall be used to protect employees and direct traffic safely.

ACCIDENT PREVENTION SIGNS AND TAGS
OSHA 1910.200

The following shall be considered for accident prevention of signs and tags:

- General
- Danger signs
- Caution signs
- Exit signs
- Safety instruction signs
- Directional signs
- Traffic signs
- Additional rules

GENERAL
OSHA 1910.200(a)

All signs and symbols required by this subpart shall be visible at all times when work is being performed, and shall only be removed or covered promptly when the hazards no longer exist.

DANGER SIGNS
OSHA 1910.200(b)

Danger signs shall be used only where an immediate hazard exists. Danger signs shall have red as the predominating color for the upper panel; black outline on the borders; and a white lower panel for additional sign wording. **(See Figure 7-1)**

Figure 7-1. Danger signs shall be used only where an immediate hazard exists.

NOTE: ALL SIGNS AND SYMBOLS SHALL BE VISIBLE AT ALL TIMES WHEN WORK IS BEING PERFORMED, AND SHALL ONLY BE REMOVED OR COVERED PROMPTLY WHEN THE HAZARDS NO LONGER EXIST.

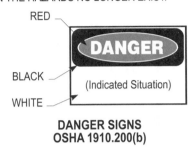

DANGER SIGNS
OSHA 1910.200(b)

CAUTION SIGNS
OSHA 1910.200(c)

Safety Tip: Caution signs must only be used to warn against potential hazards or to express caution against unsafe practices.

Caution signs shall be used only to warn against potential hazards or to caution against unsafe practices. Caution signs shall have yellow as the predominating color; black upper panel and borders: yellow lettering of "caution" on the black panel; and the lower yellow panel for additional sign wording. Black lettering shall be used for additional wording.

Standard color of the background shall be yellow; and the panel, black with yellow letters. Any letters used against the yellow background shall be black. The colors shall be those of opaque glossy samples as specified in Table 1 of American National Standard Z53.1-1967. **(See Figure 7-2)**

Figure 7-2. Caution signs shall be used only to warn against potential hazards or to caution against unsafe practices.

NOTE: STANDARD COLOR OF THE BACKGROUND SHALL BE YELLOW; AND THE PANEL, BLACK WITH YELLOW LETTERS. ANY LETTERS USED AGAINST THE YELLOW BACKGROUND SHALL BE BLACK. THE COLORS SHALL BE THOSE OF OPAQUE GLOSSY SAMPLES AS SPECIFIED IN TABLE 1 OF AMERICAN NATIONAL STANDARD Z53.1-1967.

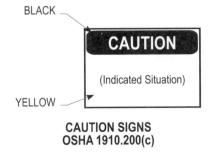

CAUTION SIGNS
OSHA 1910.200(c)

EXIT SIGNS
OSHA 1910.200(d)

Exit signs, when required, shall be lettered in legible red letters, not less than 6 in. high, on a white field and the principal stroke of the letters shall be at least 3/4 in. in width.

SAFETY INSTRUCTION SIGNS
OSHA 1910.200(e)

Safety instruction signs, when used, shall be white with green upper panel with white letters to convey the principal message. Any additional wording on the sign shall be black letters on the white background. **(See Figure 7-3)**

Figure 7-3. This illustration shows the requirements when using safety instructional safety signs to convey principal messages for safety.

NOTE: EXIT SIGNS, WHEN REQUIRED, SHALL BE LETTERED IN LEGIBLE RED LETTER, NOT LESS THAN 6 IN. HIGH, ON A WHITE FIELD AND THE PRINCIPAL STROKE OF THE LETTERS SHALL BE AT LEAST 3/4 IN. IN WIDTH.

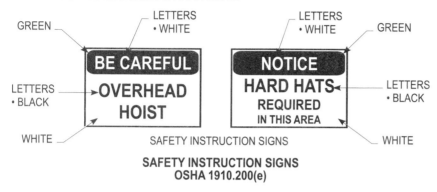

DIRECTIONAL SIGNS
OSHA 1910.200(f)

Directional signs, other than automotive traffic signs, shall be white with a black panel and a white directional symbol. Any additional wording on the sign shall be black letters on the white background.

TRAFFIC SIGNS
OSHA 1910.200(g)

Construction areas shall be posted with legible traffic signs at points of hazard. All traffic control signs or devices used for protection of construction workmen shall conform to American National Standards Institute D6.1-1971, Manual on Uniform Traffic Control Devices for Streets and Highways.

ACCIDENT PREVENTION SIGNS
OSHA 1910.200(h)

Accident prevention tags shall be used as a temporary means of warning employees of an existing hazard, such as defective tools, equipment, etc. They shall not be used in place of, or as a substitute for, accident prevention signs. **(See Figure 7-4)**

ADDITIONAL RULES

American National Standards Institute (ANSI) Z35.1-1968, Specifications for Accident Prevention Signs, and Z35.2-1968, Specifications for Accident Prevention Tags, contain rules that are additional to the rules prescribed in this section. The employer shall comply with ANSI Z35.1-1968 and Z35.2-1968 with respect to rules not specifically prescribed in this subpart.

Figure 7-4. This illustration shows the requirements when using accidental prevention signs temporarily to warn employees that a danger may exist.

NOTE 1: DIRECTIONAL SIGNS, OTHER THAN AUTOMOTIVE TRAFFIC SIGNS, SHALL BE WHITE WITH A BLACK PANEL AND A WHITE DIRECTIONAL SYMBOL. ANY ADDITIONAL WORDING ON THE SIGN SHALL BE BLACK LETTERS ON THE WHITE BACKGROUND.

NOTE 2: CONSTRUCTION AREAS SHALL BE POSTED WITH LEGIBLE SIGNS AT POINTS OF HAZARD. ALL TRAFFIC CONTROL SIGNS OR DEVICES USED FOR PROTECTION OF CONSTRUCTION WORKMEN SHALL CONFORM TO AMERICAN NATIONAL STANDARDS INSTITUTE D6.1-1971, MANUAL ON UNIFORM TRAFFIC CONTROL DEVICES FOR STREETS AND HIGHWAYS.

ACCIDENT PREVENTION TAG

**ACCIDENT PREVENTION TAGS
OSHA 1910.200(h)**

When operations are such that signs, signals and barricades do not provide the necessary protection on or adjacent to a highway or street, flagmen or other appropriate traffic controls shall be provided.

Signaling directions by flagmen shall conform to American National Standards Institute D6.1-1971, Manual on Uniform Traffic Control Devices for Streets and Highways.

Hand signaling by flagmen shall be by use of red flags at least 18 in. square or sign paddles, and in periods of darkness, red lights. **(See Figure 7-5)**

Figure 7-5. This illustration shows the requirements for additional rules when using signs, signals and barricades to warn employees of hazards.

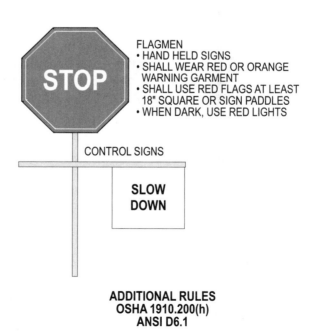

FLAGMEN
• HAND HELD SIGNS
• SHALL WEAR RED OR ORANGE WARNING GARMENT
• SHALL USE RED FLAGS AT LEAST 18" SQUARE OR SIGN PADDLES
• WHEN DARK, USE RED LIGHTS

CONTROL SIGNS

SLOW DOWN

**ADDITIONAL RULES
OSHA 1910.200(h)
ANSI D6.1**

Flagmen shall be provided with and shall wear a red or orange warning garment while flagging. Warning garments worn at night shall be made of reflective material. **(See Figure 7-6)**

NOTE: SIGNALING DIRECTIONS BY FLAGMEN SHALL CONFORM TO AMERICAN NATIONAL STANDARDS INSTITUTE D6.1-1971, MANUAL OF UNIFORM TRAFFIC CONTROL DEVICES FOR STREETS AND HIGHWAYS.

WHEN DARK, FLAGMEN SHALL USE VESTS MADE OF REFLECTIVE MATERIAL.

DOT ORANGE VESTS

ADDITIONAL RULES
OSHA 1910.200(h)

Figure 7-6. Flagmen shall wear a red or orange vest warning garment when flagging.

Traffic control devices such as barricades, cones and drums shall be used to guide or channel traffic as desired through roadway construction. See ANSI D6.1 for appropriate dimensions. **(See Figure 7-7)**

NOTE: SEE ANSI D6.1 FOR APPROPRIATE DIMENSIONS WHEN USING TRAFFIC CONTROL DEVICES SUCH AS BARRICADES, CONES AND DRUMS.

CONES   DRUM

ADDITIONAL RULES
OSHA 1910.200(h)

Figure 7-7. Control devices such as barricades, cones and drums can be used to guide or channel travel.

Three types of barricades are used to stop or control traffic. Type I and II are used for traffic control in a construction area while Type II is used as an indicator that the road is closed. See ANSI D6.1 for appropriate dimensions. **(See Figure 7-8)**

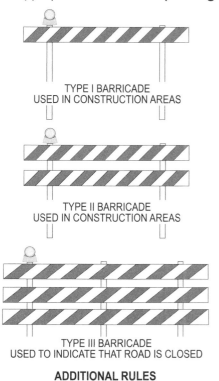

TYPE I BARRICADE
USED IN CONSTRUCTION AREAS

TYPE II BARRICADE
USED IN CONSTRUCTION AREAS

TYPE III BARRICADE
USED TO INDICATE THAT ROAD IS CLOSED

ADDITIONAL RULES
OSHA 1910.200(h)

Figure 7-8. Type I and II barricades are used in construction areas while Type III is used as an indicator that the road is closed.

OSHA Required Training
1926.201(a)(2)

Flashing warning lights shall be placed on barricades during hours of darkness. Nonflashing warning lights shall be placed on groups of barricades used in a series to channel nighttime traffic.

Caution, warning and construction information traffic signs shall be displayed as appropriate to warn or inform vehicle traffic of roadway construction activities and conditions. **(See Figure 7-9)**

Figure 7-9. The appropriate sign shall be used to display the proper warning needed to direct traffic safely.

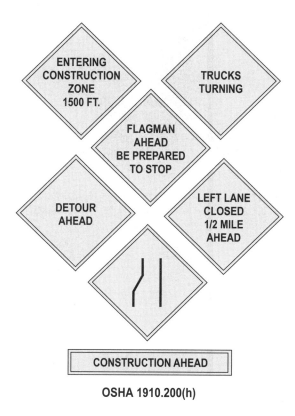

OSHA 1910.200(h)

Signs, Signals and Barricades

Section	Answer
_____	T F
_____	T F
_____	T F
_____	T F
_____	T F
_____	T F
_____	T F
_____	T F
_____	T F
_____	T F

Signs, Signals and Barricades

1. All required signs and symbols shall be visible at all times when work is being performed, and shall only be removed or covered promptly when the hazards no longer exist.

2. Danger signs shall be used only where a possible immediate hazard exists.

3. Caution signs shall be used only to warn against potential hazards or to caution against unsafe practices.

4. Exit signs, when required, shall be lettered in legible red letters, not less than 3 in. high, on a white field and the principal stroke of the letters shall be at least 3/4 in. in width.

5. Safety instruction signs, when used, shall be white with green upper panel with white letters to convey the principal message. Any additional wording on the sign shall be black letters on the white background.

6. All traffic control signs or devices used for protection of construction workmen shall conform to the Manual on Uniform Traffic Control Devices for Streets and Highways.

7. When operations are such that signs, signals and barricades do not provide the necessary protection on or adjacent to a highway or street, flagmen or other appropriate traffic controls shall be provided.

8. Hand signaling by flagmen shall be by use of red flags at least 12 in. square or sign paddles, and in periods of darkness, red lights.

9. Traffic control devices such as barricades, cones and drums shall be used to guide or channel traffic as desired through roadway construction.

10. Flashing warning lights shall be placed on barricades during hours of darkness.

Materials Handling and Storage

This chapter covers the general requirements for storage, rigging equipment for material handling and waste disposal. Employees may be required to wear personal fall arrest equipment in some instances. Slings and all fastenings and attachments shall be inspected for damage or defects by a competent person designated by the employer.

GENERAL REQUIREMENTS FOR STORAGE
OSHA 1926.250

The following general requirements shall be considered for storage:

- General
- Material storage
- Housekeeping
- Dockboards (bridge plates)

GENERAL
OSHA 1926.250(a)

Handling and storing materials involves diverse operations such as hoisting tons of steel with a crane, driving a truck loaded with concrete blocks, manually carrying bags and material and stacking drums, barrels, kegs, lumber or loose bricks.

The efficient handling and storing of materials is vital to industry. These operations provide a continuous flow of raw materials, parts and assemblies through the workplace, and ensure that materials are available when needed. Yet, the improper handling and storing of materials can cause costly injuries.

Workers frequently cite the weight and bulkiness of objects being lifted as major contributing factors to their injuries. In 1990, back injuries resulted in 400,000 workplace accidents. The second factor frequently cited by workers as contributing to their injuries was body movement. Bending, followed by twisting and turning, were the more commonly cited movements that caused back injuries. Back injuries accounted for more than 20 percent of all occupational illnesses, according to data from the National Safety Council.

In addition, workers can be injured by falling objects, improperly stacked materials or by various types of equipment. When manually moving materials, however, workers should be aware of potential injuries, including the following:

- Strains and sprains from improperly lifting loads, or from carrying loads that are either too large or too heavy

- Fractures and bruises caused by being struck by materials, or by being caught in pinch points

- Cuts and bruises caused by falling materials that have been improperly stored, or by incorrectly cutting ties or other securing devices.

Since numerous injuries can result from improperly handling and storing materials, it is important to be aware of accidents that may occur from unsafe or improperly handled equipment and improper work practices, and to recognize the methods for eliminating, or at least minimizing, the occurrence of those accidents. Consequently, employers and employees can and should examine their workplaces to detect any unsafe or unhealthful conditions, practices or equipment and take the necessary steps to correct them.

METHODS OF PREVENTION

Safety Tip: Employees must have proper training and education.

General safety principles can help reduce workplace accidents. These include work practices, ergonomic principles, and training and education. Whether moving materials manually or mechanically, employees should be aware of the potential hazards associated with the task at hand and know how to exercise control over their workplaces to minimize the danger.

Tiered material shall be stored so it will not slide, fall or collapse. Storage areas in buildings, except those located on the ground floor, shall be posted with maximum load limits in pounds per square foot (psf), and such limits shall not be exceeded.

Passageways shall be unobstructed and in good repair so that material-handling equipment may be moved safely through them. **(See Figure 8-1)**

MATERIAL STORAGE
OSHA 1926.250(b)

When manually moving materials, employees should seek help when a load is so bulky it cannot be properly grasped or lifted, when they cannot see around or over it or when a load cannot be safely handled.

When an employee is placing blocks under raised loads, the employee should ensure that the load is not released until his or her hands are clearly removed from the load. Blocking materials and timbers should be large and strong enough to support the load safely. Materials with evidence of cracks, rounded corners, splintered pieces or dry rot should not be used for blocking.

NOTE: TIERED MATERIAL SHALL BE STORED SO IT WILL NOT SLIDE, FALL OR COLLAPSE. STORAGE AREAS IN BUILDINGS, EXCEPT THOSE LOCATED ON THE GROUND FLOOR, SHALL BE POSTED WITH MAXIMUM LOAD LIMITS IN POUNDS PER SQUARE FOOT (psf), AND SUCH LIMITS SHALL NOT BE EXCEEDED.

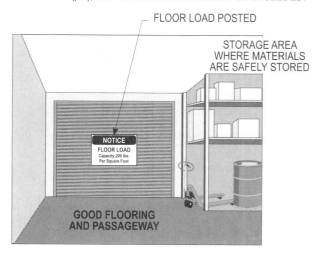

METHODS OR PREVENTION
OSHA 1926.250(a)

Figure 8-1. Passageways shall be unobstructed and in good repair so that material handling equipment can move safely through such area.

Handles and holders should be attached to loads to reduce the chances of getting fingers pinched or smashed. Workers also should use appropriate protective equipment. For loads with sharp or rough edges, wear gloves or other hand and forearm protection. To avoid injuries to the hands and eyes, use gloves and eye protection. When the loads are heavy or bulky, the mover should also wear steel-toed safety shoes or boots to prevent foot injuries if the worker slips or accidentally drops a load.

When mechanically moving materials, avoid overloading the equipment by letting the weight, size and shape of the material being moved dictate the type of equipment used for transporting it. All materials handling equipment has rated capacities that determine the maximum weight the equipment can safely handle and the conditions under which it can handle those weights. The equipment-rated capacities shall be displayed on each piece of equipment and shall not be exceeded except for load testing. When picking up items with a powered industrial truck, the load shall be centered on the forks and as close to the mast as possible to minimize the potential for the truck tipping or the load falling. A lift truck shall never be overloaded because it would be hard to control and could easily tip over. Extra weight shall not be placed on the rear of a counterbalanced forklift to offset an overload. The load shall be at the lowest position for traveling, and the truck manufacturer's operational requirements shall be followed. All stacked loads shall be correctly piled and cross-tiered, where possible. Precautions also should be taken when stacking and storing material.

Safety Tip: Mechanical equipment being used to move materials must not be overloaded.

All bound material should be stacked, placed on racks, blocked, interlocked or otherwise secured to prevent it from sliding, falling or collapsing. A load greater than that approved by a building official may not be placed on any floor of a building or other structure. Where applicable, load limits approved by the building inspector should be conspicuously posted in all storage areas.

When stacking materials, height limitations should be observed. For example, lumber shall be stacked no more than 16 ft. high if it is handled manually; 20 ft. is the maximum stacking height if a forklift is used. For quick reference, walls or posts may be painted with stripes to indicate maximum stacking heights.

Used lumber shall have all nails removed before stacking. Lumber shall be stacked and leveled on solidly supported bracing. The stacks shall be stable and self-supporting. Stacks of loose bricks should not be more than 7 ft. in height. When these stacks reach a height of 4 ft., they should be tapered back 2 in. for every foot of height above the 4 ft. level. **(See Figure 8-2)**

Figure 8-2. When stacks of brick reach a height of 4 ft., they should be tapered back 2 in. for every foot of height above the 4 ft. level.

NOTE: LUMBER SHALL BE STACKED AND LEVELED ON SOLIDLY SUPPORTED BRACING. THESE STACKS SHALL BE STABLE AND SELF-SUPPORTING. STACKS OF LOOSE BRICKS SHALL NOT BE MORE THAN 7 FT. IN HEIGHT.

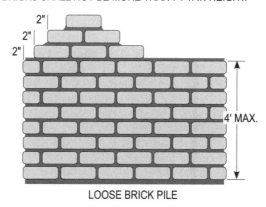

LOOSE BRICK PILE

MATERIAL STORAGE
OSHA 1926.250(b)

When masonry blocks are stacked higher than 6 ft., the stacks should be tapered back one-half block for each tier above the 6 ft. level. **(See Figure 8-3)**

Figure 8-3. When masonry blocks are stacked higher than 6 ft., they should be tapered back 1/2 block for each tier above the 6 ft. level.

NOTE 1: BAGS AND BUNDLES SHALL BE STACKED IN INTERLOCKING ROWS TO REMAIN SECURE. BAGGED MATERIAL SHALL BE STACKED BE STEPPING BACK THE LAYERS AND CROSS-KEYING THE BAGS AT LEAST EVERY 10 LAYERS.

NOTE 2: BALED PAPER AND RAGS STORED INSIDE A BUILDING SHALL NOT BE CLOSER THAN 18 IN. TO THE WALLS, PARTITIONS OR SPRINKLER HEADS.

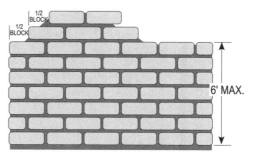

MATERIAL STORAGE
OSHA 1926.250(b)

Safety Tip: Stack drums, barrels, and kegs symmetrically. If stored on their sides, block the bottom tier.

Bags and bundles shall be stacked in interlocking rows to remain secure. Bagged material shall be stacked by stepping back the layers and cross-keying the bags at least every 10 layers. To remove bags from the stack, start from the top row first. Baled paper and rags stored inside a building shall not be closer than 18 in. to the walls, partitions or sprinkler heads. Boxed materials shall be banded or held in place using cross-ties or shrink plastic fiber.

Drums, barrels and kegs shall be stacked symmetrically. If stored on their sides, the bottom tiers shall be blocked to keep them from rolling. When stacked on end, put planks, sheets of plywood dunnage or pallets between each tier to make a firm, flat, stacking surface. When stacking materials two or more tiers high, the bottom tier shall be chocked on each side to prevent shifting in either direction.

When stacking, consider the need for availability of the material. Material that cannot be stacked due to size, shape or fragility can be safety stored on shelves or in bins. Structural steel, bar stock, poles and other cylindrical materials, unless in racks, shall be stacked and blocked to prevent spreading or tilting. Pipes and bars should not be stored in racks that face main aisles; this could create a hazard to passersby when supplies are being removed. **(See Figure 8-4)**

NOTE: DRUMS, BARRELS AND KEGS SHALL BE STACKED SYMMETRICALLY. IF STORED ON THEIR SIDES, THE BOTTOM TIERS SHALL BE BLOCKED TO KEEP THEM FROM ROLLING. WHEN STACKED ON END, PUT PLANKS, SHEETS OF PLYWOOD DUNNAGE OR PALLETS BETWEEN EACH TIER TO MAKE A FIRM, FLAT STACKING SURFACE. WHEN STACKING MATERIALS TWO OR MORE TIERS HIGH, THE BOTTOM TIER SHALL BE CHOCKED ON EACH SIDE TO PREVENT SHIFTING IN EITHER DIRECTION.

Figure 8-4. Materials shall be stacked and blocked to prevent spreading or tilting.

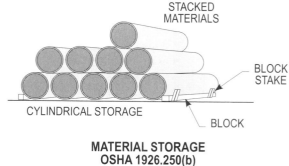

MATERIAL STORAGE
OSHA 1926.250(b)

HOUSEKEEPING
OSHA 1926.250(c)

Stored materials shall not create a hazard. Storage areas shall be kept free from accumulated materials that may cause tripping, fires, or explosions or that may contribute to the harboring of rats and other pests. When stacking and piling materials, it is important to be aware of such factors as the materials' height and weight, how accessible the stored materials are to the user and the condition of the containers where the materials are being stored.

USING MATERIAL HANDLING EQUIPMENT

To reduce potential accidents associated with workplace equipment, employees need to be trained in the proper use and limitations of the equipment they operate. This includes knowing how to effectively use equipment such as conveyors, cranes and slings.

CONVEYORS

When using conveyors, workers' hands may be caught in nip points where the conveyor runs over support members or rollers; workers may be struck by material falling off the conveyor; or they may become caught on or in the conveyor, thereby being drawn into the conveyor path.

Safety Tip: An emergency button or pullcord designed to stop the conveyor must be installed at the employee's work station.

To reduce the severity of an injury, an emergency button or pull cord designed to stop the conveyor shall be installed at the employee's work station. Continuously accessible conveyor belts should have an emergency stop cable that extends the entire length of the conveyor belt so that the cable can be accessed from any location along the belt. The emergency stop switch shall be designed to be reset before the conveyor can be restarted. Before restarting a conveyor that has stopped due to an overload, appropriate personnel shall inspect the conveyor and clear the stoppage before restarting. Employees shall never ride on a materials handling conveyor. Where a conveyor passes over work areas or aisles, guards shall be provided to keep employees from being struck by falling material. If the crossover is low enough for workers to run into, it shall be guarded to protect employees and either marked with a warning sign or painted a bright color.

Screw conveyors shall be completely covered except at loading and discharging points. At those points, guards shall protect employees against contacting the moving screw; the guards are movable, and they shall be interlocked to prevent conveyor movement when not in place.

CRANES

Safety Tip: Employees that are permitted to operate cranes must be thoroughly trained and competent.

Only thoroughly trained and competent persons shall be permitted to operate cranes. Operators should know what they are lifting and what it weighs. The rated capacity of mobile cranes varies with the length of the boom and the boom radius. When a crane has a telescoping boom, a load may be safe to lift at a short boom length and/or a short boom radius, but may overload the crane when the boom is extended and the radius increases.

All movable cranes shall be equipped with a boom angle indicator; those cranes with telescoping booms shall be equipped with some means to determine the boom length, unless the load rating is independent of the boom length. Load rating charts shall be posted in the cab of cab-operated cranes. All mobile cranes do not have uniform capacities for the same boom length and radius in all directions around the chassis of the vehicle.

Always check the crane's load chart to ensure that the crane is not going to be overloaded for the conditions under which it will operate. Plan lifts before starting them to ensure that they are safe. Take additional precautions and exercise extra care when operating around power lines.

Safety Tip: When used, outriggers must rest on firm ground timbers or be sufficiently cribbed.

Some mobile cranes cannot operate with outriggers in the traveling position. When used, the outriggers shall rest on firm ground, on timbers or be sufficiently cribbed to spread the weight of the crane and the load over a large enough area. This will prevent the crane from tipping during use. Hoisting chains and ropes shall always be free of kinks or twists and shall never be wrapped around a load. Loads should be attached to the load hook by slings, fixtures or other devices that have the capacity to support the load on the hook. Sharp edges of loads should be padded to prevent cutting slings. Proper sling angles shall be maintained so that slings are not loaded in excess of their capacity.

All cranes shall be inspected frequently by persons thoroughly familiar with the crane, the methods of inspecting the crane and what can make the crane unserviceable. Crane activity, the severity of use and environmental conditions should determine inspection schedules. Critical parts, such as crane operating mechanisms, hooks, air or hydraulic system components and other load-carrying components, should be inspected daily for any maladjustment, deterioration, leakage, deformation or other damage.

See Chapter 14 for more information on Cranes, Derricks, Hoists, Elevators and Conveyors.

SLINGS

Safety Tip: Slings must be visually inspected before use.

When working with slings, employers shall ensure that they are visually inspected before use and during operation, especially if used under heavy stress. Riggers or other knowledgeable employees should conduct or assist in the inspection because they are aware of how the sling is used and what makes a sling unserviceable. A damaged or defective sling shall be removed from service.

Slings shall not be shortened with knots or bolts or other makeshift devices; sling legs that have been kinked shall not be used. Slings shall not be loaded beyond their rated capacity, according to the manufacturer's instructions. Suspended loads shall be kept clear of all obstructions, and crane operators should avoid sudden starts and stops when moving suspended loads. Employees also shall remain clear of loads about to be lifted and suspended. All shock loading is prohibited.

POWERED INDUSTRIAL TRUCKS

Workers who must handle and store materials often use fork trucks, platform lift trucks, motorized hand trucks and other specialized industrial trucks powered by electrical motors or internal combustion engines. Affected workers, therefore, should be aware of the safety requirements pertaining to fire protection, and the design, maintenance and use of these trucks.

All new powered industrial trucks, except vehicles intended primarily for earth moving or over-the-road hauling, shall meet the design and construction requirements for powered industrial trucks established in the American National Standard for Powered Industrial Trucks, Part 11, ANSI B56.1-1969. Approved trucks shall also bear a label or some other identifying mark indicating acceptance by a nationally recognized testing laboratory.

Modifications and additions that affect capacity and safe operation of the trucks shall not be performed by an owner or user without the manufacturer's prior written approval. In these cases, capacity, operation and maintenance instruction plates and tags or decals shall be changed to reflect the new information. If the truck is equipped with front-end attachments that are not factory installed, the user should request that the truck be marked to identify these attachments and show the truck's approximate weight, including the installed attachment, when it is at maximum elevation with its load laterally centered.

Safety Tip: Modifications and additions cannot be made to a forklift without the manufacturer's approval.

There are 11 different types of industrial trucks or tractors, some having greater safeguards than others. There are also designated conditions and locations under which the vast range of industrial-powered trucks can be used. In some instances, powered industrial trucks cannot be used, and in others, they can only be used if approved by a nationally recognized testing laboratory for fire safety. For example, powered industrial trucks shall not be used in atmospheres containing hazardous concentrations of the following substances:

- Acetylene
- Butadiene
- Ethylene oxide
- Hydrogen (or gases or vapors equivalent in hazard to hydrogen, such as manufactured gas)
- Propylene oxide
- Acetaldehyde
- Cyclopropane
- Dimethyl ether
- Ethylene
- Isoprene
- Unsymmetrical dimethyl hydrazine

These trucks are not to be used in atmospheres containing hazardous concentrations of metal dust, including aluminum, magnesium and other metals of similarly hazardous characteristics or in atmospheres containing carbon black, coal or coke dust. Where dust of magnesium, aluminum or aluminum bronze dusts may be present, the fuses, switches, motor controllers and circuit breakers of trucks shall be enclosed with enclosures approved for these substances.

There also are powered industrial trucks or tractors that are designed, constructed and assembled for use in atmospheres containing flammable vapors or dusts. These include industrial-powered trucks equipped with additional safeguards to their exhaust, fuel and electrical systems; with no electrical equipment, including the ignition; with temperature limitation features; and with electric motors and all other electrical equipment completely enclosed.

These specially designed powered industrial trucks may be used in locations where volatile flammable liquids or flammable gases are handled, processed or used. The liquids, vapors or gases should, among other things, be confined within closed containers or closed systems from which they cannot escape.

Some other conditions and/or locations in which specifically designed powered industrial trucks may be used include the following:

- Only powered industrial trucks that do not have any electrical equipment, including the ignition, and have their electrical motors or other electrical equipment completely enclosed should be used in atmospheres containing flammable vapors or dust.

- Powered industrial trucks that are either powered electrically by liquefied petroleum gas or by a gasoline or diesel engine are used on piers and wharves that handle general cargo.

- Safety precautions the user can observe when operating or maintaining powered industrial trucks include:

- That high lift rider trucks be fitted with an overhead guard, unless operating conditions do not permit.

- That fork trucks be equipped with a vertical load backrest extension according to manufacturers' specifications, if the load presents a hazard.

- That battery charging installations be located in areas designated for that purpose.

- That facilities be provided for flushing and neutralizing spilled electrolytes when changing or recharging a battery to prevent fires, to protect the charging apparatus from being damaged by the truck and to adequately ventilate fumes in the charging area from gassing batteries.

- That conveyor, overhead hoist or equivalent materials handling equipment be provided for handling batteries.

- That auxiliary directional lighting is provided on the truck where general lighting is less than 2 lumens per square foot.

- That arms and legs not be placed between the uprights of the mast or outside the running lines of the truck.

- That brakes be set and wheel blocks or other adequate protection be in place to prevent movement of trucks, trailers or railroad cars when using trucks to load or unload materials onto train boxcars.

- That sufficient headroom is provided under overhead installations, lights, pipes and sprinkler systems.

- That personnel on the loading platform have the means to shut off power to the truck.

- That dockboards or bridgeplates be properly secured, so they won't move when equipment moves over them.

- That only stable or safely arranged loads be handled, and caution be exercised when handling loads.

- That trucks whose electrical systems are in need of repair have the battery disconnected prior to such repairs.

- That replacement parts of any industrial truck be equivalent in safety to the original ones.

ERGONOMIC SAFETY AND HEALTH PRINCIPLES

Ergonomics is defined as the study of work and is based on the principle that the job should be adapted to fit the person, rather than forcing the person to fit the job. Ergonomics focuses on the work environment and items such as design and function of workstations, controls, displays, safety devices, tools and lighting to fit the employees' physical requirements and to ensure their health and well being.

Ergonomics includes restructuring or changing workplace conditions to make the job easier and reducing/stressors that cause cumulative trauma disorders and repetitive motion injuries. In the area of materials handling and storing, ergonomic principles may require controls such as reducing the size or weight of the objects lifted, installing a mechanical lifting aid or changing the height of a pallet or shelf.

Although no approach has been found for totally eliminating back injuries resulting from lifting materials, a substantial number of lifting injuries can be prevented by implementing an effective ergonomics program and by training employees in appropriate lifting techniques.

In addition to using ergonomic controls, there are some basic safety principles that can be employed to reduce injuries resulting from handling and storing materials. These include taking general fire safety precautions and keeping aisles and passageways clear.

In adhering to fire safety precautions, employees should note that flammable and combustible materials shall be stored according to their fire characteristics. Flammable liquids, for example, shall be separated from other material by a fire wall. Also, other combustibles shall be stored in an area where smoking and using an open flame or a spark-producing device is prohibited. Dissimilar materials that are dangerous when they come into contact with each other shall be stored apart.

When using aisles and passageways to move materials mechanically, sufficient clearance shall be allowed for aisles at loading docks, through doorways, wherever turns shall be made, and in other parts of the workplace. Providing sufficient clearance for mechanically moved materials will prevent workers from being pinned between the equipment and fixtures in the workplace, such as walls, racks, posts or other machines. Sufficient clearance also will prevent the load from striking an obstruction and falling on an employee.

All passageways used by employees should be kept clear of obstructions and tripping hazards. Materials in excess of supplies needed for immediate operations should not be stored in aisles or passageways, and permanent aisles and passageways shall be marked appropriately.

TRAINING AND EDUCATION

OSHA recommends using a formal training program to reduce materials handling hazards. Instructors should be well-versed in matters that pertain to safety engineering and materials handling and storing. The content of the training should emphasize those factors that will contribute to reducing workplace hazards including the following:

- Alerting the employee to the dangers of lifting without proper training

- Showing the employee how to avoid unnecessary physical stress and strain

- Teaching workers to become aware of what they can comfortably handle without undue strain

- Instructing workers on the proper use of equipment

- Teaching workers to recognize potential hazards and how to prevent or correct them

Because of the high incidence of back injuries, safe lifting techniques for manual lifting should be demonstrated and practiced at the work site by supervisors as well as by employees. A training program to teach proper lifting techniques should cover the following topics:

- Awareness of the health risks to improper lifting - citing organizational case histories

- Knowledge of the basic anatomy of the spine, the muscles and the joints of the trunk and the contributions of intra-abdominal pressure while lifting.

- Awareness of individual body strengths and weaknesses - determining one's own lifting capacity.

- Recognition of the physical factors that might contribute to an accident, and how to avoid the unexpected.

- Use of safe lifting postures and timing for smooth, easy lifting and the ability to minimize the load-moment effects.

- Use of handling aids such as stages, platforms, or steps, trestles, shoulder pads, handles and wheels.

- Knowledge of body responses - warning signals - to be aware of when lifting.

A campaign using posters to draw attention to the need to do something about potential accidents, including lifting and back injuries, is one way to increase awareness of safe work practices and techniques. The plant medical staff and a team of instructors should conduct regular tours of the site to look for potential hazards and allow input from workers.

RIGGING EQUIPMENT FOR MATERIAL HANDLING OSHA 1926.251

The following shall be considered for rigging equipment for handling material:

- General
- Alloy steel chains

- Wire rope
- Fiber rope and synthetic fiber
- Synthetic webbing (nylon, polyester and polypropylene)
- Shackles and hooks

GENERAL
OSHA 1926.251(a)

The ability to handle materials and move them from one location to another, whether during transit or at the worksite, is vital to all segments of industry. Materials shall be moved, for example, in order for industry to manufacture, sell and utilize products. In short, without materials-handling capability, industry would cease to exist.

All employees in numerous workplaces take part in materials handling, to varying degrees. As a result, some employees are injured. In fact, the mishandling of materials is the single largest cause of accidents and injuries in the workplace. Most of these accidents and injuries, as well as the pain and loss of salary and productivity that often result, can be readily avoided. Whenever possible, mechanical means should be used to move materials in order to avoid employee injuries such as muscle pulls, strains and sprains. In addition, many loads are too heavy and/or bulky to be safely moved manually. Therefore, the following various types of equipment have been designed specifically to aid in the movement of materials:

- Cranes
- Derricks
- Hoists
- Powered industrial trucks
- Conveyors

Because cranes, derricks and hoists rely upon slings to hold their suspended loads, slings are the most commonly used piece of materials-handling apparatus. This discussion will offer information on the proper selection, maintenance and use of slings.

IMPORTANCE OF THE OPERATOR

The operator shall exercise intelligence, care and common sense in the selection and use of slings. Slings shall be selected in accordance with their intended use, based upon the size and type of load and the environmental conditions of the workplace. All slings shall be visually inspected before use to ensure that there is no obvious damage.

Safety Tip: Operators must exercise intelligence, care, and common sense when selecting and using slings.

A well-trained operator can prolong the service life of equipment and reduce costs by avoiding the potentially hazardous effects of overloading equipment, operating it at excessive speeds, taking up slack with a sudden jerk and suddenly accelerating or decelerating equipment. The operator can look for causes and seek corrections whenever a danger exists. He or she should cooperate with co-workers and supervisors and become a leader in carrying out safety measures, not merely for the good of the equipment and the production schedule, but, more importantly, for the safety of everyone concerned.

SLING TYPES

The dominant characteristics of a sling are determined by the components of that sling. For example, the strengths and weaknesses of a wire rope sling are essentially the same as the strengths and weaknesses of the wire rope of which it is made.

Slings are generally the following one of six types:

- Chain
- Wire rope
- Metal mesh
- Natural fiber rope
- Synthetic fiber rope
- Synthetic web

In general, use and inspection procedures tend to place these slings into the following three groups:

- Chain
- Wire rope and mesh
- Fiber rope web

Each type has its own particular advantages and disadvantages. Factors that should be taken into consideration when choosing the best sling for the job include the size, weight, shape, temperature and sensitivity of the material to be moved, as well as the environmental conditions under which the sling will be used.

ALLOY STEEL CHAINS
OSHA 1926.251(b)

Chains are commonly used because of their strength and ability to adapt to the shape of the load. Care should be taken, however, when using alloy chain slings because they are subject to damage by sudden shocks. Misuse of chain slings could damage the sling, resulting in sling failure and possible injury to an employee.

Chain slings are your best choice for lifting materials that are very hot. They can be heated to temperatures of up to 1000°F; however, when alloy chain slings are consistently exposed to service temperatures in excess of 600°F, operators shall reduce the working load limits in accordance with the manufacturer's recommendations.

Safety Tip: Chain slings must have a permanently attached and durable identification.

All sling types shall be visually inspected prior to use. When inspecting alloy steel chain slings, pay special attention to any stretching, wear in excess of the allowances made by the manufacturer and nicks and gouges. These are all indications that the sling may be unsafe and is to be removed from service. **(See Figure 8-5)**

A welded alloy-steel chain sling shall have a permanently affixed, durable identification label stating size, grade, rated capacity and manufacturer. **(See Figure 8-6)**

WIRE ROPE
OSHA 1926.251(c)

A second type of sling is made of wire rope. Wire rope is composed of individual wires that have been twisted to form strands. The strands are then twisted to form a wire rope. When wire rope has a fiber core, it is usually more flexible but is less resistant to environmental damage. Conversely, a core that is made of a wire rope strand tends to have greater strength and is more resistant to heat damage. **(See Figure 8-7)**

NOTE: WHEN INSPECTING ALLOY STEEL CHAIN SLINGS, PAY SPECIAL ATTENTION TO ANY STRETCHING, WEAR IN EXCESS OF THE ALLOWANCES MADE BY THE MANUFACTURER AND NICKS AND GOUGES.

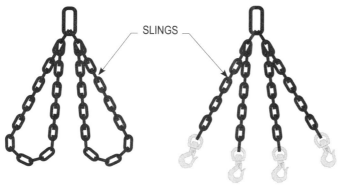

SLINGS

**ALLOY STEEL CHAINS
OSHA 1926.251(a)**

Figure 8-5. Slings shall be visually inspected prior to use and if found to unsafe, removed from service.

NOTE: CHAIN SLINGS CAN BE HEATED TO TEMPERATURES OF UP TO 1000° F; HOWEVER, WHEN ALLOY CHAIN SLINGS ARE CONSISTENTLY EXPOSED TO SERVICE TEMPERATURES IN EXCESS OF 600° F, OPERATORS SHALL REDUCE THE WORKING LOAD LIMITS IN ACCORDANCE WITH MANUFACTURER'S RECOMMENDATIONS.

WELDED ALLOY-STEEL CHAIN SLING

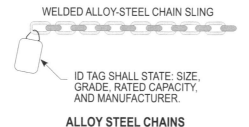

ID TAG SHALL STATE: SIZE, GRADE, RATED CAPACITY, AND MANUFACTURER.

**ALLOY STEEL CHAINS
OSHA 1926.251(b)**

Figure 8-6. Welded alloy-steel chain slings shall have a durable identification label.

NOTE: WHEN WIRE ROPE HAS A FIBER CORE, IT IS USUALLY MORE FLEXIBLE BUT IS LESS RESISTANT TO ENVIRONMENTAL DAMAGE. A CORE THAT IS MADE OF A WIRE ROPE STRAND TENDS TO HAVE GREATER STRENGTH AND IS MORE RESISTANT TO HEAT DAMAGE.

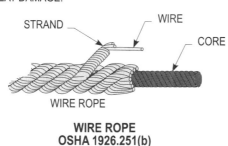

STRAND   WIRE

CORE

WIRE ROPE

**WIRE ROPE
OSHA 1926.251(b)**

Figure 8-7. This illustration shows the requirements for a sling made of rope.

ROPE LAY

Wire rope may be further defined by the "lay." The lay of a wire rope can mean any of three things:

- One complete wrap of a strand around the core: One rope lay is one complete wrap of a strand around the core. **(See Figure 8-8)**

- The direction the strands are wound around the core: Wire rope is referred to as right lay or left lay. A right lay rope is one in which the strands are wound in a right-hand direction like a conventional screw thread **(See Figure 8-9)**. A left lay rope is just the opposite.

Figure 8-8. A wire rope may be defined by the lay.

NOTE: IN REGULAR LAY ROPES, THE WIRES IN THE STRANDS ARE LAID IN ONE DIRECTION, WHILE THE STRANDS IN THE ROPE ARE LAID IN THE OPPOSITE DIRECTION. THE RESULT IS THAT THE WIRE CROWN RUNS APPROXIMATELY PARALLEL TO THE LONGITUDINAL AXIS OF THE ROPE.

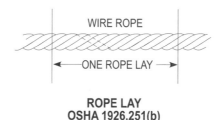

WIRE ROPE

←— ONE ROPE LAY —→

ROPE LAY
OSHA 1926.251(b)

Figure 8-9. A right lay rope has its strands wound in a right-hand direction.

NOTE: LEFT LAY ROPE HAS ITS GREATEST USAGE IN OIL FIELDS ON ROD AND TUBING LINES, BLAST HOLE RIGS AND SPUDDERS, WHERE ROTATION OF RIGHT LAY WOULD LOOSEN COUPLINGS. THE ROTATION OF A LEFT LAY ROPE TIGHTENS A STANDARD COUPLING.

RIGHT LAY ROPE

ROPE LAY
OSHA 1926.251(c)

- The direction the wires are wound in the strands in relation to the direction of the strands around the core: In regular lay rope, the wires in the strands are laid in one direction while the strands in the rope are laid in the opposite direction. In lang lay rope, the wires are twisted in the same direction as the strands. **(See Figure 8-10)**

Figure 8-10. A lang lay rope has wires that are twisted in the same direction as the strands.

NOTE: THE OUTSIDE WIRES IN LANG LAY ROPE LIE AT AN ANGLE TO THE ROPE AXIS, THE INTERNAL STRESS DUE TO BENDING OVER SHEAVES AND DRUMS IS REDUCED, CAUSING LANG LAY ROPES TO BE MORE RESISTANT TO BENDING FATIGUE.

RIGHT LAY,
REGULAR LAY ROPE

RIGHT LAY,
LANG LAY ROPE

LEFT LAY,
REGULAR LAY ROPE

LEFT LAY,
LANG LAY ROPE

ROPE LAY
OSHA 1926.251(c)

- In regular lay ropes, the wires in the strands are laid in one direction, while the strands in the rope are laid in the opposite direction. The result is that the wire crown runs approximately parallel to the longitudinal axis of the rope. These ropes have good resistance to kinking and twisting and are easy to handle. They are also able to withstand considerable crushing and distortion due to the short length of exposed wires. This type of rope has the widest range of applications.

Lang lay (where the wires are twisted in the same direction as the strands) is recommended for many excavating, construction and mining applications, including draglines, hoist lines, dredge lines and other similar lines.

Lang lay ropes are more flexible and have greater wearing surface per wire than regular lay ropes. In addition, since the outside wires in lang lay rope lie at an angle to the rope axis, internal stress due to bending over sheaves and drums is reduced, causing lang lay ropes to be more resistant to bending fatigue.

A left lay rope is one in which the strands form a left-hand helix similar to the threads of a left-hand screw thread. Left lay rope has its greatest usage in oil fields on rod and tubing lines, blast hole rigs and spudders, where rotation of right lay would loosen couplings. The rotation of a left lay rope tightens a standard coupling.

WIRE ROPE SLING SELECTION

When selecting a wire rope sling to give the best service, there are four characteristics to consider:

- Strength
- Ability to bend without distortion
- Ability to withstand abrasive wear
- Ability to withstand abuse

Strength - The strength of a wire rope is a function of its size, grade and construction. It shall be sufficient to accommodate the maximum load that will be applied. The maximum load limit is determined by means of an appropriate multiplier. This multiplier is the number by which the ultimate strength of a wire rope is divided to determine the working load limit. Thus a wire rope sling with a strength of 10,000 pounds and a total working load of 2000 pounds has a design factor (multiplier) of 5. New wire rope slings have a design factor of 5. As a sling suffers from the rigors of continued service, however, both the design factor and the sling's ultimate strength are proportionately reduced. If a sling is loaded beyond its ultimate strength, it will fail. For this reason, older slings shall be more rigorously inspected to ensure that rope conditions adversely affecting the strength of the sling are considered in determining whether or not a wire rope sling should be allowed to continue in service.

Fatigue - A wire rope shall have the ability to withstand repeated bending without the failure of the wires from fatigue. Fatigue failure of the wires in a wire rope is the result of the development of small cracks under repeated applications of bending loads. It occurs when ropes make small radius bends. The best means of preventing fatigue failure of wire rope slings is to use blocking or padding to increase the radius of the bend.

Abrasive Wear - The ability of a wire rope to withstand abrasion is determined by the size, number of wires and construction of the rope. Smaller wires bend more readily and therefore offer greater flexibility but are less able to withstand abrasive wear. Conversely, the larger wires of less flexible ropes are better able to withstand abrasion than smaller wires of the more flexible ropes.

Abuse - All other factors being equal, misuse or abuse of wire rope will cause a wire rope sling to become unsafe long before any other factor. Abusing a wire rope sling can cause serious structural damage to the wire rope, such as kinking or bird caging, which reduces the strength of the wire rope. (In bird caging, the wire rope strands are forcibly untwisted and become spread outward.) Therefore, in order to prolong the life of the sling and protect the lives of employees, the manufacturer's suggestion for safe and proper use of wire rope slings shall be strictly adhered to.

See Figure 8-11 for a detailed illustration pertaining to examples of hazardous rope conditions.

Wire Rope Life. Many operating conditions affect wire rope life. They are bending, stresses, loading conditions, speed of load application (jerking), abrasion, corrosion, sling design, materials handled, environmental conditions and history of previous problems.

Figure 8-11. This illustration shows examples of conditions when a rope is considered dangerous to use as a sling.

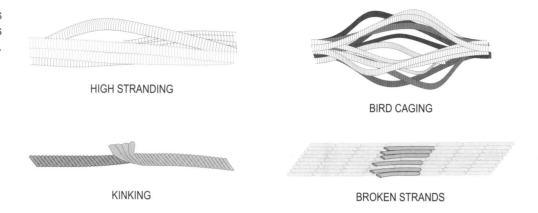

HIGH STRANDING

BIRD CAGING

KINKING

BROKEN STRANDS

SIGNS OF WIRE-ROPED DAMAGE
OSHA 1926.251(c)

Safety Tip: Wire rope slings must be visually inspected before each use.

In addition to the above operating conditions, the weight, size and shape of the loads to be handled also affect the service life of a wire rope sling. Flexibility is also a factor. Generally, more flexible ropes are selected when smaller radius bending is required. Less flexible ropes should be used when the rope must move through or over abrasive materials.

Wire Rope Sling Inspection. Wire rope slings shall be visually inspected before each use. The operator should check the twists or lay of the sling. If ten randomly distributed wires in one lay are broken, or five wires in one strand of a rope lay are damaged, the sling shall not be used. It is not sufficient, however, to check only the condition of the wire rope. End fittings and other components should also be inspected for any damage that could make the sling unsafe.

To ensure safe sling usage between scheduled inspections, all workers shall participate in a safety awareness program. Each operator shall keep a close watch on those slings he or she is using. If any accident involving the movement of materials occurs, the operator shall immediately shut down the equipment and report the accident to a supervisor. The cause of the accident shall be determined and corrected before resuming operations.

Eye splices shall have at least three tucks. Other types of eye splices shall be permitted to be used if they are not otherwise prohibited and if they have been shown to be effective (through a manufacturer's certificate of test available for examination, for example).

Safety Tip: Wire rope used for lifting shall be one continuous piece, without a knot or splice.

Each wire rope used for lifting shall be one continuous piece, without a knot or splice. (Exception may be made for end eye splices and endless slings.)

Eyes in bridles, slings or bull wires shall not be formed by wire-rope clips or knots.

Field Lubrication. Although every rope sling is lubricated during manufacture, to lengthen its useful service life it shall also be lubricated "in the field." There is no set rule on how much or how often this should be done. It depends on the conditions under which the sling is used. The heavier the loads, the greater the number of bends, or the more adverse the conditions under which the sling operates, the more frequently lubrication will be required.

Storage. Wire rope slings should be stored in a well ventilated, dry building or shed. Never store them on the ground or allow them to be continuously exposed to the elements because this will make them vulnerable to corrosion and rust. And, if it is necessary to store wire rope slings outside, make sure that they are set off the ground and protected.

Note: Using the sling several times a week, even at a light load, is a good practice. Records show that slings that are used frequently or continuously give useful service far longer than those that are idle.

DISCARDING SLINGS

Wire rope slings can provide a margin of safety by showing early signs of failure. Factors requiring that a wire sling be discarded include the following:

- Severe corrosion
- Localized wear (shiny worn spots) on the outside
- A one-third reduction in outer wire diameter
- Damage or displacement of end fittings - hooks, rings, links or collars - by overload or • Misapplication
- Distortion, kinking, bird caging or other evidence of damage to the wire rope structure
- Excessive broken wires
- U-bolt wire-rope clips

When U-bolt wire-rope clips are used to form eyes, the appropriate number of clips and the amount of space between the clips shall be determined according to Table H-1. The "U" section of a U-bolt shall always be in contact with the wire-rope dead end. **(See Figure 8-12)**

TABLE 8-12. NUMBER AND SPACING OF U-BOLT WIRE-ROPE CLIPS			
IMPROVED PLOW STEEL, ROPE DIAMETER	NUMBER OF CLIPS		MINIMUM SPACING (INCHES)
	DROP FORGED	OTHER MATERIAL	
-	3	4	3
5/8	3	4	3 -
-	4	5	4 -
7/8	4	5	5 -
1	5	6	6
1 1/8	6	7	6 -
1 -	6	6	7 -
1 3/8	7	7	8 -
1 -	7	8	9

Figure 8-12. The spacing for U-bolts on a 5/8 steel diameter rope is a minimum of 3 in.

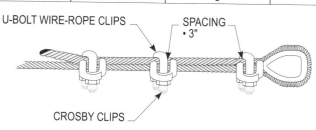

U-BOLT WIRE-ROPE CLIPS

SPACING • 3"

CROSBY CLIPS

**DISCARDING SLINGS
OSHA 1926.251(c)**

FIBER ROPE AND SYNTHETIC FIBER
OSHA 1926.251(d)

Fiber rope and synthetic web slings are used primarily for temporary work, such as construction and painting jobs, and in marine operations. They are also the best choice for use on expensive loads, highly finished parts, fragile parts and delicate equipment.

FIBER ROPE

Fiber rope slings are preferred for some applications because they are pliant, they grip the load well and they do not mar the surface of the load. They should be used only on light loads, however, and shall not be used on objects that have sharp edges capable of cutting the rope or in applications where the sling will be exposed to high temperatures, severe abrasion or acids.

The choice of rope type and size will depend upon the application, the weight to be lifted and the sling angle. Before lifting any load with a fiber rope sling be sure to inspect the sling carefully because they deteriorate far more rapidly than wire rope slings and their actual strength is very difficult to estimate. **(See Figure 8-13)**

Figure 8-13. The type of rope size depends upon the application and conditions of use.

NOTE: FIBER ROPE SLINGS SHOULD BE USED ONLY ON LIGHT LOADS AND SHALL NOT BE USED ON OBJECTS THAT HAVE SHARP EDGES CAPABLE OF CUTTING THE ROPE OR IN APPLICATIONS WHERE THE SLING WILL BE EXPOSED TO HIGH TEMPERATURES, SEVERE ABRASION OR ACIDS.

**FIBER ROPE
OSHA 1926.251(d)**

When inspecting a fiber rope sling prior to using it, look first at its surface. Look for dry, brittle, scorched or discolored fibers. If any of these conditions are found, the supervisor shall be notified and a determination made regarding the safety of the sling. If the sling is found to be unsafe, it shall be discarded.

Next, check the interior of the sling. It should be as clean as when the rope was new. A build-up of powder-like sawdust on the inside of the fiber rope indicates excessive internal wear and is an indication that the sling is unsafe.

Finally, scratch the fibers with a fingernail. If the fibers come apart easily, the fiber sling has suffered some kind of chemical damage and must be discarded.

Natural or synthetic-fiber rope shall be used only within its safe workload limits. Refer to tables in 29 CFR 1926.251.

Safety Tip: Eye splices made in synthetic rope must have at least 4 full tucks.

Splices in rope slings shall be made according to the rope manufacturer's recommendations.

Manila-rope eye splices shall have at least 3 full tucks; short splices shall have 6 full tucks (3 on each side of the center of splice).

Synthetic-rope eye splices shall have at least 4 full tucks. Short splices shall have at least 8 full tucks (4 on each side of the center of splice).

Strand ends shall not be trimmed short (flush with the surface of the rope) if they are adjacent to the full tucks. For fiber rope less than 1 in. in diameter, the tails shall project at least 6 rope diameters beyond the last full tuck; for fiber rope 1 in. in diameter or larger, the tails shall project at least 6 in. beyond the last full tuck.

For all eye splices, the eye shall be large enough to form an included angle of no more than 60 ft. at the splice when the eye is placed over the load or support.

Knots shall not be used instead of splices.

Web or rope slings shall be protected (with a blunting material) to prevent cuts from the sharp edges of material being hoisted.

SYNTHETIC WEB SLINGS
OSHA 1926.251(e)

Synthetic web slings offer a number of advantages for rigging purposes. The most commonly used synthetic web slings are made of nylon, dacron and polyester. They have the following properties in common:

- Strength - can handle load of up to 300,000 lbs.
- Convenience - can conform to any shape.
- Safety - will adjust to the load contour and hold it with a tight, non-slip grip.
- Load protection - will not mar, deface or scratch highly polished or delicate surfaces.
- Long life - are unaffected by mildew, rot or bacteria; resist some chemical action; and have • Excellent abrasion resistance.
- Economy - have low initial cost plus long service life.
- Shock absorbency - can absorb heavy shocks without damage.
- Temperature resistance are unaffected by temperatures up to 180°F.

Each synthetic material has its own unique properties. Nylon shall be used wherever alkaline or greasy conditions exist. It is also preferable when neutral conditions prevail and when resistance to chemicals and solvents is important. Dacron shall be used where high concentrations of acid solutions - such as sulfuric, hydrochloric, nitric and formic acids - and where high-temperature bleach solutions are prevalent. (Nylon will deteriorate under these conditions.) Do not use dacron in alkaline conditions because it will deteriorate; use nylon or polypropylene instead. Polyester shall be used where acids or bleaching agents are present and is also ideal for applications where a minimum of stretching is important.

Possible Defects. Synthetic web slings shall be removed from service if any of the following defects exist:

- Acid or caustic burns
- Melting or charring of any part of the surface
- Snags, punctures, tears or cuts,
- Broken or worn stitches
- Wear or elongation exceeding the amount recommended by the manufacturer
- Distortion of fittings

Each synthetic web sling shall be marked or coded to show the manufacturer's name or trademark, rated capacity and type of material.

The rated capacity of synthetic webbing shall not be exceeded.

Synthetic web slings shall be protected (with a blunting material) to prevent cuts from the sharp edges of material being hoisted. **(See Figure 8-14)**

SHACKLES AND HOOKS
OSHA 1926.251(f)

The safe workload of each shackle shall be calculated using Table H-2 or manufacturer's recommendations. **(See Figure 8-15)**

Figure 8-14. The rated capacity of synthetic web slings shall not be exceeded.

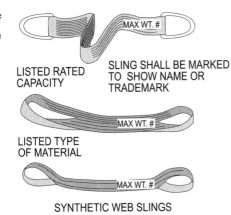

LISTED RATED CAPACITY

SLING SHALL BE MARKED TO SHOW NAME OR TRADEMARK

MAX WT. #

LISTED TYPE OF MATERIAL

MAX WT. #

SYNTHETIC WEB SLINGS

SYNTHETIC WEB SLINGS SHALL BE REMOVED FROM SERVICE IF ANY OF THE FOLLOWING DEFECTS EXIST:
• ACID OR CAUSTIC BURNS • MELTING OR CHARRING OF ANY PART OF THE SURFACE • SNAGS, PUNCTURES, TEARS OR CUTS • BROKEN OR WORN STITCHES • WEAR OR ELONGATION EXCEEDING THE AMOUNT RECOMMENDED BY THE MANUFACTURER • DISTORTION OF FITTINGS

NOTE: SYNTHETIC WEB SLINGS SHALL BE PROTECTED (WITH A BLUNTING MATERIAL) TO PREVENT CUTS FROM THE SHARP EDGES OF MATERIAL BEING HOISTED.

SYNTHETIC WEB SLINGS
OSHA 1926.251(c)

Figure 8-15. This illustration shows the safe working loads for shackles.

TABLE 8-15. SAFE WORKING LOADS FOR SHACKLES (IN TONS OF 2000 POUNDS)		
MATERIAL SIZE	PIN DIAMETER (INCHES)	SAFE WORKING LOAD
-	5/8	1.4
5/8	-	2.2
-	7/8	3.2
7/8	1	4.3
1	1 1/8	5.6
1 1/8	1 -	6.7
1 -	1 3/8	8.2
1 3/8	1 -	10.0
1 -	1 5/8	11.9
1 -	2	16.2
2	2 -	21.2

FOR EXAMPLE: A 5/8 IN. SHACKLE HAS A SAFE WORKING LOAD OF 1.4.

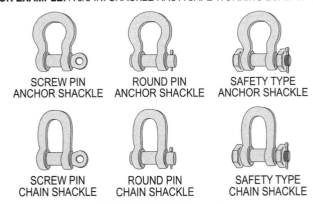

SCREW PIN ANCHOR SHACKLE

ROUND PIN ANCHOR SHACKLE

SAFETY TYPE ANCHOR SHACKLE

SCREW PIN CHAIN SHACKLE

ROUND PIN CHAIN SHACKLE

SAFETY TYPE CHAIN SHACKLE

SHACKLES AND HOOKS
OSHA 1926.251(e)

To determine the safe workload for hooks, the manufacturer's recommendations shall be followed. All hooks for which the manufacturer's recommendations are not available shall be tested to twice the intended safe workload before they are used, and records of these tests shall be maintained. **(See Figure 8-16)**

SAFETY INSTRUCTION

Workers shall be instructed in the basic safety of material storage and handling, and in the use of lifting equipment and devices necessary for their safety in their work assignment.

NOTE: ALL HOOKS FOR WHICH THE MANUFACTURER'S RECOMMENDATIONS ARE NOT AVAILABLE SHALL BE TESTED TO TWICE THE INTENDED SAFE WORKLOAD BEFORE THEY ARE USED, AND RECORDS OF THESE TESTS SHALL BE MAINTAINED.

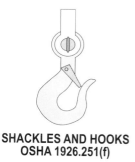

**SHACKLES AND HOOKS
OSHA 1926.251(f)**

Figure 8-16. When determining the safe workload for hooks, the manufacturer's recommendations shall be followed.

SAFE LIFTING PRACTICES

Now that the sling has been selected (based upon the characteristics of the load and the environmental conditions surrounding the lift) and inspected prior to use, the next step is learning how to use it safely. The following are four primary factors to take into consideration when safely lifting a load.

- The size, weight and center of gravity of the load
- The number of legs and the angle the sling makes with the horizontal line
- The rated capacity of the sling
- The history of the care and usage of the sling

SIZE, WEIGHT AND CENTER OF GRAVITY OF THE LOAD

The center of gravity of an object is that point at which the entire weight may be considered as concentrated. In order to make a level lift, the crane hook shall be directly above this point. While slight variations are usually permissible, if the crane hook is too far to one side of the center of gravity, dangerous tilting will result, causing unequal stresses in the different sling legs. This imbalance shall be compensated for at once.

NUMBER OF LEGS AND ANGLE WITH THE HORIZONTAL

As the angle formed by the sling leg and the horizontal line decreases, the rated capacity of the sling also decreases. In other words, the smaller the angle between the sling leg and the horizontal, the greater the stress on the sling leg and the smaller (lighter) the load the sling can safely support. Larger (heavier) loads can be safely moved if the weight of the load is distributed among more sling legs.

RATED CAPACITY OF THE SLING

The rated capacity of a sling varies depending upon the type of sling, the size of the sling and the type of hitch. Operators shall know the capacity of the sling. Charts or tables that contain this information generally are available from sling manufacturers. The values given are for new slings. Older slings shall be used with additional caution. Under no circumstances shall a sling's rated capacity be exceeded.

HISTORY OF CARE AND USAGE

The mishandling and misuse of slings are the leading causes of accidents involving their use. The majority of injuries and accidents, however, can be avoided by becoming familiar with the essentials of proper sling care and usage.

Proper care and usage are essential for maximum service and safety. Slings shall be protected from sharp bends and cutting edges by means of cover saddles, burlap padding or wood blocking, as well as from unsafe lifting procedures such as overloading.

Before making a lift, check to be certain that the sling is properly secured around the load and that the weight and balance of the load have been accurately determined. If the load is on the ground, do not allow the load to drag along the ground. This could damage the sling. If the load is already resting on the sling, ensure that there is no sling damage prior to making the lift.

Next, position the hook directly over the load and seat the sling squarely within the hook bowl. This gives the operator maximum lifting efficiency without bending the hook or overstressing the sling.

Wire rope slings are also subject to damage resulting from contact with sharp edges of the loads being lifted. These edges can be blocked or padded to minimize damage to the sling.

After the sling is properly attached to the load, there are a number of good lifting techniques that are common to all slings:

- Make sure that the load is not lagged, clamped or bolted to the floor.

- Guard against shock loading by taking up the slack in the sling slowly. Apply power cautiously so as to prevent jerking at the beginning of the lift, and accelerate or decelerate slowly.

- Check the tension on the sling. Raise the load a few inches, stop, and check for proper balance and that all items are clear of the path of travel. Never allow anyone to ride on the hood or load.

- Keep all personnel clear while the load is being raised, moved or lowered. Crane or hoist operators should watch the load at all times when it is in motion.

Finally, obey the following "nevers:"

- Never allow more than one person to control a lift or give signals to a crane or hoist operator except to warn of a hazardous situation.

- Never raise the load more than necessary.

- Never leave the load suspended in the air.

- Never work under a suspended load or allow anyone else to.

Once the lift has been completed, clean the sling, check it for damage and store it in a clean, dry airy place. It is best to hang it on a rack or wall.

Remember, damaged slings cannot lift as much as new or well-cared for older slings. Safe and proper use and storage of slings will increase their service life.

MAINTENANCE OF SLINGS

The following shall be considered for maintenance of slings:

- Chains
- Wire rope
- Fiber ropes and synthetic webs
- Formula for stress on sling legs
- Load angle factor
- Load angle factor from the chart
- How sling angles affect sling loading

CHAINS

Chain slings shall be cleaned prior to each inspection, as dirt or oil may hide damage. The operator shall be certain to inspect the total length of the sling, periodically looking for stretching, binding, wear or nicks and gouges. If a sling has stretched so that it is now more than three percent longer than it was when new, it is unsafe and shall be discarded.

Binding is the term used to describe the condition that exists when a sling has become deformed to the extent that its individual links cannot move within each other freely. It is also an indication that the sling is unsafe. Generally, wear occurs on the load-bearing inside ends of the links. Pushing links together so that the inside surface becomes clearly visible is the best way to check for this type of wear. Wear may also occur, however, on the outside of links when the chain is dragged along abrasive surfaces or pulled out from under heavy loads. Either type of wear weakens slings and makes accidents more likely.

Heavy nicks and/or gouges shall be filed smooth, measured with calipers, then compared with the manufacturer's minimum allowable safe dimensions. When in doubt, or in borderline situations, do not use the sling. In addition, never attempt to repair the welded components on a sling. If the sling needs repair of this nature, the supervisor shall be notified.

Safety Tip: Chain slings must be cleaned prior to inspection.

WIRE ROPE

Wire rope slings, like chain slings, shall be cleaned prior to each inspection because they are also subject to damage hidden by dirt or oil. In addition, they shall be lubricated according to manufacturer's instructions. Lubrication prevents or reduces corrosion and wear due to friction and abrasion. Before applying any lubricant, however, the sling user should make certain that the sling is dry. Applying lubricant to a wet or damp sling traps moisture against the metal and hastens corrosion.

Corrosion deteriorates wire rope. It may be indicated by pitting, but it is sometimes hard to detect. Therefore, if a wire rope sling shows any sign of significant deterioration, that sling shall be removed until it can be examined by a person who is qualified to determine the extent of the damage.

By following the above guidelines to proper sling use and maintenance, and by the avoidance of kinking, it is possible to greatly extend a wire rope sling's useful service life.

FIBER ROPES AND SYNTHETIC WEBS

Fiber ropes and synthetic webs are generally discarded rather than serviced or repaired. Operators shall always follow manufacturer's recommendations.

LOAD ANGLE FACTOR

The following is an example of selecting a sling using the load angle factor as shown on the chart below. **(See Figure 8-18)**

> Sling = 2 legged bridle
>
> Load = 1000 lbs.
>
> Angle with horizontal = 45°

LOAD ANGLE FACTOR FROM THE CHART

Each of the two legs would lift 500 lbs if a vertical pull were used. However, there is a 45° sling angle involved. Therefore, the 500 pound load would be multiplied by the load angle factor in the chart (1.414), giving a total of 707 lbs (500 lbs x 1.414 = 707 lbs) tension in each leg. Therefore, each leg shall have a safe working load of 707 lbs. **(See Figure 8-17)**

Figure 8-17. This illustration shows a chart for determining the load angle factor.

SLING ANGLE	LOAD ANGLE FACTOR
90°	1.000
85°	1.004
80°	1.015
75°	1.035
70°	1.064
65°	1.104
60°	1.155
55°	1.221
50°	1.305
45°	1.414
40°	1.555
35°	1.742
30°	2.000
25°	2.364
20°	3.924
15°	3.861
10°	5.747
5	11.49

LOAD ANGLE FACTOR CHART

HOW SLING ANGLES AFFECT SLING LOADING

The following formula can be used to determine how sling angles will affect sling loading:

Step 1: $\dfrac{\text{Length}}{\text{Height}}$ = Load angle factor

Step 2: Load angle x Load weight = Total tension factor

Step 3: $\dfrac{\text{Total tension}}{\text{\# of sling legs}}$ = tension per leg

See Figure 8-18 for a detailed illustration showing calculations for how sling angles will affect sling loading.

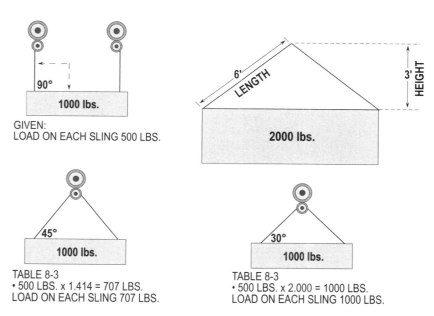

GIVEN:
LOAD ON EACH SLING 500 LBS.

TABLE 8-3
• 500 LBS. x 1.414 = 707 LBS.
LOAD ON EACH SLING 707 LBS.

TABLE 8-3
• 500 LBS. x 2.000 = 1000 LBS.
LOAD ON EACH SLING 1000 LBS.

HOW SLINGS ANGLES EFFECT SLING LOADING
OSHA 1926.251(f)

Figure 8-18. This illustration shows the type of calculations used to determine how sling angles will affect sling loading.

WASTE DISPOSAL
OSHA 1926.252

When debris must be dropped through holes in the floor without the use of a chute, the area onto which the material is dropped shall be completely enclosed with barricades not less that 42 in. high and not less than 6 ft. back from the projected edge of the opening above. At each level, signs warning of the hazard of falling materials shall be posted. Personnel shall not be allowed to remove debris from the lower area until debris handling ceases above.

All scrap material shall be removed as work progresses; waste shall be burned only in accordance with local fire and environmental regulations.

When material is dropped more than 20 ft. to a point outside a building, an enclosed chute made of wood or other equivalent material shall be used. **(See Figure 8-19)**

Figure 8-19. When material is dropped 20 ft. to a point outside the building, an enclosed chute shall be required.

NOTE: ALL SCRAP MATERIAL SHALL BE REMOVED AS WORK PROGRESSES; WASTE SHALL BE BURNED ONLY IN ACCORDANCE WITH LOCAL FIRE AND ENVIRONMENTAL REGULATIONS.

WASTE DISPOSAL
OSHA 1926.252

Solvents, oily rags and other flammable waste shall be kept in closed fire-resistant containers until they are removed from the worksite; they shall be disposed of in accordance with appropriate environmental regulations. **(See Figure 8-20)**

Figure 8-20. Oily rags and other flammable waste shall be kept in closed fire-resistant containers.

WASTE DISPOSAL
OSHA 1926.252

SUMMARY

There are good practices to follow to protect yourself while using slings to move materials. First, learn as much as you can about the materials with which you will be working. Slings come in many different types, one of which is right for your purpose. Second, analyze the load to be moved - in terms of size, weight, shape, temperature and sensitivity - then choose the sling that best meets those needs. Third, always inspect all the equipment before and after a move. Always be sure to give equipment whatever "in service" maintenance it may need. Fourth, use safe lifting practices. Use the proper lifting technique for the type of sling and the type of load.

To have an effective materials handling and storing safety and health program, managers shall take an active role in its development. First-line supervisors shall be convinced of the importance of controlling hazards associated with materials handling and storing and shall be held accountable for employee training. An ongoing safety and health program should be used to motivate employees to continue to use necessary protective gear and to observe proper job procedures.

Section	Answer	
_____	T	F
_____	T	F
_____	T	F
_____	T	F
_____	T	F
_____	T	F
_____	T	F
_____	T	F
_____	T	F
_____	T	F
_____	T	F
_____	T	F
_____	T	F
_____	T	F

Material Handling and Storage

1. When stacking materials, height limitations should be observed. For example, lumber shall be stacked no more than 12 ft. high if it is handled manually.

2. When stacking materials, 20 ft. is the maximum stacking height if a forklift is used.

3. Used lumber shall have all nails removed before stacking. Lumber shall be stacked and leveled on solidly supported bracing. The stacks shall be stable and self-supporting. Stacks of loose bricks should not be more than 6 ft. in height.

4. When masonry blocks are stacked higher than 6 ft., the stacks should be tapered back one-half block for each tier above the 6 ft. level.

5. When using conveyors, to reduce the severity of an injury, an emergency button or pull cord designed to stop the conveyor shall be installed within 10 ft. of the employee's work station.

6. Only thoroughly trained and competent persons are permitted to operate cranes.

7. All cranes shall be inspected frequently by persons thoroughly familiar with the crane, the methods of inspecting the crane and what can make the crane unserviceable.

8. When working with slings, employers shall ensure that they are visually inspected monthly.

9. A welded alloy-steel chain sling shall have a permanently affixed, durable identification label stating size, grade, rated capacity and manufacturer.

10. Wire rope is referred to as right lay or left lay. A right lay rope is one in which the strands are wound in a right-hand direction like a conventional screw thread. A left lay rope is just the opposite.

11. New wire rope slings have a design factor of 2.5.

12. Each synthetic web sling shall be marked or coded to show the manufacturer's name or trademark, rated capacity and type of material.

13. The rated capacity of synthetic webbing may be exceeded by a factor of 2.

14. When U-bolt wire-rope clips are used to form eyes, the "U" section of a U-bolt shall always be in contact with the wire-rope dead end.

_____   T   F

15. Fiber rope slings are preferred for some applications because they are pliant, they grip the load well and they do not mar the surface of the load. They should be used only on light loads.

_____   T   F

16. Manila-rope eye splices shall have at least 3 full tucks; short splices shall have 6 full tucks (3 on each side of the center of splice).

_____   T   F

17. Synthetic-rope eye splices shall have at least 3 full tucks. Short splices shall have at least 6 full tucks (3 on each side of the center of splice).

_____   T   F

18. Chain slings shall be cleaned prior to each inspection, as dirt or oil may hide damage.

_____   T   F

19. Lubrication of wire rope prevents or reduces corrosion and wear due to friction and abrasion.

_____   T   F

20. Fiber ropes and synthetic webs are generally discarded rather than serviced or repaired.

9

Hand and Power Tools

This chapter covers the general requirements, hand tools, power-operated hand tools, abrasive wheels and tools and jacks – lever and ratchet, screw and hydraulic. The following general requirements are covered for hand and power tools:

- Condition of tools
- Guarding
- Personal protective equipment
- Switches

HAND AND POWER TOOLS
OSHA 1926.300

The following shall be considered for hand and power tools:

- General requirements
- Guards
- Personal protective equipment
- Safety switches

GENERAL REQUIREMENTS
OSHA 1926.300(a)

The following shall be considered for the general requirements of hand and power tools:

- Hazard recognition
- Power tool precautions

HAZARD RECOGNITION

Tools are such a common part of our lives that it is difficult to remember that they may pose hazards. All tools are manufactured with safety in mind but, tragically, a serious accident often occurs before steps are taken to search out and avoid or eliminate tool-related hazards.

In the process of removing or avoiding the hazards, workers shall learn to recognize the hazards associated with the different types of tools and the safety precautions necessary to prevent those hazards.

The following five basic safety rules shall be used to prevent all hazards using tools:

- Keep all tools in good condition with regular maintenance
- Use the right tool for the job
- Examine each tool for damage before use
- Operate according to the manufacturer's instructions
- Provide and use the proper protective equipment

See Figure 9-1 for a detailed illustration pertaining to recognizing hazardous tools.

Figure 9-1. This illustration shows that visual inspections shall be performed on portable cord-and-plug connected equipment and flexible cord sets before they are used on any work shift.

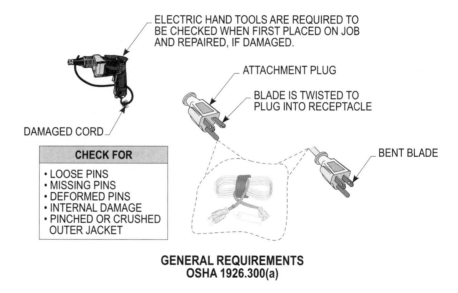

ELECTRIC HAND TOOLS ARE REQUIRED TO BE CHECKED WHEN FIRST PLACED ON JOB AND REPAIRED, IF DAMAGED.

ATTACHMENT PLUG

BLADE IS TWISTED TO PLUG INTO RECEPTACLE

DAMAGED CORD

BENT BLADE

CHECK FOR
- LOOSE PINS
- MISSING PINS
- DEFORMED PINS
- INTERNAL DAMAGE
- PINCHED OR CRUSHED OUTER JACKET

**GENERAL REQUIREMENTS
OSHA 1926.300(a)**

Employees and employers have a responsibility to work together to establish safe working procedures. If a hazardous situation is encountered, it shall be brought to the attention of the proper individual immediately.

POWER TOOL PRECAUTIONS

Safety Tip: Employees shall be trained in the proper use of all tools.

Power tools can be hazardous when improperly used. The following are several types of power tools, based on the power source they use:

- Electric
- Pneumatic
- Liquid fuel
- Hydraulic
- Powder-actuated

Employees should be trained in the use of all tools - not just power tools. They should understand the potential hazards as well as the safety precautions to prevent those hazards from occurring.

The following general precautions should be observed by power tool users:

- Never carry a tool by the cord or hose

- Never yank the cord or the hose to disconnect it from the receptacle

- Keep cords and hoses away from heat, oil and sharp edges

- Disconnect tools when not in use, before servicing and when changing accessories such as blades, bits and cutters

- All observers should be kept at a safe distance away from the work area.

- Secure work with clamps or a vise, freeing both hands to operate the tool.

- Avoid accidental starting. The worker should not hold a finger on the switch button while carrying a plugged-in tool.

- Tools should be maintained with care. They should be kept sharp and clean for the best performance. Follow instructions in the user's manual for lubricating and changing accessories.

- Be sure to keep good footing and maintain good balance. The proper apparel should be worn. Loose clothing, ties or jewelry can become caught in moving parts.

All portable electric tools that are damaged shall be removed from use and tagged "Do Not Use."

Safety Tip: All tools shall be properly maintained.

GUARDS
OSHA 1926.300(b)

Hazardous moving parts of a power tool need to be safeguarded. For example, belts, gears, shafts, pulleys, sprockets, spindles, drums, fly wheels, chains or other reciprocating, rotating or moving parts of equipment shall be guarded if such parts are exposed to contact by employees. **(See Figure 9-2)**

Figure 9-2. To protect personnel, reciprocating, rotating or moving parts of equipment shall be guarded.

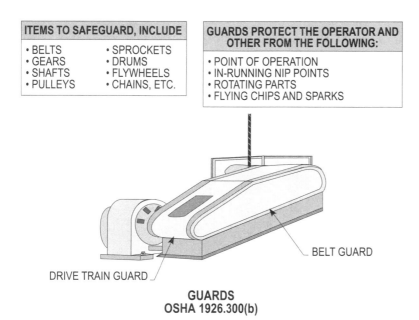

ITEMS TO SAFEGUARD, INCLUDE	
• BELTS	• SPROCKETS
• GEARS	• DRUMS
• SHAFTS	• FLYWHEELS
• PULLEYS	• CHAINS, ETC.

GUARDS PROTECT THE OPERATOR AND OTHER FROM THE FOLLOWING:
- POINT OF OPERATION
- IN-RUNNING NIP POINTS
- ROTATING PARTS
- FLYING CHIPS AND SPARKS

BELT GUARD

DRIVE TRAIN GUARD

**GUARDS
OSHA 1926.300(b)**

Safety Tip: Tools should never be operated without the proper guards.

Guards, as necessary, shall be provided to protect the operator and others from the following:

- Point of operation
- In-running nip points
- Rotating parts
- Flying chips and sparks

Safety guards shall never be removed when a tool is being used. For example, portable circular saws shall be equipped with guards. An upper guard shall cover the entire blade of the saw. A retractable lower guard shall cover the teeth of the saw, except when it makes contact with the work material. The lower guard shall automatically return to the covering position when the tool is withdrawn from the work. **(See Figure 9-3)**

Figure 9-3. When a tool is in use, safety guards shall not be removed.

NOTE: AN UPPER GUARD SHALL COVER THE ENTIRE BLADE OF THE SAW. A RETRACTABLE LOWER GUARD SHALL COVER THE TEETH OF THE SAW, EXCEPT WHEN IT MAKES CONTACT WITH THE WORK MATERIAL. THE LOWER GUARD SHALL AUTOMATICALLY RETURN TO THE COVERING POSITION WHEN THE TOOL IS WITHDRAWN FROM THE WORK.

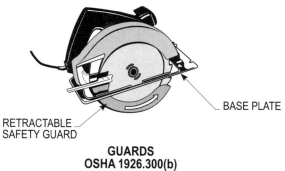

PORTABLE ELECTRIC CIRCULAR SAW

RETRACTABLE SAFETY GUARD

BASE PLATE

GUARDS
OSHA 1926.300(b)

PERSONAL PROTECTIVE EQUIPMENT
OSHA 1926.300(c)

Safety Tip: Appropriate PPE shall be worn when using portable power tools.

Appropriate personal protective equipment, e.g., safety goggles, gloves, etc., should be worn due to hazards that may be encountered while using portable power tools and hand tools. **(See Figure 9-4)**

Figure 9-4. Appropriate personnel protective equipment shall be worn by employees to protect them from hazardous.

NOTE: AROUND FLAMMABLE SUBSTANCES, SPARKS PRODUCED BY IRON AND STEEL HAND TOOLS CAN BE A DANGEROUS IGNITION SOURCE. WHERE THIS HAZARD EXISTS, SPARK-RESISTANT TOOLS MADE FROM BRASS, PLASTIC, ALUMINUM OR WOOD WILL PROVIDE FOR SAFETY.

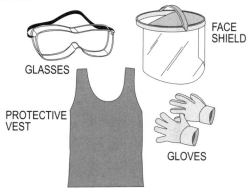

GLASSES

FACE SHIELD

PROTECTIVE VEST

GLOVES

PERSONAL PROTECTIVE EQUIPMENT (PPE)

PERSONAL PROTECTIVE EQUIPMENT
OSHA 1926.300(c)

Safety requires that floors be kept as clean and dry as possible to prevent accidental slips with or around dangerous hand tools.

Around flammable substances, sparks produced by iron and steel hand tools can be a dangerous ignition source. Where this hazard exists, spark-resistant tools made from brass, plastic, aluminum or wood will provide for safety.

SAFETY SWITCHES
OSHA 1926.300(d)

The following hand-held powered tools shall be equipped with a momentary contact "on-off" control switch:

- Drills
- Tappers
- Fastener drivers
- Horizontal, vertical and angle grinders with wheels larger than 2 in. in diameter
- Disc and belt sanders
- Reciprocating saws
- Saber saws
- Other similar tools

These tools also may be equipped with a lock-on control provided that turnoff can be accomplished by a single motion of the same finger or fingers that turn it on.

The following hand-held powered tools may be equipped with only a positive "on - off" control switch:

- Platen sanders
- Disc sanders with discs 2 in. or less in diameter
- Grinders with wheels 2 in. or less in diameter
- Routers, planers, laminate trimmers, nibblers, shears, scroll saws and jigsaws with blade shanks 2 in. wide or less.

Other hand-held powered tools such as circular saws having a blade diameter greater than 2 in., chain saws and percussion tools without positive accessory holding means shall be equipped with a constant pressure switch that will shut off the power when the pressure is released.

HAND TOOLS
OSHA 1926.301

Hand tools are non-powered. They include anything from axes to wrenches. The greatest hazards posed by hand tools result from misuse and improper maintenance. The following are some examples of hazards posed by the use of hand tools:

- Using a screwdriver as a chisel may cause the tip of the screwdriver to break and fly, hitting the user or other employees.

- If a wooden handle on a tool such as a hammer or an axe is loose, splintered or cracked, the head of the tool may fly off and strike the user or another worker.

- A wrench must not be used if its jaws are sprung, because it might slip.

See Figure 9-5 for a detailed illustration pertaining to the hazards posed by the use of hand tools.

Safety Tip: Hand tools shall not be misused and shall be properly maintained.

Figure 9-5. This illustration shows defective and non-defective tools.

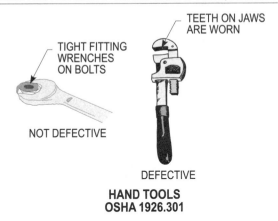

> **EXAMPLES OF HAZARDS POSED BY THE USE OF HAND TOOLS:**
> • USING A SCREWDRIVER AS A CHISEL MAY CAUSE THE TIP OF THE SCREWDRIVER TO BREAK AND FLY, HITTING THE USER OR OTHER EMPLOYEES
> • IF A WOODEN HANDLE ON A TOOL SUCH AS A HAMMER OR AN AXE IS LOOSE, SPLINTERED OR CRACKED, THE HEAD OF THE TOOL MAY FLY OFF AND STRIKE THE USER OR ANOTHER WORKER
> • A WRENCH MUST NOT BE USED IF ITS JAWS ARE SPRUNG, BECAUSE IT MIGHT SLIP

TEETH ON JAWS ARE WORN

TIGHT FITTING WRENCHES ON BOLTS

NOT DEFECTIVE

DEFECTIVE

HAND TOOLS OSHA 1926.301

Safety Tip: The employer is responsible for the safe condition of arc tools and equipment.

• Impact tools such as chisels, wedges or drift pins are unsafe if they have mushroomed heads. The heads might shatter on impact, sending sharp fragments flying.

The employer shall be responsible for the safe condition of tools and equipment used by employees, but the employees have the responsibility for properly using and maintaining tools.

Employers should caution employees that saw blades, knives or other tools be directed away from aisle areas and other employees working in close proximity. Knives and scissors must be sharp. Dull tools can be more hazardous than sharp ones. **(See Figure 9-6)**

Figure 9-6. Dull or mushroomed tools can be more hazardous to use than sharp or good condition ones.

NOTE 1: THE EMPLOYER SHALL BE RESPONSIBLE FOR THE SAFE CONDITION OF TOOLS AND EQUIPMENT USED BY EMPLOYEES, BUT THE EMPLOYEES HAVE THE RESPONSIBILITY FOR PROPERLY USING AND MAINTAINING TOOLS.

NOTE 2: EMPLOYERS SHOULD CAUTION EMPLOYEES THAT SAW BLADES, KNIVES OR OTHER TOOLS BE DIRECTED AWAY FROM AISLE AREAS AND OTHER EMPLOYEES WORKING IN CLOSE PROXIMITY.

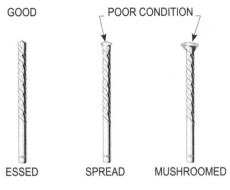

GOOD   POOR CONDITION

ESSED   SPREAD   MUSHROOMED

HAND TOOLS OSHA 1926.301

POWER-OPERATED HAND TOOLS
OSHA 1926.302

The following shall be considered for power-operated hand tools:

- Electric tools
- Pneumatic tools
- Fuel powered tools
- Powder-actuated tools

ELECTRIC TOOLS
OSHA 1926.302(a)

Employees using electric tools shall be aware of several dangers; the most serious is the possibility of electrocution.

Safety Tip: The primary hazard of electric-power tools is shock and burns.

Among the chief hazards of electric-powered tools are burns and slight shocks, which can lead to injuries or even heart failure. Under certain conditions, even a small amount of current can result in fibrillation of the heart and eventual death. A shock also can cause the user to fall off a ladder or other elevated work surface.

To protect the user from shock, tools shall either have a three-wire cord with ground and be grounded, be double insulated or be powered by a low-voltage isolation transformer. Three-wire cords contain two current carrying conductors and a grounding conductor. One end of the grounding conductor connects to the tool's metal housing. The other end is grounded through a prong on the plug. Anytime an adapter is used to accommodate a two-hole receptacle, the adapter wire shall be attached to a known ground. The third prong should never be removed from the plug. **(See Figure 9-7)**

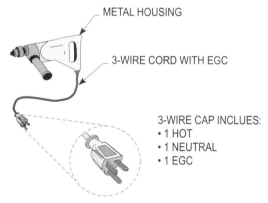

METAL HOUSING

3-WIRE CORD WITH EGC

3-WIRE CAP INCLUES:
- 1 HOT
- 1 NEUTRAL
- 1 EGC

NOTE: DO NOT EVER DAMAGE OR CUT OFF GROUNDING PIN.

ELECTRIC TOOLS
OSHA 1926.302(a)

Figure 9-7. Metal enclosed hand tools shall have their housing connected to an equipment grounding conductor.

Double insulation is more convenient. The user and the tools are protected in two ways: by normal insulation on the wires inside, and by a housing that cannot conduct electricity to the operator in the event of a malfunction. These following general practices should be followed when using electric tools:

- Electric tools should be operated within their design limitations
- Gloves and safety footwear are recommended during use of electric tools
- When not in use, tools should be stored in a dry place
- Electric tools should not be used in damp or wet locations
- Work areas should be well lighted

See Figure 9-8 for a detailed illustration pertaining to a double insulated tool.

Figure 9-8. Double insulated tools do not require a third wire to serve as an equipment grounding conductor.

THE FOLLOWING GENERAL PRACTICES SHOULD BE WHEN USING ELECTRIC TOOLS:
• ELECTRIC TOOLS SHOULD BE OPERATED WITHIN THEIR DESIGN LIMITATIONS • GLOVES AND SAFETY FOOTWEAR ARE RECOMMENDED DURING USE OF ELECTRIC TOOLS • WHEN NOT IN USE, TOOLS SHOULD BE STORED IN DRY PLACE • ELECTRIC TOOLS SHOULD NOT BE USED IN WET LOCATIONS • WORK AREAS SHOULD BE WELL LIGHTED

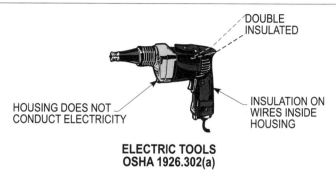

DOUBLE INSULATED

HOUSING DOES NOT CONDUCT ELECTRICITY

INSULATION ON WIRES INSIDE HOUSING

ELECTRIC TOOLS
OSHA 1926.302(a)

PNEUMATIC TOOLS
OSHA 1926.302(b)

Safety Tip: Eye and face protection shall be used when working with pneumatic tools.

Pneumatic tools are powered by compressed air and include chippers, drills, hammers and sanders.

There are several dangers encountered in the use of pneumatic tools. The main one is the danger of getting hit by one of the tool's attachments or by some kind of fastener the worker is using with the tool.

Eye protection is required and face protection is recommended for employees working with pneumatic tools.

Noise is another hazard. Working with noisy tools such as jackhammers requires proper, effective use of hearing protection.

When using pneumatic tools, employees shall check to see that they are fastened securely to the hose to prevent them from becoming disconnected. A short wire or positive locking device attaching the air hose to the tool will serve as an added safeguard. **(See Figure 9-9)**

Figure 9-9. When using pneumatic tools, employees shall check to verify they are securely fastened and locking safeguard is in place.

NOTE 1: A SAFETY CLIP OR RETAINER SHALL BE INSTALLED TO PREVENT ATTACHMENTS, SUCH AS CHISELS ON A CHIPPING HAMMER, FROM BEING UNINTENTIONALLY SHOT FROM THE BARREL.

NOTE 2: SCREENS SHALL BE SET UP TO PROTECT NEARBY WORKERS FROM BEING STRUCK BY FLYING FRAGMENTS AROUND CHIPPERS, RIVETING GUNS, STAPLERS OR AIR DRILLS.

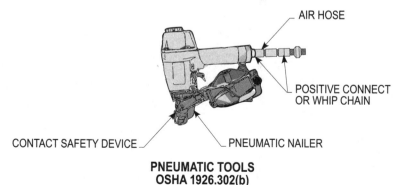

AIR HOSE

POSITIVE CONNECT OR WHIP CHAIN

CONTACT SAFETY DEVICE

PNEUMATIC NAILER

PNEUMATIC TOOLS
OSHA 1926.302(b)

A safety clip or retainer shall be installed to prevent attachments, such as chisels on a chipping hammer, from being unintentionally shot from the barrel.

Screens shall be set up to protect nearby workers from being struck by flying fragments around chippers, riveting guns, staplers or air drills.

Compressed air guns should never be pointed toward anyone. Users should never "dead-end" it against themselves or anyone else.

FUEL POWERED TOOLS
OSHA 1926.302(c)

All fuel powered tools shall be shut off before being refueled.

HYDRAULIC POWER TOOLS
OSHA 1926.302(d)

The fluid used in hydraulic power tools shall be an approved fire-resistant fluid and shall retain its operating characteristics at the most extreme temperatures to which it will be exposed.

Safety Tip: Hydraulic fluid, under pressure, can puncture the skin.

The manufacturer's recommended safe operating pressure for hoses, valves, pipes, filters and other fittings shall not be exceeded.

Do not check for leaks using your hands because fluid under pressure may puncture the skin.

POWDER-ACTUATED TOOLS
OSHA 1926.302(e)

Powder-actuated tools operate like a loaded gun and should be treated with the same respect and precautions. In fact, they are so dangerous that they shall be operated only by specially trained employees.

Safety Tip: Powder-actuated tools should be treated with the same respect and precaution as a loaded gun.

Safety precautions to remember include the following:

• These tools should not be used in an explosive or flammable atmosphere.

• Before using the tool, the worker should inspect it to determine that it is clean, that all moving parts operate freely and that the barrel is free from obstructions.

• The tool should never be pointed at anybody.

• The tool should not be loaded unless it is to be used immediately.

OSHA Required Training
1926.302(e)(1) and (12)

Note: A loaded tool should not be left unattended, especially where it would be available to unauthorized persons. **(See Figure 9-10)**

• Hands should be kept clear of the barrel end.

Note: To prevent the tool from firing accidentally, two separate motions are required for firing: one to bring the tool into position, and another to pull the trigger. The tools shall not be able to operate until they are pressed against the work surface with a force of at least 5 pounds greater than the total weight of the tool.

Figure 9-10. Certain requirements shall be observed by employees before using power-actuated tools.

SAFETY PRECAUTIONS FOR POWER-ACTUATED TOOLS
• TOOLS SHOULD NOT BE USED IN AN EXPLOSIVE OR FLAMMABLE ATMOSPHERE
• BEFORE USING THE TOOL, THE WORKER SHOULD INSPECT IT TO DETERMINE THAT IS CLEAN, THAT ALL MOVING PARTS OPERATE FREELY AND THAT THE BARREL IS FREE FROM OBSTRUCTIONS
• THE TOOL SHOULD NEVER BE POINTED AT ANYBODY
• THE TOOL SHOULD NOT BE LOADED UNLESS IT IS TO BE USED IMMEDIATELY
• HANDS SHOULD BE KEPT CLEAR OF BARREL END
• IF A POWDER-ACTUATED TOOL MISFIRES THE EMPLOYEE SHOULD WAIT AT LEAST 30 SECONDS, THEN TRY FIRING IT AGAIN. IF IT STILL WILL NOT FIRE, THE USER SHOULD WAIT ANOTHER 30 SECONDS SO THAT THE FAULTY CARTRIDGE IS LESS LIKELY TO EXPLODE, THEN CAREFULLY REMOVE THE LOAD. THE BAD CARTRIDGE SHOULD BE PUT IN WATER
• THE MUZZLE END OF THE TOOL SHALL HAVE A PROTECTIVE SHIELD OR GUARD CENTERED PERPENDICULARLY ON THE BARREL TO CONFINE ANY FLYING FRAGMENTS OR PARTICLES THAT MIGHT OTHERWISE CREATE A HAZARD WHEN THE TOOL IS FIRED. THE TOOL SHALL BE DESIGNED SO THAT IT WILL NOT FIRE UNLESS IT HAS THIS KIND OF SAFETY DEVICE
• ALL POWDER-ACTUATED TOOLS SHALL BE DESIGNED FOR VARYING POWDER CHANGES SO THAT THE USER CAN SELECT A POWDER LEVEL NECESSARY TO DO THE WORK WITHOUT EXCESSIVE FORCE
• IF THE TOOL DEVELOPS A DEFECT DURING USE, IT SHALL BE TAGGED AND TAKEN OUT OF SERVICE IMMEDIATELY UNTIL IT IS PROPERLY REPAIRED

POWER-ACTUATED TOOLS
OSHA 1926.302(e)

Safety Tip: If a powder-actuated tool misfires employees shall wait at least 30 seconds before trying to fire again.

• If a powder-actuated tool misfires, the employee should wait at least 30 seconds, then try firing it again. If it still will not fire, the user should wait another 30 seconds so that the faulty cartridge is less likely to explode, then carefully remove the load. The bad cartridge should be put in water.

• Suitable eye and face protection are essential when using a powder-actuated tool.

• The muzzle end of the tool shall have a protective shield or guard centered perpendicularly on the barrel to confine any flying fragments or particles that might otherwise create a hazard when the tool is fired. The tool shall be designed so that it will not fire unless it has this kind of safety device.

• All powder-actuated tools shall be designed for varying powder charges so that the user can select a powder level necessary to do the work without excessive force.

• If the tool develops a defect during use, it shall be tagged and taken out of service immediately until it is properly repaired.

FASTENERS

When using powder-actuated tools to apply fasteners, there are some precautions to consider. Fasteners shall not be fired into material that would let them pass through to the other side. The fastener shall not be driven into materials like brick or concrete any closer than 3 in. to an edge or corner. In steel, the fastener shall not come any closer than 1/2 in. from a corner or edge. Fasteners shall not be driven into very hard or brittle materials that might chip or splatter or make the fastener ricochet.

An alignment guide shall be used when shooting a fastener into an existing hole. A fastener shall not be driven into a spalled area caused by an unsatisfactory fastening.

POWERED ABRASIVE WHEEL TOOLS
OSHA 1926.303

The following shall be considered for powered abrasive wheel tools:

- Power
- Guarding
- Use of abrasive wheels

POWER
OSHA 1926.303(a)

Powered abrasive grinding, cutting, polishing and wire buffing wheels create special safety problems because they may throw off flying fragments. Sufficient power shall be supplied to all grinding machines in order to maintain the spindle speed at safe levels.

GUARDING
OSHA 1926.303(b)

Grinding machines shall be guarded to conform to American National Institute (ANSI) B 7.1-1970.

USE OF ABRASIVE WHEELS
OSHA 1926.303(c)

Floor- or bench-mounted abrasive wheels shall be equipped with guards on the spindle, nut and wheel. Maximum wheel exposure shall be 90°. The guards shall be strong enough to withstand the effects of and contain a bursting wheel.

Safety Tip: Guards for abrasive wheels must be strong enough to withstand and contain a bursting wheel.

Floor- or bench-mounted abrasive wheels shall be equipped with work rests that are rigidly supported and easily adjusted. Work rests shall be adjusted to within 1/8 in. of the surface of the abrasive wheel.

Cup-type wheels used for external grinding shall be protected by either a revolving cup guard or a band-type guard, per ANSI B7.1-1970. Guards shall be set so that the worker is protected from wheel contact. **(See Figure 9-11)**

All workers using abrasive wheels shall be provided with and use safety-approved eye wear, safety glasses with side shields, goggles or face shield. See Section E for more details on eye and face protection.

Before an abrasive wheel is mounted, it should be inspected closely and sound or ring-tested to be sure that it is free from cracks or defects. This test can be used for both light or heavy grinding disks or wheels. To conduct the test, a light disk or wheel should be suspended from its hole on a small pin or finger, and a heavy one should be placed vertically on a hard floor. The wheel should be tapped gently with a light tool, such as a wooden screwdriver handle; a mallet may be used for heavy wheels. The tap should be made at a point 45° from the vertical center line and about 1 to 2 in. from the periphery. A wheel or disk in good condition will give a clear, metallic "ping" indicates good condition, not the pitch wheels and disks of various grades and sizes give different pitches. **(See Figure 9-12)**

Safety Tip: An abrasive wheel shall be inspected and sound-o-ring tested before it is mounted.

Figure 9-11. Requirements to adhere to when floor or bench-mounted abrasive wheels are used.

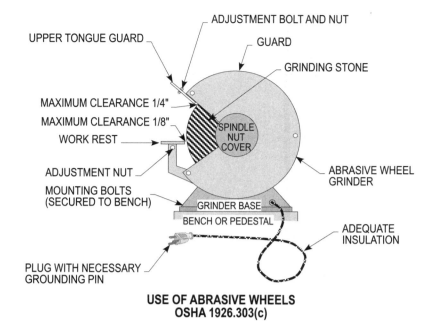

USE OF ABRASIVE WHEELS
OSHA 1926.303(c)

Figure 9-12. Before use, visually inspect an abrasive wheel and sound or-ring test to verify that it is free from cracks or defects.

NOTE 1: ALL WORKERS USING ABRASIVE WHEELS SHALL BE PROVIDED WITH AND USE SAFETY-APPROVED EYE WEAR, SAFETY GLASSES WITH SIDE SHIELDS, GOGGLES OR FACE SHIELD.

NOTE 2: DUE TO THE POSSIBILITY OF A WHEEL DISINTEGRATING (EXPLODING) DURING START-UP, THE EMPLOYEE SHOULD NEVER STAND DIRECTLY IN FRONT OF THE WHEEL AS IT ACCELERATES TO FULL OPERATING SPEED.

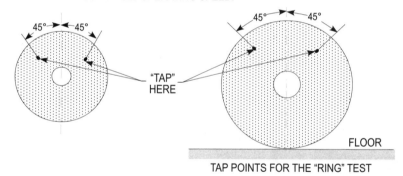

TAP POINTS FOR THE "RING" TEST

USE OF ABRASIVE WHEELS
OSHA 1926.303(c)

Safety Tip: Employees should never stand directly in front of a wheel during start-up.

To prevent the wheel from cracking, the user should be sure it fits freely on the spindle. The spindle nut shall be tightened enough to hold the wheel in place, without distorting the flange. Follow the manufacturer's recommendations. Care shall be taken to assure that the spindle wheel will not exceed the abrasive wheel specifications.

Due to the possibility of a wheel disintegrating (exploding) during start-up, the employee should never stand directly in front of the wheel as it accelerates to full operating speed.

Portable grinding tools shall be equipped with safety guards to protect workers not only from the moving wheel surface, but also from flying fragments in case of breakage. **(See Figure 9-13)**

In addition, when using a powered grinder:

- Always use eye protection
- Turn off the power when not in use
- Never clamp a hand-held grinder in a vise

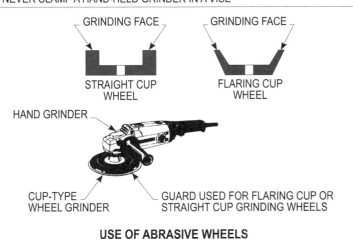

THE FOLLOWING SHALL BE APPLIED WHEN USING A POWERED GRINDER:
• ALWAYS USE EYE PROTECTION
• TURN OFF THE POWER WHEN NOT IN USE
• NEVER CLAMP A HAND-HELD GRINDER IN A VISE

GRINDING FACE — STRAIGHT CUP WHEEL

GRINDING FACE — FLARING CUP WHEEL

HAND GRINDER

CUP-TYPE WHEEL GRINDER

GUARD USED FOR FLARING CUP OR STRAIGHT CUP GRINDING WHEELS

USE OF ABRASIVE WHEELS
OSHA 1926.303(c)

Figure 9-13. This illustration shows requirements pertaining to the use of portable grinding tools.

ELECTRIC WOODWORKING TOOLS
OSHA 1926.304

The following shall be considered for electric woodworking tools:

- General requirements
- Disconnect switches
- Speeds
- Self-feed
- Guarding
- Personal protective equipment
- Radial arm saws
- Hand-fed rip saws (table saws)

GENERAL REQUIREMENTS

Workers who are allowed to operate woodworking machines shall be instructed in the machine hazards and safe use of the machine.

Safety Tip: Employees should be properly trained before being allowed to operate woodworking machines.

DISCONNECT SWITCHES
OSHA 1926.304(a)

All fixed electric woodworking tools shall be equipped with a disconnect switch that can be locked or tagged in the off position. **(See Figure 9-14)**

A manual restart switch or button shall be installed on all woodworking machines to prevent them from automatically starting when power is restored following a power failure. **(See Figure 9-15)**

SPEEDS
OSHA 1926.304(b)

All circular saws over 20 in. in diameter shall be permanently marked with their operating speed.

Figure 9-14. Fixed electric woodworking tools shall be equipped with a disconnect switch.

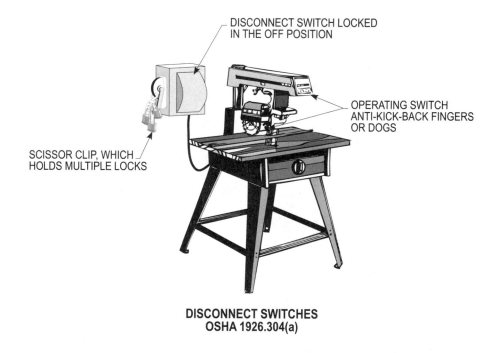

DISCONNECT SWITCH LOCKED IN THE OFF POSITION

OPERATING SWITCH ANTI-KICK-BACK FINGERS OR DOGS

SCISSOR CLIP, WHICH HOLDS MULTIPLE LOCKS

DISCONNECT SWITCHES
OSHA 1926.304(a)

Figure 9-15. To prevent woodworking machines from restarting after an overload condition, a manual restart switch or button shall be provided.

NOTE: WORKERS WHO ARE ALLOWED TO OPERATE WOODWORKING MACHINES SHALL BE INSTRUCTED IN THE MACHINE HAZARDS AND SAFE USE OF THE MACHINE.

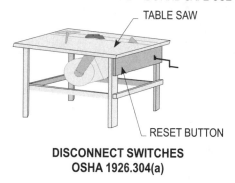

TABLE SAW

RESET BUTTON

DISCONNECT SWITCHES
OSHA 1926.304(a)

SELF-FEED
OSHA 1926.304(c)

When the nature of the work permits, automatic feeding devices shall be installed on the machine.

GUARDING
OSHA 1926.304(d)

Portable, electric circular saws shall be equipped with guards above and below the base plates or shoes.

The lower guard on portable electric circular saws shall automatically and instantly return to the covered position when the saw is withdrawn from the material. **(See Figure 9-16)**

The upper and lower guards on portable electric circular saws shall cover the saw to the depth of the teeth. **(See Figure 9-17)**

NOTE: PORTABLE ELECTRIC CIRCULAR SAWS SHALL BE EQUIPPED WITH GUARDS ABOVE AND BELOW THE BASE PLATES OR SHOES.

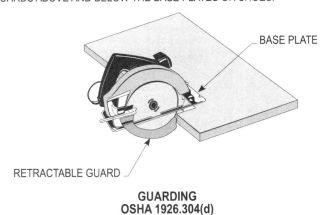

GUARDING
OSHA 1926.304(d)

Figure 9-16. The lower guard of a saw shall automatically and instantly return to the covered position after sawing is complete.

NOTE: THE LOWER GUARD ON PORTABLE ELECTRIC CIRCULAR SAWS SHALL AUTOMATICALLY AND INSTANTLY RETURN TO THE COVERED POSITION WHEN THE SAW IS WITHDRAWN FROM THE MATERIAL.

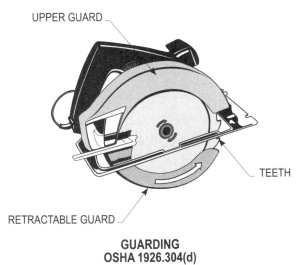

GUARDING
OSHA 1926.304(d)

Figure 9-17. The upper and lower guards the circular saw shall cover the depth of the blade's teeth.

PERSONAL PROTECTIVE EQUIPMENT
OSHA 1926.304(e)

Employees who use hand and power tools and who are exposed to the hazards of falling, flying, abrasive and splashing objects, or exposed to harmful dusts, fumes, mists, vapors or gases shall be provided with the particular personal equipment necessary to protect them from the hazard.

Safety Tip: Employees shall be provided with and must use the proper PPE.

RADIAL ARM SAWS
OSHA 1926.304(g)

The sides of the lower exposed portion of radial arm saw blades shall have guards that cover the entire blade.

Radial arm saws used for ripping shall have anti-kick-back fingers or dogs located on the out-feed side. **(See Figure 9-18)**

OSHA Required Training
1926.304(f)

Figure 9-18. When ripping is done, anti-kick-back fingers or dogs shall be used.

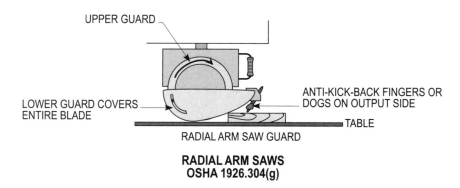

UPPER GUARD

ANTI-KICK-BACK FINGERS OR DOGS ON OUTPUT SIDE

LOWER GUARD COVERS ENTIRE BLADE

TABLE

RADIAL ARM SAW GUARD

RADIAL ARM SAWS
OSHA 1926.304(g)

In repetitive operations, to prevent the blade from traveling beyond the position necessary to complete the cut, radial arm saws shall have adjustable stops.

Radial arm saws shall be installed so that the cutting head of the blade returns to the starting position when it is released by the operator. **(See Figure 9-19)**

Figure 9-19. When released by the operator, the cutting head of the blade shall return to the starting position.

NOTE: IN REPETITIVE OPERATIONS, TO PREVENT THE BLADE FROM TRAVELING BEYOND THE POSITION NECESSARY TO COMPLETE THE CUT, RADIAL ARM SAWS SHALL HAVE ADJUSTABLE STOPS.

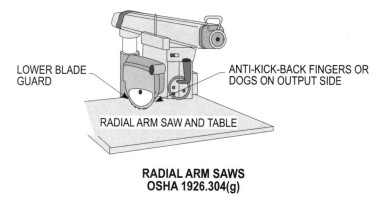

LOWER BLADE GUARD

ANTI-KICK-BACK FINGERS OR DOGS ON OUTPUT SIDE

RADIAL ARM SAW AND TABLE

RADIAL ARM SAWS
OSHA 1926.304(g)

HAND-FED RIP SAWS (TABLE SAWS)
OSHA 1926.304(h)

Circular hand-fed rip saws shall have guards that automatically adjust and completely enclose the portion of the saw that is above the table. The blade shall also be guarded below the table.

Circular hand-fed rip saws shall have spreaders to prevent the stock from squeezing the saw blade or being thrown back on the operator. **(See Figure 9-20)**

Circular hand-fed rip saws shall have anti-kick-back fingers or dogs located to prevent the blade from throwing stock back at the operator.

JACKS
OSHA 1926.305

All jacks - lever and rachet jacks, screw jacks and hydraulic jacks - shall have a device that stops them from jacking up too high. Also, the manufacturer's load limit shall be permanently marked in a prominent place on the jack and should not be exceeded.

A jack should never be used to support a lifted load. Once the load has been lifted, it shall immediately be blocked up.

Use wooden blocking under the base if necessary to make the jack level and secure. If the lift surface is metal, place a 1 in. thick hardwood block or equivalent between it and the metal jack head to reduce the danger of slippage.

To set up a jack, make certain of the following:

- The base rests on a firm level surface
- The jack is correctly centered
- The jack head bears against a level surface
- The lift force is applied evenly

Proper maintenance of jacks is essential for safety. All jacks shall be inspected before each use and lubricated regularly. If a jack is subjected to an abnormal load or shock, it should be thoroughly examined to make sure it has not been damaged.

Hydraulic jacks exposed to freezing temperatures shall be filled with an adequate antifreeze liquid.

NOTE 1: CIRCULAR HAND-FED SAWS SHALL HAVE GUARDS THAT AUTOMATICALLY ADJUST AND COMPLETELY ENCLOSE THE PORTION OF THE SAW THAT IS ABOVE THE TABLE. THE BLADE SHALL BE GUARDED BELOW THE TABLE.

NOTE 2: CIRCULAR HAND-FED SAWS SHALL HAVE SPREADERS TO PREVENT THE STOCK FROM SQUEEZING THE SAW BLADE OR BEING THROWN BACK ON THE OPERATOR.

Figure 9-20. Anti-kick-back fingers or dogs shall be located to prevent the blade from throwing stock back at the operator.

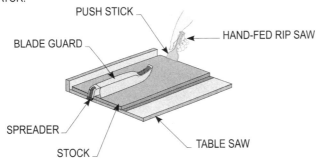

HAND-FED RIP SAWS (TABLE SAWS)
OSHA 1926.304(h)

Hand and Power Tools

Section	Answer
_____	T F
_____	T F
_____	T F
_____	T F
_____	T F
_____	T F
_____	T F
_____	T F
_____	T F
_____	T F
_____	T F
_____	T F

Hand and Power Tools

1. Safety guards shall never be removed when a tool is being used.

2. The following hand-held powered tools shall be equipped with a positive contact "on-off" control switch: drills, tappers, fastener drivers, horizontal, vertical and angle grinders with wheels larger than 2 in. in diameter, disc and belt sanders, reciprocating saws, saber saws and other similar tools.

3. The following hand-held powered tools may be equipped with only a momentary "on off" control switch: platen sanders, disc sanders with discs 2 in. or less in diameter; grinders with wheels 2 in. or less in diameter; routers, planers, laminate trimmers, nibblers, shears, scroll saws and jigsaws with blade shanks 1/4 in. wide or less.

4. Hand-held powered tools such as circular saws having a blade diameter greater than 2 in., chain saws and percussion tools without positive accessory holding means shall be equipped with a constant pressure switch that will shut off the power when the pressure is released.

5. The employer is responsible for the safe condition of tools and equipment used by employees but the employees have the responsibility for properly using and maintaining tools.

6. To protect the user from shock, tools shall either have a three-wire cord with ground and be grounded, be double insulated or be powered by a low-voltage isolation transformer.

7. There are several dangers encountered in the use of pneumatic tools. The main one is the danger of getting hit by one of the tool's attachments or by some kind of fastener the worker is using with the tool.

8. To prevent a powder-actuated tool from firing accidentally, two separate motions are required for firing: one to bring the tool into position and another to pull the trigger. The tools shall not be able to operate until they are pressed against the work surface with a force of at least 3 pounds greater than the total weight of the tool.

9. Floor- or bench-mounted abrasive wheels shall be equipped with work rests that are rigidly supported and easily adjusted. Work rests shall be adjusted to within 1/8 in. of the surface of the abrasive wheel.

10. All fixed electric woodworking tools shall be equipped with a disconnect switch that can be locked or tagged in the off position.

11. All circular saws over 15 in. in diameter shall be permanently marked with their operating speed.

12. Portable, electric circular saws shall be equipped with guards above and below the base plates or shoes.

_____	T F
_____	T F
_____	T F

13. The sides of the lower exposed portion of radial arm saw blades shall have guards that cover at least 6 in. of the blade.

14. In repetitive operations, to prevent the blade from traveling beyond the position necessary to complete the cut, radial arm saws shall have adjustable stops.

15. All jacks - lever and ratchet jacks, screw jacks and hydraulic jacks - shall have a device that stops them from jacking up too high. Also, the manufacturer's load limit shall be permanently marked in a prominent place on the jack and should not be exceeded.

10

Welding and Cutting

This chapter covers gas welding and cutting, arc welding and cutting, fire prevention, ventilation and protection in welding, cutting and heating, welding, cutting and heating in way of preservative coating and conveyors.

GAS WELDING AND CUTTING
OSHA 1926.350

The following shall be considered for gas welding and cutting:

- Transporting, moving and storing gas cylinders
- Placing cylinders
- Treatment of cylinders
- Use of fuel gas
- Fuel gas and oxygen manifolds
- Hose
- Torches
- Regulators and gauges
- Oil and grease hazards
- Additional rules

TRANSPORTING, MOVING AND STORING GAS CYLINDERS
OSHA 1926.350(a)

When transporting, moving or storing compressed gas cylinders, valve protection caps shall be in place and secured. **(See Figure 10-1)**

Figure 10-1. This illustration shows the requirements for compressed gas cylinders when transporting, moving or storing.

NOTE: WHEN CYLINDERS ARE HOISTED, THEY SHALL BE SECURED ON A CRADLE, SLINGBOARD OR PALLET.

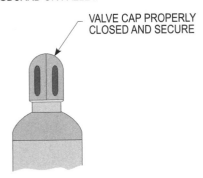

VALVE CAP PROPERLY CLOSED AND SECURE

TRANSPORTING, MOVING AND STORING GAS CYLINDERS
OSHA 1926.350

When cylinders are hoisted, they shall be secured on a cradle, slingboard or pallet. They shall not be hoisted or transported by means of magnets or choker slings.

Cylinders shall be moved by tilting and rolling them on their bottom edges. They shall not be intentionally dropped, struck or permitted to strike each other violently.

When cylinders are transported by powered vehicles, they shall be secured in a vertical position. **(See Figure 10-2)**

Figure 10-2. This illustration shows the requirements for cylinders when transported by powered vehicles.

NOTE: CYLINDERS SHALL NOT BE HOISTED OR TRANSPORTED BY MEANS OF MAGNETS OR CHOKER SLINGS.

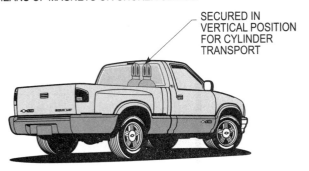

SECURED IN VERTICAL POSITION FOR CYLINDER TRANSPORT

TRANSPORTING, MOVING AND STORING GAS CYLINDERS
OSHA 1926.350

Safety Tip: A cylinder must not be lifted by the valve protection cap.

Valve protection caps shall not be used for lifting cylinders from one vertical position to another. Bars shall not be used under valves or valve protection caps to pry cylinders loose when frozen. Warm, not boiling, water shall be used to thaw cylinders loose. **(See Figure 10-3)**

Unless cylinders are firmly secured on a special carrier intended for this purpose, regulators shall be removed and valve protection caps put in place before cylinders are moved. **(See Figure 10-4)**

A suitable cylinder truck, chain or other steadying device shall be used to keep cylinders from being knocked over while in use. **(See Figure 10-5)**

When work is finished, when cylinders are empty or when cylinders are moved at any time, the cylinder valve shall be closed.

Compressed gas cylinders shall be secured in an upright position at all times, if necessary, for short periods of time while cylinders are actually being hoisted or carried. **(See Figure 10-6)**

NOTE: BARS SHALL NOT BE USED UNDER VALVES OR VALVE PROTECTION CAPS TO PRY CYLINDERS LOOSE WHEN FROZEN. WARM, NOT BOILING, WATER SHALL BE USED TO THAW CYLINDERS LOOSE.

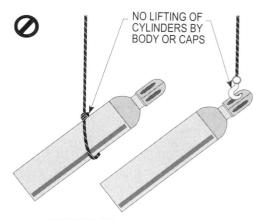

IMPROPER HOISTING OF CYLINDERS

TRANSPORTING, MOVING AND STORING GAS CYLINDERS
OSHA 1926.350

Figure 10-3. The valve protection caps or the body of cylinders shall not be used for lifting cylinders from one vertical position to another.

NOTE: REGULATORS SHALL BE REMOVED AND VALVE PROTECTION CAPS PUT IN PLACE BEFORE CYLINDERS ARE MOVED.

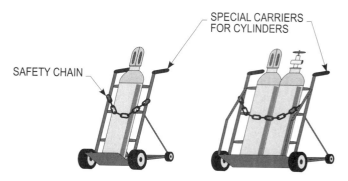

TRANSPORTING, MOVING AND STORING GAS CYLINDERS
OSHA 1926.350

Figure 10-4. A special carrier shall be used to move cylinders from one location to another.

NOTE: OXYGEN CYLINDERS IN STORAGE SHALL BE SEPARATED FROM FUEL-GAS CYLINDERS OR COMBUSTIBLE MATERIALS (ESPECIALLY OIL OR GREASE), A MINIMUM DISTANCE OF 20 FT. OR BY A NONCOMBUSTIBLE BARRIER AT LEAST 5 FT. HIGH HAVING A FIRE-RESISTANCE RATING OF AT LEAST ONE-HALF HOUR.

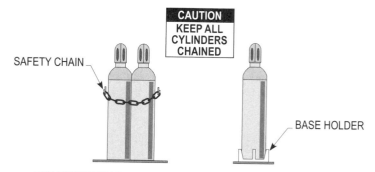

TRANSPORTING, MOVING AND STORING GAS CYLINDERS
OSHA 1926.350

Figure 10-5. A suitable cylinder truck, chain or other steadying device shall be used to prevent the cylinder from being knocked over.

Figure 10-6. A means shall be provided to prevent cylinders from falling and striking one another.

Safety Tip: A distance of 20 ft. or a noncombustible barrier shall separate oxygen cylinders from fuel-gas cylinders.

NOTE: INSIDE OF BUILDINGS, CYLINDERS SHALL BE STORED IN A WELL-PROTECTED, WELL-VENTILATED, DRY LOCATION, AT LEAST 20 FT. FROM HIGHLY COMBUSTIBLE MATERIALS SUCH AS OIL OR EXCELSIOR, CYLINDERS SHOULD BE STORED IN DEFINITELY ASSIGNED PLACES AWAY FROM ELEVATORS, STAIRS OR GANGWAYS.

CYLINDERS SHALL NOT BE PERMITTED TO STRIKE ONE ANOTHER

TRANSPORTING, MOVING AND STORING GAS CYLINDERS
OSHA 1926.350

Oxygen cylinders in storage shall be separated from fuel-gas cylinders or combustible materials (especially oil or grease), a minimum distance of 20 ft. (6.1 m) or by a noncombustible barrier at least 5 ft. (1.5 m) high having a fire-resistance rating of at least one-half hour.

Inside of buildings, cylinders shall be stored in a well-protected, well-ventilated, dry location, at least 20 ft. (6.1 m) from highly combustible materials such as oil or excelsior cylinders should be stored in definitely assigned places away from elevators, stairs or gangways. Assigned storage places shall be located where cylinders will not be knocked over or damaged by passing or falling objects, or subject to tampering.

The in-plant handling, storage and utilization of all compressed gases in cylinders, portable tanks, rail tank cars or motor vehicle cargo tanks shall be in accordance with Compressed Gas Association Pamphlet P-1-1965.

PLACING CYLINDERS
OSHA 1926.350(b)

Cylinders shall be kept far enough away from the actual welding or cutting operation so that sparks, hot slag or flame will not reach them. When this is impractical, fire resistant shields shall be provided.

Cylinders shall be placed where they cannot become part of an electrical circuit. Electrodes shall not be struck against a cylinder to strike an arc.

Fuel gas cylinders shall be placed with valve end up whenever they are in use. They shall not be placed in a location where they would not be subject to open flame, hot metal or other sources of artificial heat.

Cylinders containing oxygen or acetylene or other fuel gas shall not be taken into confined spaces.

TREATMENT OF CYLINDERS
OSHA 1926.350(c)

Cylinders, whether full or empty, shall not be used as rollers or supports.

No person other than the gas supplier shall attempt to mix gases in a cylinder. No one except the owner of the cylinder, or person authorized by him, shall refill a cylinder. No one shall use a cylinder's contents for purposes other than those intended by the supplier. All cylinders used shall meet the Department of Transportation requirements published in 49 CFR Part 178, Subpart C, Specification for Cylinders.

No damaged or defective cylinder shall be used.

USE OF FUEL GAS
OSHA 1926.350(d)

The employer shall thoroughly instruct employees in the safe use of fuel gas, as follows:

- Fuel gas shall not be used from cylinders through torches or other devices that are equipped with shutoff valves without reducing the pressure through a suitable regulator attached to the cylinder valve or manifold.

- Before a regulator to a cylinder valve is connected, the valve shall be opened slightly and closed immediately. (This action is generally termed "cracking" and is intended to clear the valve of dust or dirt that might otherwise enter the regulator.)

Note: The person cracking the valve shall stand to one side of the outlet, not in front of it. The valve of a fuel gas cylinder shall not be cracked where the gas would reach welding work, sparks, flame or other possible sources of ignition.

- The cylinder valve shall always be opened slowly to prevent damage to the regulator; for quick closing, valves of fuel gas cylinders shall not be opened more than 1 1/2 turns. **(See Figure 10-7)** When a special wrench is required, it shall be left in position on the stem of the valve while the cylinder is in use so that the fuel gas flow can be shut off quickly in case of an emergency. In the case of manifolded or coupled cylinders, at least one such wrench shall always be available for immediate use. Nothing shall be placed on top of a fuel gas cylinder, when in use, that may damage the safety device or interfere with the quick closing of the valve. **(See Figure 10-8)**

Safety Tip: All employees shall be thoroughly instructed in the safe use of liquid gas.

OSHA Required Training
1926.350(d)(1) thru (5)

NOTE 1: FUEL GAS SHALL NOT BE USED FROM CYLINDERS THROUGH TORCHES OR OTHER DEVICES THAT ARE EQUIPPED WITH SHUTOFF VALVES WITHOUT REDUCING THE PRESSURE THROUGH A SUITABLE REGULATOR ATTACHED TO THE CYLINDER VALVE OR MANIFOLD.

NOTE 2: BEFORE A REGULATOR TO A CYLINDER IS CONNECTED, THE VALVE SHALL BE OPENED SLIGHTLY AND CLOSED IMMEDIATELY.

Figure 10-7. For quick closing of fuel gas cylinders, the valve shall be open as shown in illustration.

NO MORE THAN 1 1/2 TURNS WHEN OPENING VALVE

USE OF FUEL GAS
OSHA 1926.350(d)

Figure 10-8. When cracking open the valve of a cylinder, a special wrench shall be used, so the fuel gas flow can be shut off quickly if an emergency occurs.

NOTE: BEFORE A REGULATOR IS REMOVED FROM A CYLINDER VALVE, THE CYLINDER VALVE SHALL ALWAYS BE CLOSED AND THE GAS RELEASED FROM THE REGULATOR.

NO SPECIAL TOOL WAS USED ON VALVE

IMPROPER METHOD FOR CRACKING A CYLINDER

**USE OF FUEL GAS
OSHA 1926.350(d)**

• Before a regulator is removed from a cylinder valve, the cylinder valve shall always be closed and the gas released from the regulator.

If, when the valve on a fuel gas cylinder is opened, there is found to be a leak around the valve stem, the valve shall be closed and the gland nut tightened. If this action does not stop the leak, the use of the cylinder shall be discontinued, and it shall be properly tagged and removed from the work area. In the event that fuel gas should leak from the cylinder valve, rather than from the valve stem, and the gas cannot be shut off, the cylinder shall be properly tagged and removed from the work area. If a regulator attached to a cylinder valve will effectively stop a leak through the valve seat, the cylinder need not be removed from the work area.

If a leak should develop at a fuse plug or other safety device, the cylinder shall be removed from the work area.

FUEL GAS AND OXYGEN MANIFOLDS
OSHA 1926.350(e)

Safety Tip: Manifold and header hose connections when not in use shall be capped.

Fuel gas and oxygen manifolds shall bear the name of the substance they contain in letters at least 1 in. high, which shall be either painted on the manifold or on a sign permanently attached to it. These manifolds shall be placed in safe, well ventilated and accessible locations and not be located within enclosed spaces.

Manifold hose connections, including both ends of the supply hose that lead to the manifold, shall be such that the hose cannot be interchanged between fuel gas and oxygen manifolds and supply header connections. Adapters shall not be used to permit the interchange of hose. Hose connections shall be kept free of grease and oil.

When not in use, manifold and header hose connections shall be capped.

Nothing shall be placed on top of a manifold, when in use, which will damage the manifold or interfere with the quick closing of the valves.

HOSE
OSHA 1926.350(f)

Fuel gas and oxygen hose shall be easily distinguishable from each other. The contrast may be made by different colors or by surface characteristics readily distinguishable by the sense of touch. Oxygen and fuel gas hoses shall not be interchangeable. A single

hose having NOTE: OXYGEN AND FUEL GAS HOSES SHALL NOT BE USED. (See Figure 10-9)
SINGLE HOSE HAVING MORE THAN ONE GAS PASSAGE SHALL NOT BE USED.

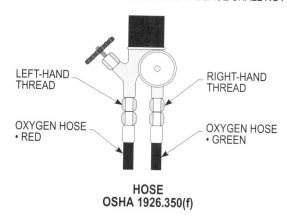

LEFT-HAND THREAD

RIGHT-HAND THREAD

OXYGEN HOSE • RED

OXYGEN HOSE • GREEN

HOSE
OSHA 1926.350(f)

Figure 10-9. For identification, fuel gas and oxygen hoses may color coded as shown in illustration.

When parallel sections of oxygen and fuel gas hose are taped together, not more than 4 in. out of 12 in. shall be covered by tape.

All hoses in use, carrying acetylene, oxygen, natural or manufactured fuel gas, or any gas or substance that may ignite or enter into combustion or be in any way harmful to employees, shall be inspected at the beginning of each working shift. Defective hose shall be removed from service.

Hose which has been subject to flashback, or that shows evidence of severe wear or damage, shall be tested to twice the normal pressure to which it is subject, but in no case less than 300 p.s.i. Defective hose, or hose in doubtful condition, shall not be used. **(See Figure 10-10)**

NOTE: HOSE COUPLINGS SHALL BE OF THE TYPE THAT CANNOT BE UNLOCKED OR DISCONNECTED BY MEANS OF A STRAIGHT PULL WITHOUT ROTARY MOTION.

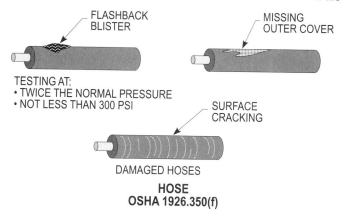

FLASHBACK BLISTER

MISSING OUTER COVER

TESTING AT:
• TWICE THE NORMAL PRESSURE
• NOT LESS THAN 300 PSI

SURFACE CRACKING

DAMAGED HOSES
HOSE
OSHA 1926.350(f)

Figure 10-10. This illustration shows the requirements for testing hoses subjected to flashback or shows evidence of severe wear or damage.

Hose couplings shall be of the type that cannot be unlocked or disconnected by means of a straight pull without rotary motion.

Boxes used for the storage of gas hose shall be ventilated.

Hoses, cables and other equipment shall be kept clear of passageways, ladders and stairs.

TORCHES
OSHA 1926.350(g)

Clogged torch tip openings shall be cleaned with suitable cleaning wires, drills or other devices designed for such purpose.

Torches in use shall be inspected at the beginning of each working shift for leaking shutoff valves, hose couplings and tip connections. Defective torches shall not be used.

Torches shall be lighted by friction lighters or other approved devices, and not by matches or from hot work.

REGULATORS AND GAUGES
OSHA 1926.350(h)

Oxygen and fuel gas pressure regulators, including their related gauges, shall be in proper working order while in use. **(See Figure 10-11)**

Figure 10-11. Regulators and gauges shall be functional when in use.

NOTE: CLOGGED TORCH TIPS OPENINGS SHALL BE CLEANED WITH SUITABLE CLEANING WIRES, DRILLS OR OTHER DEVICES DESIGNED FOR SUCH PURPOSE.

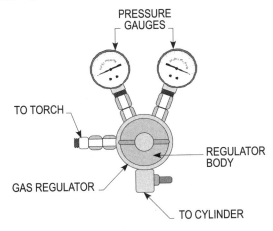

PRESSURE GAUGES

TO TORCH

REGULATOR BODY

GAS REGULATOR

TO CYLINDER

REGULATORS AND GAUGES
OSHA 1926.350(f)

OIL AND GREASE HAZARDS
OSHA 1926.350(i)

Oxygen cylinders and fittings shall be kept away from oil or grease. Cylinders, cylinder caps and valves, couplings, regulators, hose and apparatus shall be kept free from oil or greasy substances and shall not be handled with oily hands or gloves. Oxygen shall not be directed at oily surfaces, greasy clothes or within a fuel oil or other storage tank or vessel.

ADDITIONAL RULES
OSHA 1926.350(j)

For additional details not covered in this subpart, applicable technical portions of American National Standards Institute, Z49.1-1967, Safety in Welding and Cutting, shall apply.

ARC WELDING AND CUTTING
OSHA 1926.351

The following shall be considered for arc welding and cutting:

- Manual electrode holders
- Welding cables and connectors
- Ground returns and machine groundings
- Operating instructions
- Shielding

MANUAL ELECTRODE HOLDERS
OSHA 1926.351(a)

Only manual electrode holders that are specifically designed for arc welding and cutting, and are of a capacity capable of safely handling the maximum rated current required by the electrodes, shall be used.

Any current-carrying parts passing through the portion of the holder that the arc welder or cutter grips in his hand, and the outer surfaces of the jaws of the holder, shall be fully insulated against the maximum voltage encountered to ground. **(See Figure 10-12)**

Safety Tip: Current-carrying parts that the welder is exposed to must be properly insulated.

Figure 10-12. Only manual electrodes holders that are specifically designed for arc welding and cutting shall be used.

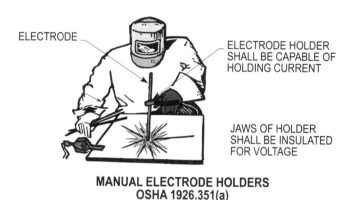

ELECTRODE

ELECTRODE HOLDER
SHALL BE CAPABLE OF
HOLDING CURRENT

JAWS OF HOLDER
SHALL BE INSULATED
FOR VOLTAGE

**MANUAL ELECTRODE HOLDERS
OSHA 1926.351(a)**

WELDING CABLES AND CONNECTORS
OSHA 1926.351(b)

All arc welding and cutting cables shall be of the completely insulated, flexible type, capable of handling the maximum current requirements of the work in progress, taking into account the duty cycle under which the arc welder or cutter is working.

Safety Tip: Cables that are in need of repair shall not be used.

Only cable free from repair or splices for a minimum distance of 10 ft. from the cable end to which the electrode holder is connected shall be used, except that cables with standard insulated connectors or with splices whose insulating quality is equal to that of the cable are permitted.

Cables in need of repair shall not be used. When a cable, other than the cable lead referred to above, becomes worn to the extent of exposing bare conductors, the portion thus exposed shall be protected by means of rubber and friction tape or other equivalent insulation. **(See Figure 10-13)**

When it becomes necessary to connect or splice lengths of cable one to another, substantial insulated connectors of a capacity at least equivalent to that of the cable shall be used. If connections are effected by means of cable lugs, they shall be securely fastened together to give good electrical contact, and the exposed metal parts of the lugs shall be completely insulated.

Figure 10-13. Other than welding lead, cables worn or damaged to the extent that bare conductors are present, the exposed portion shall be protected with rubber and friction tape or equivalent insulation.

NOTE: ONLY CABLE FREE FROM REPAIR OR SPLICES FOR A MINIMUM DISTANCE OF 10 FT. FROM THE CABLE END TO WHICH THE ELECTRODE HOLDER IS CONNECTED SHALL BE USED, EXCEPT THAT CABLE WITH STANDARD INSULATED CONNECTORS OR WITH SPLICES WHOSE INSULATING QUALITY IS EQUAL TO THAT OF THE CABLE ARE PERMITTED.

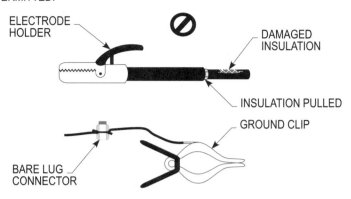

WELDING CABLES AND CONNECTORS
OSHA 1926.351(b)

GROUND RETURNS AND MACHINE GROUNDINGS
OSHA 1926.351(c)

Safety Tip: Ground return cables shall have adequate current-carrying capacity.

A ground return cable shall have a safe current-carrying capacity equal to or exceeding the specified maximum output capacity of the arc welding or cutting unit which it services. When a single ground return cable services more than one unit, its safe current-carrying shall exceed the total specified maximum output capacities of the all the units which it services.

Pipelines containing gases or flammable liquids, or conduits containing electrical circuits, shall not be used as a ground return.

When a structure or pipeline is employed as a ground return circuit, it shall be determined that the required electrical contact exists at all joints. The generation of an arc, sparks or heat at any point shall cause rejection of the structures as a ground circuit.

When a structure or pipeline is continuously employed as a ground return circuit, all joints shall be bonded, and periodic inspections shall be conducted to ensure that no condition of electrolysis or fire hazard exists by virtue of such use.

The frames of all arc welding and cutting machines shall be grounded either through a third wire in the cable containing the circuit conductor or through a separate wire which is grounded at the source of the current. Grounding circuits, other than by means of the structure, shall be checked to ensure that the circuit between the ground and the grounded power conductor has resistance low enough to permit sufficient current to flow to cause the fuse or circuit breaker to interrupt the current.

All ground connections shall be inspected to ensure that they are mechanically strong and electrically adequate for the required current.

OPERATING INSTRUCTIONS
OSHA 1926.351(d)

Employers shall instruct employees in the safe means of arc welding and cutting as follows:

• When electrode holders are to be left unattended, the electrodes shall be removed and the holders shall be so placed or protected that they cannot make electrical contact with employees or conducting objects. **(See Figure 10-14)**

> NOTE 1: HOT ELECTRODE HOLDERS SHALL NOT BE DIPPED IN WATER, TO DO SO MAY EXPOSE THE ARC WELDER OR CUTTER TO ELECTRIC SHOCK.
>
> NOTE 2: WHEN THE ARC WELDER OR CUTTER HAS OCCASION TO LEAVE HIS WORK OR TO STOP WORK FOR ANY APPRECIABLE LENGTH OF TIME, OR WHEN THE ARC WELDING OR CUTTING MACHINE IS TO BE MOVED, THE POWER SUPPLY SWITCH TO THE EQUIPMENT SHALL BE OPENED.

Figure 10-14. When left unattended, electrodes shall be removed and holders placed or protected from making contact with employees.

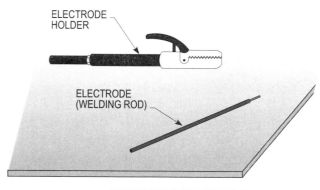

OPERATING INSTRUCTIONS
OSHA 1926.351(d)

• Hot electrode holders shall not be dipped in water; to do so may expose the arc welder or cutter to electric shock.

• When the arc welder or cutter has occasion to leave his work or to stop work for any appreciable length of time, or when the arc welding or cutting machine is to be moved, the power supply switch to the equipment shall be opened.

• Any faulty or defective equipment shall be reported to the supervisor.

• A disconnecting means shall be provided in the supply circuit for each motor generated arc welder, and for each AC transformer and DC rectifier arc welder that is not equipped with a disconnect mounted as an integral part of the welder. **(See Figure 10-15)**

> NOTE 1: ANY FAULTY OR DEFECTIVE EQUIPMENT SHALL BE REPORTED TO THE SUPERVISOR.
>
> NOTE 2: A SWITCH OR CIRCUIT BREAKER SHALL BE PROVIDED BY WHICH EACH RESISTANCE WELDER AND ITS CONTROL EQUIPMENT CAN BE ISOLATED FROM THE SUPPLY CIRCUIT. THE AMPERE RATING OF THIS DISCONNECTING MEANS SHALL NOT BE LESS THAN THE SUPPLY CONDUCTOR AMPACITY.

Figure 10-15. Motor and nonmotor generated arc welders shall be provided with a disconnecting means.

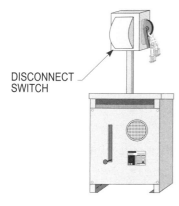

OPERATING INSTRUCTIONS
OSHA 1926.351(d)

• A switch or circuit breaker shall be provided by which each resistance welder and its control equipment can be isolated from the supply circuit. The ampere rating of this disconnecting means shall not be less than the supply conductor ampacity.

SHIELDING
OSHA 1926.351(e)

Whenever practicable, all arc welding and cutting operations shall be shielded by noncombustible or flameproof screen that will protect employees and other persons working in the vicinity from the direct rays of the arc. **(See Figure 10-16)**

Figure 10-16. Whenever practicable, during arc welding and cutting operations, a noncombustible or flameproof screen (shield) shall be used.

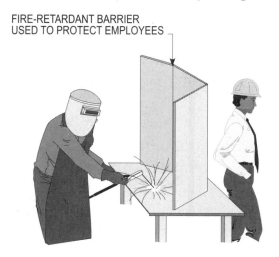

FIRE-RETARDANT BARRIER
USED TO PROTECT EMPLOYEES

SHIELDING
OSHA 1926.351(e)

FIRE PREVENTION
OSHA 1926.352

Safety Tip: Objects to be welded, cut, or heated shall be moved to a safe designated location.

When practical, objects to be welded, cut or heated shall be moved to a designated safe location or, if these objects cannot be readily moved, all movable fire hazards in the vicinity shall be taken to a safe place, or otherwise protected. If these objects cannot be moved and if all the fire hazards cannot be removed, positive means shall be taken to confine the heat, sparks and slag, and to protect the immovable fire hazards from them.

OSHA Required Training
1926.352(e)

No welding, cutting or heating shall be done where the application of flammable paints, or the presence of other flammable compounds or heavy dust concentrations creates a hazard.

Suitable fire extinguishing equipment shall be immediately available in the work area and shall be maintained in a state or readiness for instant use. **(See Figure 10-17)**

When the welding, cutting or heating operation is such that normal fire prevention precautions are not sufficient, additional personnel shall be assigned to guard against fire while the actual welding, cutting or heating operation is being performed, and for a sufficient period of time after completion of the work to ensure that no possibility of fire exists. Such personnel shall be instructed as to the specific anticipated fire hazards and how the fire fighting equipment provided is to be used.

When welding, cutting or heating is performed on walls, floors and ceilings, since direct penetration of sparks or heat transfer may introduce a fire hazard to an adjacent area, the same precautions shall be taken on the opposite side as are taken on the side on which the welding is being performed.

FIRE EXTINGUISHER

NOTE 1: NO WELDING, CUTTING OR HEATING SHALL BE DONE WHERE THE APPLICATION OF FLAMMABLE PAINTS, OR THE PRESENCE OF OTHER FLAMMABLE COMPOUNDS OR HEAVY DUST CONCENTRATIONS CREATE A HAZARD.

NOTE 2: WHEN THE WELDING, CUTTING OR HEATING OPERATION IS SUCH THAT NORMAL FIRE PREVENTION PRECAUTIONS ARE NOT SUFFICIENT, ADDITIONAL PERSONNEL SHALL BE ASSIGNED TO GUARD AGAINST FIRE WHILE THE ACTUAL WELDING, CUTTING OR HEATING OPERATION IS BEING PERFORMED, AND FOR A SUFFICIENT PERIOD OF TIME AFTER COMPLETION OF THE WORK TO ENSURE THAT NO POSSIBILITY OF FIRE EXISTS.

Figure 10-17. Work areas shall be provided with a suitable free extinguisher.

FIRE PREVENTION
OSHA 1926.352

For the elimination of possible fire in enclosed spaces as a result of gas escaping through leaking or improperly closed torch valves, the gas supply to the torch shall be positively shut off at some point outside the enclosed space whenever the torch is not to be used or whenever the torch is left unattended for a substantial period of time, such as during the lunch period. Overnight and at the change of shifts, the torch and hose shall be removed from the confined space. Open end fuel gas and oxygen hoses shall be immediately removed from enclosed spaces when they are disconnected from the torch or other gas-consuming device.

Except when the contents are being removed or transferred, drums, pails and other containers that contain or have contained flammable liquids shall be kept closed. Empty containers shall be removed to a safe area apart from hot work operations or open flames.

Safety Tip: Containers that have previously contained toxic or flammable substances shall be filled with water or thoroughly cleaned before employees are allowed to weld or cut on them.

Drums, containers or hollow structures that have contained toxic or flammable substances shall, before welding, cutting or heating is undertaken on them, either be filled with water or thoroughly cleaned of such substances and ventilated and tested.

Before heat is applied to a drum, container or hollow structure, a vent or opening shall be provided for the release of any built-up pressure during the application of heat.

VENTILATION AND PROTECTION IN WELDING, CUTTING AND HEATING
OSHA 1926.353

The following shall be considered for ventilation and protection in welding, cutting and heating:

- Mechanical ventilation
- Welding, cutting and heating in confined spaces
- Welding, cutting and heating of metals of toxic significance
- Inert-gas metal-arc welding
- General welding, cutting and heating

MECHANICAL VENTILATION
OSHA 1926.353(a)

Mechanical ventilation shall consist of either general mechanical ventilation systems or local exhaust systems.

Ventilation shall be deemed adequate if it is of sufficient capacity and so arranged as to remove fumes and smoke at the source and keep their concentration in the breathing zone within safe limits as defined in Subpart D of Part 1926, Occupational Health and Environmental Controls. **(See Figure 10-18)**

Figure 10-18. Adequate mechanical ventilation or a local exhaust system shall be provided.

NOTE 1: CONTAMINATED AIR EXHAUSTED FROM A WORKING SPACE SHALL BE DISCHARGED CLEAR OF THE SOURCE OF INTAKE AIR.

NOTE 2: OXYGEN SHALL NOT BE USED FOR VENTILATION PURPOSES, COMFORT COOLING, BLOWING DUST FROM CLOTHING OR FOR CLEANING THE WORK AREA.

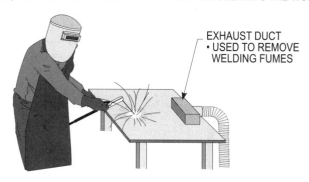

EXHAUST DUCT
• USED TO REMOVE WELDING FUMES

MECHANICAL VENTILATION
OSHA 1926.353(a)

Contaminated air exhausted from a working space shall be discharged clear of the source of intake air.

All air replacing that withdrawn shall be clean and respirable.

Oxygen shall not be used for ventilation purposes, comfort cooling, blowing dust from clothing or for cleaning the work area.

WELDING, CUTTING AND HEATING IN CONFINED SPACES
OSHA 1926.353(b)

Except where air line respirators are required or allowed as described below, adequate mechanical ventilation meeting the requirements described above shall be provided whenever welding, cutting or heating is performed in a confined space.

Safety Tip: A means to quickly remove employees fro a confined space shall be provided.

When sufficient ventilation cannot be obtained without blocking the means of access, employees in the confined space shall be protected by air line respirators in accordance with the requirements of Subpart E of Part 1926, Personal Protective and Life Saving Equipment. An employee on the outside of the confined space shall be assigned to maintain communication with those working within it and to aid them in an emergency.

Where a welder must enter a confined space through a small opening, means shall be provided for quickly removing him in case of emergency. When safety belts and lifelines are used for this purpose they shall be so attached to the welder's body that his body cannot be jammed in a small exit opening. An attendant with a preplanned rescue procedure shall be stationed outside to observe the welder at all times and be capable of putting rescue operations into effect.

WELDING, CUTTING OR HEATING OF METALS OF TOXIC SIGNIFICANCE
OSHA 1926.353(c)

Welding, cutting or heating in any enclosed spaces involving the following metals shall be performed with adequate mechanical ventilation as described above:

• Zinc-bearing base or filler metals or metals coated with zinc-bearing materials.

- Lead base metals.

- Cadmium-bearing filler materials.

- Chromium-bearing metals or metals coated with chromium-bearing materials.

- Welding, cutting or heating in any enclosed spaces involving the following metals shall be performed with adequate local exhaust ventilation as described above or employees shall be protected by air line respirators in accordance with the requirements of Subpart E.

- Metals containing lead, other than as an impurity, or metals coated with lead-bearing materials.

- Cadmium-bearing or cadmium-coated base metals.

- Metal coated with mercury-bearing metals.

- Beryllium-containing base or filler metals. Because of its high toxicity, work involving beryllium shall be done with both local exhaust ventilation and air line respirators.

- Employees performing such operations in the open air shall be protected by filter type respirators in accordance with the requirements of Subpart E, except that employees performing such operations on beryllium-containing base or filler metals shall be protected by air line respirators in accordance with the requirements of Subpart E. **(See Figure 10-19)**

Figure 10-19. In open air welding, cutting or heating operations, employees shall be protected by wearing filter type respirators.

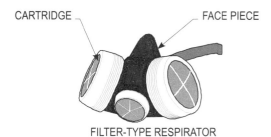

NOTE: OTHER EMPLOYEES EXPOSED TO THE SAME ATMOSPHERE AS THE WELDERS OR BURNERS SHALL BE PROTECTED IN THE SAME MANNER AS THE WELDER OR BURNER.

CARTRIDGE — FACE PIECE

FILTER-TYPE RESPIRATOR

WELDING, CUTTING OR HEATING OF
METALS OF TOXIC SIGNIFICANCE
OSHA 1926.353(c)

- Other employees exposed to the same atmosphere as the welders or burners shall be protected in the same manner as the welder or burner.

INERT-GAS METAL-ARC WELDING
OSHA 1926.353(d)

Since the inert-gas metal-arc welding process involves the production of ultraviolet radiation of intensities of 5 to 30 times that produced during shielded metal-arc welding, the decomposition of chlorinated solvents by ultraviolet rays, and the liberation of toxic fumes and gases, employees shall not be permitted to engage in, or be exposed to the process, until the following special precautions have been taken:

- The use of chlorinated solvents shall be kept at least 200 ft., unless shielded, from the exposed arc, and surfaces prepared with chlorinated solvents shall be thoroughly dry before welding is permitted on such surfaces. **(See Figure 10-20)**

Figure 10-20. Welding operations shall not take place near chlorinated solvents.

NOTE: EMPLOYEES IN THE AREA NOT PROTECTED FROM THE ARC BY SCREENING SHALL BE PROTECTED BY FILTER LENSES. WHEN TWO OR MORE WELDERS ARE EXPOSED TO EACH OTHER'S ARC, FILTER LENS GOGGLES OF A SUITABLE TYPE SHALL BE WORN UNDER WELDING HELMETS. HAND SHIELDS TO PROTECT THE WORKER AGAINST FLASHES AND RADIANT ENERGY SHALL BE USED WHEN EITHER THE HELMET IS LIFTED OR THE SHIELD IS REMOVED.

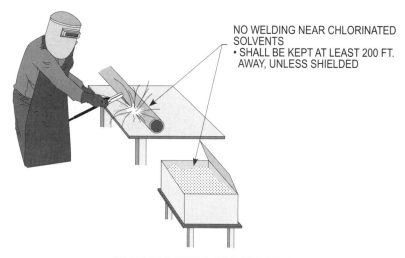

NO WELDING NEAR CHLORINATED SOLVENTS
• SHALL BE KEPT AT LEAST 200 FT. AWAY, UNLESS SHIELDED

INERT-GAS METAL-ARC WELDING
OSHA 1926.353(d)

Safety Tip: Employees shall be protected by filter lenses if not protected from the arc by screening.

• Employees in the area not protected from the arc by screening shall be protected by filter lenses meeting the requirements of Subpart E. When two or more welders are exposed to each other's arc, filter lens goggles of a suitable type, meeting the requirements of Subpart E, shall be worn under welding helmets. Hand shields to protect the welder against flashes and radiant energy shall be used when either the helmet is lifted or the shield is removed.

• Welders and other employees who are exposed to radiation shall be suitably protected so that the skin is covered completely to prevent burns and other damage by ultraviolet rays. Welding helmets and hand shields shall be free of leaks and openings, and highly reflective surfaces. **(See Figure 10-21)**

Figure 10-21. Employees shall be protected from the welding arc or from radiation exposure by wearing the appropriate protective equipment and clothing.

NOTE: WHEN INERT-GAS METAL-ARC WELDING IS BEING PERFORMED ON STAINLESS STEEL, ADEQUATE LOCAL EXHAUST VENTILATION SHALL BE USED TO PROTECT AGAINST DANGEROUS CONCENTRATIONS OF NITROGEN DIOXIDE.

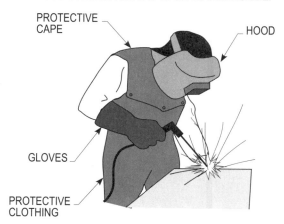

PROTECTIVE CAPE

HOOD

GLOVES

PROTECTIVE CLOTHING

INERT-GAS METAL-ARC WELDING
OSHA 1926.353(d)

• When inert-gas metal-arc welding is being performed on stainless steel, adequate local exhaust ventilation as described above or air line respirators in accordance with the requirements of Subpart E shall be used to protect against dangerous concentrations of nitrogen dioxide.

GENERAL WELDING, CUTTING AND HEATING
OSHA 1926.353(e)

Welding, cutting or heating not involving conditions or toxic materials described above may normally be done without mechanical ventilation or respiratory protective equipment. These protections shall be provided, however, where an unsafe accumulation of contaminants exists because of unusual physical or atmospheric conditions.

Employees performing any type of welding, cutting or heating shall be protected by suitable eye protective equipment in accordance with the requirements of Subpart E.

Safety Tip: Welders exposed to an unsafe accumulation of contaminants because of unused physical or atmospheric conditions shall be provided with mechanical ventilation or respiratory protective equipment.

WELDING, CUTTING AND HEATING IN WAY OF PRESERVATIVE COATINGS
OSHA 1926.354

Before welding, cutting, or heating is commenced on any surface covered by a preservative coating whose flammability is not known, a test shall be made by a competent person to determine its flammability. Preservative coatings shall be considered to be highly flammable when scrapings burn with extreme rapidity.

OSHA Required Training
1926.354(a)

When coatings are determined to be highly flammable, they shall be stripped from the area to be heated to prevent ignition.

The following shall be used for protection against toxic preservative coatings:

- In enclosed spaces, all surfaces covered with toxic preservatives shall be stripped of all toxic coatings for a distance of at least 4 in. from the area of heat application, or the employees shall be protected by air line respirators meeting the requirements of Subpart E.

- In the open air, employees shall be protected by a respirator, in accordance with the requirements of Subpart E.

- The preservative coatings shall be removed a sufficient distance from the area to be heated to ensure that the temperature of the unstripped metal will not be appreciably raised. Artificial cooling of the metal surrounding the heating area may be used to limit the size of the area required to be cleaned.

Welding and Cutting

Section	Answer	
_____	T	F
_____	T	F
_____	T	F
_____	T	F
_____	T	F
_____	T	F
_____	T	F
_____	T	F
_____	T	F
_____	T	F

Welding and Cutting

1. When transporting, moving or storing compressed gas cylinders valve protection caps shall be in place and secured.

2. When cylinders are transported by powered vehicles, they shall be secured in a horizontal position.

3. Unless cylinders are firmly secured on a special carrier intended for this purpose, regulators shall be removed and valve protection caps put in place before cylinders are moved.

4. When work is finished, when cylinders are empty or when cylinders are moved at any time, the cylinder valve shall be open.

5. Oxygen cylinders in storage shall be separated from fuel-gas cylinders or combustible materials (especially oil or grease), a minimum distance of 15 ft. or by a noncombustible barrier at least 2.5 ft. high having a fire-resistance rating of at least one hour.

6. Cylinders containing oxygen or acetylene or other fuel gas shall not be taken into confined spaces.

7. Fuel gas and oxygen manifolds shall bear the name of the substance they contain in letters at least 1 in. high which shall be either painted on the manifold or on a sign permanently attached to it.

8. Fuel gas and oxygen hose shall be easily distinguishable from each other. The contrast may be made by different colors or by surface characteristics readily distinguishable by the sense of touch. Oxygen and fuel gas hoses shall not be interchangeable.

9. Torches in use shall be inspected at the beginning of each working shift for leaking shutoff valves, hose couplings, and tip connections. Defective torches shall not be used.

10. Suitable fire extinguishing equipment shall be within 200 ft. of the work area and shall be maintained in a state or readiness for instant use.

Electrical

Electricity has long been recognized as a serious workplace hazard, exposing employees to such dangers as electric shock, electrocution, fires and explosions.

Experts in electrical safety have traditionally looked toward the widely used National Electrical Code (NEC) for help in the practical safeguarding of persons from these hazards. The Occupational Safety and Health Administration (OSHA) recognized the important role of the NEC in defining basic requirements for safety in electrical installations by including the entire 1971 NEC by reference in Subpart K of 29 Code of Federal Regulations Part 1926 (Construction Safety and Health Standards).

In a final rule dated July 11, 1986, OSHA updated, simplified and clarified Subpart K, 29 CFR 1926. The revisions serve these objectives:

• NEC requirements that directly affect employees in construction workplaces have been placed in the text of the OSHA standard, eliminating the need for the NEC to be incorporated by reference.

• Certain requirements that supplemented the NEC have been integrated in the new format.

• Performance language is utilized and superfluous specifications omitted and changes in technology accommodated.

In addition, the standard is easier for employers and employees to use and understand. Also, the OSHA revision of the electrical standards has been made more flexible, eliminating the need for constant revision to keep pace with the NEC, which is revised every three years.

SUBPART K

The NEC provisions directly related to employee safety are included in the body of the standard itself - making it unnecessary to continue the adoption by reference of the NEC. Subpart K is divided into four major groups plus a general definitions section:

- Installation safety requirements - OSHA 1926.402 - 1926.415
- Safety-related work practices - OSHA 1926.416 - 1926.430
- Safety-related maintenance and environmental considerations - OSHA 1926.431 - 1926.440
- Safety Requirements for Special Equipment - OSHA 1926.441 - 1926.448
- Definitions - OSHA 1926.449

INSTALLATION SAFETY REQUIREMENTS - COVERED
OSHA 1926.402(a)

Part I of the standard is very comprehensive. Only some of the major topics and brief summaries of these requirements are included in this discussion.

Sections 29 CFR 1926.402 through 1926.408 contain installation safety requirements for electrical equipment and installations used to provide electric power and light at the jobsite. These sections apply to installations, both temporary and permanent, used on the jobsite; but they do not apply to existing permanent installations that were in place before the construction activity commenced.

If an installation is made in accordance with the 1984 National Electrical Code, it will be considered to be in compliance with Sections 1926.403 through 1926.408, except for:

- Ground-fault protection for employees - 1926.404(b)(1)
- Protection of lamps on temporary wiring - 1926.405(a)(2)(ii)(E)
- Suspension of temporary lights by cords - 1926.405(a)(2)(ii)(F)
- Portable lighting used in wet or conductive locations - 1926.405(a)(2)(ii)(G)
- Extension cord sets and flexible cords - 1926.405(a)(2)(ii)(J)

GENERAL REQUIREMENTS
OSHA 1926.403

The following shall be considered when applying the general requirements:

- Approval
- Examination, installation and use of equipment
- Interrupting rating
- Mounting and cooling of equipment
- Splices
- Arcing parts
- Marking
- Identification of disconnecting means
- 600 volts nominal or less
- Over 600 volts, nominal

APPROVAL
OSHA 1926.403(a)

OSHA requires that all electrical conductors and equipment shall be free of hazards that may cause death or serious physical harm and that conductors and equipment be accepted only when approved. The definition of approved is to be acceptable to the authority having jurisdiction (AHJ). The authority having jurisdiction will usually require third party certification of all equipment. Third party certification is where a qualified testing laboratory perform tests to verify if the equipment will do what it is designed to do and still be safe to the user. The only problem with accepting equipment that is not certified by a third party is that the authority having jurisdiction takes the sole responsibility that the equipment is safe. **(See Figure 11-1)**

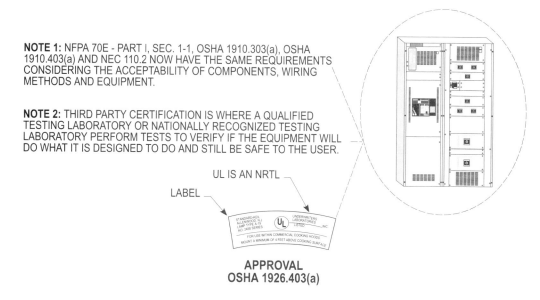

Figure 11-1. Conductors and equipment installed or used on construction sites shall be approved by the authority having jurisdiction.

NOTE 1: NFPA 70E - PART I, SEC. 1-1, OSHA 1910.303(a), OSHA 1910.403(a) AND NEC 110.2 NOW HAVE THE SAME REQUIREMENTS CONSIDERING THE ACCEPTABILITY OF COMPONENTS, WIRING METHODS AND EQUIPMENT.

NOTE 2: THIRD PARTY CERTIFICATION IS WHERE A QUALIFIED TESTING LABORATORY OR NATIONALLY RECOGNIZED TESTING LABORATORY PERFORM TESTS TO VERIFY IF THE EQUIPMENT WILL DO WHAT IT IS DESIGNED TO DO AND STILL BE SAFE TO THE USER.

UL IS AN NRTL

LABEL

APPROVAL
OSHA 1926.403(a)

EXAMINATION, INSTALLATION AND USE OF EQUIPMENT OSHA 1926.403(b)

All electrical equipment and work shall be related to the Occupational Safety and Health Administration (OSHA) regulations and comply to the provisions of standards and installation instructions of qualified testing laboratories that have tested and certified the equipment. Such Nationally Recognized Testing Laboratories (NRTL) and certifying organizations are:

Safety Tip: Electrical equipment shall be approved by a NRTL.

- Underwriters Laboratories (UL)
- Factory Mutual Engineering Corp. (FM)
- Electrical Testing Laboratories (ETL)*
- MET Testing Company, Inc.*
- Dash, Straus and Goodhue, Inc.*
- Canadian Standards Association (CSA)*
- Other laboratories approved by OSHA
*(Limited Certification)

In judging and evaluating electrical equipment that is installed in the employee workplace, the following steps shall be followed:

- Judge the wiring installation to ensure that it complies with the provisions of 1926 Subpart K. Evaluate the equipment for suitability for installation and use according to nameplate ratings and instructions. List the appropriate standard for such equipment. Every product listed in the UL Electrical Construction Materials Directory (Green Book) shall be installed as described in the application data given with the listing in the book. It becomes the responsibility of the authority having jurisdiction (AHJ) to decide and determine the suitability of electrical equipment that has not been tested and certified by a qualified testing laboratory.

• Mechanical strength and durability of enclosures, raceways, etc. enclosing electrical parts and wiring shall be adequate to provide proper protection of such elements.

• Electrical insulation shall be installed by Table 310.13 in the NEC and loaded according to the ampacities of Table 310.16 in the NEC for systems rated at 2000 volts or less. Insulation of conductors can be checked with any good insulation resistance tester.

• Equipment and insulations shall be sized and selected to operate safely under normal conditions of use and also under abnormal conditions without adverse heating effects. (See Section 310.10 in the NEC)

• Electrical equipment shall be equipped with shields, etc. to disguise the arcing effect and limit the burning effect of the arc. (See Section 110.18 in the NEC)

• Electrical equipment with components shall be classified by its type, size, voltage, current capacity and specific use. Evaluation to verify that the type and size of equipment is being supplied with the proper voltage and is loaded to a value no greater than it is designed to carry. (See Section 90.7 and 110.3(B) in the NEC)

• All listed, labeled or certified equipment shall be installed and used in accordance with the instructions included in the listing, labeling or certification. **(See Figure 11-2)**

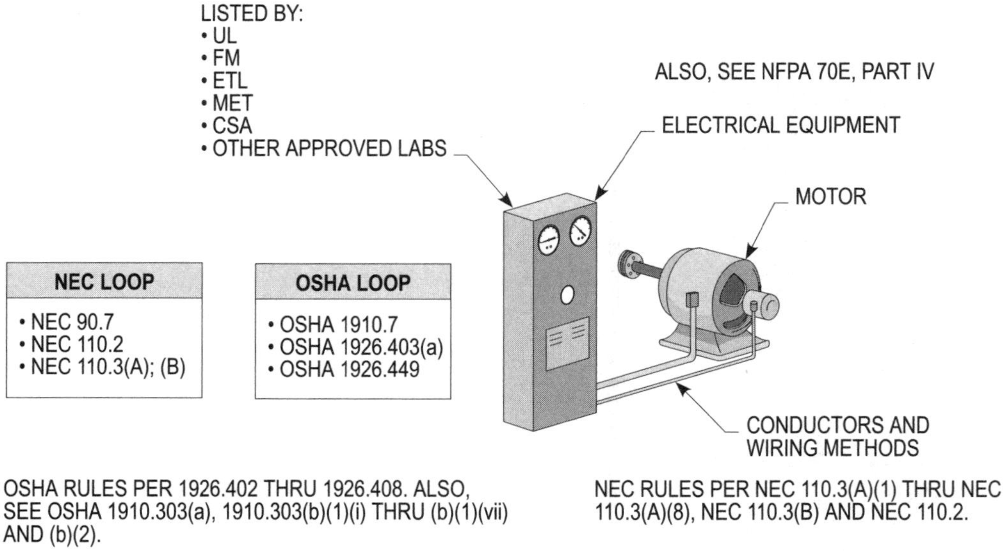

Figure 11-2. This illustration pertains to the requirements for acceptance of electrical equipment per OSHA.

Safety Tip: Equipment shall be rated in current carrying capacity as well as interrupting current capacity.

LISTED BY:
• UL
• FM
• ETL
• MET
• CSA
• OTHER APPROVED LABS

ALSO, SEE NFPA 70E, PART IV

ELECTRICAL EQUIPMENT

MOTOR

NEC LOOP
• NEC 90.7
• NEC 110.2
• NEC 110.3(A); (B)

OSHA LOOP
• OSHA 1910.7
• OSHA 1926.403(a)
• OSHA 1926.449

CONDUCTORS AND WIRING METHODS

OSHA RULES PER 1926.402 THRU 1926.408. ALSO, SEE OSHA 1910.303(a), 1910.303(b)(1)(i) THRU (b)(1)(vii) AND (b)(2).

NEC RULES PER NEC 110.3(A)(1) THRU NEC 110.3(A)(8), NEC 110.3(B) AND NEC 110.2.

**EXAMINATION, INSTALLATION AND USE OF EQUIPMENT
OSHA 1926.403(b)**

INTERRUPTING RATING
OSHA 1926.403(c)

Interrupting capacity is different from the rating of the amperes that is required to supply a load. The electrical system is faced with what is known as fault currents. A fault current is the amount of current that might develop under a direct short condition. In the past, this was not much of a problem, but with increased electrical use and larger generating and distribution capacities, the problem of fault currents have become a serious problem. Such destructive currents shall be taken into consideration more than in the past. If a piece of equipment is rated at a certain number of amperes, this does not necessarily mean that it can be disconnected under load or faults without damage. Equipment shall be rated in carrying capacity as well as interrupting capacity.

Switches and circuit breakers shall have a voltage rating at least equal to the circuit voltage. The interrupting rating shall be high enough to handle the maximum short-circuit current of the circuit per NEC 110.9 and 110.10.

MOUNTING AND COOLING OF EQUIPMENT
OSHA 1926.403(d)

MOUNTING

Mounting of equipment is directly related to workmanship. Wooden plugs driven into holes in masonry, plaster, concrete, etc., will rot and deteriorate, thereby allowing the equipment to become loose and dangerous for someone turning it off or on. Therefore, only approved methods such as special anchoring devices may be used to mount equipment.

COOLING

Electricity produces heat and shall be provided with some kind of cooling means. Electrical equipment shall be installed so that circulation of air and convection methods of cooling will be provided. If mounted too close to walls, ceilings, floors, etc. which will interfere with the cooling of such electric equipment, ventilation openings in the electric equipment will not be free to allow natural circulation.

Safety Tip: Electrical equipment shall be installed so as to allow for proper cooling.

The total space of the room shall be evaluated where the equipment is mounted. If such space is inadequate to permit a low enough ambient temperature, means shall be taken to allow the lowering of high ambient temperatures by either natural or mechanical ventilation.

SPLICES
OSHA 1926.403(e)

Splices in conductors shall be accessible for maintenance and splices are not allowed to be made in raceways. Splices shall be accessible where they are made in junction boxes, auxiliary gutters, wireways, metal surface raceways, etc.

When making a splice, the wires shall be cleaned and a good mechanical connection made. The wires can be soldered, provided a suitable solder and flux are applied. The temperature should be controlled since a cold solder joint has no value. Where the wires become too hot, the heat will damage the insulation. Soldering shall not be permitted on conductors used for grounding.

Listed connectors can be used for splices where the wires are clean and free from corrosion. Insulation at least equivalent to that on the wire shall be applied to the splice. This requirement shall be applied to all types of splices. For the splicing of high-voltage cables, the specifications with the high-voltage cables are to be applied. It is very important to apply inhibitors with aluminum conductors. Brush the aluminum conductor to remove the oxide film and then immediately apply the inhibitor to prevent the reoccurrence of the oxide film. Conductors shall be cleaned and spliced with a good mechanical connection and be accessible.

Many electrical connections have failed because the splicer made an improper splice. The splicer should make certain that the connectors are listed for the use when conductors are being spliced.

ARCING PARTS
OSHA 1926.403(f)

The making and breaking of electrical contacts normally produce sparking or arcing. Any parts that usually cause arcing or sparking are enclosed unless they are isolated or separated from combustible material.

For example: There are enclosures in the field that are not total enclosures. Arcing parts creating sparks or arcs could easily ignite any nearby combustible materials. Inspectors require combustible material to be kept well away from such equipment to help prevent costly and life threatening fires.

Note: Motors with commutators, collectors and brushes housed within the motor are not required to be guarded unless there are exposed rotating parts. Motors having electrical contacts with voltage in supply circuits operating at over 150 volts-to-ground shall be guarded.

MARKING
OSHA 1926.403(g)

Equipment shall be installed in such a manner so the marking can be examined by the inspector without removing the installed equipment from a hard-wired (permanent) position. The marking on the equipment shall be durable enough to withstand the environment in which the equipment is exposed.

For example: Equipment located in wet locations has to be rated with wet-proof enclosures to prevent the entrance of water.

IDENTIFICATION OF DISCONNECTING MEANS
OSHA 1926.403(f)

Safety Tip: Circuit directories shall be fully and clearly filledout to indicate the load served and its location.

Panelboard circuit directories shall be fully and clearly filled out to indicate the load served and its location. Marking on equipment shall be painted lettering or another suitable identification. The circuit breakers in panelboards and fuses in fusible disconnects shall be marked to match the markings on the equipment they serve. This is a mandatory requirement by OSHA, and it is to be applied to both new and old electrical systems. Adequate circuit mapping and proper identification shall be provided for expanding or alteration of all electrical systems. It is extremely important that the right disconnecting means can be opened to deenergize a circuit quickly and safely when there is a threat of injury to personnel working, repairing or pulling maintenance on the equipment supplied.

Painted labeling or embossed plates fixed to the equipment complies with the requirement that disconnects be "legibly marked" and that the "marking shall be durable". Glue-on paper labels or marking with crayon, ink or chalk most likely will be rejected as not being suitable permanent markings.

For example, a circuit breaker in a panelboard is marked No. 1 and supplies power to a disconnecting means in sight of a motor controller and motor.

The disconnecting means, motor controller and motor shall be legibly marked No. 1 to correspond to the marking on the circuit directory on the panelboard enclosure. Effective identification of disconnecting means and protective devices is a must by OSHA to ensure the necessary safety for employees in the workplace.

600 VOLTS NOMINAL OR LESS
OSHA 1926.403(i)

The following general requirements shall be considered for systems of 600 volts nominal or less:

- Working space
- Working clearances
- Clear spaces
- Access and entrance to working space
- Headroom
- Guarding of live parts

WORKING SPACE

All electrical equipment shall be provided with sufficient working space to allow safe access for servicing the equipment without exposing electrical workers to shock hazards. The workspace is required and shall be maintained around and about all electrical equipment where parts can be serviced while energized.

Note: Equipment of equal depth can be mounted below or above another piece of equipment, if they comply with the height requirements.

Safety Tip: Sufficient working space shall be provided to allow safe access to electrical equipment.

WORKING CLEARANCES

A minimum working space of 30 in. width is required in front of electrical equipment operating at 600 volts or less. This space permits sufficient room to avoid body contact or elbows from contacting live parts and metal parts at the same time while working on the equipment. Equipment doors and hinged panels shall have at least a 90° opening provided in the workspace. This opening allows an electrical worker to have adequate room to repair, adjust or reset overcurrent protection devices without placing his or her body between the panel door and panelboard. **(See Figure 11-3)**

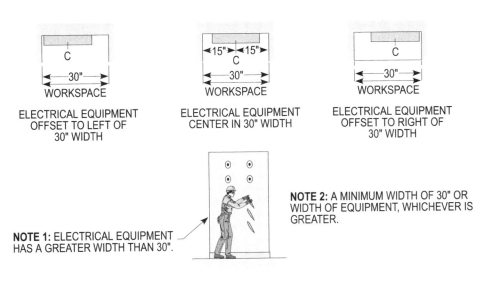

NOTE 1: ELECTRICAL EQUIPMENT HAS A GREATER WIDTH THAN 30".

NOTE 2: A MINIMUM WIDTH OF 30" OR WIDTH OF EQUIPMENT, WHICHEVER IS GREATER.

**WORKSPACE
OSHA 1926.403(i)**

Figure 11-3. This illustration shows a minimum workspace of 30 in. is required in front of electrical equipment. The workspace can be centered or offset, if water pipe or other types of piping, etc. are present.

Working space is not required in back of electrical equipment where there are not any removable or adjustable parts such as circuit breakers, fuses or switches mounted on the back of the equipment. All connections and service areas for maintenance shall be accessible from other locations other than the back of the equipment. If rear access is necessary to work on deenergized parts on the back of enclosed equipment, a minimum working space of 30 in. horizontally is required. **(See Figure 11-4)**

See Figures 11-5(a) through (c) for a detailed illustration pertaining to work clearance in front of electrical equipment.

Figure 11-4. When installing electrical equipment, both front and back of such equipment shall be evaluated to verify if live parts and deenergized parts can be worked on.

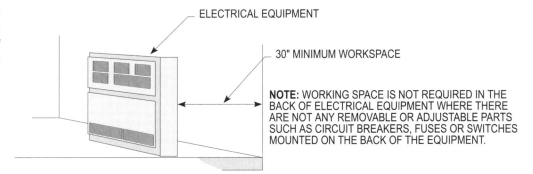

WORKING SPACES IN BACK OF EQUIPMENT
OSHA 1926.403(f)

Figure 11-5(a). This illustration shows work clearances in front of electrical equipment with ungrounded wall opposite equipment.

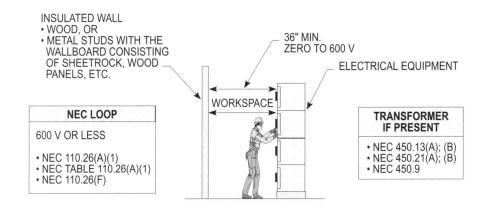

WORKSPACE IN FRONT OF ELECTRICAL EQUIPMENT
OSHA 1926.403(i)
TABLE K-1

Figure 11-5(b). This illustration shows work clearances in front of electrical equipment with grounded wall opposite equipment.

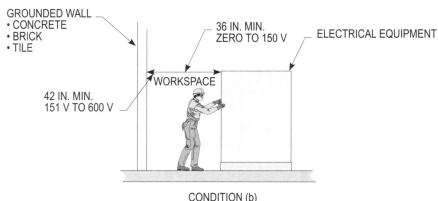

WORKSPACE IN FRONT OF ELECTRICAL EQUIPMENT
OSHA 1926.403(i)
TABLE K-1

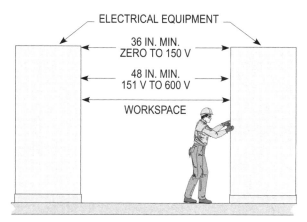

ELECTRICAL EQUIPMENT

36 IN. MIN.
ZERO TO 150 V

48 IN. MIN.
151 V TO 600 V

WORKSPACE

CONDITION (c)
EXPOSED ENERGIZED PARTS IN
FRONT AND BACK OF WORKER

WORKSPACE IN FRONT OF ELECTRICAL EQUIPMENT
OSHA 1926.403(i)
TABLE K-1

Figure 11-5(c). This illustration shows work clearances in front of electrical equipment opposite from other electrical equipment.

CLEAR SPACES

Electrical equipment shall be located where adequate working space is accessible to take voltage measurements, check continuity of circuits, adjust or replace defective overcurrent protection devices and tighten loose connections. Working space in front of electrical equipment shall be free from storage of materials, etc. Mains and overcurrent protection devices shall be accessible to the users in case of emergencies as well as disconnecting means of equipment. Any and all exposed live parts located in a passageway or open space are required to be guarded. **(See Figures 11-6(a) and (b))**

Safety Tip: Working sapce in front of equipment shall be free from storage of materials.

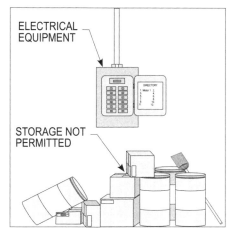

ELECTRICAL
EQUIPMENT

STORAGE NOT
PERMITTED

CLEAR SPACES
OSHA 1926.403(i)

Figure 11-6(a). Storage shall not be permitted in front or below electrical equipment.

ACCESS AND ENTRANCE TO WORKING SPACE

Electrical equipment shall be designed and installed to have at least one entrance to the work space. Easy and fast access to electrical devices is essential. **(See Figure 11-7)**

Special consideration should be given to electrical equipment that is over 6 ft. wide with 1200 amps or more of bus containing overcurrent devices, switching devices or control devices. Such equipment shall have a clearance of 24 in. wide and 6 1/2 ft. high at each end for safe exit in case of a ground fault. If there are not two ways to enter and leave the workspace of such equipment, a worker could be trapped at one end of the workspace by a ground fault with fire between him and the exit. **(See Figure 11-8)**

Figure 11-6(b). A 6 in. clearance from the face of the gutter to the front face of the panel enclosure shall be considered adequate working space.

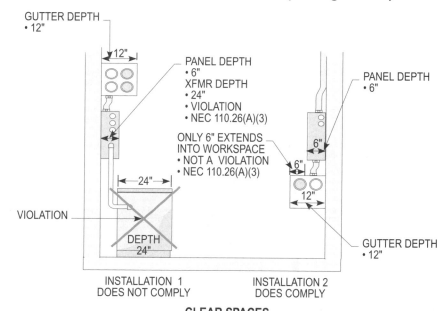

**CLEAR SPACES
OSHA 1926.403(i)**

Figure 11-7. Access and working space for electrical equipment shall be provided.

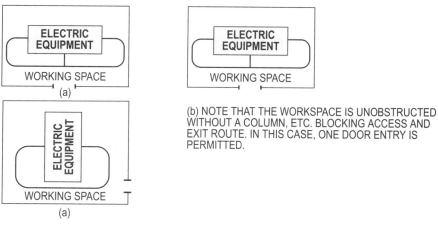

(b) NOTE THAT THE WORKSPACE IS UNOBSTRUCTED WITHOUT A COLUMN, ETC. BLOCKING ACCESS AND EXIT ROUTE. IN THIS CASE, ONE DOOR ENTRY IS PERMITTED.

(a) AT LEAST ONE ENTRANCE IS REQUIRED TO PROVIDE ACCESS TO THE WORKING SPACE AROUND ELECTRIC EQUIPMENT

**ACCESS AND ENTRANCE TO WORKING SPACE
OSHA 1926.403(i)**

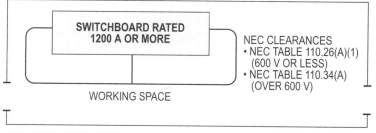

FOR EQUIPMENT RATED 1200 A OR MORE, AND OVER 6 FT. WIDE, THERE MUST BE ONE ENTRANCE NOT LESS THAN 24 IN. WIDE AND 6 FT.-6 IN. HIGH AT EACH END OR WORKSPACE IS REQUIRED TO BE DOUBLED.

**ACCESS AND ENTRANCE TO WORKING SPACE
OSHA 1926.403(i)**

However, the workspace can be doubled per Table K-1 and based on conditions 1, 2 or 3 and only one entrance is required. With the workspace doubled in front of the equipment a worker can move back out of the endangered workspace and exit along the length of the equipment. The deeper workspace in front of the equipment provides a safe route to exit without providing two entrances. **(See Figure 11-9)**

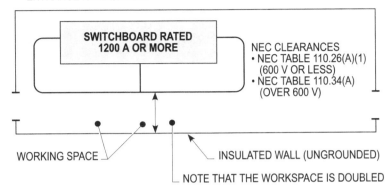

EXAMPLE: IF THE WORKSPACE OF 3 FT. PER NEC TABLE 110.26(A)(1) IS DOUBLED (3' x 2 = 6') PROVIDING A WORKSPACE OF 6', ONLY ONE ENTRANCE IS REQUIRED.

SWITCHBOARD RATED 1200 A OR MORE

NEC CLEARANCES
• NEC TABLE 110.26(A)(1) (600 V OR LESS)
• NEC TABLE 110.34(A) (OVER 600 V)

WORKING SPACE

INSULATED WALL (UNGROUNDED)

NOTE THAT THE WORKSPACE IS DOUBLED

ACCESS AND ENTRANCE TO WORKING SPACE
OSHA 1926.403(i)

Figure 11-9. Only one entrance shall be required where the workspace is doubled between the switchboard and wall.

HEADROOM

A minimum headroom clearance of 6 1/2 ft. shall be maintained from the floor or platform up to the luminaire (lighting fixture) or any overhead obstruction. This overhead workspace is mandatory and applies especially to service equipment, switchboards, panelboards and motor control centers. The purpose of the overhead workspace is to protect employees from accidentally contacting grounded objects with their bodies (head, hands, etc.) while touching live parts and completing a circuit to ground that could cause a fatal electric shock. Electricians or maintenance workers should never have to stoop or bend down to gain access to service, repair or modify components inside electrical equipment, as previously mentioned. **(See Figures 11-10(a) and (b))**

Safety Tip: Minimum headrooom clearance of 6 1/2 ft. shall be maintained for service equipment, switchboards, panelboards, and MCC.

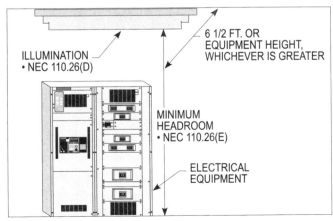

NOTE 1: ADJACENT LIGHTING UNITS PROVIDING PROPER LIGHTING WILL SATISFY THIS RULE PER NEC 110.26(D).

ILLUMINATION
• NEC 110.26(D)

6 1/2 FT. OR EQUIPMENT HEIGHT, WHICHEVER IS GREATER

MINIMUM HEADROOM
• NEC 110.26(E)

ELECTRICAL EQUIPMENT

NOTE 2: LIGHTING FIXTURES CAN BE INCANDESCENT OR FLUORESCENT AS LONG AS THEY PROVIDE THE PROPER LIGHTING FOR PARTS TO BE SERVICED.

HEADROOM
OSHA 1926.403(i)

Figure 11-10(a). Headroom workspace shall be at least 6 1/2 ft. or the height of the electrical equipment, whichever is greater.

Figure 11-10(b). This illustration shows the guidelines for installing electrical equipment per NEC.

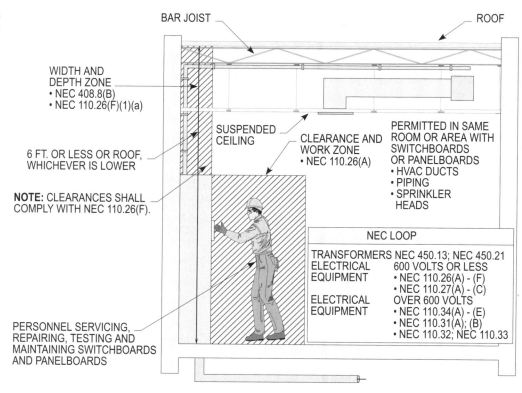

**INSTALLATION REQUIREMENTS PER NEC
NEC 110.26(A) THRU (F)**

GUARDING OF LIVE PARTS

The general rule of protecting live parts (energized) from accidental contact is by installing them in a complete enclosure that provides a dead front. Sometimes it is not practical to construct enclosures to house large control panels, etc., and in such cases if the apparatus is rated at 50 volts or more, then suitable guards or isolation should be provided by one of the following rules. **(See Figure 11-11)**

Figure 11-11. Live parts in electrical equipment equipped with dead front hardware shall be considered guarded.

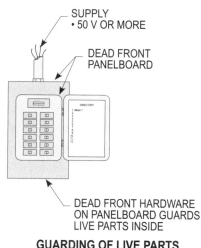

**GUARDING OF LIVE PARTS
OSHA 1926.403(iii)(2)**

Live parts in electrical equipment that are not mounted in a completely enclosed enclosure can be installed in a room, vault or similar enclosure that is accessible only to qualified personnel. **(See Figure 11-12)**

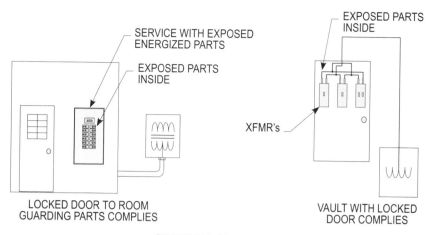

SERVICE WITH EXPOSED ENERGIZED PARTS

EXPOSED PARTS INSIDE

EXPOSED PARTS INSIDE

XFMR's

LOCKED DOOR TO ROOM GUARDING PARTS COMPLIES

VAULT WITH LOCKED DOOR COMPLIES

GUARDING OF LIVE PARTS
OSHA 1926.403(iii)(2)(A)

Figure 11-12. Exposed live parts installed in locked rooms or vaults shall be considered guarded.

Live parts may be separated by permanent partitions or screens so located that only qualified persons have access and reach to the live parts. Openings in partitions or screens shall be designed to prevent accidental contact with live parts or bring conductive objects into contact with live parts. **(See Figure 11-13)**

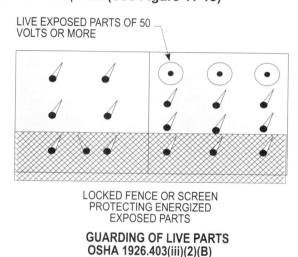

LIVE EXPOSED PARTS OF 50 VOLTS OR MORE

LOCKED FENCE OR SCREEN PROTECTING ENERGIZED EXPOSED PARTS

GUARDING OF LIVE PARTS
OSHA 1926.403(iii)(2)(B)

Figure 11-13. Live parts separated by permanent partitions or screens shall be considered guarded.

Exposed live parts in equipment may be located on a suitable balcony, gallery or platform that is high enough or designed in such a manner to keep unqualified personnel out. **(See Figure 11-14)**

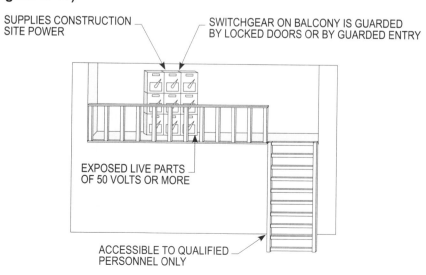

SUPPLIES CONSTRUCTION SITE POWER

SWITCHGEAR ON BALCONY IS GUARDED BY LOCKED DOORS OR BY GUARDED ENTRY

EXPOSED LIVE PARTS OF 50 VOLTS OR MORE

ACCESSIBLE TO QUALIFIED PERSONNEL ONLY

GUARDING OF LIVE PARTS
OSHA 1926.403(iii)(2)(C)

Figure 11-14. A gallery or platforms shall be permitted to guard exposed live parts of electrical equipment.

Live exposed parts shall be permitted to be elevated at least 8 ft. above the floor. **(See Figure 11-15)**

Figure 11-15. Live exposed parts elevated 8 ft. or greater above grade shall be considered guarded.

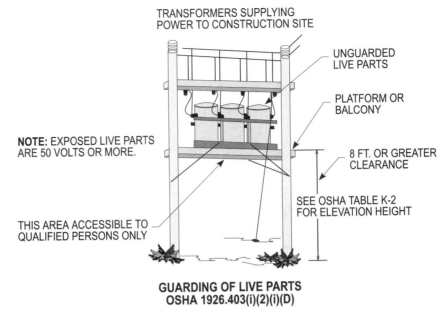

TRANSFORMERS SUPPLYING POWER TO CONSTRUCTION SITE

UNGUARDED LIVE PARTS

PLATFORM OR BALCONY

8 FT. OR GREATER CLEARANCE

NOTE: EXPOSED LIVE PARTS ARE 50 VOLTS OR MORE.

SEE OSHA TABLE K-2 FOR ELEVATION HEIGHT

THIS AREA ACCESSIBLE TO QUALIFIED PERSONS ONLY

GUARDING OF LIVE PARTS
OSHA 1926.403(i)(2)(i)(D)

At times, electrical equipment is located in work areas where the work activity around it might damage the equipment. In such cases, the equipment shall be properly protected with enclosures or guards that provide the necessary strength to prevent any damage to the electrical equipment.

All guarded locations shall be posted with warning signs, giving warning that only qualified personnel are allowed to enter and service the electrical components. Any dangers that might exist should be posted to warn employees to be careful. **(See Figure 11-16)**

Figure 11-16. This illustration shows guarded locations posted with signs to prohibit the entry of unqualified personnel and warn employees to be careful.

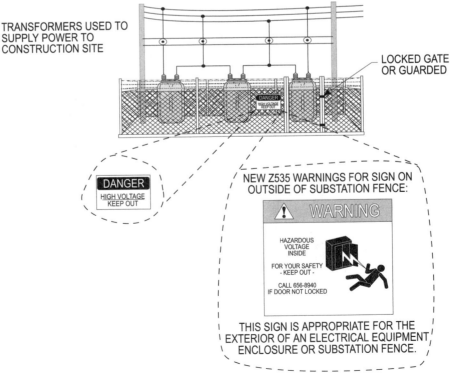

TRANSFORMERS USED TO SUPPLY POWER TO CONSTRUCTION SITE

LOCKED GATE OR GUARDED

DANGER
HIGH VOLTAGE
KEEP OUT

NEW Z535 WARNINGS FOR SIGN ON OUTSIDE OF SUBSTATION FENCE:

⚠ WARNING

HAZARDOUS VOLTAGE INSIDE

FOR YOUR SAFETY - KEEP OUT -

CALL 656-8940 IF DOOR NOT LOCKED

THIS SIGN IS APPROPRIATE FOR THE EXTERIOR OF AN ELECTRICAL EQUIPMENT ENCLOSURE OR SUBSTATION FENCE.

GUARDING OF LIVE PARTS
OSHA 1926.403(i)(2)(iii)

OVER 600 VOLTS, NOMINAL
OSHA 1926.403(j)

The following requirements shall be considered for systems over 600 volts, nominal:

- General
- Enclosure for electrical installations
- Installation accessible to qualified persons
- Installation accessible to unqualified persons
- Workspace about equipment
- Lighting outlets and points of control
- Elevation of unguarded live parts
- Entrance and access to workspace

GENERAL

The installation requirements of conductors, overcurrent protection devices and equipment are usually more stringent for systems rated over 600 volts. Conductors are normally shielded and special terminations are required where they terminate to the equipment, etc. Greater clearances around electrical equipment is required due to the threat of flash over voltage from live parts to grounded metal.

ENCLOSURE FOR ELECTRICAL INSTALLATIONS

In locations where entrance and accessibility is controlled by lock and key or other acceptable and approved means, these locations are considered as accessible to only qualified personnel. Doors and covers of enclosures used solely as pull boxes, splice boxes or junction boxes shall be locked, bolted or screwed on. Ungrounded box covers that weigh over 100 lbs. are considered as complying with this requirement. Such locations that are involved are as follows:

- Electrical installations in vaults,
- Electrical installations in rooms,
- Electrical installations in closets and
- Locations surrounded by a:
    - **(a)** Wall,
    - **(b)** Screen, or
    - **(c)** Fence

The design and construction of enclosures shall be suitable to the nature and degree of the hazards involved.

Note: A wall, screen or fence that is less than 8 ft. tall is not considered as preventing access to equipment unless other features are provided with a degree of isolation equivalent to an 8 ft. fence. A fence less than 8 ft. in height is equipped with barbed wire at the top of the fence. This addition of the barbed wire is the extra feature needed to prevent easy access of unqualified personnel. **(See Figure 11-17)**

INSTALLATION ACCESSIBLE TO QUALIFIED PERSONS

Electrical installations installed outdoors containing exposed live parts shall be accessible to qualified personnel only. It is essential that the live parts of high-voltage systems be totally enclosed in the equipment or be isolated to prevent access by the general public or untrained employees. **(See Figure 11-18)**

Figure 11-17. A wall, screen or fence installed properly prevent easy access of unqualified personnel.

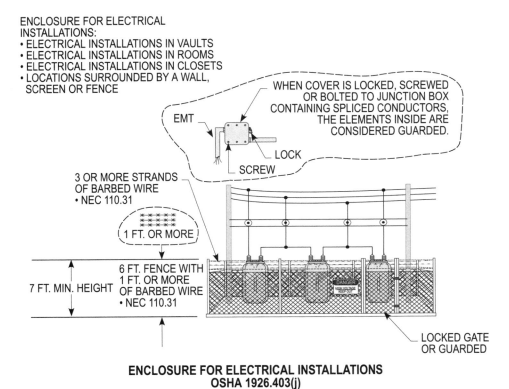

ENCLOSURE FOR ELECTRICAL INSTALLATIONS:
• ELECTRICAL INSTALLATIONS IN VAULTS
• ELECTRICAL INSTALLATIONS IN ROOMS
• ELECTRICAL INSTALLATIONS IN CLOSETS
• LOCATIONS SURROUNDED BY A WALL, SCREEN OR FENCE

WHEN COVER IS LOCKED, SCREWED OR BOLTED TO JUNCTION BOX CONTAINING SPLICED CONDUCTORS, THE ELEMENTS INSIDE ARE CONSIDERED GUARDED.

EMT
LOCK
SCREW

3 OR MORE STRANDS OF BARBED WIRE
• NEC 110.31

1 FT. OR MORE

7 FT. MIN. HEIGHT

6 FT. FENCE WITH 1 FT. OR MORE OF BARBED WIRE
• NEC 110.31

DANGER
HIGH VOLTAGE
KEEP OUT

LOCKED GATE OR GUARDED

**ENCLOSURE FOR ELECTRICAL INSTALLATIONS
OSHA 1926.403(j)**

Figure 11-18. Live energized parts behind locked doors entering rooms or vaults or totally enclosed enclosures shall be considered accessible to qualified persons only.

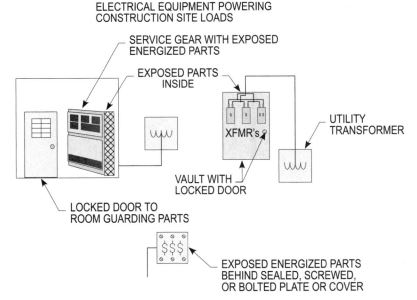

ELECTRICAL EQUIPMENT POWERING CONSTRUCTION SITE LOADS

SERVICE GEAR WITH EXPOSED ENERGIZED PARTS

EXPOSED PARTS INSIDE

XFMR's

UTILITY TRANSFORMER

VAULT WITH LOCKED DOOR

LOCKED DOOR TO ROOM GUARDING PARTS

$$$

EXPOSED ENERGIZED PARTS BEHIND SEALED, SCREWED, OR BOLTED PLATE OR COVER

**INSTALLATION ACCESSIBLE TO QUALIFIED PERSONS
OSHA 1926.443(j)(2)(i)**

INSTALLATION ACCESSIBLE TO UNQUALIFIED PERSONS

Ventilating or similar openings in equipment shall be so designed that foreign objects inserted through these openings are deflected from energized parts. When exposed to physical damage from vehicular traffic, suitable guards and caution signs shall be provided. Metal enclosed equipment located outdoors accessible to the general public shall be designed so that exposed nuts or bolts are not readily removed, permitting access to live parts. Where metal enclosure equipment is accessible to the general public and the bottom of the enclosure is less than 8 ft. above the finished floor or grade level, the enclosure door or hinged cover shall be kept locked. **(See Figure 11-19)**

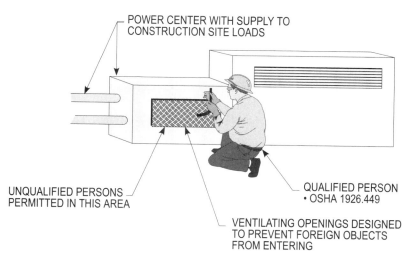

POWER CENTER WITH SUPPLY TO
CONSTRUCTION SITE LOADS

UNQUALIFIED PERSONS
PERMITTED IN THIS AREA

QUALIFIED PERSON
• OSHA 1926.449

VENTILATING OPENINGS DESIGNED
TO PREVENT FOREIGN OBJECTS
FROM ENTERING

INSTALLATIONS ACCESSIBLE TO UNQUALIFIED PERSONS
OSHA 1926.443(j)(2)(ii)

Figure 11-19. Unqualified persons shall be permitted to have access to electrical equipment with openings designed to deflect foreign objects from live parts.

WORKSPACE ABOUT EQUIPMENT

There shall be sufficient clear working space about high-voltage equipment to permit ready and safe operation of such equipment. There have been minimum clearances set for workspace about and around electrical equipment, and they are:

• Where energized parts are exposed, the minimum clear working space shall not be less than 6 1/2 ft. high measured vertically from the floor or platform or less than 3 ft. wide measured parallel to the equipment. See Table K-1 for a detailed procedure on clearances about electrical equipment supplied with voltage over 600 volts.

• Space shall be adequate to permit doors and hinged panels to open a minimum of 90° so personnel can stand in front of the equipment without making contact with their body and the doors or hinged panels while touching live parts.

WORKING SPACE

Table K-2 lists the minimum clear working space in front of electrical equipment. This can be such equipment as switchboards, control panels, switches, circuit breakers, motor control centers, relays, etc.

In measuring the distances given in Table K-2, they shall be measured from exposed live parts or if in enclosures, the distance is measured from the enclosure front or opening. **(See Figure 11-20)**

Working space is not required in back of electrical equipment where there are not any renewable or adjustable parts such as fuses and circuit breakers on the back and all connections are accessible from locations other than the back. If rear access is required to work on energized parts on the back of enclosed equipment, a minimum distance of 30 in. horizontally is required for the working space. **(See Figure 11-21)**

Figure 11-20. This illustration shows requirements for working space when the voltage is over 600 volts.

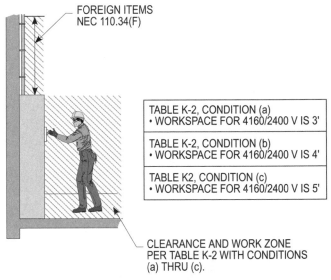

FOREIGN ITEMS
NEC 110.34(F)

TABLE K-2, CONDITION (a)
• WORKSPACE FOR 4160/2400 V IS 3'

TABLE K-2, CONDITION (b)
• WORKSPACE FOR 4160/2400 V IS 4'

TABLE K2, CONDITION (c)
• WORKSPACE FOR 4160/2400 V IS 5'

CLEARANCE AND WORK ZONE
PER TABLE K-2 WITH CONDITIONS
(a) THRU (c).

**WORKING SPACE
OSHA TABLE K-2, CONDITIONS (a) THRU (c)**

Figure 11-21. When installing electrical equipment, both front and back of such equipment shall be evaluated to verify if live parts and deenergized parts can be worked on.

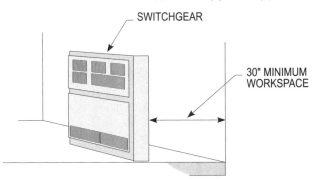

SWITCHGEAR

30" MINIMUM
WORKSPACE

NOTE: WORKING SPACE IS NOT REQUIRED IN THE BACK OF ELECTRICAL EQUIPMENT WHERE THERE ARE NOT ANY REMOVABLE OR ADJUSTABLE PARTS SUCH AS CIRCUIT BREAKERS, FUSES OR SWITCHES MOUNTED ON THE BACK OF THE EQUIPMENT.

**WORKING SPACE
OSHA 1926.403(j)(3)(i)**

LIGHTING OUTLETS AND POINTS OF CONTROL

Lighting outlets shall be arranged so that those changing lamps or making repairs will not be exposed to the hazard of contacting exposed live parts.

ELEVATION OF UNGUARDED LIVE PARTS

Safety Tip: There shall be at least one entrance with a maximum of 24 in. x 6 1/2 ft. to provide access to the working space of high-voltage equipment.

It is permissible to install unguarded live parts at elevated heights of at least 8 ft. 6 in. from any and all finished grade areas. The elevated height is based and selected from the supply voltage supplying the live pans. **(See Figure 11-22)**

ENTRANCE AND ACCESS TO WORKSPACE

There shall be at least one entrance a minimum of 24 in. by 6 1/2 ft. to provide access to the working space of high-voltage equipment (over 600 volts). This is a minimum and more should be provided when possible. Where the equipment is over 6 ft. in width, there shall be one entrance at each end per NEC 110.33(A)(1). **(See Figure 11-23)**

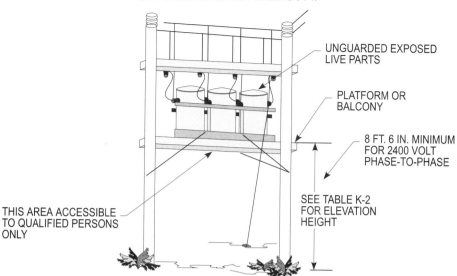

NOTE: THE ELEVATIONS HEIGHT FOR 13,200 VOLTS PHASE-TO-PHASE IS 9 FT.

UNGUARDED EXPOSED LIVE PARTS

PLATFORM OR BALCONY

8 FT. 6 IN. MINIMUM FOR 2400 VOLT PHASE-TO-PHASE

SEE TABLE K-2 FOR ELEVATION HEIGHT

THIS AREA ACCESSIBLE TO QUALIFIED PERSONS ONLY

ELEVATION OF UNGUARDED LIVE PARTS
OSHA 1926.403(j)(3)(iii)

Figure 11-22. Electrical equipment elevated at 8 ft. 6 in. above grade shall be considered guarded by height.

NOTE 1: AT LEAST ONE ENTRANCE IS REQUIRED AT EACH END OF SWITCHBOARDS AND CONTROL PANELS WHERE THE EQUIPMENT IS OVER 6 FT. IN WIDTH.

NOTE 2: ONE ENTRANCE AT EACH END IS NOT REQUIRED WHERE THE WORKSPACE IS DOUBLED OR THE LOCATION PERMITS A CONTINUOUS AND UNOBSTRUCTED WAY OF EXIT TRAVEL.

NOTE 3: SUITABLE MEANS IS REQUIRED TO BE DESIGNED TO GUARD BARE OR INSULATED PARTS OF MORE THAN 600 VOLTS, NOMINAL, LOCATED ADJACENT TO SUCH ENTRANCES.

Figure 11-23. Based on conditions of use, electrical equipment shall be provided with proper exits for personnel servicing such equipment.

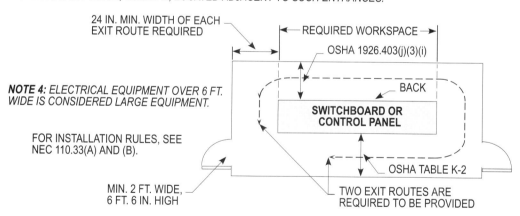

24 IN. MIN. WIDTH OF EACH EXIT ROUTE REQUIRED

REQUIRED WORKSPACE

OSHA 1926.403(j)(3)(i)

NOTE 4: ELECTRICAL EQUIPMENT OVER 6 FT. WIDE IS CONSIDERED LARGE EQUIPMENT.

FOR INSTALLATION RULES, SEE NEC 110.33(A) AND (B).

BACK

SWITCHBOARD OR CONTROL PANEL

OSHA TABLE K-2

MIN. 2 FT. WIDE, 6 FT. 6 IN. HIGH

TWO EXIT ROUTES ARE REQUIRED TO BE PROVIDED

ENTRANCE AND ACCESS TO WORKSPACE
OSHA 1926.403(j)(4)

On switchboards and control panels exceeding 48 in. in width, there shall be one entrance at each end of the board where practical.

If bare or insulated parts of more than 600 volts, nominal, are located adjacent to such entrances, there shall be suitable means designed to guard them.

WIRING DESIGN AND PROTECTION
OSHA 1926.404

The following shall be considered for wiring design and protection:

• Use and identification of grounded and grounding conductors
• Branch-circuits
• Outside conductors and lamps
• Services
• Overcurrent protection 600 volts nominal or less
• Grounding

USE AND IDENTIFICATION OF GROUNDED AND GROUNDING CONDUCTORS
OSHA 1926.404(a)

Conductors used in electrical wiring systems shall be identified properly to protect personnel working on such systems. It is essential to know which conductors by color represent the ungrounded (phase) conductor, the grounded (neutral) conductor and equipment grounding conductor in an electrical circuit, because it is by color coding that conductors are connected to color coded terminals of equipment. **(See Figure 11-24)**

Figure 11-24. Conductors in an electrical system shall be identified by color, based on an ungrounded (phase) conductor, grounded (neutral) conductor or equipment grounding conductor.

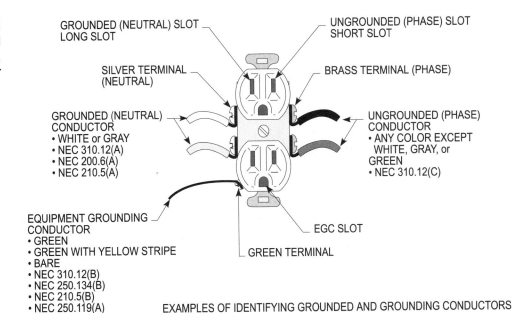

GROUNDED (NEUTRAL) SLOT LONG SLOT

UNGROUNDED (PHASE) SLOT SHORT SLOT

SILVER TERMINAL (NEUTRAL)

BRASS TERMINAL (PHASE)

GROUNDED (NEUTRAL) CONDUCTOR
• WHITE or GRAY
• NEC 310.12(A)
• NEC 200.6(A)
• NEC 210.5(A)

UNGROUNDED (PHASE) CONDUCTOR
• ANY COLOR EXCEPT WHITE, GRAY, or GREEN
• NEC 310.12(C)

EQUIPMENT GROUNDING CONDUCTOR
• GREEN
• GREEN WITH YELLOW STRIPE
• BARE
• NEC 310.12(B)
• NEC 250.134(B)
• NEC 210.5(B)
• NEC 250.119(A)

EGC SLOT

GREEN TERMINAL

EXAMPLES OF IDENTIFYING GROUNDED AND GROUNDING CONDUCTORS

**USE AND IDENTIFICATION OF GROUNDED AND GROUNDING CONDUCTORS
OSHA 1926.404(a)**

POLARITY OF CONNECTIONS

Conductors shall be connected to terminals, lugs or leads that have the same color codings or markings. This rule shall be applied in all electrical installations to prevent the polarity of the electrical circuit from being reversed.

For example: If the ungrounded (phase) conductor was connected by accident to the white lead (grounded (neutral) conductor) of a light socket, the shell of the socket would be energized with a ungrounded (phase) conductor. An employee changing a light bulb could receive a fatal electric shock.

Grounding terminals on equipment shall not be permitted for any purpose other than grounding.

BRANCH CIRCUITS
OSHA 1926.404(b)

The following shall be considered for wiring design and protection of branch circuits:

- Ground-fault protection
- Ground-fault circuit-interrupters
- Assured equipment grounding conductor program
- Outlet devices

GROUND-FAULT PROTECTION

OSHA Required Training
1926.404(b)(iii)(B)

OSHA ground-fault protection rules and regulations have been determined necessary and appropriate for employee safety and health. Therefore, it is the employer's responsibility to provide either:

- Ground-fault circuit interrupters on construction sites for receptacle outlets in use and not part of the permanent wiring of the building or structure; or

Safety Tip: All receptacle outlets on construction sites that are not a part of the permanent wiring of the building shall be provided with GFCI's.

- A scheduled and recorded assured equipment grounding conductor program on construction sites, covering all cord sets, receptacles that are not part of the permanent wiring of the building or structure and equipment connected by cord and plug that are available for use or used by employees. **(See Figures 11-25(a) and (b))**

GROUND-FAULT CIRCUIT INTERRUPTERS

The employer shall provide approved ground-fault circuit interrupters for all 120-volt, single-phase, 15- and 20-ampere receptacle outlets on construction sites that are not a part of the permanent wiring of the building or structure and that are in use by employees. Receptacles on the ends of extension cords are not part of the permanent wiring and, therefore, shall be protected by GFCIs whether or not the extension cord is plugged into permanent wiring. These GFCIs monitor the current-to-the-load for leakage to ground. When this leakage exceeds 5 mA _+ 1 mA, the GFCI interrupts the current. They are rated to trip quickly enough to prevent electrocution. This protection is required in addition to, not as a substitute for, the grounding requirements of OSHA safety and health rules and regulations, 29 CFR 1926. The requirements that employers shall meet, if they choose the GFCI option, are stated in 29 CFR 1926.404(b)(1)(ii).

ASSURED EQUIPMENT GROUNDING CONDUCTOR PROGRAM

The assured equipment grounding conductor program covers all cord sets, receptacles that are not a part of the permanent wiring of the building or structure, and equipment connected by cord and plug that are available for use or used by employees. The requirements that the program shall meet are stated in 29 CFR 1926.404(b)(1)(iii), but employers may provide additional tests or procedures. OSHA requires that a written description of the employer's assured equipment grounding conductor program, including the specific procedures adopted, be kept at the jobsite. This program should outline the employer's specific procedures for the required equipment inspections, tests and test schedule.

The required tests shall be recorded, and the record maintained until replaced by a more current record. The written program description and the recorded tests shall be made available, at the jobsite, to OSHA and to any affected employee upon request. The employer shall designate one or more competent persons to implement the program.

Figure 11-25(a). Circuits supplying extension cord sets shall be protected by a GFCI.

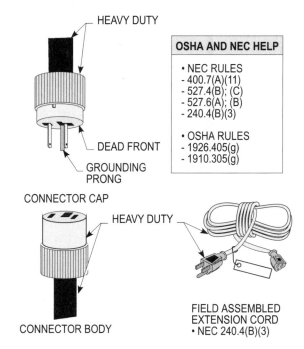

GFCI PROTECTION FOR EXTENSION CORD
OSHA 1926.404(b)(1)(i)

Figure 11-25(b). Circuits supplying electric hand tools used by employees performing construction work shall be protected by a GFCI.

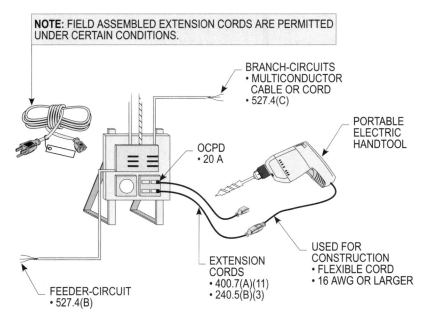

GFCI PROTECTION OF HAND TOOLS
OSHA 1926.404(b)(1)(i)

Electrical equipment noted in the assured equipment grounding conductor program shall be visually inspected for damage or defects before each day's use. Any damaged or defective equipment shall not be used by the employee until repaired.

Two tests are required by OSHA. One is a continuity test to ensure that the equipment grounding conductor is electrically continuous. It shall be performed on all cord sets, receptacles that are not part of the permanent wiring of the building or structure and on cord- and plug-connected equipment that is required to be grounded. This test may be performed using a simple continuity tester, such as a lamp and battery, a bell and battery, an ohmmeter or a receptacle tester.

The other test shall be performed on receptacles and plugs to ensure that the equipment grounding conductor is connected to its proper terminal. This test can be performed with the same equipment used in the first test.

These tests are required before first use, after any repairs, after damage is suspected to have occurred, and at 3-month intervals. Cord sets and receptacles that are essentially fixed and not exposed to damage shall be tested at regular intervals not to exceed 6 months. Any equipment that fails to pass the required tests shall not be made available or used by employees. **(See Figure 11-26)**

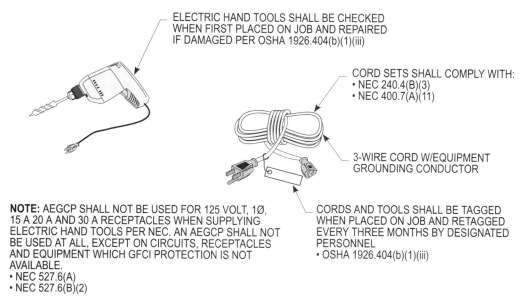

ELECTRIC HAND TOOLS SHALL BE CHECKED WHEN FIRST PLACED ON JOB AND REPAIRED IF DAMAGED PER OSHA 1926.404(b)(1)(iii)

CORD SETS SHALL COMPLY WITH:
• NEC 240.4(B)(3)
• NEC 400.7(A)(11)

3-WIRE CORD W/EQUIPMENT GROUNDING CONDUCTOR

CORDS AND TOOLS SHALL BE TAGGED WHEN PLACED ON JOB AND RETAGGED EVERY THREE MONTHS BY DESIGNATED PERSONNEL
• OSHA 1926.404(b)(1)(iii)

NOTE: AEGCP SHALL NOT BE USED FOR 125 VOLT, 1Ø, 15 A 20 A AND 30 A RECEPTACLES WHEN SUPPLYING ELECTRIC HAND TOOLS PER NEC. AN AEGCP SHALL NOT BE USED AT ALL, EXCEPT ON CIRCUITS, RECEPTACLES AND EQUIPMENT WHICH GFCI PROTECTION IS NOT AVAILABLE.
• NEC 527.6(A)
• NEC 527.6(B)(2)

AEGCP
OSHA 1926.404(B)(1)(iii)

Figure 11-26. Under certain conditions of use, the assured equipment grounding conductor program shall be permitted for construction site circuitry.

OUTLET DEVICES

Outlet devices shall have an ampere rating not less than the load to be served. Outlet devices shall be sized and selected based on the following requirements:

• Single receptacle installed on an individual branch circuit shall have an amp rating of not less than that of the branch-circuit. **(See Figure 11-27)**

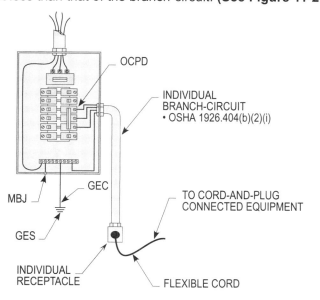

OCPD

INDIVIDUAL BRANCH-CIRCUIT
• OSHA 1926.404(b)(2)(i)

TO CORD-AND-PLUG CONNECTED EQUIPMENT

GEC

MBJ

GES

INDIVIDUAL RECEPTACLE

FLEXIBLE CORD

INDIVIDUAL BRANCH CIRCUIT AND RECEPTACLE
OSHA 1926.404(b)(2)(ii)

Figure 11-27. A single outlet on an individual branch-circuit shall have an amp rating equal to such circuit.

• Where connected to a branch-circuit supplying two or more receptacles or outlets, a receptacle shall not supply a total cord-and-plug connected load in excess of the maximum specified in Table K-4. **(See Figure 11-28)**

Figure 11-28. When there are receptacles and other outlets on the same branch-circuit, the total cord-and-plug connected load shall not be in excess of the maximum specified in Table K-4.

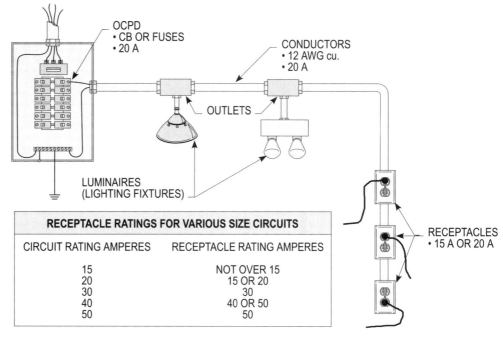

RECEPTACLE RATINGS FOR VARIOUS SIZE CIRCUITS	
CIRCUIT RATING AMPERES	RECEPTACLE RATING AMPERES
15	NOT OVER 15
20	15 OR 20
30	30
40	40 OR 50
50	50

TWO OR MORE RECEPTACLES OR OUTLETS ON A BRANCH CIRCUIT
OSHA 1926.404(b)(2)(ii)
OSHA TABLE K-4

• The rating of an attachment plug or receptacle used for cord-and-plug connection of a motor to a branch circuit shall not exceed 15 amps at 125 volts or 10 amps at 250 volts if individual overload protection is omitted. **(See Figure 11-29)**

Figure 11-29. This illustration shows the cord-and-plug connected requirements for a motor.

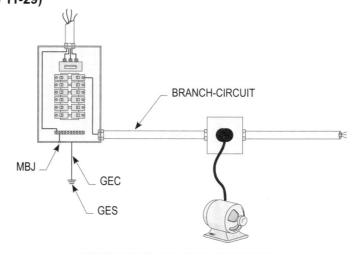

MOTOR CORD-AND-PLUG CONNECTED
OSHA 1926.404(b)(2)(iii)

OUTSIDE CONDUCTORS AND LAMPS <600 VOLTS
OSHA 1926.404(c)

Conductor wiring runs outside shall have a minimum spacing between individual conductors. Proper distance between the spacing of the insulator supports on the surface of the building or pole shall be measured and maintained to provide safe clearances between each conductor in the run span.

CONDUCTORS ON POLES

Conductors supported on poles shall be supported with horizontal clearances that will provide climbing space. This climbing space is required to allow the workman access to service conductor connections, supports, tighten span, etc. A horizontal climbing space shall be provided, where conductors are mounted on poles, as follows:

- Power conductors, below communication conductors .... 30 in.
- Power conductors alone or above communication conductors:
    (a) Less than 300 volts ....24 in.
    (b) Exceeding 300 volts ....30 in.
- Communication conductors below power conductors .....same as power conductors

See Figure 11-30 for a detailed illustration pertaining to conductors on poles.

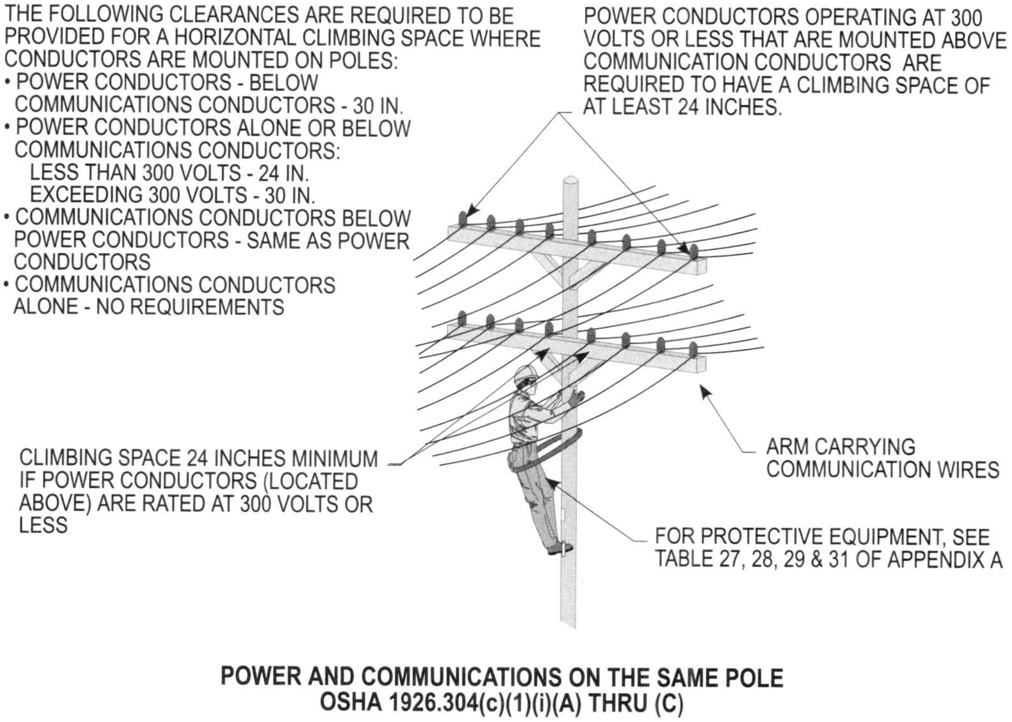

THE FOLLOWING CLEARANCES ARE REQUIRED TO BE PROVIDED FOR A HORIZONTAL CLIMBING SPACE WHERE CONDUCTORS ARE MOUNTED ON POLES:
- POWER CONDUCTORS - BELOW COMMUNICATIONS CONDUCTORS - 30 IN.
- POWER CONDUCTORS ALONE OR BELOW COMMUNICATIONS CONDUCTORS:
    LESS THAN 300 VOLTS - 24 IN.
    EXCEEDING 300 VOLTS - 30 IN.
- COMMUNICATIONS CONDUCTORS BELOW POWER CONDUCTORS - SAME AS POWER CONDUCTORS
- COMMUNICATIONS CONDUCTORS ALONE - NO REQUIREMENTS

POWER CONDUCTORS OPERATING AT 300 VOLTS OR LESS THAT ARE MOUNTED ABOVE COMMUNICATION CONDUCTORS ARE REQUIRED TO HAVE A CLIMBING SPACE OF AT LEAST 24 INCHES.

CLIMBING SPACE 24 INCHES MINIMUM IF POWER CONDUCTORS (LOCATED ABOVE) ARE RATED AT 300 VOLTS OR LESS

ARM CARRYING COMMUNICATION WIRES

FOR PROTECTIVE EQUIPMENT, SEE TABLE 27, 28, 29 & 31 OF APPENDIX A

POWER AND COMMUNICATIONS ON THE SAME POLE
OSHA 1926.304(c)(1)(i)(A) THRU (C)

Figure 11-30. This illustration shows safe climbing spaces for power conductors located over communications conductors.

CLEARANCES FROM GROUND

Conductors routed overhead outside must be installed to protect personnel from contacting them. They must provide enough height above ground, grade, platforms, etc. to keep personnel from touching them. There are two requirements that determine the minimum height. Are there people under the overhead conductor or vehicles, trucks, etc. Each situation requires a different height based on voltage-to-ground.

A minimum of 10 ft. is required above finished grade, sidewalks, or from any platform or projection where overhead open conductors could be contacted. This clearance pertains to the conductors where only persons having access under the span of overhead conductors are involved.

Overhead conductors must be installed at a height of at least 12 ft. in areas that are subject to vehicular traffic other than truck travel. Such travel is cars, pickups, and similar height in small vehicles.

Safety Tip: Outside overhead conductors shall be installed to protect personnel from contactin them.

Where truck travel is present, overhead conductors are to be installed at a height of at least 15 ft. This allows trucks with loads to have more height to pass under without pulling the overhead conductors down.

A height of 18 ft. is required for overhead conductors passing over public streets, alleys, roads and driveways. **(See Figure 11-31)**

Figure 11-31. The chart in the illustration shows the clearances from ground for overhead conductors based on which condition is underneath.

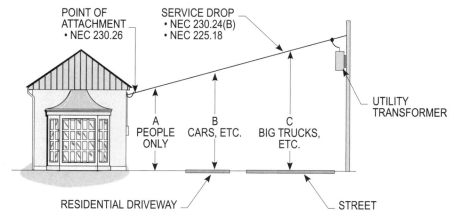

NOMINAL VOLTAGE	SERVICE ENTRANCE (A)	RESIDENTIAL PROPERTY (B)	PUBLIC STREETS (C)
0 - 150 V	10 FT.	12 FT.	18 FT.
151 - 300 V	12 FT.	12 FT.	18 FT.
301 - 600 V	15 FT.	15 FT.	18 FT.

CLEARANCES FROM GROUND OF OVERHEAD CONDUCTORS OSHA 1926.404(c)(ii)(A) THRU (C)

CLEARANCE FROM BUILDING OPENINGS

Conductors are required to have a clearance of least 3 ft. from windows, porches, platforms, etc. Branch-circuit, feeder, and service conductors are out of reach where they are attached above the window. When a clearance of 3 ft. is maintained, the conductors are difficult to reach and are considered to be at a safe distance. **(See Figures 11-32(a) and (b))**

CLEARANCE OVER ROOFS

Safety Tip: Because workmen walking on roofs could accidentally contact overhead conductors, it is important to maintain conductor clearances over flat roofs.

It is extremely important to maintain the clearances of conductors crossing over flat roofs. Workmen walking on the roof to location of equipment requiring servicing could accidentally contact overhead conductors if safe heights are not provided. Overhead conductors are required to have a clearance of at least 8 ft. for the safety of personnel walking, repairing, or servicing equipment on a flat roof. The clearance is 18 ft. where vehicle traffic has access to a flat roof such as the top floor of a garage. **(See Figure 11-33)**

The minimum clearance for insulated conductors is 8 ft., for covered conductors it is 10 ft., and for bare conductors it is 15 ft., except that:

Conductors with a voltage between conductors of 300 volts or less may pass 3 ft. over roofs that have a slope of not less than 4 in. x 12 in. **(See Figure 11-34)**

Conductors may be attached to a conduit system extending through the roof provided conductors do not pass over more than 4 ft. of the roof. The attachment of the conductor to the conduit must be at least 18 in. from the roof line measured vertically. **(See Figure 11-35)**

LOCATIONS OF OUTDOOR LAMPS

In some types of outdoor lighting it would be difficult to keep all electrical equipment above the lamps, and hence a lockable disconnecting means may be required. A disconnecting means should be provided for the equipment on each individual pole,

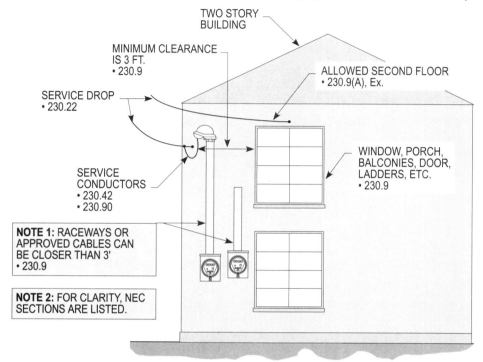

CLEARANCE FROM BUILDING OPENINGS
OSHA 1926.404(c)(1)(iii)

Figure 11-32(a). During construction (or permanent installations), conductors shall have a clearance of 3 ft. maintained from windows designed to be opened, including porches, platforms, etc.

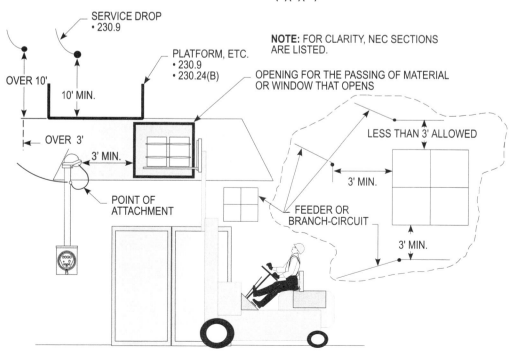

CLEARANCE FROM BUILDING OPENINGS
OSHA 1926.404(c)(1)(iii)

Figure 11-32(b). During construction stages (not necessarily permanent), conductors from building openings such as for storage shall have a clearance of 3 ft.

Figure 11-33. During construction operations, service-drop conductors shall have minimum clearances that comply with OSHA 1926.404(c)(1)(iv) where they are installed above roosf subject to accessibility.

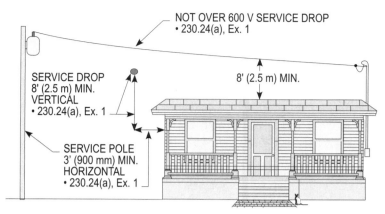

ROOFS SUBJECT TO PEDESTRIAN TRAVEL
OSHA 1926.404(c)(1)(iv)

Figure 11-34. Service-drop conductors shall be permitted to pass above roof where voltages between conductors does not exceed 300 volts and are installed at a height of 3 ft. (900 mm), where the roof has a slope of at least 4 in. (100 mm) by 12 in. (300 mm).

NOTE 1: SERVICE-DROP CONDUCTORS ARE PERMITTED TO PASS ABOVE ROOFS WHERE THE VOLTAGE BETWEEN CONDUCTORS DOES NOT EXCEED 300 VOLTS AND ARE INSTALLED AT A HEIGHT OF 3 FT., WHERE THE ROOF HAS A SLOPE OF AT LEAST 4 IN. x 12 IN.

NOTE 2: SERVICE-DROP CONDUCTORS ARE REQUIRED TO BE INSTALLED AT A CLEARANCE OF 8 FT. WHEN THE VOLTAGE BETWEEN CONDUCTORS EXCEEDS 300 VOLTS, REGARDLESS OF THE SLOPE OF THE ROOF.

NFPA 70E - PART I, SEC. 2-3.4.1, Ex. 2
OSHA 1910.304(c)(4)(i)

Figure 11-35. Service-drop conductors shall be permitted to be installed for roof overhangs provided no more than 6 ft. (1.8 m) of such conductors does not pass over more than 4 ft. (1.2 m) of the roof.

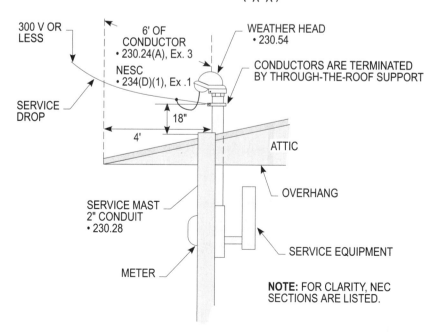

OVERHANG PORTION OF THE ROOF
OSHA 1926.404(c)(1)(iv)

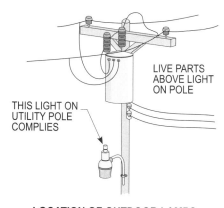

LIVE PARTS ABOVE LIGHT ON POLE

THIS LIGHT ON UTILITY POLE COMPLIES

LOCATION OF OUTDOOR LAMPS
OSHA 1926.404(c)(iv)(2)

Figure 11-36. Lamps in fixtures that are installed outdoors shall be located below energized (live) conductors and equipment.

tower, or other structure where the conditions are such that lamp replacements can be necessary while lighting systems are in use. Other adequate clearances or safeguards are permitted.

Lamps for outdoor lighting shall be located below energized (live) conductors, transformers, and other electrical equipment. This is for the safety of personnel replacing lamps in and around wiring systems and equipment. **(See Figure 11-36)**

SERVICES
OSHA 1926.404(d)

Services consist of the conductors and equipment that delivers energy from the electrical supply systems to the wiring systems of the premises served. The disconnecting means must be located in a readily accessible location and be identified as a disconnect, that will disconnect the service-entrance conductors.

DISCONNECTING MEANS

The service disconnecting means has to be installed in a readily accessible location, either outside of a building or structure or as close as possible to the point where the service entrance conductors enter the premise. The service disconnecting means is required to disconnect all the ungrounded (hot) service entrance conductors supplying power to the service equipment. This disconnecting means must plainly indicate that it is either in the open or closed position. The handle of the main circuit breaker or switch must be marked (off) in the open position and marked (on) in the closed position to qualify as the service disconnecting means. **(See Figure 11-37)**

Safety Tip: Service entrance conductors shall have a disconnecting means installed in a readily accessible location.

SIMULTANEOUS OPENING OF POLES

All ungrounded service conductors must be simultaneously disconnected by each service equipment disconnecting means. The service disconnecting means can be one disconnect located in one single enclosure or two to six disconnects that are installed in one enclosure. It's not uncommon to see two to six disconnects mounted below a meter base and an auxiliary gutter. It doesn't matter how the disconnects are arranged and installed as long as they are grouped in one location. **(See Figures 11-38(a) and (b))**

SERVICES OVER 600 VOLTS

Wiring methods and elements supplied with systems rated over 600 volts have more stringent installation rules than components served by 600 volts or less. The stricter requirements are to provide better protection for personnel from the higher voltages.

Figure 11-37. The service disconnecting means (disconnect) shall be located in a readily accessible location for all dwelling units.

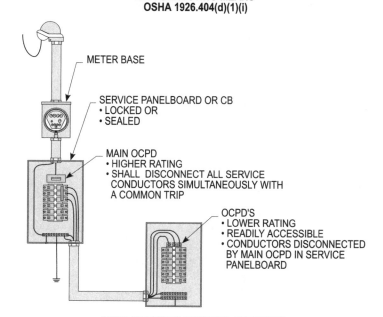

SERVICE DROP INCLUDES SPLICES

SERVICE-ENTRANCE CONDUCTORS IN CONDUIT

SERVICE DISCONNECT
• 230.70(A)
• 230.79

READILY ACCESSIBLE LOCATION

FEEDER CONDUCTORS IN SE CABLE

MAIN LUGS ONLY
• 408.16(A)

SERVICE DISCONNECT INSTALLED WHERE SERVICE RACEWAY ENTERS BUILDING THROUGH WALL
• 230.70(a)

READILY ACCESSIBLE LOCATION

COMBINATION METER WITH SERVICE DISCONNECT
• 230.70(A)
• 230.79
• 408.16(A)

DISCONNECTING MEANS
OSHA 1926.404(d)(1)(i)

Figure 11-38(a). The disconnecting means (main OCPD) shall disconnect all service-entrance conductors from the power supply simultaneously.

METER BASE

SERVICE PANELBOARD OR CB
• LOCKED OR
• SEALED

MAIN OCPD
• HIGHER RATING
• SHALL DISCONNECT ALL SERVICE CONDUCTORS SIMULTANEOUSLY WITH A COMMON TRIP

OCPD'S
• LOWER RATING
• READILY ACCESSIBLE
• CONDUCTORS DISCONNECTED BY MAIN OCPD IN SERVICE PANELBOARD

SIMULTANEOUS OPENING OR POLES
OSHA 1926.404(d)(1)(ii)

Figure 11-38(b). No more than six fusible switches or circuit breakers mounted in a single enclosure, a group of separate enclosures or in or on a panelboard or switchboard shall be installed for each set of service-entrance conductors.

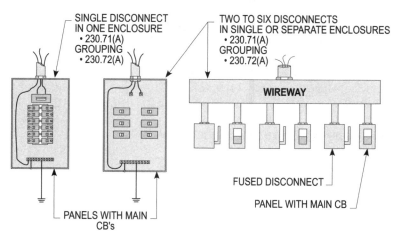

SINGLE DISCONNECT IN ONE ENCLOSURE
• 230.71(A) GROUPING
• 230.72(A)

TWO TO SIX DISCONNECTS IN SINGLE OR SEPARATE ENCLOSURES
• 230.71(A) GROUPING
• 230.72(A)

WIREWAY

FUSED DISCONNECT

PANEL WITH MAIN CB

PANELS WITH MAIN CB's

GROUPED IN ONE LOCATION
OSHA 1926.404(d)(1)(i); (d)(1)(ii)

GUARDING

Only qualified persons are allowed to service open conductors used as service entrance conductors. The open service conductors are required to be guarded so that only qualified persons and not the general public or untrained personnel have access to them. Fences or other suitable barriers are considered suitable means to isolate the conductors from unauthorized personnel. **(See Figure 11-39)**

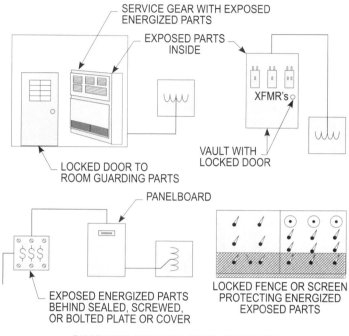

Figure 11-39. Live parts of electrical equipment rated over 600 volts shall be protected by a suitable guarding means.

SERVICE GEAR WITH EXPOSED ENERGIZED PARTS

EXPOSED PARTS INSIDE

XFMR's

VAULT WITH LOCKED DOOR

LOCKED DOOR TO ROOM GUARDING PARTS

PANELBOARD

EXPOSED ENERGIZED PARTS BEHIND SEALED, SCREWED, OR BOLTED PLATE OR COVER

LOCKED FENCE OR SCREEN PROTECTING ENERGIZED EXPOSED PARTS

GUARDING SYSTEMS OVER 600 VOLTS
OSHA 1926.404(d)(2)(i)

WARNING SIGNS

Live parts of electrical systems rated over 600 volts are required to be posted with warning signs to prevent unauthorized persons from coming in contact with such parts. High-voltage systems are very dangerous due to the threat of arcing or flashing across and making contact with grounded objects such as a person's body. When an individual is shocked with high-voltage, there is usually an explosion at the point the current leaves the body. **(See Figures 11-40(a) and (b))**

Safety Tip: Warning signs shall be posted to prevent unauthorized persons from coming into contact with live parts of electrical systems rated at or over 600 volts.

OVERCURRENT PROTECTION 600 VOLTS NOMINAL OR LESS
OSHA 1926.404(e)

The following shall be considered for overcurrent protection 600 volts nominal or less:

- Protection of conductors and equipment
- Grounded conductors
- Disconnection of fuses and thermal cutouts
- Arcing or suddenly moving parts
- Circuit breakers
- Over 600 volts nominal

Figure 11-40(a). This illustration shows the types of warning signs per Z 535 standard, National Electrical Code and National Electrical Safety Code.

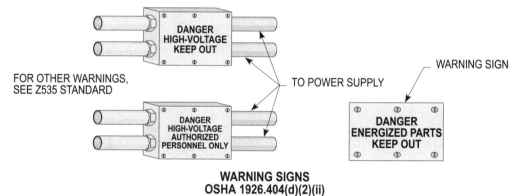

WARNING SIGNS
OSHA 1926.404(d)(2)(ii)

Figure 11-40(b). This illustration shows warning signs alerting employees about high-voltage electrical systems per National Electrical Code requirements.

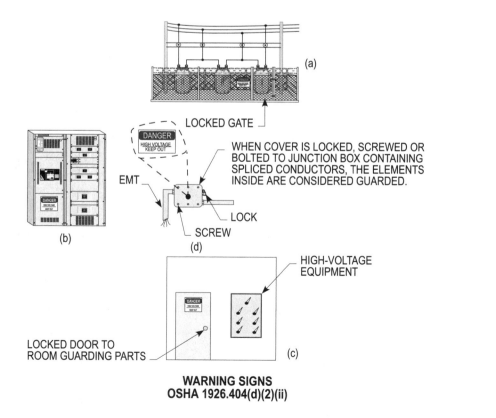

WARNING SIGNS
OSHA 1926.404(d)(2)(ii)

PROTECTION OF CONDUCTORS AND EQUIPMENT

Overcurrent protection devices shall be sized and selected to protect conductors and equipment per NEC 210.20 and NEC 240.1 and 2. Overcurrent protection devices must not allow current to flow through conductors and equipment that will exceed their current-carrying ability. A 20 amp overcurrent protection device protecting a 12 AWG THWN copper conductor must trip if the current flow exceeds 20 amps.

GROUNDED CONDUCTORS

The NEC does not allow an overcurrent protection device to be placed in a grounded (neutral) conductor unless the overcurrent protection device operates to open the ungrounded (phase) conductors simultaneously with the grounded (neutral) conductor. In other words, a fuse shall not be permitted in the grounded (neutral) conductor because the neutral fuse might blow with the ungrounded (phase) conductor fuses not blowing. Then the grounded (neutral) conductor would be open while the ungrounded (phase) conductors are still alive, and this creates a hazard. Only one exception is made, which permits a fuse to be placed in a grounded (neutral) conductor. This is for running overload protection of a three-wire, three-phase circuit with one leg grounded. To provide running overload protection, a fuse is required in the grounded (neutral) conductor where fuses are used in the ungrounded (phase) conductors. **(See Figure 11-41)**

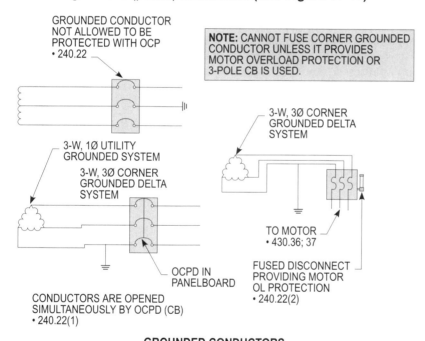

Figure 11-41. The grounded (neutral) conductor shall not be permitted to be protected by a CB unless it opens simultaneously (common trip) with all circuit conductors. To be fused, it shall provide overload protection to a motor.

GROUNDED CONDUCTORS
OSHA 1926.404(e)(1)(ii)

DISCONNECTION OF FUSES AND THERMAL CUTOUTS

Each set of cartridge fuses shall have an individual disconnect on the supply side, so that any one set of fuses can be individually disconnected from the load. For other types of fuses and for thermal cutouts, the rule applies only for voltage of over 150 volts-to-ground. **(See Figure 11-42)**

Where the fuses or cutouts are accessible only to qualified persons, such as electrical workers or maintenance electricians, these rules do not apply. One disconnect shall be permitted to serve a number of fused circuits; a disconnect for each circuit is not required.

Figure 11-42. A single disconnect can deenergize a single electric apparatus, or a single disconnect shall be permitted to serve more than one motor or heating unit protected by more than one set of fuses.

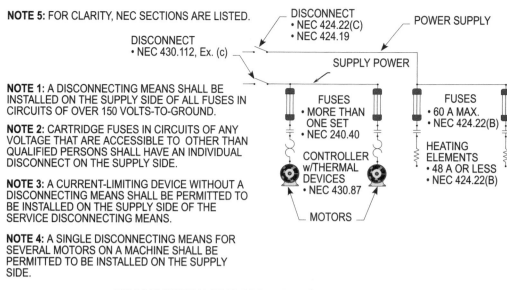

NOTE 5: FOR CLARITY, NEC SECTIONS ARE LISTED.

NOTE 1: A DISCONNECTING MEANS SHALL BE INSTALLED ON THE SUPPLY SIDE OF ALL FUSES IN CIRCUITS OF OVER 150 VOLTS-TO-GROUND.

NOTE 2: CARTRIDGE FUSES IN CIRCUITS OF ANY VOLTAGE THAT ARE ACCESSIBLE TO OTHER THAN QUALIFIED PERSONS SHALL HAVE AN INDIVIDUAL DISCONNECT ON THE SUPPLY SIDE.

NOTE 3: A CURRENT-LIMITING DEVICE WITHOUT A DISCONNECTING MEANS SHALL BE PERMITTED TO BE INSTALLED ON THE SUPPLY SIDE OF THE SERVICE DISCONNECTING MEANS.

NOTE 4: A SINGLE DISCONNECTING MEANS FOR SEVERAL MOTORS ON A MACHINE SHALL BE PERMITTED TO BE INSTALLED ON THE SUPPLY SIDE.

DISCONNECTION OF FUSES AND THERMAL CUTOUTS
OSHA 1926.404(e)(1)(iii)

ARCING OR SUDDENLY MOVING PARTS

Where overcurrent protection devices are installed it is essential that care be taken to prevent the possible injury to personnel by the flashing, arcing or sudden movement of parts. If a large overcurrent protection device should unexpectedly open up, the flash or arc can be a source of danger to the person serving the equipment. Overcurrent protection devices (CB) with internal contacts that open an overload without moving the circuit breaker handle should be utilized. This type of circuit breaker with a trip free handle eliminates the danger of suddenly moving parts creating arcs that can burn or blind maintenance personnel while servicing or repairing electrical equipment. **(See Figure 11-43)**

Figure 11-43. Enclosures housing electrical parts that can arc or suddenly move shall be installed to provide safety.

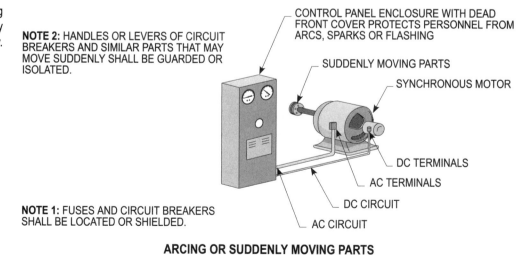

NOTE 2: HANDLES OR LEVERS OF CIRCUIT BREAKERS AND SIMILAR PARTS THAT MAY MOVE SUDDENLY SHALL BE GUARDED OR ISOLATED.

NOTE 1: FUSES AND CIRCUIT BREAKERS SHALL BE LOCATED OR SHIELDED.

ARCING OR SUDDENLY MOVING PARTS
OSHA 1926.404(e)(1)(v)

CIRCUIT BREAKERS

Any and all circuit breakers with their various ratings shall be capable of opening (off) and closing (on) by hand. Operation by hand simply means that the circuit breaker has a handle that the operator can grasp and open or close the circuit. Circuit breakers that are pneumatic or electrically controlled can become inoperative, or the contacts might stick in the make position and without a handle to deenergize the circuit breaker, there would be no way to open the circuit. Should the electrical or pneumatic control fail to operate the circuit, it can be opened manually by hand using the handle on the circuit breaker.

Circuit breakers operated by remote control shall be equipped with a handle so they can be open (off) or closed (on) by hand operation. Circuit breaker handles that operate vertically instead of rotationally or horizontally shall be installed with the on position of the handle in the up position. Circuit breakers used to switch 120 or 277 volt fluorescent lighting circuits shall be marked SWD. **(See Figure 11-44)**

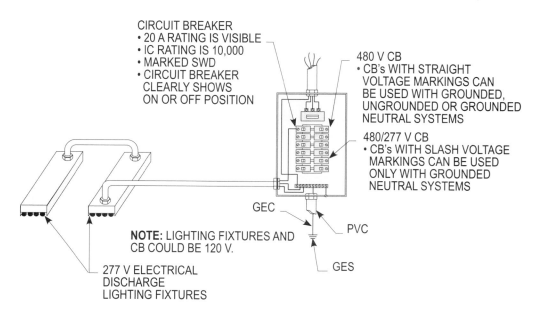

CIRCUIT BREAKER
• 20 A RATING IS VISIBLE
• IC RATING IS 10,000
• MARKED SWD
• CIRCUIT BREAKER CLEARLY SHOWS ON OR OFF POSITION

480 V CB
• CB's WITH STRAIGHT VOLTAGE MARKINGS CAN BE USED WITH GROUNDED, UNGROUNDED OR GROUNDED NEUTRAL SYSTEMS

480/277 V CB
• CB's WITH SLASH VOLTAGE MARKINGS CAN BE USED ONLY WITH GROUNDED NEUTRAL SYSTEMS

GEC

PVC

GES

NOTE: LIGHTING FIXTURES AND CB COULD BE 120 V.

277 V ELECTRICAL DISCHARGE LIGHTING FIXTURES

Figure 11-44. Circuit breakers used as switches shall be marked, switching duty (SWD), either for HID or fluorescent type. Circuit breakers can be turned ON and OFF by a remote button if they are also equipped with a manual control.

OVER 600 VOLTS NOMINAL

Short-circuit protection for high-voltage feeders can be provided using fuses or circuit breakers. Fuses shall be permitted to be sized and selected up to three times the ampacity of the conductor. Circuit breakers shall be permitted to be set up to six times the ampacity of the conductor. These values are not to be applied for overload protection. They are intended to be utilized for short-circuit protection only. Branch-circuits operating on high-voltage systems shall have an overcurrent protection device provided in each ungrounded (phase) conductor. Circuit breakers shall be permitted to be equipped with two overcurrent relays operated by two current transformers installed in each of the two phases. **(See Figure 11-45)**

GROUNDING
OSHA 1926.404(f)

There are three main steps that are required to be applied by designers, installers, and inspectors to ground and bond electrical systems safely. The three steps in grounding a system are as follows:

- Circuit and system grounding,
- Equipment grounding and
- Bonding.

Circuit and system grounding consists of connecting one of the current-carrying conductors of a premises wiring system to ground.

Figure 11-45. This illustration shows maximum overcurrent protection for high-voltage feeder conductors (over 600 volts) using circuit breakers and fuses.

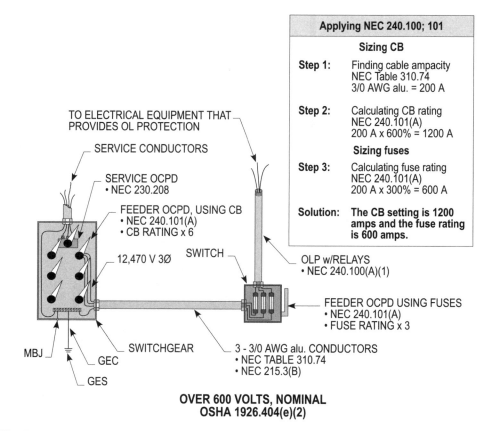

OVER 600 VOLTS, NOMINAL
OSHA 1926.404(e)(2)

Equipment grounding consists of connecting non-current-carrying metal parts of a wiring electrical system to ground. Metal parts include conduit, motor frames, the enclosing metal case of switches, appliances and other equipment. **(See Figure 11-46)**

Figure 11-46. This illustration shows the procedures for connecting the grounded (neutral) conductor and equipment grounding conductor to the grounding electrode conductor.

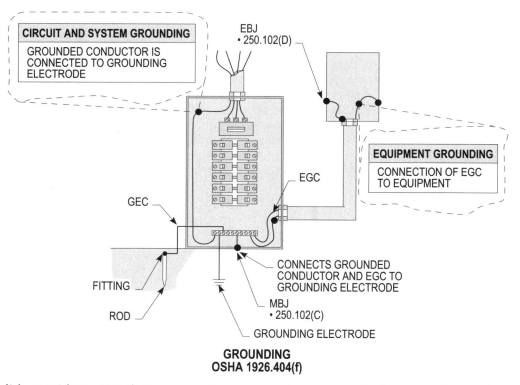

GROUNDING
OSHA 1926.404(f)

It is most important that construction personnel never modify the grounding system once it has been designed, sized and installed. If it becomes necessary to modify the grounding system due to an upgrade in the service, feeder or branch-circuit only qualified personnel should size and install the new grounding system.

SYSTEMS TO BE GROUNDED

All three-wire, DC systems are required to have their grounded (neutral) conductor grounded. A three-wire system allows phase-to-phase voltage and phase-to-ground voltage to be obtained.

All two-wire, DC systems operating over 50 volts through 300 volts or less between conductors shall be grounded except:

- A system equipped with a ground detector and supplying only industrial equipment in limited areas.

- A rectifier-derived DC system supplied from an AC system, which shall be grounded as required by Section 250.5 of the NEC.

- The grounding electrode conductors to earth ground DC systems shall be sized and selected per NEC 250.93. DC fire-protective signaling circuits having a maximum current of .03 amps.

AC circuits of less than 50 volts shall be grounded only if the circuit is derived from a transformer having an ungrounded primary circuit or a primary circuit of over 150 volts-to-ground.

A low-voltage bell ringing circuit in a building supplied from a 120 volt circuit shall not be required to be grounded, since the circuit is derived from a transformer with a grounded primary circuit of less than

150 volts-to-ground. Low-voltage AC circuits routed outdoors and overhead shall be grounded in all types of installations. If the low-voltage bell ringing circuit is derived from a transformer whose primary supply is an ungrounded circuit, the secondary low-voltage circuit shall be grounded. Transformers with primaries supplied by a 277 volt circuit shall have the secondary low-voltage circuit grounded. The reason for this is that the primary supply is over 150 volts-to-ground.

All AC systems of 50 to 1000 volts-to-ground shall be grounded as follows:

- The maximum voltage to ground on the ungrounded (phase) conductors does not exceed 150 volts. A 120/240 volt, three-wire system or a 120/208 volt, four-wire system shall be grounded due to the grounded (neutral) conductor being less than 150 volts-to-ground.

- Wye connected 480/277 volt, three phase, four-wire systems shall be grounded. The voltage-to-ground is 277 volts.

- Three-phase, delta connected, 240 volt systems shall be grounded if one phase has a grounded (neutral) conductor. In this case the grounded (neutral) conductor of the one phase would be grounded with a voltage-to-ground of less than 150 volts.

- Any uninsulated service conductor shall be grounded. In this case, the grounded (neutral) conductor is bare in junction boxes, at the weatherhead, in the meter base, service equipment, etc.

PORTABLE AND VEHICLE-MOUNTED GENERATORS

Portable Generators. The frame of a portable generator shall not be required to be grounded if the generator supplies only equipment mounted on the generator, or if the generator supplies cord-and-plug connected equipment through receptacles mounted on the generator, provided they are as follows:

- The equipment grounding conductor for the receptacles is bonded to the generator frame, and

- Exposed metal parts of the equipment served are bonded to the generator frame. This requires the use of grounding type receptacles for any equipment that is cord-and-plug connected. The exposed metal parts of equipment is bonded to the frame through the equipment grounding conductor in the cord.

Vehicle-Mounted Generators. The frame of the vehicle shall be permitted to be used to ground a circuit supplied by a generator on the vehicle provided it complies with the following:

- The generator frame is bonded to the vehicle frame,

- The generator serves equipment mounted on the vehicle, or if the generator supplies cord-and-plug connected equipment through receptacles mounted on the vehicle, and

- Exposed metal parts of the equipment served are bonded to the generator frame either directly or through the receptacles.

Neutral Conductor Bonding. If the generator is a component of a separately derived system, a grounded (neutral) conductor shall be bonded to the generator frame.

GROUNDING CONNECTIONS

Grounding connections shall be made at the service equipment grounding bar and each grounding electrode in the grounding electrode system. These connections shall be made proper with fittings designed for the purpose. The connection points shall be clean, with fittings and conductors tightened wrench-tight for good continuity. The connection of the grounding electrode conductor to the grounding electrode shall be installed in a place that is accessible where the electrode is a metallic water pipe or building steel. Other electrodes and connections shall be permitted to be inaccessible if they are buried in the ground or slab foundation. **(See Figure 11-47)**

For a grounded system, a grounding electrode conductor shall be utilized to connect the equipment grounding conductor and the grounded circuit conductor (neutral) to the grounding electrode. The equipment grounding conductor and grounding electrode conductor shall be connected to the grounded circuit conductor on the supply side of the service disconnecting means or on the supply line side of the system disconnecting means or overcurrent protection devices if the system is separately derived. **(See Figure 11-48)**

UNGROUNDED SYSTEMS

For an ungrounded service-supplied system from an ungrounded transformer, the equipment grounding conductor shall be terminated to the grounding electrode conductor at the service equipment grounding terminal bar. For an ungrounded separately derived system, the equipment grounding conductor shall be terminated to the grounding electrode conductor at or ahead of the system disconnecting means or overcurrent protection devices. This connection shall be made at a common grounding terminal bar, etc. in the separately derived system or in the equipment being supplied. **(See Figure 11-48)**

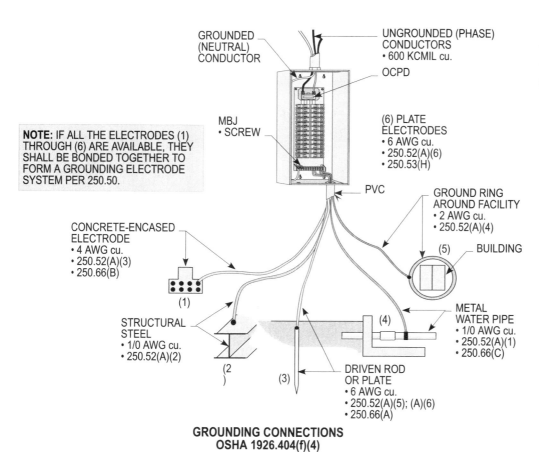

Figure 11-47. This illustration shows the types of grounding electrodes that shall be bonded together to form the grounding electrode system, if available.

NOTE: IF ALL THE ELECTRODES (1) THROUGH (6) ARE AVAILABLE, THEY SHALL BE BONDED TOGETHER TO FORM A GROUNDING ELECTRODE SYSTEM PER 250.50.

GROUNDED (NEUTRAL) CONDUCTOR

UNGROUNDED (PHASE) CONDUCTORS
• 600 KCMIL cu.

OCPD

MBJ
• SCREW

(6) PLATE ELECTRODES
• 6 AWG cu.
• 250.52(A)(6)
• 250.53(H)

PVC

CONCRETE-ENCASED ELECTRODE
• 4 AWG cu.
• 250.52(A)(3)
• 250.66(B)

GROUND RING AROUND FACILITY
• 2 AWG cu.
• 250.52(A)(4)

(5)   BUILDING

(1)

STRUCTURAL STEEL
• 1/0 AWG cu.
• 250.52(A)(2)

(4)

METAL WATER PIPE
• 1/0 AWG cu.
• 250.52(A)(1)
• 250.66(C)

(2)

(3)

DRIVEN ROD OR PLATE
• 6 AWG cu.
• 250.52(A)(5); (A)(6)
• 250.66(A)

GROUNDING CONNECTIONS
OSHA 1926.404(f)(4)

Figure 11-48. This illustration shows the procedures for installing a grounded or ungrounded electrical system.

SERVICE EQUIPMENT

GROUNDED SERVICE CONDUCTOR

NO GROUNDED SERVICE CONDUCTOR PRESENT

NOTE: AN EGC MAY BE PULLED IN CONDUIT FOR GROUNDING EQUIPMENT

TO ACCESSIBLE MWP OR SBS

EQUIPMENT GROUNDING CONDUCTORS

MAIN BONDING JUMPER

MAIN BONDING JUMPER

GROUNDED SYSTEM

UNGROUNDED SYSTEM

GROUNDED AND UNGROUNDED SYSTEMS
OSHA 1926.404(f)(5)(ii); (5)(iii)

GROUNDING PATH

The path to ground from circuits, equipment and enclosures shall be permanent, continuous and effective. The path to ground mentioned above includes the equipment grounding conductor, which shall be permitted to be a wire, metal conduit or other metal-clad wiring and the grounding electrode conductor.

The path to ground shall be permanent, continuous, of low resistance and be sized with enough current-carrying capacity to safely conduct all fault currents. **(See Figures 11-49(a) and (b))**

Figure 11-49(a). This illustration shows that an effective ground-fault current path from circuits, equipment and conductor enclosures shall be installed in a manner that creates a permanent, low-impedance circuit capable of safely carrying the maximum ground-fault current likely to be imposed on it from any point on the wiring system.

NOTE: FOR CLARITY, NEC SECTIONS ARE LISTED.

SUPPLEMENTARY GROUNDING
• 250.54
• 250.136(A)

MOTOR
• 430.7
• 430.6

DISCONNECT
• 430.102

GROUNDED (NEUTRAL) CONDUCTOR SIZED TO CARRY FAULT-CURRENT
• 250.24(B)(1)
• 250.4(A)(5)

COMPUTER
• 645.15
• 250.146(D)

OCPD
• 200 A

CIRCUITS RUN TOGETHER TO LOWER IMPEDANCE
• 250.4(A)(5)

NEUTRAL BUS
• 408.20

OCPD
• 20 A

CORD-AND-PLUG

MBJ
• 250.102(C)

METAL CONDUIT OR EGC OR BOTH SIZED TO CARRY FAULT
• TABLE 250.122
• 250.4(A)(5)

EBJ
• 250.102(D); (E)

GES
• 250.50
• 250.52(A)(1) - (A)(7)

**EFFECTIVE GROUND FAULT CURRENT PATH
OSHA 1926.404(f)(6)**

Figure 11-49(b). This illustration shows the elements necessary to provide a low-impedance path for the current to flow over to clear a ground-fault condition.

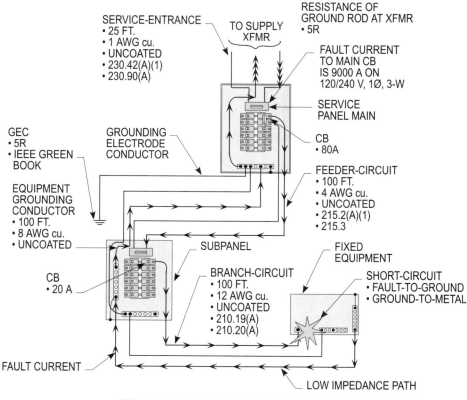

SERVICE-ENTRANCE
• 25 FT.
• 1 AWG cu.
• UNCOATED
• 230.42(A)(1)
• 230.90(A)

TO SUPPLY XFMR

RESISTANCE OF GROUND ROD AT XFMR
• 5R

FAULT CURRENT TO MAIN CB IS 9000 A ON 120/240 V, 1Ø, 3-W

SERVICE PANEL MAIN

GEC
• 5R
• IEEE GREEN BOOK

GROUNDING ELECTRODE CONDUCTOR

CB
• 80A

EQUIPMENT GROUNDING CONDUCTOR
• 100 FT.
• 8 AWG cu.
• UNCOATED

FEEDER-CIRCUIT
• 100 FT.
• 4 AWG cu.
• UNCOATED
• 215.2(A)(1)
• 215.3

SUBPANEL

FIXED EQUIPMENT

CB
• 20 A

BRANCH-CIRCUIT
• 100 FT.
• 12 AWG cu.
• UNCOATED
• 210.19(A)
• 210.20(A)

SHORT-CIRCUIT
• FAULT-TO-GROUND
• GROUND-TO-METAL

FAULT CURRENT

LOW IMPEDANCE PATH

**EFFECTIVE GROUND FAULT CURRENT PATH
OSHA 1926.404(f)(6)**

SUPPORTS, ENCLOSURES AND EQUIPMENT TO BE GROUNDED

Enclosures for enclosing conductors shall be grounded except for the following:

Metal enclosures such as sleeves and similar enclosures used to protect cable assemblies from physical damage shall not be required to be grounded. Installation of short runs used as extensions from existing nonmetallic sheathed cable shall be permitted without grounding where there is little likelihood of an accidental connection to ground or of a person touching the conduit, raceway or armor and any grounded metal or other grounded surface at the same time.

If a short piece of metal conduit, EMT or AC cable is added to a nonmetallic wiring system, such as romex or rope, where it is less than 25 ft. in length and free from probable contact with ground, metal piping or conductive material it shall not be required to be grounded. If the short piece is 25 ft. or more in length, it shall always be grounded. Short pieces of conduit, 10 ft. or less in length, shall not be required to be grounded. Where it is within reach of ground or a grounded object, it shall be grounded regardless of length.

SERVICE EQUIPMENT ENCLOSURES

All metal enclosures enclosing the elements of the service equipment shall be grounded. Metal raceways and metal enclosures utilized to house the elements of the service equipment shall be grounded. Grounding of such items is accomplished by bonding the metal raceway, metal sheath or armor to the grounding terminal bar located in the service entrance cabinet. **(See Figure 11-50)**

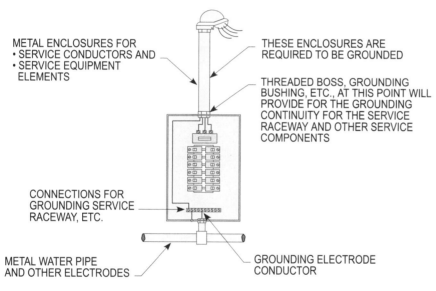

METAL ENCLOSURES FOR
• SERVICE CONDUCTORS AND
• SERVICE EQUIPMENT ELEMENTS

THESE ENCLOSURES ARE REQUIRED TO BE GROUNDED

THREADED BOSS, GROUNDING BUSHING, ETC., AT THIS POINT WILL PROVIDE FOR THE GROUNDING CONTINUITY FOR THE SERVICE RACEWAY AND OTHER SERVICE COMPONENTS

CONNECTIONS FOR GROUNDING SERVICE RACEWAY, ETC.

METAL WATER PIPE AND OTHER ELECTRODES

GROUNDING ELECTRODE CONDUCTOR

**SERVICE EQUIPMENT ENCLOSURES
OSHA 1926.404(f)(7)(ii)**

Figure 11-50. All elements of the service wiring methods and equipment shall be bonded and grounded together with fittings and enclosures approved for such use.

FIXED EQUIPMENT

Examples of fixed equipment are service switches, motor cases, outlet boxes, switch boxes, motor controllers or any other exposed metal parts of electrical equipment. Exposed noncurrent-carrying metal parts of fixed equipment likely to become energized shall be grounded under any of the following conditions:

Safety Tip: All exposed noncurrent-carrying metal parts of fixed equipment shall be grounded.

• Where equipment is within reach of a person touching the equipment and at the same time is in contact with ground or a grounded object. The grounded object can be a water faucet in a bathroom, kitchen or outside. Where the electrical equipment is close to a grounded object (within reach), the equipment shall always be grounded. Exposed metal parts that are within 8 ft. vertically or 5 ft. horizontally of ground or a grounded object are considered to be within reach and shall be grounded.

- Where equipment is located in wet or damp locations and is not isolated, it is considered readily accessible to persons, then regardless of the type of wiring used, such equipment shall be required to be grounded for safety from electrical shock.

- Where metal equipment is mounted on metal or metal lathe, it shall be grounded.

- Where the equipment is located in hazardous (classified) locations.

- Where the wiring method is metal conduit, EMT, AC, MC cable or any other "metal-clad" type of wiring, grounding shall be required.

- Where the voltage is over 150 volts-to-ground, the equipment shall be grounded and bonded. This includes equipment on three-phase, four-wire, 240 volt circuits with a 208 volt high leg, and 480/277 volt circuits as well. Exceptions are as follows:

(a) Switches and circuit breakers not used as service equipment shall not be required to be grounded if accessible only to qualified persons.

(b) Electrically heated appliances shall be permitted to be left ungrounded by special permission. Special permission is the intent not to cover house or commercial heating in general.

(c) Equipment installed on wooden poles and mounted higher than 8 ft. above ground. Here again care must be taken to verify that this equipment is high enough to avoid persons touching it while standing on grounded objects, etc.

EQUIPMENT CONNECTED BY CORD AND PLUG

Under any of the following conditions, exposed noncurrent-carrying metal parts or cord-and-plug connected equipment likely to become energized shall be grounded to protect the user from electrical shock:

- In hazardous (classified) locations, all metal electrical equipment shall be grounded.

- Where operated at over 150 volts-to-ground, except guarded motors and metal frames of electrically heated appliances if the appliance frames are permanently and effectively insulated from ground.

- Handheld electrically motor operated tools.

- Cord-and-plug connected appliances utilized in damp or wet locations or by personnel standing on the ground or on metal floors or working inside of metal tanks or boilers.

- Portable and mobile X-ray and associated equipment.

- Tools likely to be used in wet and conductive locations.

- Portable hand lamps.

- Tools used in wet and conductive locations shall not be required to be grounded if they are supplied through an isolating transformer with an ungrounded secondary 50 volts or less.

• Listed portable tools and appliances protected by an approved system of double insulation or its equivalent, shall not be required to be grounded. Where such a system is utilized, the equipment shall be distinctively marked to indicate that the tool or appliance is using an approved system of double insulation.

NONELECTRICAL EQUIPMENT

The noncurrent-carrying metal parts of the following nonelectrical equipment shall be grounded:

• Operator's cab and tracks of electric cranes.

• Metal frame of an elevator operated nonelectrically where there is wiring to the elevator that is used to supply power to lights or other equipment.

• Shifting cables on electric elevators that are hand operated. Cables that raise and lower the elevator are not meant to be included.

• Metal enclosures around high-voltage systems rated at 1 kV between conductors.

METHODS OF GROUNDING EQUIPMENT

Noncurrent-carrying metal parts of fixed equipment, where required to be grounded, shall be grounded with an equipment grounding conductor that is enclosed within the same raceway, cable or cord as the circuit conductors. For DC circuits, the equipment grounding conductor shall be permitted to be run separately from the circuit conductors.

GROUNDING CONDUCTOR

Grounding conductors shall be sized to conduct safely any fault current that may be imposed on it.

EQUIPMENT CONSIDERED EFFECTIVELY GROUNDED

Electric equipment secured to and in electrical equipment in contact with a grounded metal rack or structure provided for its support shall be considered to be effectively grounded. The structural metal frame of a building shall not be permitted to be used as the equipment grounding conductor for AC electrical equipment. Metal car frames supported by metal hoisting cables connected to or running over metal sleeves or drums of grounded elevator machines shall be considered to be effectively grounded.

BONDING

Bonding shall be provided where necessary to assure electrical continuity and the capacity to conduct safely any fault current likely to be imposed.

"Bonding" means joining together the metal parts of the wiring system in such a manner as to provide good metallic contact between the different pans. This must be done to join the exposed metal parts of the system together in a continuous metallic system, thus assuring a good electrical path to ground throughout the system for boxes, cabinets and other items that need grounding to ensure safety. **(See Figure 11-51)**

Figure 11-51. This illustration shows the bonding of electrical equipment and other electrical conductive materials and parts.

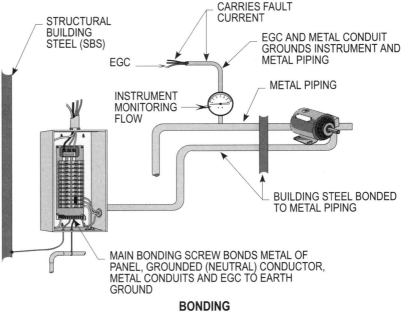

STRUCTURAL BUILDING STEEL (SBS)

CARRIES FAULT CURRENT

EGC AND METAL CONDUIT GROUNDS INSTRUMENT AND METAL PIPING

EGC

METAL PIPING

INSTRUMENT MONITORING FLOW

BUILDING STEEL BONDED TO METAL PIPING

MAIN BONDING SCREW BONDS METAL OF PANEL, GROUNDED (NEUTRAL) CONDUCTOR, METAL CONDUITS AND EGC TO EARTH GROUND

**BONDING
OSHA 1926.404(f)(9)**

MADE ELECTRODES

If made electrodes are used, they shall meet the following requirements:

- They shall be free from nonconductive coatings (paint, enamel).

- If practical, they shall be installed below permanent moisture level.

- A single electrode consisting or a rod, pipe or plate that has a resistance to ground of more than 25 ohms shall be augmented by one additional electrode installed no closer than 6 ft. to the first electrode.

See Figure 11-52 for a detailed illustration pertaining to rod, pipe or plate electrodes.

Figure 11-52. Rod, pipe and plate electrodes shall have a resistance to ground of 25 ohms or less, wherever practicable. When the resistance is greater than 25 ohms, two or more electrodes shall be permitted to be connected in parallel or extended to a greater length.

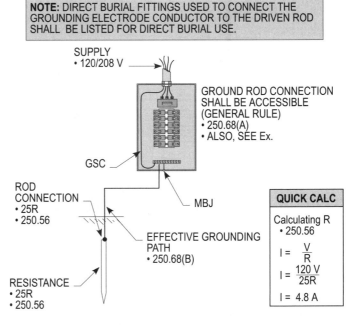

NOTE: DIRECT BURIAL FITTINGS USED TO CONNECT THE GROUNDING ELECTRODE CONDUCTOR TO THE DRIVEN ROD SHALL BE LISTED FOR DIRECT BURIAL USE.

SUPPLY
- 120/208 V

GROUND ROD CONNECTION SHALL BE ACCESSIBLE (GENERAL RULE)
- 250.68(A)
- ALSO, SEE Ex.

GSC

ROD CONNECTION
- 25R
- 250.56

MBJ

EFFECTIVE GROUNDING PATH
- 250.68(B)

RESISTANCE
- 25R
- 250.56

QUICK CALC

Calculating R
- 250.56

$I = \dfrac{V}{R}$

$I = \dfrac{120\ V}{25R}$

$I = 4.8\ A$

**MADE ELECTRODES (RODS, PIPES OR PLATES)
OSHA 1926.404(f)(10)**

GROUNDING OF SYSTEMS SUPPLYING PORTABLE OR MOBILE EQUIPMENT

The following grounding procedures shall be applied when grounding electrical systems rated at 1000 volts and over:

- Portable equipment shall be supplied from a system having an impedance ground.

- Exposed metal parts of portable equipment shall be grounded with an equipment grounding conductor to the point at which the neutral impedance is grounded.

- A ground fault detection system with relaying shall be provided that disconnects a high-voltage system that has developed a ground fault. The detection system and relaying must disconnect the high-voltage circuit to the portable equipment if a break should develop in the equipment grounding conductor.

- The grounding electrode for the high-voltage circuit feeding portable equipment shall be separated at least 20 ft. from any other electrode. There shall not be any connection directly to a buried pipe, fence, etc.

See Figures 11-53(a) and (b) for a detailed illustration pertaining to grounding systems supplying portable or mobile equipment.

GROUNDING OF EQUIPMENT

Any and all noncurrent-carrying metal parts of fixed and portable equipment, including associated fences, housings, etc. shall be grounded. Equipment isolated from ground and located to prevent personnel from making contact with ground and making contact with ground while touching metal parts of equipment that is energized, shall be considered to be guarded. Pole-mounted distribution apparatus that is elevated at least 8 ft. above grade shall be considered grounded.

WIRING METHODS, COMPONENTS AND EQUIPMENT FOR GENERAL USE OSHA 1926.405

The following shall be considered for wiring methods, components and equipment for general use:

- Wiring methods
- Cabinets, boxes and fittings
- Knife switches
- Switchboards and panelboards
- Enclosures for damp or wet locations
- Conductors for general wiring
- Flexible cords and cables
- Portable cables over 600 volts
- Fixture wires
- Equipment for general use

Figure 11-53(a). This illustration shows the procedures for grounding noncurrent-carrying metal parts.

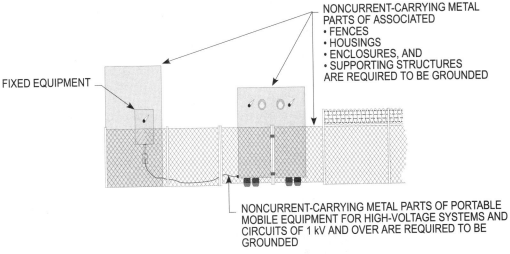

NONCURRENT-CARRYING METAL PARTS OF ASSOCIATED
• FENCES
• HOUSINGS
• ENCLOSURES, AND
• SUPPORTING STRUCTURES
ARE REQUIRED TO BE GROUNDED

FIXED EQUIPMENT

NONCURRENT-CARRYING METAL PARTS OF PORTABLE MOBILE EQUIPMENT FOR HIGH-VOLTAGE SYSTEMS AND CIRCUITS OF 1 kV AND OVER ARE REQUIRED TO BE GROUNDED

GROUNDING PROCEDURES FOR SYSTEMS AND CIRCUITS OF 1000 VOLTS AND OVER OSHA 1926.404(f)(11)(ii)(A) THRU (D)

Figure 11-53(b). This illustration shows the grounding procedures of the supply circuit, noncurrent-carrying parts and methods of installing ground rods.

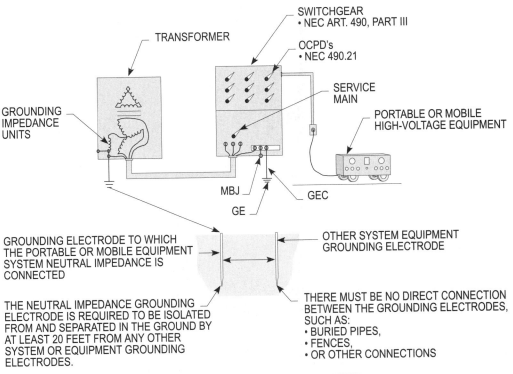

TRANSFORMER

SWITCHGEAR
• NEC ART. 490, PART III

OCPD's
• NEC 490.21

SERVICE MAIN

PORTABLE OR MOBILE HIGH-VOLTAGE EQUIPMENT

GROUNDING IMPEDANCE UNITS

MBJ
GEC
GE

GROUNDING ELECTRODE TO WHICH THE PORTABLE OR MOBILE EQUIPMENT SYSTEM NEUTRAL IMPEDANCE IS CONNECTED

THE NEUTRAL IMPEDANCE GROUNDING ELECTRODE IS REQUIRED TO BE ISOLATED FROM AND SEPARATED IN THE GROUND BY AT LEAST 20 FEET FROM ANY OTHER SYSTEM OR EQUIPMENT GROUNDING ELECTRODES.

OTHER SYSTEM EQUIPMENT GROUNDING ELECTRODE

THERE MUST BE NO DIRECT CONNECTION BETWEEN THE GROUNDING ELECTRODES, SUCH AS:
• BURIED PIPES,
• FENCES,
• OR OTHER CONNECTIONS

GROUNDING OF SYSTEMS SUPPLYING PORTABLE OR MOBILE EQUIPMENT OSHA 1926.404(f)(11)(ii)(A) THRU (D)

WIRING METHODS
OSHA 1926.405(a)

The provisions of this section are not to be applied to the conductors that are an integral part of factory assembled equipment such as motors, controllers, motor controller centers and like equipment. Properly designed electrical systems utilize approved wiring materials. Such materials include various types of conductors, conduit, cable trays, motors, transformers, switches, lighting fixtures, etc. Wiring methods or equipment designed and inspected by the provisions of the OSHA and the NEC assure safety for employees who are around, near or work in places where they are installed.

ELECTRICAL CONTINUITY OF METAL RACEWAYS AND ENCLOSURES

All metal enclosures enclosing the elements of the service equipment shall be grounded. Metal raceways and metal enclosures utilized to house the elements of the service equipment shall be grounded. Grounding of such items is accomplished by bonding the metal raceway, metal sheath or armor to the grounding terminal bar located in the service entrance cabinet.

Where metal raceways or cable armor is used as the grounding means for equipment, care must be exercised to see that there are no parcel breaks, or breaks in the metallic continuity of the metal raceways, cables and enclosures throughout the wiring systems.

Any nonconducting paint, enamel or similar coating shall be removed at threads, contact points and contact surfaces or be connected by means of fittings so designed as to make such removal unnecessary.

The bonding methods between metal raceways and metal enclosures or between sections of such raceways are not specified. Electrical continuity is simply required and shall be assured. **(See Figure 11-54)**

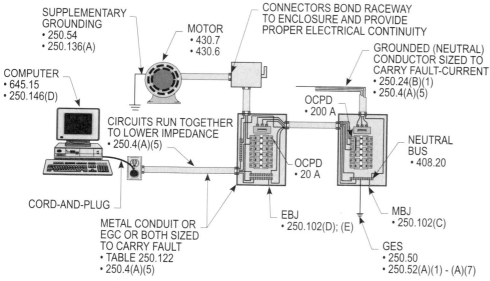

ELECTRICAL CONTINUITY OF METAL RACEWAYS AND ENCLOSURES
OSHA 1926.405(a)(1)(i)

Figure 11-54. An effective ground-fault current path from circuits, equipment and conductor enclosures shall be installed in a manner that creates a permanent, low-impedance circuit capable of safely carrying the maximum ground-fault current likely to be imposed on it from any point on the wiring system.

WIRING IN DUCTS

Wiring methods shall not be installed in ducts that handle dust, loose stock or flammable vapors. No wiring methods of any type shall be installed in any duct used for vapor removal.

TEMPORARY WIRING

Temporary electrical power and lighting wiring methods shall be permitted to be of a class less than required for a hard-wired installation. Except as modified in this section, all other rules of this standard for hard-wired electrical systems shall be applied to temporary wiring installations. All temporary wiring shall be immediately removed upon completion of construction or the purpose for which it was installed.

GENERAL REQUIREMENTS FOR TEMPORARY WIRING

Temporary wiring methods shall be permitted to be used for feeders and branch-circuits if they conform to the following requirements:

Feeders. Feeders shall originate in an approved distribution center. The conductors shall be permitted within multiconductor cord or cable assemblies; if not subject to physical damage, they shall be permitted to be run using open conductors on insulators located and supported at intervals not more than 10 ft. apart. **(See Figure 11-55)**

Figure 11-55. Conductor for feeders shall be permitted within multiconductor cord or cable assemblies per NEC. Check with OSHA when using open conductors on insulators.

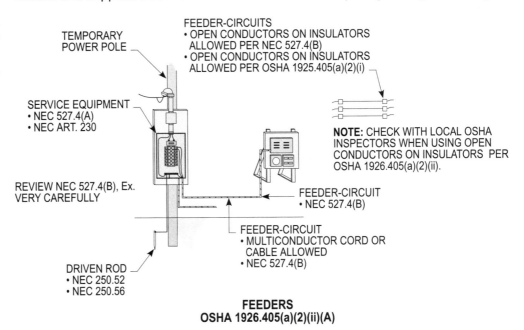

FEEDERS
OSHA 1926.405(a)(2)(ii)(A)

Safety Tip: All branch-circuits shall originate in an approved power outlet or panelboard.

Branch-circuits. All branch-circuits shall originate in an approved power outlet or panelboard. Conductors shall be permitted to be within multiconductor cord or cable assemblies or used as open conductors. Where run as open conductors, they shall be attached at intervals not exceeding 10 ft. Branch-circuit conductors of the open type shall not be to be laid on the floor. Branch-circuits that supply receptacles or fixed equipment shall contain a separate equipment grounding conductor if routed as open conductors. **(See Figure 11-56)**

Receptacles. Receptacles shall be of the grounding type. Unless installed in a complete metallic raceway, each branch-circuit shall be equipped with a separate equipment grounding conductor if routed as open conductors. **(See Figure 11-57)**

Disconnecting means. Suitable disconnecting switches or plug connectors shall be installed to allow the disconnection of the ungrounded (phase) conductors of each temporary circuit. This role shall be required to limit the possibility of personnel receiving electrical shock while working on the elements of temporary wiring systems. **(See Figure 11-58)**

Lamp protection. Lamps used for general illumination shall be protected from accidental contact or breakage, Protection shall be provided by using a suitable fixture or lampholder with a guard. Metal case sockets shall be grounded, and guards are to be of the totally enclosed type. Temporary lights shall not be suspended by their electric cords unless approved for such use. **(See Figure 11-59)**

Portable lighting. Portable lighting used in wet or conductive locations shall be operated at 12 volts or less. If 120 volts are utilized, they shall be GFCI-protected. **(See Figure 11-60)**

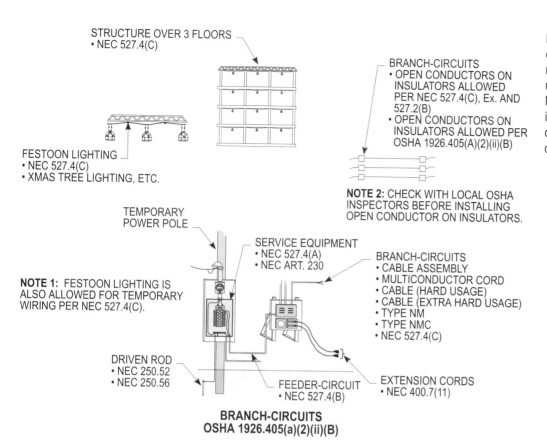

STRUCTURE OVER 3 FLOORS
• NEC 527.4(C)

FESTOON LIGHTING
• NEC 527.4(C)
• XMAS TREE LIGHTING, ETC.

BRANCH-CIRCUITS
• OPEN CONDUCTORS ON INSULATORS ALLOWED PER NEC 527.4(C), Ex. AND 527.2(B)
• OPEN CONDUCTORS ON INSULATORS ALLOWED PER OSHA 1926.405(A)(2)(ii)(B)

NOTE 2: CHECK WITH LOCAL OSHA INSPECTORS BEFORE INSTALLING OPEN CONDUCTOR ON INSULATORS.

TEMPORARY POWER POLE

SERVICE EQUIPMENT
• NEC 527.4(A)
• NEC ART. 230

BRANCH-CIRCUITS
• CABLE ASSEMBLY
• MULTICONDUCTOR CORD
• CABLE (HARD USAGE)
• CABLE (EXTRA HARD USAGE)
• TYPE NM
• TYPE NMC
• NEC 527.4(C)

NOTE 1: FESTOON LIGHTING IS ALSO ALLOWED FOR TEMPORARY WIRING PER NEC 527.4(C).

DRIVEN ROD
• NEC 250.52
• NEC 250.56

FEEDER-CIRCUIT
• NEC 527.4(B)

EXTENSION CORDS
• NEC 400.7(11)

BRANCH-CIRCUITS
OSHA 1926.405(a)(2)(ii)(B)

Figure 11-56. Conductors for branch-circuits shall be permitted to be within multiconductor or cable assemblies or used as open conductors. Type NM and NMC cable shall be permitted to be installed for temporary wiring in any dwelling, building or structure under construction, regardless of their height.

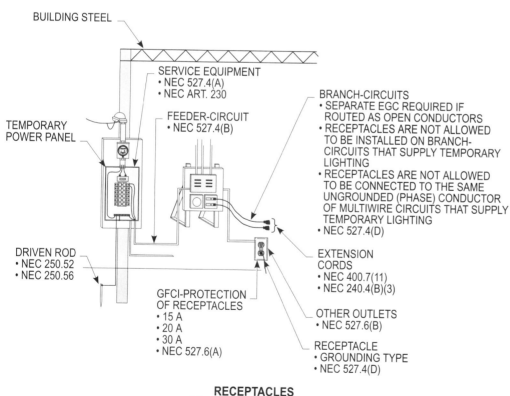

BUILDING STEEL

SERVICE EQUIPMENT
• NEC 527.4(A)
• NEC ART. 230

FEEDER-CIRCUIT
• NEC 527.4(B)

TEMPORARY POWER PANEL

BRANCH-CIRCUITS
• SEPARATE EGC REQUIRED IF ROUTED AS OPEN CONDUCTORS
• RECEPTACLES ARE NOT ALLOWED TO BE INSTALLED ON BRANCH-CIRCUITS THAT SUPPLY TEMPORARY LIGHTING
• RECEPTACLES ARE NOT ALLOWED TO BE CONNECTED TO THE SAME UNGROUNDED (PHASE) CONDUCTOR OF MULTIWIRE CIRCUITS THAT SUPPLY TEMPORARY LIGHTING
• NEC 527.4(D)

DRIVEN ROD
• NEC 250.52
• NEC 250.56

EXTENSION CORDS
• NEC 400.7(11)
• NEC 240.4(B)(3)

GFCI-PROTECTION OF RECEPTACLES
• 15 A
• 20 A
• 30 A
• NEC 527.6(A)

OTHER OUTLETS
• NEC 527.6(B)

RECEPTACLE
• GROUNDING TYPE
• NEC 527.4(D)

RECEPTACLES
OSHA 1926.405(a)(2)(ii)(C)

Figure 11-57. Receptacles used for temporary wiring shall be of the grounding type. Receptacles shall not be permitted to be installed on branch-circuits or be connected to the same ungrounded (phase) conductor of multiwire circuits that supply temporary lighting.

Splices. On construction sites, a box shall not be required for splices or junction terminations if the circuit conductors are multiconductor cord, cable assemblies or open run conductors. A box shall be used wherever a change is made to a raceway or a cable that is metal clad or metal sheathed.

Figure 11-58. Suitable disconnecting switches or plug connectors shall be provided to disconnect the ungrounded (phase) conductors of each temporary circuit.

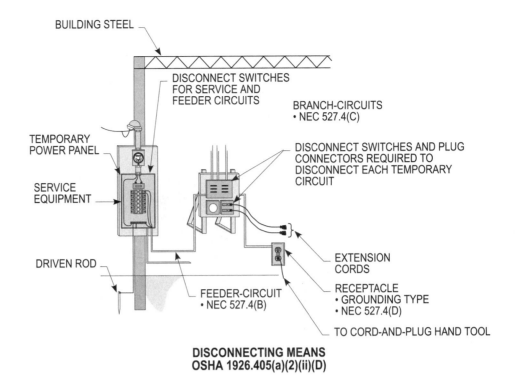

BUILDING STEEL

DISCONNECT SWITCHES FOR SERVICE AND FEEDER CIRCUITS

BRANCH-CIRCUITS
• NEC 527.4(C)

DISCONNECT SWITCHES AND PLUG CONNECTORS REQUIRED TO DISCONNECT EACH TEMPORARY CIRCUIT

TEMPORARY POWER PANEL

SERVICE EQUIPMENT

EXTENSION CORDS

DRIVEN ROD

RECEPTACLE
• GROUNDING TYPE
• NEC 527.4(D)

FEEDER-CIRCUIT
• NEC 527.4(B)

TO CORD-AND-PLUG HAND TOOL

**DISCONNECTING MEANS
OSHA 1926.405(a)(2)(ii)(D)**

Figure 11-59. This illustration shows that lamps used for general illumination shall be protected from accidental contact or breakage.

NOTE 1: METAL CASE SOCKETS SHALL BE GROUNDED, AND GUARDS ARE TO BE OF THE TOTALLY ENCLOSED TYPE.

NOTE 2: TEMPORARY LIGHTS SHALL NOT BE PERMITTED TO BE SUSPENDED BY THEIR ELECTRIC CORDS UNLESS APPROVED FOR SUCH USE.

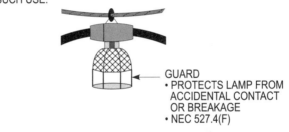

GUARD
• PROTECTS LAMP FROM ACCIDENTAL CONTACT OR BREAKAGE
• NEC 527.4(F)

**LAMP PROTECTION
OSHA 1926.405(a)(2)(ii)(E)**

Figure 11-60. Portable lighting units located in a wet or conductive location require a specific protection scheme.

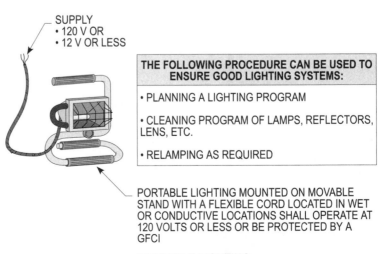

SUPPLY
• 120 V OR
• 12 V OR LESS

THE FOLLOWING PROCEDURE CAN BE USED TO ENSURE GOOD LIGHTING SYSTEMS:
• PLANNING A LIGHTING PROGRAM
• CLEANING PROGRAM OF LAMPS, REFLECTORS, LENS, ETC.
• RELAMPING AS REQUIRED

PORTABLE LIGHTING MOUNTED ON MOVABLE STAND WITH A FLEXIBLE CORD LOCATED IN WET OR CONDUCTIVE LOCATIONS SHALL OPERATE AT 120 VOLTS OR LESS OR BE PROTECTED BY A GFCI

**PORTABLE LIGHTING
OSHA 1926.405(a)(2)(ii)(G)**

Flexible cords and cables. All flexible cords and cables shall be protected from accidental damage. It is essential that sharp corners and projections be avoided. Where wiring methods are passing through doorways or other areas where pinch points can occur, care of cords and cables shall be required. Proper protection shall always be provided to avoid damaging the wiring system.

Extension Cord Sets. Extension cord sets used with portable electric tools and appliances shall be of three-wire type and shall be designed for hard or extra-hard usage. Flexible cords used with temporary and portable lights shall be designed for hard or extra-hard usage.

NOTE: The National Electrical Code, ANSI/NFPA 70, in Article 400, Table 400-4, lists various types of flexible cords, some of which are noted as being designed for hard or extra-hard usage. Examples of these types of flexible cords include hard service cord (types S, ST, SO, STO) and junior hard service cord (Types SJ, SJO, SJT, SJTO).

Over 600 volts. When wiring methods of over 600 volts are utilized, suitable fencing, barriers or other effective means shall be provided to prevent access of unqualified persons.

The following are approved methods of guarding live parts of electrical equipment from accidental contact by persons:

• By enclosing the live parts in a cabinet, box, etc.

• By locating in a room accessible only to "qualified persons." Locations intended only for "unqualified persons" shall be posted with Danger - High-Voltage-Keep Out warning signs.

• By locating behind a permanent partition or screen. An example would be a screened-in space in back of a switchboard having a locked door operable only by qualified persons.

• By elevating at least 8 ft. 6 in. above the floor.

See Figure 11-61 for a detailed illustration pertaining to the guarding of exposed live parts of systems over 600 volts.

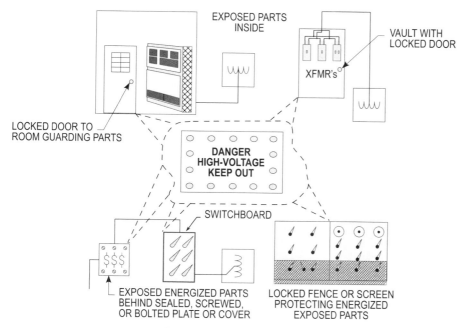

EXPOSED PARTS INSIDE

VAULT WITH LOCKED DOOR

XFMR's

LOCKED DOOR TO ROOM GUARDING PARTS

DANGER HIGH-VOLTAGE KEEP OUT

SWITCHBOARD

EXPOSED ENERGIZED PARTS BEHIND SEALED, SCREWED, OR BOLTED PLATE OR COVER

LOCKED FENCE OR SCREEN PROTECTING ENERGIZED EXPOSED PARTS

SYSTEMS OVER 600 VOLTS
OSHA 1926.405(a)(2)(iii)

Figure 11-61. Live parts of electrical equipment shall be protected and guarded against accidental contact by unqualified personnel.

CABINETS, BOXES AND FITTINGS
OSHA 1926.405(b)

A cabinet or cutout box is a box or enclosure that is fitted with swinging doors. Pull boxes and junction boxes have covers rather than swinging doors. Fittings are accessories that are used to perform mechanical connections such as to connect a raceway to a junction box, etc. An example of a fitting is a locknut or bushing, etc.

Adequate closing of a box or cabinet is accomplished by a connector and locknut installed around a cable or locknut and bushing installed around a conduit. Unused openings in cabinets, boxes and fittings shall also be effectively closed.

COVERS AND CANOPIES

All pull boxes, junction boxes and fittings shall be provided with covers approved for the purpose.

Covers of outlet boxes having holes through which flexible cord pendants pass shall be provided with bushings that are designed for the purpose. If no bushing is used, then the box cover shall have rounded, smooth surfaces on which the cords can bear without causing abrasion. Metal boxes and covers shall be grounded to prevent a fatal electrical shock.

PULL AND JUNCTION BOXES FOR
SYSTEMS OVER 600 VOLTS

The requirements in this standard apply to all pull and junction boxes housing conductors over 600 volts, including the following rules:

- All boxes shall have a complete enclosure for containing conductors or cables.

- All covers on boxes shall be securely fastened in place to prevent easy access of unauthorized personnel. Covers on underground boxes that weigh over 100 lbs. shall be considered to be complying with this requirement. Boxes shall have covers mounted properly, and all openings to have seals to prevent the entering of dirt, lint, etc. Covers for boxes shall be permanently marked, "DANGER! HIGH-VOLTAGE". The marking shall be placed on the outside of the box cover and must be readily visible and legible.

KNIFE SWITCHES
OSHA 1926.405(c)

Single-throw knife switches shall be mounted vertically if possible but shall be permitted to be mounted horizontally as long as they are installed so gravity won't tend to close the blades. Where they are mounted vertically, they shall be installed so the hinge is located at the bottom. For both horizontal and vertical mounting, the line side conductors shall be connected to the line side lugs that are opposite the blades. The blades will always be dead when the switch is in the open position. Electricians should always take voltage readings to ensure no blade has been broken. This procedure can prevent a fatal electrical shock.

Double-throw switches shall be mounted horizontally. However, they shall be permitted to be mounted vertically, but they shall be provided with a locking device that ensures the blades will remain open when in the off position. The power line shall be permitted to be connected to the blade part of the double-throw switch.

SWITCHBOARDS AND PANELBOARDS
OSHA 1926.405(d)

A panelboard shall be permitted to be mounted in cabinet or cutout box that is installed in a wall or partition or mounted against a wall or partition. Panelboards are used to serve buildings with small capacity circuits and loads.

Switchboards are used to serve buildings with larger loads and circuits of higher capacity. Switchboards shall be permitted to be mounted on the wall or installed to stand on the floor. Switchboards with exposed live parts shall be located in dry locations and accessible only to qualified persons who are trained to service such equipment.

Panelboards shall be installed in cabinets, cutout boxes or enclosures that are approved for the purpose. They shall be dead front except for panelboards of the dead front externally operable type and that are permitted to be installed in places accessible only to qualified persons. Exposed blades of knife switches shall be dead when in the open position to deenergize power to the load and allow maintenance to be performed on the equipment.

ENCLOSURES FOR DAMP OR WET LOCATIONS
OSHA 1926.405(e)

In a damp or wet location, cutout boxes and surface mounted cabinets shall be set off the wall at least 1/4 in. or more to prevent condensation between the wall and the enclosure. It is essential in wet locations, that cabinets and cutout boxes be of the weatherproof type. In damp locations, they shall not be required to be weatherproof. However, if they are not weatherproof type, then they shall be placed or equipped to prevent water from entering and accumulating within the enclosure. Water accumulation can be prevented by drilling drain holes in the bottom of the enclosure. **(See Figure 11-62)**

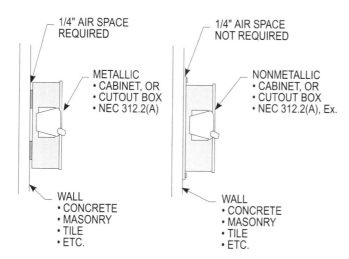

Figure 11-62. Allow 1/4 in. space between enclosure and wall for surface mounted cutout boxes and cabinets in wet or damp locations.

1/4" AIR SPACE REQUIRED

1/4" AIR SPACE NOT REQUIRED

METALLIC
• CABINET, OR
• CUTOUT BOX
• NEC 312.2(A)

NONMETALLIC
• CABINET, OR
• CUTOUT BOX
• NEC 312.2(A), Ex.

WALL
• CONCRETE
• MASONRY
• TILE
• ETC.

WALL
• CONCRETE
• MASONRY
• TILE
• ETC.

NOTE 1: CABINETS AND CUTOUT BOXES SHALL BE OF THE WEATHERPROOF TYPE IN WET LOCATIONS.

NOTE 2: CABINETS AND CUTOUT BOXES NOT OF THE WEATHERPROOF TYPE SHALL BE PLACED OR EQUIPPED TO PREVENT WATER FROM ENTERING AND ACCUMULATING WITHIN THE ENCLOSURE IN DAMP LOCATIONS.

ENCLOSURES FOR DAMP OR WET LOCATIONS
OSHA 1926.405(e)(1)

CONDUCTORS FOR GENERAL WIRING
OSHA 1926.405(f)

Conductors used for general wiring shall be insulated unless they are permitted to be otherwise due to the type of installation. The conductor insulation shall be of a type that is approved for the voltage, operating temperature and location of use for safety. Insulated conductors shall be distinguishable by appropriate color or other suitable means to identify them as being grounded (neutral) conductors, ungrounded (phase) conductors or equipment grounding conductors.

Grounded or grounded (neutral) conductors shall be white or gray in color per NEC 310.12 (A). Equipment grounding conductors shall be green or green with one or more yellow stripes per NEC 310.12(B). Grounded (neutral) conductors and equipment grounding conductors in sizes 6 AWG and smaller shall be color coded as mentioned above. However, grounded (neutral) conductors 4 AWG and larger shall be permitted to be taped white or gray or painted white or gray per NEC 200.6. The grounded (neutral) conductor shall be permitted to be stripped bare on the line side of the service equipment per NEC 230.41, Ex. (a). Equipment grounding conductors shall be permitted to be stripped bare for all sizes on the load side of the service equipment per NEC 250.119. Ungrounded (phase) conductors shall be permitted to be any color in any combination except those colors listed above. **(See Figure 11-63)**

Figure 11-63. The recommendation to color code ungrounded (phase) conductors was deleted in the 1975 NEC.

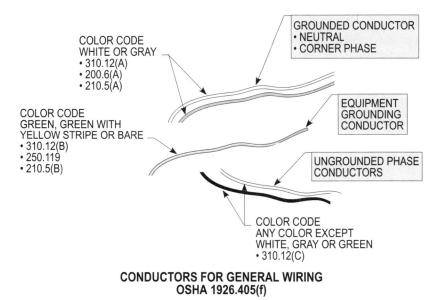

**CONDUCTORS FOR GENERAL WIRING
OSHA 1926.405(f)**

FLEXIBLE CORDS AND CABLES
OSHA 1926.405(g)

Cords shall be suitable for the conditions of use and location. The different types of cords are listed in NEC Table 400.4 and the service for which each cord is approved to be utilized. There are three designations indicating the roughness of cords, and they are as follows:

- Extra hard usage,
- Hard usage, and
- Not hard usage.

Flexible cords and cables shall be approved and suitable for conditions of use and location. Flexible cords and cables shall be used only for the following:

- Pendants
- Wiring of fixtures
- Connection of portable lamps or appliances
- Elevator cable
- Wiring of cranes and hoists
- Connection of stationary equipment to facilitate their frequent interchange
- Prevention of the transmission of noise or vibration
- Appliances where the fastening means and mechanical connections are designed to allow removal for maintenance and repair

Prohibited Uses

Unless specifically permitted, flexible cords and cables shall not be permitted to be used in the following manner:

- As a suitable method for the permanent wiring of a structure
- Where run through holes in walls, ceilings or floors
- Where run through doorways, windows or similar openings
- Where attached to building surfaces
- Where concealed behind building walls, ceilings or floors
- Identification, splices and terminations

Conductors of a flexible cord or cable used as a grounded (neutral) conductor or an equipment grounding conductor shall be identifiable from other conductors. Types SJ, SJO, SJT, SJTO, S, SO, ST, STO, HSJ, HSJO and AFS shall be marked on the surface with the type designation, size and number of conductors in the cord. Flexible cords shall be used in continuous lengths without a splice or tap when first installed. The repair of hard-usage and extra hard-usage flexible cords, 12 AWG or larger, shall be permitted if spliced so the splice retains the insulations and usage characteristics of the cord that is spliced. Flexible cords shall be connected to devices and fittings so strain relief is provided that prevents pull from being directly transmitted to joints or terminal screws.

PORTABLE CABLES OVER 600 VOLTS
OSHA 1926.405(h)

Multiconductor portable cables that are for use in servicing power to portable or mobile equipment rated over 600 volts, nominal, shall consist of 8 AWG or larger conductors equipped with flexible standing. Cables operated greater than 2000 volts shall be shielded for the purpose of containing the voltage stresses to the insulation. Grounding conductors shall be provided. Connectors for these cables shall be the locking type with provisions to prevent opening or closing while they are energized. Strain relief shall always be provided at connections and terminations. Portable cables shall not be operated without splices unless the splices are of the permanent molded vulcanized or other suitable type. Termination enclosures shall be marked with a high-voltage hazard warning, and terminations are to be accessible to authorized and qualified personnel only.

FIXTURE WIRES
OSHA 1926.405(i)

Fixture wires shall be a type approved for the voltage, temperature and condition of use. A fixture wire that is used as a grounded (neutral) conductor shall be identified as such.

USES PERMITTED

Fixture wires shall be permitted for installation in lighting fixtures and in similar equipment where they are enclosed or protected and are not subjected to bending or twisting while they are in use. Fixture wires shall be permitted to be used for connecting lighting fixtures to the branch circuit conductors supplying power to the fixtures.

USES NOT PERMITTED

Use of fixture wires as branch-circuit conductors is prohibited unless they are used as permitted as Class 1 power limited circuits and fire protection signaling circuits.

EQUIPMENT FOR GENERAL USE
OSHA 1926.405(j)

It is mandatory that fixtures, lampholders, lamps, rosettes and receptacles have no live parts exposed to contact except rosettes and cleat-type lampholders and receptacles that are located at 8 ft. or more above the finished floor. Under these conditions of use, they shall be permitted to have exposed parts. The only time personnel should have contact with the lampholder is in a maintenance condition of changing a light bulb, etc. with the toggle switch (wall switch) in the closed position and locked out. **(See Figure 11-64)**

Figure 11-64. Fixtures shall be located at least 8 ft. above the floor with exposed live parts.

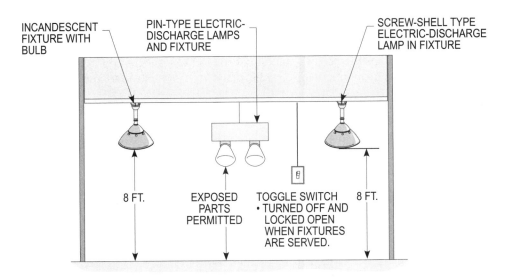

EQUIPMENT FOR GENERAL USE - LIGHTING FIXTURES, LAMPHOLDERS, LAMPS AND RECEPTACLES
OSHA 1926.405(j)(1)(i)

Handlamps of the portable type served through flexible cords shall be equipped with a handle of molded composition or other material approved for such purpose. A substantial guard shall be attached to the lampholder or the handle for the protection and safety of the user. This rule prevents the user from touching the hot bulb or coming into contact with live parts.

It is essential that lampholders of the screw-shell type be installed for use as lampholders only. Lampholders installed in wet or damp locations shall be of the weatherproof type. The ungrounded (phase) conductor shall be terminated to the inside tit, and the grounded (neutral) conductor shall be connected to the screw shell of the lampholder.

Fixtures that are installed in wet or damp locations shall be approved for the purpose and are to be constructed or installed so that water will not enter or accumulate in wireways, lampholders or other electrical parts. **(See Figure 11-65)**

NOTE: FIXTURES THAT ARE INSTALLED IN WET OR DAMP LOCATIONS SHALL BE APPROVED FOR THE PURPOSE AND ARE TO BE CONSTRUCTED OR INSTALLED SO THAT WATER WILL NOT ENTER OR ACCUMULATE IN WIREWAYS, LAMPHOLDERS OR OTHER ELECTRICAL PARTS.

Figure 11-65. Fixtures shall be suitable for the location where they are installed.

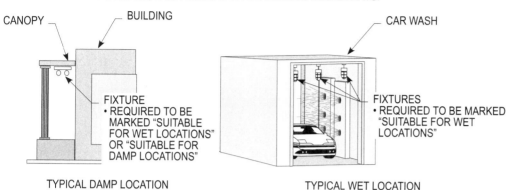

CANOPY — — BUILDING

CAR WASH

FIXTURE
• REQUIRED TO BE MARKED "SUITABLE FOR WET LOCATIONS" OR "SUITABLE FOR DAMP LOCATIONS"

FIXTURES
• REQUIRED TO BE MARKED "SUITABLE FOR WET LOCATIONS"

TYPICAL DAMP LOCATION

TYPICAL WET LOCATION

EQUIPMENT FOR GENERAL USE
OSHA 1926.405(j)(1)(iii)(v)

RECEPTACLES, CORD CONNECTORS AND ATTACHMENT PLUGS

Receptacles, cord connectors and attachment plugs shall be constructed so the receptacle or cord connectors will not accept an attachment plug with a different voltage or current rating than that for which the device is intended to be used. A 20 ampere T-slot type receptacle or cord connector shall not be permitted to accept a 15 ampere attachment plug of the same voltage rating.

All receptacles that are installed in a wet or damp location shall be approved and suitable for the location. This rule limits the corrosion effect that water can have on the components in the receptacle. **(See Figure 11-66)**

RECEPTACLE OUTLET

Figure 11-66. This illustration shows that receptacles and cover plates installed in wet and damp locations shall be approved and suitable for the location.

NOTE 1: RECEPTACLES THAT ARE INSTALLED IN WET LOCATIONS ARE ALSO CONSIDERED SUITABLE FOR DAMP LOCATIONS.

NOTE 2: RECEPTACLES THAT ARE LOCATED UNDER ROOFED OPEN PORCHES, CANOPIES, MARQUEES AND THE LIKE ARE CONSIDERED AS BEING INSTALLED IN A LOCATION THAT IS PROTECTED FROM THE WEATHER.

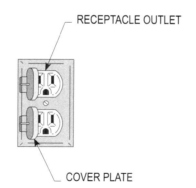

COVER PLATE

RECEPTACLES, CORD CONNECTORS
AND ATTACHMENT PLUGS (CAPS)
OSHA 1926.405(j)(2)(i)

APPLIANCES

Appliances are not normally supposed to have live parts exposed to contact by personnel using them. However, some appliances operating with current carrying parts at high temperatures shall be permitted to have exposed live parts where it is necessary for cool and safe operation. Some appliances are designed with open type screens to allow heat from components to escape easily and prevent heating problems.

All appliances shall be equipped with a disconnecting means or a disconnecting means shall be provided.

Appliances shall be marked with their rating in volts and amperes or in volts and watts (VA).

MOTORS

Where it is required that one piece of equipment is required to be "in sight from" another piece of equipment, one shall be located in a place that is invisible and not more than 50 ft. from each other. **(See Figure 11-67)**

Figure 11-67. Certain requirements shall be adhered to when locating the motor, load and controller from the disconnecting means.

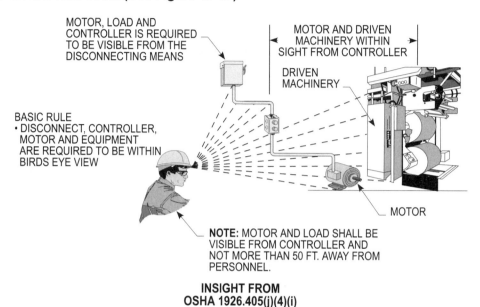

MOTOR, LOAD AND CONTROLLER IS REQUIRED TO BE VISIBLE FROM THE DISCONNECTING MEANS

MOTOR AND DRIVEN MACHINERY WITHIN SIGHT FROM CONTROLLER

DRIVEN MACHINERY

BASIC RULE
• DISCONNECT, CONTROLLER, MOTOR AND EQUIPMENT ARE REQUIRED TO BE WITHIN BIRDS EYE VIEW

MOTOR

NOTE: MOTOR AND LOAD SHALL BE VISIBLE FROM CONTROLLER AND NOT MORE THAN 50 FT. AWAY FROM PERSONNEL.

INSIGHT FROM
OSHA 1926.405(j)(4)(i)

Safety Tip: A disconnecting means shall be located within sight from the controller location.

A disconnecting means shall be located in sight from the controller location. A single disconnecting means shall be permitted to be located adjacent to a group of coordinated controllers mounted adjacent one to another such as on a multi-motor continuous process machine. The controller disconnecting means for a motor branch-circuit over 600 volts shall be permitted to be located out of sight of the motor branch-circuit controller where the controller has a warning label and is marked by listing the location and identification of the disconnecting means that is locked in the open position.

The disconnecting means shall be required to disconnect the motor and the controller from all ungrounded (phase) conductors and shall be arranged so that each pole is operated independently.

Where a motor and the driven machinery are not located in sight from the controller location, the installation shall be designed to comply with one of the following:

• The controller disconnecting means shall be capable of being locked in the open position.

• A switch that disconnects for the motor from its source of supply shall be installed within sight from the motor location.

See Figures 11-68(a), (b), (c) and (d) for a detailed illustration pertaining to the requirements for motor disconnecting means.

After the disconnecting means has been installed, it shall indicate very plainly to the user that it is in the open (off) or closed (on) position. This rule is to help personnel working on the motor to know for sure that the disconnecting means in truly in the open position and all ungrounded (phase) conductors are deenergized. **(See Figure 11-69)**

NOTE 1: ALL OF THE UNGROUNDED (PHASE) CONDUCTORS SHALL BE DISCONNECTED FROM BOTH THE MOTOR AND CONTROLLER SUPPLYING THE MOTOR CIRCUIT.

NOTE 2: THE CONTROLLER HAS A DIRECT RELATIONSHIP TO THE DISCONNECTING MEANS AND SHALL BE INSTALLED WITHIN SIGHT AND WITHIN 50 FT. OF THE DISCONNECTING MEANS.

Figure 11-68. This illustration shows the requirements for the disconnecting means in (a), (b), (c) and (d).

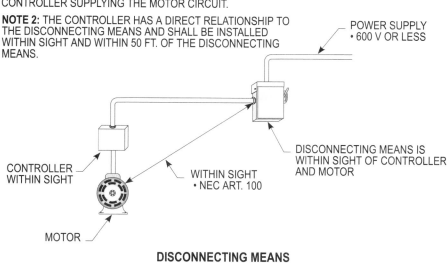

POWER SUPPLY
• 600 V OR LESS

DISCONNECTING MEANS IS WITHIN SIGHT OF CONTROLLER AND MOTOR

CONTROLLER WITHIN SIGHT

WITHIN SIGHT
• NEC ART. 100

MOTOR

DISCONNECTING MEANS
OSHA 1926.405(j)(4)(i)

NOTE: THE CONTROLLER SHALL HAVE A WARNING LABEL THAT MARKS AND LISTS THE LOCATION AND IDENTIFICATION OF THE DISCONNECTING MEANS.

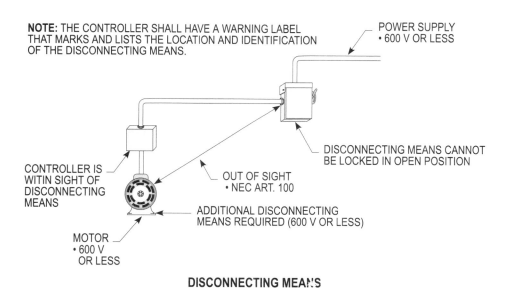

POWER SUPPLY
• 600 V OR LESS

DISCONNECTING MEANS CANNOT BE LOCKED IN OPEN POSITION

CONTROLLER IS WITIN SIGHT OF DISCONNECTING MEANS

OUT OF SIGHT
• NEC ART. 100

ADDITIONAL DISCONNECTING MEANS REQUIRED (600 V OR LESS)

MOTOR
• 600 V OR LESS

DISCONNECTING MEANS
OSHA 1926.405(j)(4)(i)

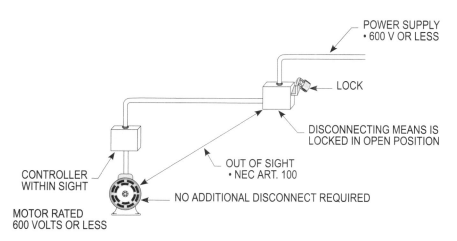

POWER SUPPLY
• 600 V OR LESS

LOCK

DISCONNECTING MEANS IS LOCKED IN OPEN POSITION

CONTROLLER WITHIN SIGHT

OUT OF SIGHT
• NEC ART. 100

NO ADDITIONAL DISCONNECT REQUIRED

MOTOR RATED 600 VOLTS OR LESS

DISCONNECTING MEANS
OSHA 1926.405(j)(4)(i)

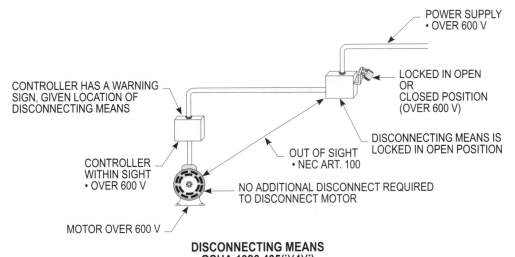

DISCONNECTING MEANS
OSHA 1926.405(j)(4)(i)

Figure 11-69. This illustration shows the requirements for switching and locking the disconnecting means.

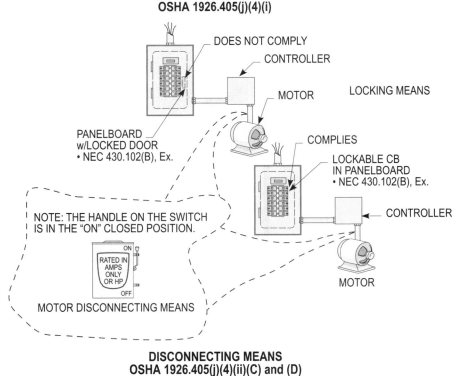

DISCONNECTING MEANS
OSHA 1926.405(j)(4)(ii)(C) and (D)

At least one of the disconnecting means shall be readily accessible so the user can have easy and fast access to disconnect power to the motor or other elements that need servicing. OSHA is very strict on this requirement because so many people have been shocked, burned, hurt or killed for not observing this rule.

An individual disconnecting means shall be provided for each motor. However, a single disconnecting means shall be permitted to be utilized for a group of motors with any one of the following being applied:

• Where a number of motors drive specific parts of a single machine or piece of electrical apparatus, such as metal and woodworking machines, cranes, hoists, etc.

• Where a group of motors is protected by one set of branch-circuit protective overcurrent devices.

• Where a group of motors is installed in a single room that are within sight from the location of the disconnecting means.

See Figures 11-70(a) and (b) for a detailed illustration pertaining to a single disconnecting means being utilized for a group of motors.

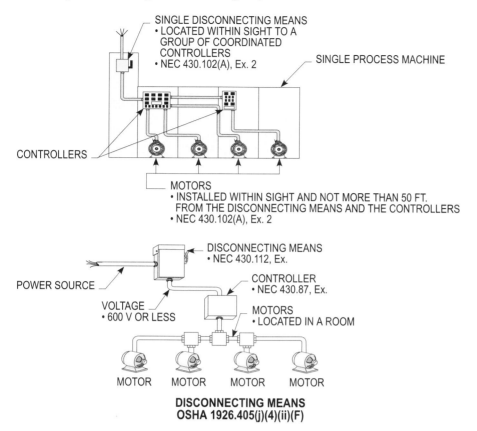

Figures 11-70(a) and (b). A single disconnect shall be permitted to be used as a disconnecting means for several motors located in a room and within sight or located within sight to a group of coordinated controllers.

MOTOR OVERLOAD, SHORT-CIRCUIT AND GROUND FAULT PROTECTION

All and any motors, motor control apparatus and motor branch-circuit conductors shall be protected against overheating due to motor overloads or failure to start to drive the load. Protection devices shall be sized and selected to protect the motor from short-circuits and ground faults. Overload protection can be limited where a shutdown might introduce hazards, as in the case of fire pumps.

Where continued operation of a motor is necessary for a safe shutdown of equipment of a process and motor overload sensing devices are terminated to a supervised alarm, overload protection shall not be required under these conditions of use.

PROTECTION OF LIVE PARTS

Stationary motor with commutators, collectors and brush rigging located inside of motor end brackets and not conductively connected to supply circuits that operate at 150 volts or less to ground shall not be required to be guarded. Exposed live parts of motors and controllers operating at 50 volts or greater between terminals shall be guarded from accidental contact by one of the following:

• By installation in a room or enclosure accessible to qualified personnel only.

• By installation on a suitable balcony, gallery or platform, so the motor and components are elevated and arranged to exclude unauthorized personnel.

• By elevation of at least 8 ft. above the finished floor.

Where live parts of motors and controllers that operate at over 150 volts-to-ground are guarded against accidental contact by location and adjustments and other maintenance is necessary during the operation, suitable insulating mats or platforms shall be provided. The attendant shall be standing on the mat or platform before he or she touches the live parts. **(See Figure 11-71)**

Figure 11-71. This illustration shows that a suitable mat or platform shall be provided for employees working on motors or controllers having exposed live parts operating at over 150 volts-to-ground.

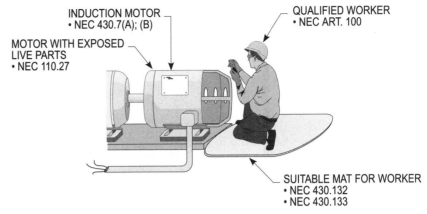

INDUCTION MOTOR
• NEC 430.7(A); (B)

MOTOR WITH EXPOSED LIVE PARTS
• NEC 110.27

QUALIFIED WORKER
• NEC ART. 100

SUITABLE MAT FOR WORKER
• NEC 430.132
• NEC 430.133

PROTECTION OF LIVE PARTS
OSHA 1926.405(j)(4)(ii)(F)(3)

TRANSFORMERS

The rules of OSHA 1926.405(j)(5) apply to all transformers except the following:

- Current transformers.

- Dry-type transformers that are installed as a component part or other electrical apparatus.

- Transformers that are an integral part of an X-ray, high frequency or electrostatic-coating electrical apparatus.

- Transformers utilized with Class 2 and Class 3 circuits, sign and outline lighting, electric discharge lighting and power-limited fire-protective signaling circuits.

OPERATING VOLTAGE

The operating voltage of exposed live parts of transformer installations shall be indicated with warning signs or visible markings shall be provided on the equipment or structure.

TRANSFORMERS OVER 35 kV, DRY TYPE

Transformers of the dry type, high fire point liquid, insulated and askarel insulated shall be permitted to be installed indoors, and if rated over 35 kV they shall be installed in a vault.

OIL INSULATED TRANSFORMERS

Oil filled transformers installed indoors that present a fire hazard to employees shall be installed in a vault.

FIRE PROTECTION

Combustible material, combustible buildings and parts of buildings, fire escapes and door and window openings shall be safeguarded from fires which may originate in oil-insulated transformers attached to or adjacent to a building or combustible material.

TRANSFORMER VAULTS

Transformer vaults shall be constructed so as to prevent access of unauthorized personnel from entering. Locks and latches shall be arranged so that a vault door can be readily or quickly opened from the inside using panic bars, pressure plates, etc.

Any and all pipe or duct systems, foreign to the vault installation, shall not be permitted to enter or pass through a transformer vault.

Materials shall not be permitted to be stored in transformer vaults. **(See Figure 11-72)**

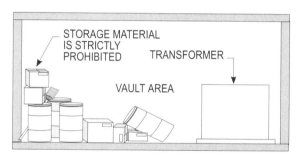

TRANSFORMER VAULTS
OSHA 1926.405(j)(5)(vi)

Figure 11-72. No storage of any kind may be put in a vault other than the transformers and equipment necessary for their operation.

CAPACITORS

Capacitors are used to improve power factor and aid a motor with additional starting torque. Capacitors shall be provided with the proper size conductors, overcurrent protection devices and disconnecting means. **(See Figure 11-73)**

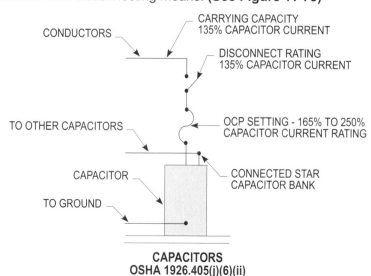

CAPACITORS
OSHA 1926.405(j)(6)(ii)

Figure 11-73. Capacitor elements shall be sized and protected properly based on the type of installation.

All capacitors, except surge capacitors or capacitors included as a component part of other electrical apparatus, shall be provided with an automatic means of draining the stored charge when the capacitor is disconnected from its power supply. If the windings in the motor are not used to accomplish this purpose, other means shall be provided.

Capacitors that are rated over 600 volts shall comply with the following rules:

- Isolating or disconnecting switches without an interrupting rating shall be interlocked with the load interrupting device, or they shall be provided with caution signs to prevent switching the load current.

- Series capacitors shall be switched by using one of the following methods:

(a) Mechanically sequenced isolating and bypass switches.

(b) By using proper interlocks.

(c) Switching procedure that is properly displayed at the switching location.

SPECIFIC PURPOSE EQUIPMENT AND INSTALLATIONS OSHA 1926.406

The following shall be considered for specific purpose equipment and installations:

- Cranes and hoists
- Elevators, escalators and moving walks
- Electric welders

CRANES AND HOISTS OSHA 1926.406(a)

Wiring methods to cranes and hoists shall be designed and installed for the moving of the elements and parts required for proper operation and safety. The wiring methods on cranes and hoists should be rigid conduit or EMT, which provides excellent protection of conductors. Motors, brakes, magnets and other devices shall be permitted to be connected to the power supply utilizing short lengths of flexible conduit or metal-clad cable for flexibility. This rule provides easy and fast access for interchanging parts and equipment. The wiring method used has to be approved for the location where the crane or hoist is installed and used.

DISCONNECTING MEANS

The disconnecting means shall comply with the following rules:

- A readily accessible disconnecting means shall be provided between the runway contact conductors and the power supply to the crane or hoist.

- A disconnecting means shall be arranged to be locked in the open position, and shall be provided in the power leads from the runway contact conductors or other power supply on any and all cranes and monorail hoist systems.

- Where the disconnecting means is not readily accessible from the crane or monorail hoist operating station, a means shall be provided at the operating station to open the power supply circuit to any and all motors of the crane or monorail hoist.

See Figure 11-74 for a detailed illustration pertaining to cranes and hoists with a platform.

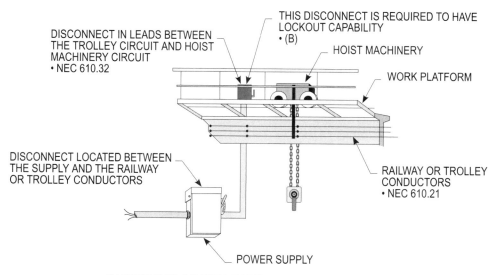

DISCONNECT IN LEADS BETWEEN THE TROLLEY CIRCUIT AND HOIST MACHINERY CIRCUIT
• NEC 610.32

THIS DISCONNECT IS REQUIRED TO HAVE LOCKOUT CAPABILITY
• (B)

HOIST MACHINERY

WORK PLATFORM

DISCONNECT LOCATED BETWEEN THE SUPPLY AND THE RAILWAY OR TROLLEY CONDUCTORS

RAILWAY OR TROLLEY CONDUCTORS
• NEC 610.21

POWER SUPPLY

AN ADDITIONAL CONTROL SWITCH OR A REMOTE CONTROL SWITCH IS NECESSARY IF THE SECOND DISCONNECTING MEANS (B) IS NOT ACCESSIBLE TO THE OPERATOR.

DISCONNECTING MEANS
OSHA 1926.405(a)(1)(i) and (a)(1)(ii)

Figure 11-74. This illustration shows the disconnecting means for the crane at grade level as well as on the platform.

Exception: Except for a monorail hoist or hand-propelled crane bridge installation that meets all of the following:

- If the unit is ground or floor controlled, then the disconnect shall be permitted to be omitted.

- If the unit is within view of the power supply disconnecting means, then the disconnect shall be permitted to be omitted.

- If no fixed work platform has been provided for servicing the unit, then the disconnect shall be permitted to be omitted.

See Figure 11-75 for a detailed illustration pertaining to cranes and hoists without a platform.

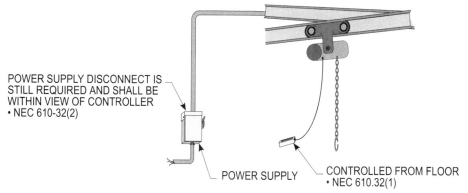

POWER SUPPLY DISCONNECT IS STILL REQUIRED AND SHALL BE WITHIN VIEW OF CONTROLLER
• NEC 610-32(2)

POWER SUPPLY

CONTROLLED FROM FLOOR
• NEC 610.32(1)

Figure 11-75. This illustration shows the disconnecting means for the crane at grade level.

SECOND DISCONNECT NOT REQUIRED. A MONORAIL HOIST DOES NOT REQUIRE A DISCONNECTING MEANS IN THE LEADS TO THE HOIST MACHINERY IF IT IS CONTROLLED FROM THE GROUND OR FLOOR, IF IT IS WITHIN VIEW OF THE POWER SUPPLY DISCONNECT AND IF THERE IS NO WORK PLATFORM PROVIDED TO SERVICE THE HOIST MACHINERY.

DISCONNECTING MEANS
OSHA 1926.406(a)(1)(i) and (a)(1)(ii)

CONTROL

A limit switch or other device shall be provided to prevent the load block from passing the safe upper limit of travel of any and all hoisting mechanisms.

CLEARANCE

The working space in the direction of access to live parts that are likely to require examination, adjustment, servicing or maintenance shall have a minimum clearance of 2 1/2 ft. Where the controls are enclosed in cabinets, the door(s) shall be open at least 90° or be removable for easy access so the wiring systems can be serviced safely without exposing personnel to shock hazards.

GROUNDING

All exposed metal parts of cranes, monorail hoists, hoists and accessories including pendant controls shall be metallically joined together into a continuous electrical conductor so that the entire crane or hoist will be grounded in accordance with 1926.404(f). Moving parts, other than removable accessories or attachments, metal-to-metal bearing surfaces shall be considered to be electrically connected to each other through the bearing surfaces for grounding purposes. The trolley frame and bridge frame shall be considered as electrically grounded through the bridge and trolley wheels and their respective tracks unless conditions such as paint or other insulating materials prevent reliable metal-to-metal contact. In this case a separate bonding conductor shall be provided.

ELEVATORS, ESCALATORS AND MOVING WALKS
OSHA 1926.406(b)

The following shall be considered for elevators, escalators and moving walks:

- Disconnecting means
- Control panels

DISCONNECTING MEANS

Elevators, dumbwaiters, escalators and moving walks shall have an individual disconnecting means to disconnect all ungrounded main power supply conductors to each unit. **(See Figure 11-76)**

CONTROL PANELS

Control panels that are not located in the same location as the drive machine shall be located in cabinets with doors or panels that are capable of being locked in the closed position.

ELECTRIC WELDERS – DISCONNECTING MEANS
OSHA 1926.406(c)

A disconnecting means shall be provided in the supply line for each motor generator arc welder. Each AC transformer and DC rectifier arc welder that is not equipped with a disconnect mounted as an integral part of the welder shall have such disconnecting means installed within sight of the welder.

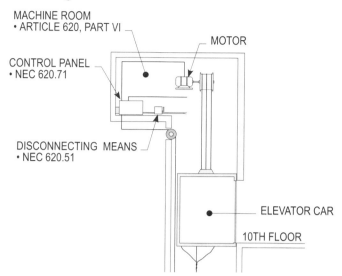

NOTE: THE DISCONNECTING MEANS SHALL BE LOCATED WHERE IT IS READILY ACCESSIBLE TO QUALIFIED PERSONNEL.

DISCONNECTING MEANS
OSHA 1926.406(b)(1)

Figure 11-76. The disconnecting means for an elevator shall be located where it is readily accessible to qualified personnel.

RESISTANCE WELDERS

A switch or circuit breaker shall be provided by each resistance welder and its control equipment, which can be isolated from the power supply circuit. The ampere rating of the disconnecting means shall not be less than the ampacity of the power supply conductors. **(See Figure 11-77)**

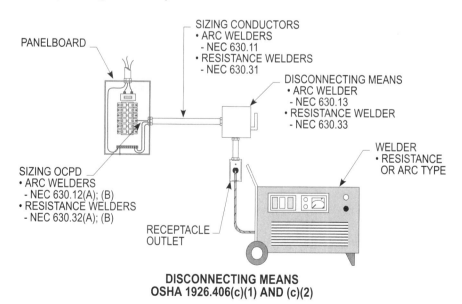

DISCONNECTING MEANS
OSHA 1926.406(c)(1) AND (c)(2)

Figure 11-77. Properly sized overcurrent protection devices, conductors and disconnects shall be provided for electric welders.

X-RAY EQUIPMENT

A disconnecting means shall be provided in the power supply circuit. The disconnecting means shall be capable of being operated from a location readily accessible from the X-ray control. Where equipment is connected to a 120 volt branch-circuit of not more than 30 amps, a grounding-type attachment plug cap and receptacle of the proper rating shall be permitted to be used as a disconnecting means.

If more than one piece of equipment is supplied from the same high-voltage circuit, each piece or each group of equipment as a unit shall be provided with a high voltage switch or equivalent disconnecting means. The disconnecting means shall be constructed, enclosed or located in such a manner so as to avoid contact by personnel with its live or energized parts.

All radiographic and fluoroscopic type equipment shall be effectively enclosed. As an alternate they shall be permitted to be equipped with interlocks that deenergize the equipment automatically to prevent ready access to live and energized current-carrying parts.

HAZARDOUS (CLASSIFIED) LOCATIONS
OSHA 1926.407

The following shall be considered for hazardous (classified) locations:

- Scope
- Electrical installations
- Conduits

SCOPE
OSHA 1926.407(a)

A hazardous (classified) location is any location where a potential hazard, either a fire or an explosion, can exist due to presence of flammable, combustible or ignitable materials. These materials can consist of gases, vapors, liquids, dust, fibers, etc. Hazardous (classified) locations are classified according to the properties and quantities of the hazardous material that can be present. Hazardous locations are divided into three classes, two divisions, and seven classified groups as follows:

- Class I, II, and III; Division 1 and 2
- Groups A, B, C, D, E, F and G.

Wiring methods used in hazardous (classified) locations shall comply with more stringent requirements than wiring methods used in other locations.

The 1996 NEC also recognizes Class I, Zones 0, 1, and 2. The concept of using Zones per the International Electrotechnical Commission (IEC), instead of Division per the NEC, has been used internationally for many years. In the past, the IEC has regulated the requirements for such systems.

The IEC system uses two groups to identify the hazards involved. Group I is used for mining, while Group II is used for surface industries and offshore installations.

Group II consists of three subgroups, which are A, B and C. Subgroups A, B and C are designed to represent categories of flammable gases or vapors that are based upon the minimum ignition energy of the hazard. Subgroup A represents the most difficult flammable gas or vapor to ignite, while Subgroup C is the easiest to ignite.

It must be understood that each room, section or area shall be considered individually in determining its classification. These classified locations are designated as follows:

- Class I, Division 1 Class I, Division 2
- Class II, Division 1 Class II, Division 2
- Class III, Division 1 Class III, Division 2

See OSHA 1926.449 for definitions of these locations.

ELECTRICAL INSTALLATIONS
OSHA 1926.407(b)

Any and all equipment, wiring methods and installations that are installed in hazardous (classified) locations shall be of a type and design that provides protection from hazards. Hazards that are created by combustible and flammable vapors, liquids, gases, dusts or fibers that are exposed to an arc or spark shall be contained by using an explosion proof enclosure or other approved means. The NEC contains requirements that are appropriate for determining the type and design of equipment and installations that provide protection from these types of hazards.

All electrical equipment installed in hazardous locations shall be approved as intrinsically safe or approved for the hazardous location

INTRINSICALLY SAFE

Intrinsically safe equipment and its associated wiring shall be permitted to be installed in any hazardous (classified) location for which it has been approved.

An electrical system is intrinsically safe when a spark is so weak or the temperature is so low that it cannot ignite flammable atmospheres. The useful power of intrinsically safe circuits is approximately one watt. This energy level is sufficient to power most instruments.

APPROVED FOR THE HAZARDOUS LOCATION

Equipment shall be approved not only for the class of location but also for the ignitable or combustible properties of the specific gas, vapor, dust or fiber that is present. Chapter 5 in the NEC lists and defines hazardous gases, vapors and dusts by class, division and group, characterized by their ignitable or combustible properties.

MARKING

Equipment that is not listed in (a) through (c) below shall be marked to show the class, group and operating temperatures or temperature range, which is based in a 40°C ambient. The temperature marking shall not be permitted to exceed the ignition temperature of the specific gas or vapor to be encountered.

The nameplate on approved equipment shall be required to show class, group and operating temperature based on operation in a 40°C ambient. Where temperature range is given, it shall be indicated by identification numbers per NEC Table 500.8(B).

Equipment that is not marked to indicate a division, or is marked "Division 1" or "Div. 1", shall be suitable for either Division 1 and 2 locations. However, equipment marked "Division 2" or "Div. 2" shall be suitable for only Division 2 locations.

Equipment of the non-heat-producing type, such as junction boxes, conduit and fittings, and equipment of the heat-producing type having a maximum temperature of 100°C or less (212°F) shall not be required to have a marked operating temperature or temperature range. Junction boxes containing only termination points for conductors (splices) are good examples of non-heat-producing equipment. **(See Figure 11-78)**

Fixed lighting fixtures that are marked for use in Class I, Division 2, or Class II, Division 2 locations only, shall not be required to be marked to indicate the group.

Figure 11-78. This illustration shows that general-purpose enclosures shall be permitted to be used to house splices, taps and terminations due to the fact they are considered non-heat-producing type boxes, etc.

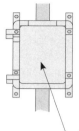

GENERAL PURPOSE
ENCLOSURE

NOTE 1: EQUIPMENT OF THE NON-HEAT-PRODUCING TYPE, SUCH AS JUNCTION BOXES, CONDUIT AND FITTINGS AND EQUIPMENT OF THE HEAT-PRODUCING TYPE HAVING TEMPERATURES OF 100° C OR LESS (212°F) IS NOT REQUIRED TO HAVE A MARKED OPERATING TEMPERATURE OR TEMPERATURE RANGE.

NOTE 2: JUNCTION BOXES CONTAINING ONLY SPLICES, TAPS OR TERMINATIONS ARE GOOD EXAMPLES OF NON-HEAT-PRODUCING EQUIPMENT.

BOX CONTAINS SPLICES, TAPS OR TERMINATIONS

NON-HEAT PRODUCING EQUIPMENT
OSHA 1926.407(b)(2)(ii)(A)

The fixture surface temperature of Class I, Division 1 fixtures shall not be permitted to exceed the ignition temperature of the gases or vapors present per NEC 500.3(E). For Class I, Division 2 fixtures, it is the lamp surface temperature that shall not be permitted to exceed the ignition temperature of the gases or vapors involved. **(See Figure 11-79)**

Figure 11-79. This illustration shows the temperature ratings of lighting fixtures based on lamp or surface temperature.

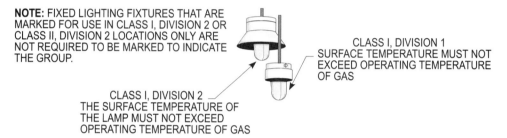

NOTE: FIXED LIGHTING FIXTURES THAT ARE MARKED FOR USE IN CLASS I, DIVISION 2 OR CLASS II, DIVISION 2 LOCATIONS ONLY ARE NOT REQUIRED TO BE MARKED TO INDICATE THE GROUP.

CLASS I, DIVISION 1
SURFACE TEMPERATURE MUST NOT EXCEED OPERATING TEMPERATURE OF GAS

CLASS I, DIVISION 2
THE SURFACE TEMPERATURE OF THE LAMP MUST NOT EXCEED OPERATING TEMPERATURE OF GAS

FIXTURE MARKING
OSHA 1926.407(b)(2)(ii)(B)

Fixed general-purpose equipment installed in Class I locations not including fixed lighting fixtures, that can be used in Class I, Division 2 locations shall not be required to be marked with the class, group, division or operating temperature.

Fixed dust-tight equipment not including fixed lighting fixtures that can be used in Class II, Division 2 and Class III locations shall not be required to be marked with the Class, Group, Division or operating temperature.

Note: The National Electrical Code, NFPA 70, contains guidelines for determining the type and design of equipment and installations that will meet this requirement. The guidelines of this document address electric wiring, equipment and systems installed in hazardous (classified) locations and contain specific provisions for the following: wiring methods, wiring connections, conductor insulation, flexible cords, sealing and drainage, transformers, capacitors, switches, circuit breakers, fuses, motor controllers, receptacles, attachment plugs, meters, relays, instruments, resistors, generators, motors, lighting fixtures, storage battery charging equipment, electric cranes, electric hoists and similar equipment, utilization equipment, signaling systems, alarm systems, remote control systems, local loud speaker and communications systems, ventilation piping, live parts, lightning surge protection and grounding. Compliance with these guidelines will constitute one means, but not the only means, of compliance with this paragraph.

CONDUITS
OSHA 1926.407(c)

All conduits shall be threaded and shall be made up wrenchtight with at least five fully engaged threads. Where it is impractical to make a threaded joint tight, a bonding jumper shall be used.

SPECIAL SYSTEMS
OSHA 1926.408

This chapter covers special requirements for installations with voltages over 600 volts. It suggests that information concerning the particular equipment or the conductors utilized be obtained from the manufacturer of the product and that their recommendations be carefully followed. Equipment and conductors used with voltages over 600 volts are not usually listed by Testing Laboratories. In designing, installing, and inspecting high-voltage equipment and conductors, NEMA specifications should be secured and studied. These specifications will list extremely useful information that is most helpful for installation purposes.

Included also in this chapter are low-voltage systems such as Class 1, Class 2, and Class 3 circuits. Communication systems and fire protective signaling systems are discussed along with their installation requirements.

SYSTEMS OVER 600 VOLTS
OSHA 1926.408(a)

General requirements for all wiring methods, equipment and circuits operating at over 600 volts shall comply with the NEC and OSHA regulations.

WIRING METHODS FOR FIXED INSTALLATIONS

Above ground conductors shall be installed in the following wiring methods:

  • Rigid metal conduit
  • IMC
  • Cable trays
  • Cable bus
  • Other suitable raceways
  • Metal clad cable

Where qualified personnel is in the location, open runs of MV cable or bare cable or bare conductors or busbars shall be permitted to be utilized. All metallic shielding components for conductors such as tapes, wires or braids shall be grounded. Open runs of insulated wires and cables equipped with a bare lead sheath or a braided outer covering shall be supported in a manner to prevent physical damage to the braid or sheath.

INTERRUPTING AND ISOLATING DEVICES

CIRCUIT BREAKERS

Indoor installations shall be enclosed in fire-resistant cell-mounted units, except that open mounting of circuit breakers shall be permitted in locations accessible only to qualified personnel.

Exception. Open mounting of circuit breakers shall be permitted in locations accessible only to qualified personnel.

FUSED CUTOUTS

Fused cutouts that are installed in buildings or transformer vaults shall be of a type suitable for the purpose, in other words approved. They shall be readily accessible for fuse replacement by qualified personnel.

EQUIPMENT ISOLATING MEANS

A means shall be provided to isolate the equipment for inspection and repairs. Isolating means not suitable to interrupt the load current of the circuit shall be interlocked with an approved circuit interrupter or be provided with a sign warning not to open under load. **(See Figure 11-80)**

Figure 11-80. Isolating switches not capable of interrupting the load current shall be interlocked with an approved circuit interrupter or be equipped with a warning sign not to open under load.

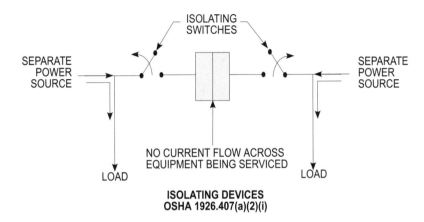

MOBILE AND PORTABLE EQUIPMENT

Mobile and portable equipment shall be designed and installed in a safe manner. Enclosures shall be grounded and live parts shall be enclosed to protect personnel from electrical shock.

POWER CABLE CONNECTIONS TO MOBILE MACHINES

A metallic enclosure shall be provided on the mobile machine to enclose the terminals of the power cable. The enclosure shall be equipped with provisions for the connection of the ground wire(s) terminal to properly ground the frame of the machine. The method used to terminate the cable from stressing the electrical connections. The enclosure is to have provisions for locking, to allow only qualified personnel to open it. The enclosure shall be marked with a sign warning of the presence of energized parts operating at over 600 volts.

GUARDING LIVE PARTS

Energized switching and control parts shall be enclosed in properly grounded metal cabinets or enclosures. Circuit breakers and protective equipment shall have the operating means projecting through the metal cabinet or enclosure in such a manner so the units can be reset without opening locked doors. Enclosures and metal cabinets shall be locked to allow only qualified personnel to have access. The enclosure shall be marked with a sign warning of the presence of energized parts. Collector ring assemblies on revolving-type machines such as shovels, draglines, etc. shall be guarded.

TUNNEL INSTALLATIONS

Wiring methods installed in tunnels that operate at over 600 volts shall be installed in a safe manner. Conductors and live parts shall be enclosed with metal enclosures, raceways, etc. being properly grounded to protect personnel from a fatal electrical shock.

The provisions of this Part shallbe applied to the installations and use of high-voltage power distribution and utilization equipment that is portable, mobile or both such as:

- Substations
- Trailers or cars
- Mobile shovels
- Draglines
- Hoists
- Drills
- Dredges
- Compressors
- Pumps
- Conveyors
- Underground excavators

CONDUCTORS

Conductors in tunnels shall be designed and installed in one or more of the following wiring methods:

- Metal conduit or other metal raceways
- Type MC cable
- Other approved multiconductor cable

Conductors shall be located or guarded to protect them from physical damage. Multiconductor portable cable shall be permitted to supply mobile equipment. An equipment grounding conductor shall be run with circuit conductors inside the metal raceway or inside the multiconductor cable jacket. The equipment grounding conductor shall be permitted to be insulated or bare.

GUARDING LIVE PARTS

Any and all bare terminals of transformers, switches, motor controllers and other equipment shall be enclosed to prevent personnel from accidentally contacting energized parts. All enclosures used in tunnels shall be drip-proof, weatherproof or of the submersible type as required by the conditions of use.

DISCONNECTING MEANS

A disconnecting means that opens all ungrounded (phase) conductors at the same time shall be installed at each transformer or motor location for the protection of personnel serving such equipment.

GROUNDING AND BONDING

Any and all non-current-carrying metal parts of electrical equipment and metal raceways and cable sheaths shall be grounded and bonded to all metal pipes and rails. This grounding and bonding shall be done at the portal and at intervals not exceeding 1000 ft. throughout the tunnel. **(See Figure 11-81)**

Figure 11-81. All metal parts of noncurrent-carrying enclosures, raceways, etc. in tunnels shall be grounded and bonded to metal pipes and rails at the portal and at intervals not exceeding 1000 ft. throughout the tunnel.

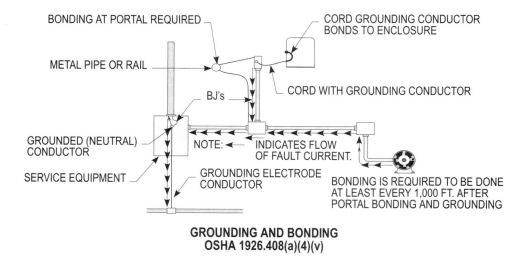

GROUNDING AND BONDING
OSHA 1926.408(a)(4)(v)

CLASS 1, CLASS 2 AND CLASS 3 REMOTE CONTROL, SIGNALING AND POWER-LIMITED CIRCUITS
OSHA 1926.408(b)

Class 1, Class 2 or Class 3 remote-control, signaling or power-limited circuits are characterized by their usage and electrical power limitation, which differentiates them from light and power circuits. These circuits are classified with their respective voltage and power limitations. They shall comply with the following provisions.

CLASS 1 CIRCUITS

A Class 1 power-limited circuit is supplied from a source having a rated output of 30 volts or less and 1000 volt amps or less.

A Class 1 nonpower-limited remote-control circuit or a Class 1 signaling circuit is one whose voltage is rated at 600 volts or less. However, the power output of the source does not have to be limited. **(See Figure 11-82)**

Figure 11-82. Class 1 circuits can be provided with outputs that are power-limited or nonpower-limited.

NOTE 1: A CLASS 1 POWER-LIMITED CIRCUIT IS SUPPLIED FROM A SOURCE HAVING A RATED OUTPUT OF 30 VOLTS AND 1000 VOLT-AMPS.

NOTE 2: A CLASS 1 NONPOWER-LIMITED REMOTE CONTROL CIRCUIT OR A CLASS 1 SIGNALING CIRCUIT IS ONE WHOSE VOLTAGE IS RATED AT 600 VOLTS OR LESS.

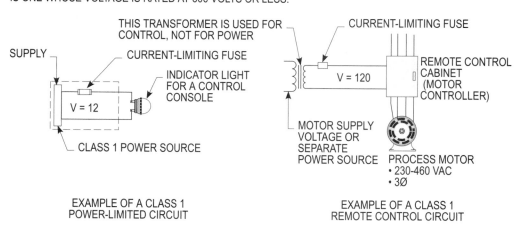

POWER LIMITED OR NONPOWER LIMITED CIRCUITS
OSHA 1926.408(b)(1)(i)(A); (B)

CLASS 2 AND CLASS 3 CIRCUITS

Class 2 and Class 3 circuits shall meet the provisions of the following:

Power for Class 2 and Class 3 circuits is limited either inherently, in which case no overcurrent protection device is required, or by a combination of a power source and overcurrent protection device. **(See Figure 11-83)**

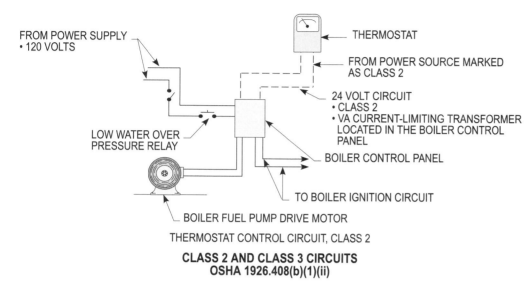

FROM POWER SUPPLY
• 120 VOLTS

THERMOSTAT

FROM POWER SOURCE MARKED AS CLASS 2

24 VOLT CIRCUIT
• CLASS 2
• VA CURRENT-LIMITING TRANSFORMER LOCATED IN THE BOILER CONTROL PANEL

LOW WATER OVER PRESSURE RELAY

BOILER CONTROL PANEL

TO BOILER IGNITION CIRCUIT

BOILER FUEL PUMP DRIVE MOTOR

THERMOSTAT CONTROL CIRCUIT, CLASS 2

**CLASS 2 AND CLASS 3 CIRCUITS
OSHA 1926.408(b)(1)(ii)**

Figure 11-83. This illustration shows that Class 2 circuits usually operate at 30 volts or less and 100 volts-amps or less.

The maximum circuit voltage is 150 volts or less for a Class 3 inherently limited power source, and 100 volts or less for a Class 2 inherently limited power source.

The maximum circuit supply voltage for a Class 2 power supply is 30 volts or less for AC and 60 volts or less for DC that is limited by overcurrent protection devices. A Class 3 power supply that is limited by overcurrent protection devices shall not exceed 150 volts for AC or DC. **(See Figure 11-84)**

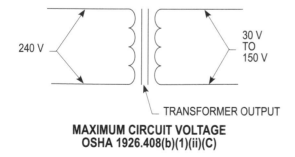

240 V

30 V
TO
150 V

TRANSFORMER OUTPUT

**MAXIMUM CIRCUIT VOLTAGE
OSHA 1926.408(b)(1)(ii)(C)**

Figure 11-84. This illustration shows that Class 3 circuits usually operate at 30 volts up to 150 volts.

MARKING

A Class 2 or Class 3 power supply unit shall be durably marked and plainly visible to indicate the class of supply circuit and its electrical rating used to supply the system.

COMMUNICATIONS SYSTEMS
OSHA 1926.408(c)

The provisions listed apply basically to systems that are connected to a central station and operate as part of a central station system.

SCOPE

These requirements for communication systems apply to systems such as central station connected and non-central station connected telephone circuits, radio and television receiving and transmitting equipment. Other apparatus included are community antenna television and radio distribution systems, telegraph, district messenger, and outside wiring for fire and burglar alarm and central station systems that are installed in premises.

PROTECTIVE DEVICES

Protective devices shall comply with the following:

- Communication circuits that are located so as to be exposed to accidental contact with light or power conductors operating at over 300 volts shall be provided with additional protection. Each circuit exposed shall be provided with a protector that is listed.

- Each conductor of a lead-in for outdoor antennas shall be provided with an antenna discharge unit or other means that will safely drain static charges from the antenna system.

CONDUCTOR LOCATION

The location of conductors shall comply with the following:

Outside of Buildings

- Receiving distribution lead-in or aerial drop cables attached to buildings and lead-in conductors to radio transmitters shall be installed to avoid the possibility of accidental contact with electric light or power conductors.

- The clearance between lead-in conductors and any lightning protection conductors shall be at least 6 ft. This is to prevent crossover flash overvoltage from a lightning strike, which can damage the components of such systems.

See Figure 11-85 for a detailed illustration pertaining to the location of conductors outside of buildings.

Figure 11-85. The clearance between lead-in conductors and any lightning protection conductors shall be at least 6 ft. to help prevent a lightning strike from flashing across to circuit elements.

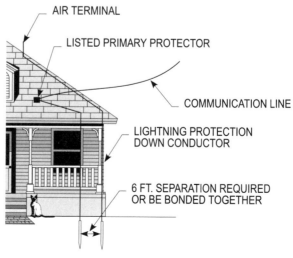

AIR TERMINAL

LISTED PRIMARY PROTECTOR

COMMUNICATION LINE

LIGHTNING PROTECTION DOWN CONDUCTOR

6 FT. SEPARATION REQUIRED OR BE BONDED TOGETHER

**SEPARATION OF ELECTRODES
OSHA 1926.408(c)(3)**

On Poles

- Where it is possible to install communications conductors on poles, they shall be located below the light or power conductors and shall not to be attached to a cross arm that carries light or power conductors. **(See Figure 11-86)**

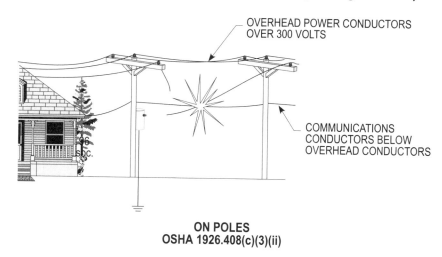

OVERHEAD POWER CONDUCTORS OVER 300 VOLTS

COMMUNICATIONS CONDUCTORS BELOW OVERHEAD CONDUCTORS

ON POLES
OSHA 1926.408(c)(3)(ii)

Figure 11-86. Communications conductors, if possible, shall be installed below power conductors where they are supported on poles.

Inside of Buildings

- Indoor antennas, lead-ins and other communications conductors attached as open conductors to the inside of buildings shall be located at least 2 in. from conductors of any light or power or Class 1 circuits. Where a method of conductor separation is listed to do the job properly, the 2 in. separation shall not be required. **(See Figure 11-87)**

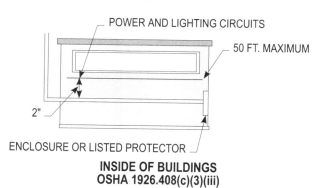

POWER AND LIGHTING CIRCUITS

50 FT. MAXIMUM

2"

ENCLOSURE OR LISTED PROTECTOR

INSIDE OF BUILDINGS
OSHA 1926.408(c)(3)(iii)

Figure 11-87. This illustration shows that lead-in conductors for antennas communications circuits that enter the premises shall be separated at least 2 in. from power and light conductors. (See NEC 800.50, Ex. 3 and 800.52(A)(2)).

EQUIPMENT LOCATION

Outdoor metal structures supporting antennas, as well as self-supporting antennas such as vertical rods or dipole structures, shall be located well away from overhead conductors. Such overhead conductors are electric light and power circuits that are rated over 150 volts-to-ground. Antenna or structures shall be located to avoid the possibility of falling into or making contact with these circuits rated over 150 volts to-ground or 250 volts between conductors.

GROUNDING

Communications systems shall be grounded and bonded into the grounding electrode system to create an equipotential plane. This grounding procedure will help all the systems to elevate with as much equal voltage rise as possible in case of a lightning strike. The system shall be grounded as follows:

LEAD-IN CONDUCTORS

Where exposed to contact with electric light and power conductors, the metal sheath of aerial cables entering buildings shall be grounded or be interrupted close to the entrance to the building by an insulating joint or equivalent device. Where protective devices are installed, they shall be grounded in a suitable manner.

ANTENNA STRUCTURES

All masts and metal structures supporting antennas shall be grounded without splice or connection between the grounding conductor and the grounding electrode. **(See Figure 11-88)**

Figure 11-88. This illustration shows the bonding and grounding requirements for communications circuits, radio and television equipment and CATV circuits.

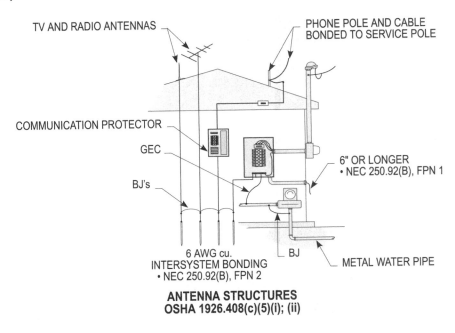

TV AND RADIO ANTENNAS

PHONE POLE AND CABLE BONDED TO SERVICE POLE

COMMUNICATION PROTECTOR

GEC

BJ's

6" OR LONGER
• NEC 250.92(B), FPN 1

6 AWG cu.
INTERSYSTEM BONDING
• NEC 250.92(B), FPN 2

BJ

METAL WATER PIPE

ANTENNA STRUCTURES
OSHA 1926.408(c)(5)(i); (ii)

EQUIPMENT ENCLOSURES

Transmitters shall be enclosed in a metal frame or grill or be separated from the operating space by a barrier or other equivalent means. All noncurrent-carrying metallic parts shall be connected to ground (earth). All external metal handles and controls that are accessible to operating personnel shall be grounded. Unpowered equipment and enclosures shall be considered grounded where connected to an attached coaxial cable with its metallic shield grounded properly.

SAFETY RELATED WORK PRACTICES
OSHA 1926.416

The following shall be considered for safety related work practices:

- Protection of employees
- Passageways and open spaces
- Load ratings
- Fuses
- Cords and cables

PROTECTION OF EMPLOYEES
OSHA 1926.409(a)(1)

No employer shall permit an employee to work in such proximity to any part of an electric power circuit that the employee could contact the electric power circuit in the course of work, unless the employee is protected against electric shock by deenergizing the circuit and grounding it or by guarding it effectively by insulation or other means. **(See Figures 11-89(a), (b) and (c))**

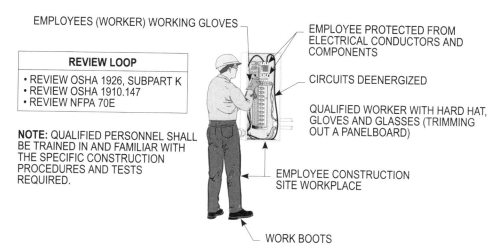

Figure 11-89(a). An employee (worker) shall follow specific requirements when performing a job task in a construction site environment.

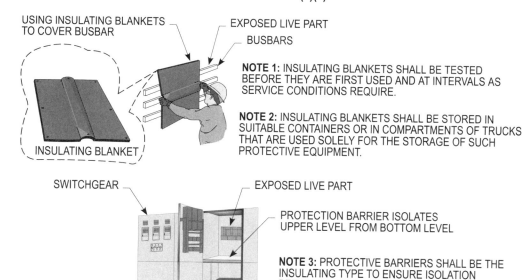

Figure 11-89(b). This illustration shows the procedures to be applied when working on exposed live parts in electrical equipment.

In work areas where the exact location of underground electric power lines is unknown, employees using jackhammers, bars, or other hand tools that may contact a line shall be provided with insulated protective gloves.

Figure 11-89(c). This illustration shows the procedures to be applied when grounding out deenergized parts for safe work conditions.

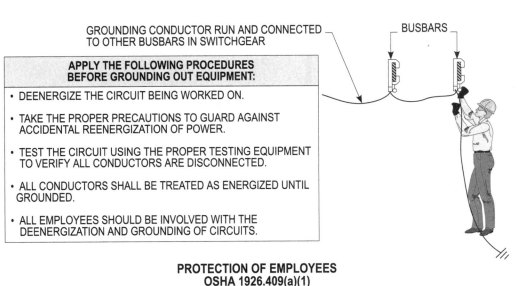

GROUNDING CONDUCTOR RUN AND CONNECTED TO OTHER BUSBARS IN SWITCHGEAR

BUSBARS

> **APPLY THE FOLLOWING PROCEDURES BEFORE GROUNDING OUT EQUIPMENT:**
>
> • DEENERGIZE THE CIRCUIT BEING WORKED ON.
>
> • TAKE THE PROPER PRECAUTIONS TO GUARD AGAINST ACCIDENTAL REENERGIZATION OF POWER.
>
> • TEST THE CIRCUIT USING THE PROPER TESTING EQUIPMENT TO VERIFY ALL CONDUCTORS ARE DISCONNECTED.
>
> • ALL CONDUCTORS SHALL BE TREATED AS ENERGIZED UNTIL GROUNDED.
>
> • ALL EMPLOYEES SHOULD BE INVOLVED WITH THE DEENERGIZATION AND GROUNDING OF CIRCUITS.

PROTECTION OF EMPLOYEES
OSHA 1926.409(a)(1)

Before work is begun the employer shall ascertain by inquiry or direct observation, or by instruments, whether any part of an energized electric power circuit, exposed or concealed, is so located that the performance of the work may bring any person, tool, or machine into physical or electrical contact with the electric power circuit. The employer shall post and maintain proper warning signs where such a circuit exists. The employer shall advise employees of the location of such lines, the hazards involved, and the protective measures to be taken.

PASSAGEWAYS AND OPEN SPACES
OSHA 1926.416(b)

Barriers or other means of guarding shall be provided to ensure that workspace for electrical equipment will not be used as a passageway during periods when energized parts of electrical equipment are exposed.

Working spaces, walkways and similar locations shall be kept clear of cords so as not to create a hazard to employees.

LOAD RATINGS
OSHA 1926.416(c)

In existing installations, no changes in circuit protection shall be made to increase the load in excess of the load rating of the circuit wiring.

FUSES
OSHA 1926.416(d)

When fuses are installed or removed with one or both terminals energized, special tools insulated for the voltage shall be used.

CORDS AND CABLES
OSHA 1926.416(e)

Worn or frayed electric cords or cables shall not be used.

Extension cords shall not be fastened with staples, hung from nails, or suspended by wire.

LOCKOUT AND TAGGING OF CIRCUITS
OSHA 1926.417

The following shall be considered for lockout and tagging of circuits:

- Controls
- Equipment and circuits
- Tags

CONTROLS
OSHA 1926.417(a)

Controls that are to be deactivated during the course of work on energized or deenergized equipment or circuits shall be tagged.

EQUIPMENT AND CIRCUITS
OSHA 1926.417(b)

Equipment or circuits that are deenergized shall be rendered inoperative and shall have tags attached at all points where such equipment or circuits can be energized.

TAGS
OSHA 1926.417(c)

Tags shall be placed to identify plainly the equipment or circuits being worked on.

The employer shall ensure that all wiring components and utilization equipment in hazardous locations are maintained in a dust-tight, dust-ignition-proof or explosion-proof condition, as appropriate.

MAINTENANCE OF EQUIPMENT
OSHA 1926.431

There shall be no loose or missing screws, gaskets, threaded connections, seals or other impairments to a tight condition.

ENVIRONMENTAL DETERIORATION OF EQUIPMENT
OSHA 1926.432

The following shall be considered for environmental deterioration of equipment:

- Deteriorating agents
- Protection against corrosion

DETERIORATING AGENTS
OSHA 1926.432(a)

Unless identified for use in the operating environment, no conductors or equipment shall be located:

- In damp or wet locations
- Where exposed to gases, fumes, vapors, liquids or other agents having a deteriorating effect on the conductors or equipment
- Where exposed to excessive temperatures

Control equipment, utilization equipment and busways approved for use in dry locations only shall be protected against damage from the weather during building construction.

PROTECTION AGAINST CORROSION
OSHA 1926.416(b)

Metal raceways, cable armor, boxes, cable sheathing, cabinets, elbows, couplings, fittings, supports and support hardware shall be of materials appropriate for the environment in which they are to be installed.

BATTERIES AND BATTERY CHARGING
OSHA 1926.441

The following shall be considered for batteries and battery charging:

- General requirements
- Charging

GENERAL REQUIREMENTS
OSHA 1926.441(a)

Batteries of the unsealed type shall be located in enclosures with outside vents or in well ventilated rooms and shall be arranged so as to prevent the escape of fumes, gases or electrolyte spray into other areas.

Ventilation shall be provided to ensure diffusion of the gases from the battery and to prevent the accumulation of an explosive mixture.

Racks and trays shall be substantial and shall be treated to make them resistant to the electrolyte.

Floors shall be of acid resistant construction unless protected from acid accumulations.

Face shields, aprons, and rubber gloves shall be provided for workers handling acids or batteries.

Facilities for quick drenching of the eyes and body shall be provided within 25 ft. (7.62 m) of battery handling areas.

Facilities shall be provided for flushing and neutralizing spilled electrolyte and for fire protection.

CHARGING
OSHA 1926.441(b)

Battery charging installations shall be located in areas designated for that purpose.

Charging apparatus shall be protected from damage by trucks.

When batteries are being charged, the vent caps shall be kept in place to avoid electrolyte spray. Vent caps shall be maintained in functioning condition.

Electrical

Section	Answer	
_____	T	F
_____	T	F
_____	T	F
_____	T	F
_____	T	F
_____	T	F
_____	T	F
_____	T	F
_____	T	F
_____	T	F
_____	T	F
_____	T	F
_____	T	F
_____	T	F
_____	T	F

Electrical

1. Conductors and equipment installed or used on construction sites shall be approved by the authority having jurisdiction.

2. Glue-on paper labels are permitted to be used for identification of the disconnecting means.

3. A minimum working space of 36 in. width is required in front of electrical equipment operating at 600 volts or less.

4. If rear access is necessary to work on deenergized parts on the back enclosed equipment, a minimum working space of 30 in. horizontally is required.

5. A minimum headroom clearance of 6 ft. 7 in. shall be maintained from the floor or platform up to the luminaire (lighting fixture) or any overhead obstruction.

6. A platform shall be permitted to guard exposed live parts of electrical equipment.

7. There shall be at least one entrance a minimum of 24 in. x 6 1/2 ft. to provide access to the working space of high-voltage equipment (over 600 volts).

8. Overhead conductors shall be installed a height of at least 12 ft. in areas that are subject to vehicular traffic other than truck travel.

9. Conductors shall have a clearance of at least 5 ft. from windows, porches, platforms, etc.

10. All ungrounded service conductors shall be simultaneously disconnected by each service equipment disconnecting means.

11. Live parts of electrical systems rated over 600 volts shall be posted with warning signs.

12. The frame of a vehicle shall not be permitted to be used to ground a circuit supplied by a generator on the vehicle.

13. Short pieces of conduit, 25 ft. or less in length shall not be required to be grounded.

14. Rod, pipe and plate electrodes shall have a resistance to ground of 25 ohms or less, wherever practicable.

15. The grounding electrode for a high-voltage circuit feeding portable equipment shall be separated at least 6 ft. from any other electrode.

Section	Answer
_____	T F
_____	T F
_____	T F
_____	T F
_____	T F
_____	T F
_____	T F
_____	T F
_____	T F
_____	T F
_____	T F
_____	T F
_____	T F
_____	T F
_____	T F

16. Where branch-circuits are run as open conductors for temporary wiring, they shall be attached at intervals not exceeding 20 ft.

17. Extension cords sets used with portable electric tools and appliances shall be of three-wire type and shall be designed for hard or extra-hard usage.

18. In a damp or wet location, cutout boxes and surface mounted cabinets shall be set off the wall at least 1/2 in. or more to prevent condensation between the wall and the enclosure.

19. Luminaires (lighting fixtures) shall be located at least 8 ft. above the floor with exposed live parts.

20. Where it is required that one piece of equipment be "in sight from" another piece of equipment, one shall be located in a place that is visible and not more than 50 ft. from each other.

21. Materials shall be permitted to be stored in transformer vaults.

22. The working space for cranes and hoists in the direction of access to live parts that are likely to require examination, adjustment, servicing or maintenance shall have a minimum clearance of 2 1/2 ft.

23. Elevators shall have an individual disconnecting means to disconnect all ungrounded main power supply conductors to each unit.

24. Equipment that is not marked to indicate a division, or is marked "Division 1" or "Div. 1", shall be suitable for either Division 1 and 2 locations.

25. Fixed lighting fixtures that are marked for use in Class I, Division 2, or Class II, Division 2 locations only, shall be marked to indicate the group.

26. All conduits used in hazardous locations shall be threaded and shall be made up wrenchtight with at least five fully engaged threads.

27. The clearance for communication lead-in conductors and any lightning protection conductors shall be at least 3 ft.

28. Indoor antennas, lead-ins and other communications conductors attached as open conductors to the inside of buildings shall be located at least 1 in. from conductors of any light or power or Class 1 circuits.

29. Batteries of the unsealed type shall be located in enclosures with outside vents or in well ventilated rooms.

30. Facilities for quick drenching of the eyes and body shall be provided within 50 ft. of battery handling areas.

12

Scaffolding

Scaffolding is a vital part of the construction process. To ensure the safety of workers and the public, scaffolding must be handled properly. A competent person should be present at the worksite during the entire scaffold construction process to ensure that all scaffolding is erected, moved, used, and dismantled safely.

The five most serious scaffold hazards are:

- Falls
- Unsafe access
- Struck by falling objects
- Electrocution
- Scaffold collapse

SCOPE, APPLICATION AND DEFINITIONS
OSHA 1926.450

OSHA 1926.450(a) applies to all scaffolds used in construction, alteration, repair (including painting and decorating) and demolition operations covered under 29 CFR Part 1926, except that crane or derrick suspended personnel platforms will continue to be regulated under OSHA 1926.550(g). In addition, aerial lifts are covered exclusively in OSHA 1926.453, as noted in OSHA 1926.450(a).

OSHA 1926.450(b) lists and defines all major terms used in Subpart L.

CAPACITY
OSHA 1926.451(a)

OSHA 1926.451(a) sets the minimum strength criteria for all scaffold components and connections.

Each scaffold and scaffold component shall be capable of supporting, without failure, its own weight and at least 4 times the maximum intended load applied or transmitted to it. OSHA 1926.451(a)(2), (a)(3), (a)(4), (a)(5) and OSHA 1926.451(g) provide exceptions to this general rule, and are discussed below.

OSHA provides that the 4 to 1 factor for a component applies only to the load that is actually applied or transmitted to that component, and not to the total load placed on the scaffold. OSHA requires that each component be adequate to meet the 4 to 1 factor, but only for the portion of the MIL applied or transmitted to that component. The MIL for each component depends on the type and configuration of the scaffold system.

Direct connections to roofs and floors and counterweights used to balance adjustable suspension scaffolds shall be capable of resisting at least 4 times the tipping moment imposed by the scaffold operating at the rated load of the UL rated hoist or at 1.5 (minimum) times the tipping moment imposed by the scaffold operating at the stall load for hoist not UL rated, whichever is greater.

Suspension rope, including its connecting hardware, used on non-adjustable suspension scaffolds shall be capable of supporting, without failure, at least 6 times the maximum intended load applied or transmitted to that rope.

Each suspension rope, including connecting hardware, used on adjustable suspension scaffolds shall be capable of supporting, without failure, at least 6 times the maximum intended load applied or transmitted to that rope with the scaffold operating at either (a) the rated load of the hoist or (b) 2 (minimum) times the stall load of the hoist, whichever is greater.

The stall load of any scaffold hoist cannot exceed 3 times its rated load. OSHA finds that this requirement is reasonably necessary to prevent accidental overloading of suspension scaffold support systems. UL standard 1323 limits the output force of a scaffold hoist to three times the rated load of the hoist.

All scaffolds shall be designed by a qualified person and constructed and loaded in accordance with that design.

SCAFFOLD PLATFORM CONSTRUCTION
OSHA 1926.451(b)

All platforms, except walkways and those platforms used by employees performing scaffold erection and dismantling operations, shall be fully decked or planked.

All platform units shall be placed so that spaces between units do not exceed 1 in., except where employers establish that more space is needed.

For example: This would be necessary to fit around uprights when using side brackets to extend platform width.

If this exception applies, employers shall place platform units as close together as possible, with the space between the platform and uprights not to exceed 9 in. OSHA sets 9 in. as the maximum space allowed, because the minimum width for scaffold units that could be expected to sustain a working load is just over 9 in.

In a situation where no work, other than erecting or dismantling the scaffold, is being done at intermediate levels, OSHA requires only that the planking established by the employer as necessary to provide safe working conditions for employees erecting or dismantling the scaffold be used. On the other hand, if scaffold erection or dismantling is being performed from an intermediate level platform that is being or will be used as a work area, that platform shall be fully planked in accordance with OSHA 1926.451(b)(1).

OSHA believes that platforms used solely as walkways or solely by employees erecting or dismantling scaffolds should be at least 2 planks wide.

All scaffold platforms and walkways shall be at least 18 in. wide, with lesser widths allowed for ladder jack scaffolds, top plate bracket scaffolds, pump jack scaffolds, roof bracket scaffolds and boatswains' chairs, and for scaffolds in areas shown to be too narrow to accommodate an 18 in. wide surface.

The rationale for setting a 12 in. minimum width for ladder jack scaffolds was the difficulty of handling one 18 in. wide plank or two 9 in. planks on a ladder, which is considered more hazardous than working on a 12 in. wide plank. Pump jack scaffolds are the exception to OSHA 1926.451(b)(2), for which a minimum platform width of 12 in. is permitted. In addition, top plate bracket scaffolds are permitted to have platforms not less than 12 in. in width.

The space between the front edge of a platform and the face of the structure where the scaffold is being used shall be no more than 14 in. from the face of the structure, unless the employer implements guardrail systems or personal fall arrest systems that comply with OSHA 1926.451(g) to protect employees from falling between the platform and the structure. The following are exceptions to this rule:

- Front edges of outrigger scaffolds shall be no more than 3 in. from the face of the structure.

- Front edges of scaffolds used for plastering and lathing operations shall be no more than 18 in. from the face of the structure.

- Each end of a platform unit, unless cleated or otherwise restrained by hooks or equivalent means, shall extend over the center line of its support at least 6 in. The use of cleats, hooks and similar securing devices is allowed as an alternative to the 6 in. extension, because of their ability to restrain movement of platform units.

- Each end of a platform unit 10 ft. or less in length shall not extend over its support more than 12 in. unless the unit is designed, and installed so that the cantilevered portion of the unit is able to support employees or material without tipping or has guardrails that prevent employee access to the cantilevered end. **(See Figure 12-1)**

- Each platform unit greater than 10 ft. in length shall not extend over its support more than 18 in., unless the unit is designed and installed so that the cantilevered portion of the unit is able to support employees without tipping, or that the unit has guardrails that block employee access to the cantilevered end.

- Where platform units are abutted to create a long platform, each abutted end shall rest on a separate support surface. Abutted platform units do not rest one on another, but instead are end-to-end. Consequently, one unit does not support the other, and proper support can only be provided by separate support surfaces.

Safety Tip: All platforms and walkways shall be at least 18 in. wide.

Figure 12-1. This illustration shows the distance requirements for each end of a platform unit 10 ft. or less in length.

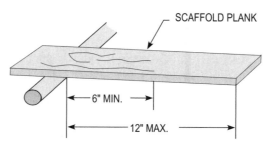

SCAFFOLD PLANK

6" MIN.

12" MAX.

NOTE: EACH PLATFORM UNIT GREATER THAN 10 FT. IN LENGTH SHALL NOT EXTEND OVER ITS SUPPORT MORE THAN 18 IN., UNLESS THE UNIT IS DESIGNED AND INSTALLED SO THAT THE CANTILEVERED PORTION OF THE UNIT IS ABLE TO SUPPORT EMPLOYEES WITHOUT TRIPPING, OR THAT THE UNIT HAS GUARDRAIL THAT BLOCK EMPLOYEE ACCESS TO THE CANTILEVERED END.

**GENERAL REQUIREMENTS
OSHA 1926.451(b)(5)(i)**

• Where platforms are overlapped to create a long platform, the overlap shall occur only over supports, and shall not be less than 12 in. unless the platforms are nailed together or otherwise restrained to prevent movement. **(See Figure 12-2)**

Figure 12-2. This illustration shows the planking requirements for platforms that are overlapped.

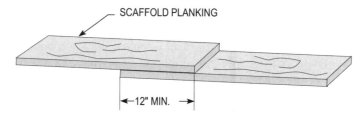

SCAFFOLD PLANKING

12" MIN.

NOTE: ALL POINTS OF A SCAFFOLD WHERE THE PLATFORM CHANGES DIRECTION, SUCH AS TURNING A CORNER, AND ANY PLATFORM THAT RESTS ON A BEARER AT AN ANGLE OTHER THAN A RIGHT ANGLE SHALL BE LAID FIRST AND PLATFORMS THAT REST AT RIGHT ANGLES OVER THE SAME BEARER SHALL BE LAID SECOND, ON TOP OF THE FIRST PLATFORM.

**GENERAL REQUIREMENTS
OSHA 1926.451(b)(7)**

• All points of a scaffold where the platform changes direction, such as turning a corner, any platform that rests on a bearer at an angle other than a right angle shall be laid first and platforms which rest at right angles over the same bearer shall be laid second, on top of the first platform.

Wood platforms shall not be covered with opaque finishes, except that platform edges may be covered or marked for purposes of identification. Platforms may be coated periodically with wood preservatives, fire-retardant finishes and slip-resistant finishes, but the coating may not obscure the top or bottom wood surfaces. This requirement is intended to ensure that structural defects in platforms are not covered from view by the use of an opaque coating or finish. Hairline cracks can significantly reduce the strength of a wood member, so early detection of structural defects is important. Opaque finishes can cover such cracks and make them difficult to discover. The edges of platform units are excepted from this rule to allow identification marks, grading marks or other similar type of marks to be placed on the unit edges.

Safety Tip: Components by different manufacturers cannot be intermixed, unless the parts fit without using force and the scaffold's integrity is maintained.

Scaffold components manufactured by different manufacturers cannot be intermixed unless the component parts fit together without force and the scaffold's structural integrity is maintained by the user. Scaffold components manufactured by different manufacturers shall not be modified in order to intermix them unless the resulting scaffold is determined by a competent person to be structurally sound. OSHA expects that the competent person who evaluates the scaffold will have the appropriate knowledge, skill and experience regarding scaffold systems and components. **(See Figure 12-3)**

NOTE 1: SCAFFOLD COMPONENTS BY DIFFERENT MANUFACTURERS CANNOT BE MODIFIED IN ORDER TO INTERMIX THEM UNLESS THE RESULTING SCAFFOLD IS DETERMINED BY A COMPETENT PERSON TO BE STRUCTURALLY SOUND.

NOTE 2: SCAFFOLD COMPONENTS BY DIFFERENT MANUFACTURERS CANNOT BE INTERMIXED UNLESS COMPONENT PARTS FIT TOGETHER WITHOUT FORCE.

Figure 12-3. This illustration shows various scaffold components available by manufacturers.

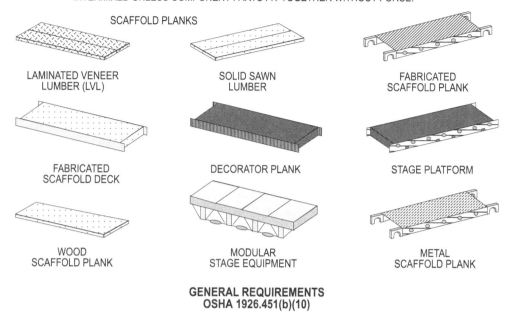

SCAFFOLD PLANKS

LAMINATED VENEER LUMBER (LVL)	SOLID SAWN LUMBER	FABRICATED SCAFFOLD PLANK
FABRICATED SCAFFOLD DECK	DECORATOR PLANK	STAGE PLATFORM
WOOD SCAFFOLD PLANK	MODULAR STAGE EQUIPMENT	METAL SCAFFOLD PLANK

GENERAL REQUIREMENTS
OSHA 1926.451(b)(10)

Scaffold components made of dissimilar metals shall not be used together unless a competent person has determined that galvanic action will not reduce the strength of any component to a level below that required by OSHA 1926.451(a).

CRITERIA FOR SUPPORTED SCAFFOLDS
OSHA 1926.451(c)

Scaffolds with a height to base width ratio of more than 4 to 1 (including outrigger supports, if used) shall be restrained from tipping by guying, tying, bracing or equivalent means. **(See Figures 12-4(a) and (b))**

Guys, ties and braces shall be installed at locations where horizontal members support both inner and outer legs. OSHA 1926.451(c)(1)(ii) requires the following:

- Guys, ties and braces shall be installed according to the scaffold manufacturer's recommendations or at the closest horizontal member to the 4:1 height and be repeated vertically at locations of horizontal members every 20 ft. or less thereafter for scaffolds 3 ft. wide or less and every 26 ft. or less thereafter for scaffolds greater than 3 ft. wide.

- The top tie, guy or brace of a completed scaffold shall be placed no further than the 4:1 height from the top.

- Guys, ties and braces shall be installed at each end of the scaffold and at horizontal intervals not to exceed 30 ft. (measured from one end (not both) towards the other).

- Scaffolds with eccentric loads (such as cantilevered work platforms) shall be restrained from tipping through the use of ties, guys, braces or outriggers.

Figure 12-4(a). This illustration shows the requirements for maximum vertical tie spacing wider than 3 ft. bases.

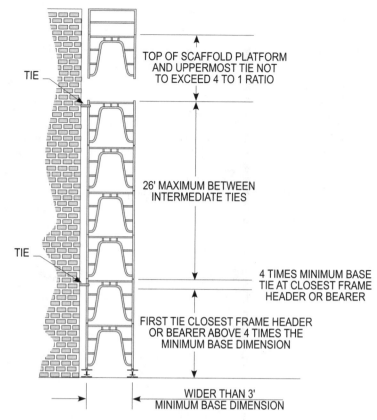

MAXIMUM VERTICAL TIE SPACING WIDER THAN 3' BASES
OSHA 1926.451(c)

Figure 12-4(b). This illustrations shows the requirements for maximum vertical tie spacing for 3 ft. and narrower bases.

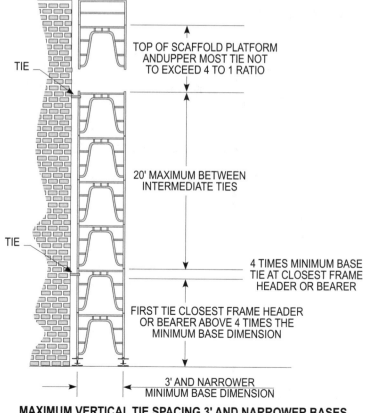

MAXIMUM VERTICAL TIE SPACING 3' AND NARROWER BASES
OSHA 1926.451(c)

OSHA 1926.451(c)(2) requires the following:

- Supported scaffold poles, legs, posts, frames and uprights bear on base plates and mud sills or other adequate firm foundation.

- Footings shall be level, sound, rigid and capable of supporting the scaffold in a loaded condition without settling or displacement.

- Unstable objects shall neither be used to support scaffolds or platform units, nor be used as working platforms, respectively.

- Front-end loaders and similar pieces of equipment shall not be used as scaffold supports unless specifically designed by the manufacturer for such use.

- Forklifts shall not be used to support scaffold platforms unless the entire platform is attached to the fork and the forklift is not moved horizontally while the platform is occupied. Both these requirements relate to the need for solid support for scaffold platforms and reflect the fact that front-end loaders, forklifts and other such equipment are not generally designed for this purpose.

- Supported scaffold poles, legs, posts, frames and uprights shall be plumb and braced to prevent swaying and displacement.

CRITERIA FOR SUSPENSION SCAFFOLDS
OSHA 1926.451(d)

All suspension scaffold support devices, such as outrigger beams, cornice hooks, parapet clamps and similar devices, shall rest on surfaces capable of supporting at least 4 times the loads imposed on them by the scaffold operating at the rated load of the hoist (or at least 1.5 times the loads imposed on them by the scaffold operating at the stall load of the hoist, whichever is greater).

Suspension scaffold outrigger beams, when used, shall be made of structural metal, or equivalent strength material, and be restrained to prevent movement.

Outrigger beams shall be secured directly to the supporting surface or be stabilized using counterweights, except that masons' multi-point adjustable suspension scaffolds shall not be stabilized by counterweights. The rule does not allow counterweights for stabilizing such masons' suspension scaffolds because, with the large loads often placed on masons' multi-point adjustable suspension scaffolds and the large counterweights that would be necessary to anchor such systems, OSHA is concerned that the supporting roof or floor would become dangerously overloaded.

Safety Tip: A competent person shall evaluate direct connections.

Direct connections shall be evaluated by a competent person who affirms, based on that evaluation, that supporting surfaces can support the anticipated loads. In addition, masons' multi-point adjustable suspension scaffold connections shall be designed by an engineer experienced in such scaffold design. OSHA anticipates that compliance with these provisions will ensure that roof or floor decks are capable of supporting the loads to be imposed.

Counterweights shall meet the following requirements:

- Made of non-flowable material
- Specifically designed for use as scaffold counterweights
- Secured to outrigger beams to prevent accidental displacement
- Not be removed from an outrigger beam until the scaffold is disassembled

See Figure 12-5 for a detailed illustration pertaining to counterweights.

Figure 12-5. This illustration shows the requirements for counterweights.

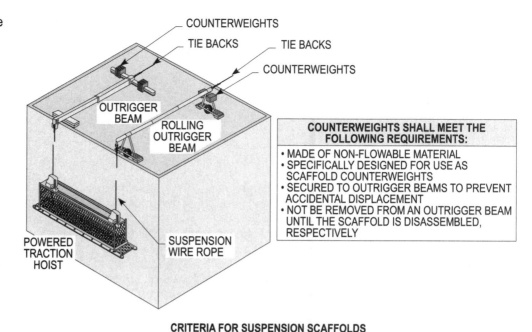

COUNTERWEIGHTS SHALL MEET THE FOLLOWING REQUIREMENTS:
- MADE OF NON-FLOWABLE MATERIAL
- SPECIFICALLY DESIGNED FOR USE AS SCAFFOLD COUNTERWEIGHTS
- SECURED TO OUTRIGGER BEAMS TO PREVENT ACCIDENTAL DISPLACEMENT
- NOT BE REMOVED FROM AN OUTRIGGER BEAM UNTIL THE SCAFFOLD IS DISASSEMBLED, RESPECTIVELY

CRITERIA FOR SUSPENSION SCAFFOLDS
OSHA 1926.451(d)(3)

These requirements are necessary to ensure that counterweights are used only for their intended purpose and are not displaced or removed prematurely.

The following requirements apply to securing outrigger beams.

- Outrigger beams not stabilized by direct connections to the supporting surface shall be secured by tiebacks.

- Tiebacks shall be as strong as the suspension ropes, be secured to a structurally sound anchorage, and be installed perpendicular to the structure unless opposing angle tiebacks are installed.

- Outrigger beams shall be placed perpendicular to their bearing support, with the exception described more fully below.

Safety Tip: Counterweights shall be designed for no other purpose than to counterweight the system.

OSHA has determined that it is reasonably necessary to require that counterweights be designed for no other purpose than to counterweight the system, and to prohibit the use of construction materials as counterweights. In addition, OSHA has determined that it is appropriate to require the marking of counterweights with their weights because that information is needed for the proper design, selection and installation of counterweights.

Outrigger beams used with suspension scaffolds shall be:

- Provided with stop bolts or shackles at both ends
- Securely fastened together with the flanges turned out when channel iron beams are used in place of I-beams
- Installed with all bearing supports perpendicular to the beam center line; and set and maintained with the web in a vertical position.

See Figure 12-6 for a detailed illustration pertaining to outrigger beams used with suspension scaffolds.

In addition, when an outrigger beam is used, the shackle or clevis with which the suspension rope is attached to the outrigger beam shall be placed directly over the hoisting machine, i.e., over the center line of the stirrup.

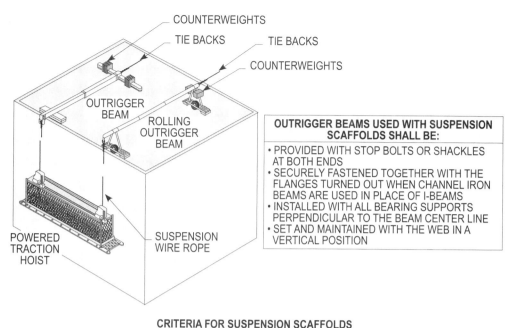

COUNTERWEIGHTS

TIE BACKS   TIE BACKS

COUNTERWEIGHTS

OUTRIGGER
BEAM   ROLLING
            OUTRIGGER
            BEAM

OUTRIGGER BEAMS USED WITH SUSPENSION SCAFFOLDS SHALL BE:
- PROVIDED WITH STOP BOLTS OR SHACKLES AT BOTH ENDS
- SECURELY FASTENED TOGETHER WITH THE FLANGES TURNED OUT WHEN CHANNEL IRON BEAMS ARE USED IN PLACE OF I-BEAMS
- INSTALLED WITH ALL BEARING SUPPORTS PERPENDICULAR TO THE BEAM CENTER LINE
- SET AND MAINTAINED WITH THE WEB IN A VERTICAL POSITION

POWERED
TRACTION
HOIST

SUSPENSION
WIRE ROPE

CRITERIA FOR SUSPENSION SCAFFOLDS
OSHA 1926.451(d)(4)

Figure 12-6. This illustration shows the requirements for outrigger beams used with suspension scaffolds.

The following requirements apply to suspension scaffold support devices other than outrigger beams. (These devices include cornice hooks, roof irons, parapet clamps or similar devices.) The devices shall be:

- Made of steel, wrought iron, or materials of equivalent strength

- Supported by bearing blocks

- Secured against movement by tiebacks installed at right angles to the face of the building or structure unless opposing angle tiebacks are installed and secured to a structurally sound point of anchorage on the building or structure (sound points of anchorage include structural members, but do not include standpipes, vents, other piping systems, or electrical conduit)

- Tiebacks shall be equivalent in strength to the strength of the hoisting rope

See Figure 12-7 for a detailed illustration pertaining to scaffold support devices other than outrigger beams.

Winding drum hoists shall have at least four wraps of suspension rope at the lowest point of scaffold travel. All other types of hoists shall have suspension rope long enough to lower scaffolds to the level below, without having the rope end pass through the hoist, or to have the rope end configured or provided with means so that the end does not pass through the hoist.

Wire rope shall meet the following requirements:

- The use of repaired wire rope as suspension rope is prohibited

- Wire suspension ropes shall not be joined together except through the use of eye splice thimbles connected with shackles or cover plates and bolts

- The load end of wire suspension ropes shall be equipped with proper size thimbles and secured by eye splicing or equivalent means

See Figure 12-8 for a detailed illustration pertaining to requirements for wire rope.

Safety Tip: Winding drum hoists shall have at least four wraps of suspension rope at the lowest point of travel.

Figure 12-7. This illustration shows the requirements for scaffold support devices other than outrigger beams.

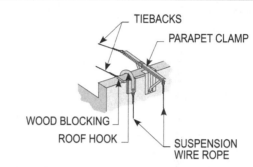

SUSPENSION SCAFFOLD SUPPORT DEVICES OTHER THAN OUTRIGGER BEAMS SHALL COMPLY WITH THE FOLLOWING REQUIREMENTS:

- BE MADE OF STEEL, WROUGHT IRON OR MATERIALS OF EQUIVALENT STRENGTH
- BE SUPPORTED BY BEARING BLOCKS
- SECURED AGAINST MOVEMENT BY TIEBACKS INSTALLED AT RIGHT ANGLES TO THE FACE OF THE BUILDING OR STRUCTURE UNLESS OPPOSING ANGLE TIEBACKS ARE INSTALLED AND SECURED TO A STRUCTURALLY SOUND POINT OF ANCHORAGE ON THE BUILDING OR STRUCTURE
- TIEBACKS SHALL BE EQUIVALENT IN STRENGTH TO THE STRENGTH OF THE HOISTING ROPE

CRITERIA FOR SUSPENSION SCAFFOLDS
OSHA 1926.451(d)(5)

Figure 12-8. This illustration shows the requirements for wire rope.

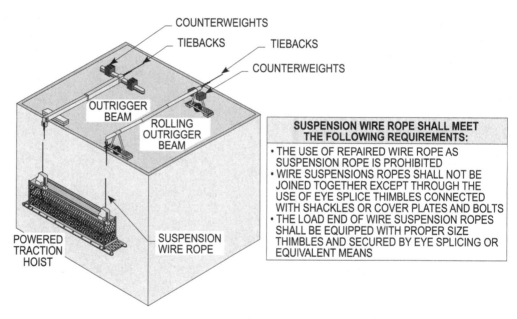

SUSPENSION WIRE ROPE SHALL MEET THE FOLLOWING REQUIREMENTS:

- THE USE OF REPAIRED WIRE ROPE AS SUSPENSION ROPE IS PROHIBITED
- WIRE SUSPENSIONS ROPES SHALL NOT BE JOINED TOGETHER EXCEPT THROUGH THE USE OF EYE SPLICE THIMBLES CONNECTED WITH SHACKLES OR COVER PLATES AND BOLTS
- THE LOAD END OF WIRE SUSPENSION ROPES SHALL BE EQUIPPED WITH PROPER SIZE THIMBLES AND SECURED BY EYE SPLICING OR EQUIVALENT MEANS

CRITERIA FOR SUSPENSION SCAFFOLDS
OSHA 1926.451(d)(6)

Safety Tip: Wire ropes shall be inspected by a competent person prior to each work shift.

Wire ropes shall be inspected for defects by a competent person prior to each work shift and after every occurrence that could affect a rope's integrity. The wire rope shall be replaced if the rope has any physical damage that impairs its function and strength; any kinks that might impair the tracking or wrapping of rope around the drum(s) or sheave(s); six randomly distributed broken wires in one rope lay or three broken wires in one strand in one rope lay; abrasion, corrosion, scrubbing, flattening or peening causing loss of more than one-third of the original diameter of the outside wires; evidence of any heat damage resulting from a torch or any damage caused by contact with electrical wires; or evidence that a secondary brake has been activated during an overspeed condition and engages the suspension rope.

Swagged attachments or spliced eyes on wire suspension ropes shall not be used unless they are made by the wire rope manufacturer or a qualified person. This provision is essential to ensure the strength and integrity of such attachments as eyes.

When wire rope clips are used on suspension scaffolds, there shall be a minimum of 3 wire rope clips installed, with the clips a minimum of 6 rope diameters apart; employers shall follow the manufacturer's recommendations when installing clips, retightening clips after initial loading and inspecting and retightening clips at the start of each work shift; U-bolt clips (a variety of wire rope clip) shall not be used at the point of suspension for any scaffold hoist; and when U-bolt clips are used, the U-bolt shall be placed over the dead end of the rope, and the saddle shall be placed over the live end of the rope.

Safety Tip: If wire rope clips are used on suspension scaffolds, there shall be a minimum of 3 wire rope clips installed.

Suspension scaffold power-operated hoists and manually operated hoists shall meet the following:

- Be of a type tested and listed by a qualified testing laboratory

- Gasoline-powered equipment and hoists must not be used on suspension scaffolds

- Gears and brakes of power operated hoists used on suspension scaffolds must be enclosed

- In addition to the normal operating brake, suspension scaffold power operated hoists and manually operated hoists shall have a braking device or locking pawl that engages automatically when a hoist makes either of the following uncontrolled movements: an instantaneous change in momentum or an accelerated overspeed

- Manually operated hoists shall require a positive crank force to descend

See Figure 12-9 for a detailed illustration pertaining to suspended scaffold power-operated hoists and manually operated hoists.

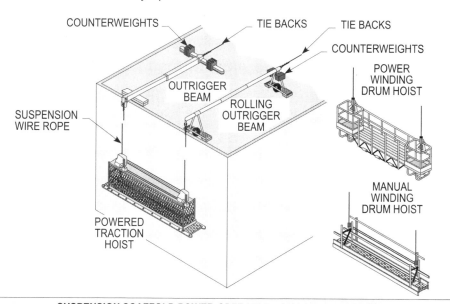

Figure 12-9. This illustration shows the requirements for suspended scaffold power-operated hoists and manually operated hoists.

SUSPENSION SCAFFOLD POWER-OPERATED HOISTS AND MANUALLY OPERATED HOISTS SHALL MEET THE FOLLOWING:
• BE OF A TYPE TESTED AND LISTED BY A QUALIFIED TESTING LABORATORY • GASOLINE-POWERED EQUIPMENT AND HOISTS MUST NOT BE USED ON SUSPENSION SCAFFOLDS • GEARS AND BRAKES OF POWER OPERATED HOISTS USED ON SUSPENSION SCAFFOLDS SHALL BE ENCLOSED • IN ADDITION TO THE NORMAL OPERATING BRAKE, SUSPENSION SCAFFOLD POWER OPERATED HOISTS AND MANUALLY OPERATED HOISTS SHALL HAVE A BRAKING DEVICE OR LOCKING PAWL THAT ENGAGES AUTOMATICALLY WHEN A HOIST MAKES EITHER OF THE FOLLOWING MOVEMENTS: AN INSTANTANEOUS CHANGE IN MOMENTUM OR AN ACCELERATED OVERSPEED • MANUALLY OPERATED HOISTS SHALL REQUIRE A POSITIVE CRANK FORCE TO DESCEND

CRITERIA FOR SUSPENSION SCAFFOLES
OSHA 1926.451(d)(13)

Safety Tip: Two-point and multi-point suspension scaffolds shall be tied or secured to prevent them from swaying.

Two-point and multi-point suspension scaffolds shall be tied or otherwise secured to prevent them from swaying, as determined necessary based on an evaluation by a competent person. In addition, window cleaners' anchors shall not be used for the purpose of preventing swaying. This prohibition is based on the fact that window cleaners' anchors are not designed for the load that could be imposed.

Single function emergency escape and rescue devices shall not be used as working platforms. This prohibition does not apply to systems that are designed to function both as working platforms and as emergency systems.

ACCESS
OSHA 1926.451(e)

Employers shall provide scaffold access for each affected employee. It also specifies that the access requirements for employees erecting or dismantling supported scaffolds are prescribed in OSHA 1926.451(e)(9).

Access to and between scaffold platforms more than 2 ft. above or below the point of access shall be by portable ladders, hook-on ladders, attachable ladders, scaffold stairways, stairway-type ladders (such as ladder stand), ramps, walkways, integral prefabricated scaffold access or equivalent means, or by direct access from another scaffold, structure, personnel hoist or similar surface. In addition, the final rule requires that cross braces not be used as a means of access.

Additional requirements for the proper construction and use of portable ladders are contained in Subpart X (Stairways and Ladders) of the construction standards.

Portable, hook-on and attachable ladders shall be positioned so as not to tip the scaffold.

Hook-on and attachable ladders shall meet the following:

- Have bottom rungs positioned not more than 24 in. above the scaffold supporting level

- Have rest platforms at 35 ft. maximum vertical intervals on all supported scaffolds more than 35 ft. high

- Be specifically designed for use with the manufactured type of scaffold to be used

- Have a minimum rung length of 11 1/2 in.

- Have uniformly spaced rungs with a maximum spacing between rungs of 16 3/4 in., respectively

Stairway-type ladders shall meet the following:

- Be positioned so that the bottom step is not more than 24 in. above the scaffold supporting level

- Be provided with rest platforms at 12 ft. maximum vertical intervals

- Have a minimum step width of 16 in. (except for mobile scaffold stairway-type ladders, which are permitted to have a minimum step width of 11 in.)

- Have slip-resistant treads on all steps and landings

Scaffold stairway towers used for access to scaffolds and other elevated work surfaces shall meet the following:

- A stairrail consisting of a toprail and a midrail shall be provided on each side of each scaffold stairway.

- The toprail of each stairrail system shall be capable of serving as a handrail, unless a separate handrail is provided.

- Handrails, and toprails that serve as handrails, shall provide a handhold for employees grasping them to avoid falling.

- Stairrail systems and handrails shall be surfaced in a manner that prevents injury to employees from punctures or lacerations, and to prevent snagging of clothing.

- Ends of stairrail systems and handrails shall be constructed in a manner that does not constitute a projection hazard.

- Scaffold stairway handrails, and toprails that are used as handrails, shall have a minimum clearance of 3 in. between the handrail or toprail and other objects. Inadequate hand clearances can render handrails essentially useless.

- Stairrails can be no less than 28 in. or more than 37 in. from the upper surface of the stairrail to the surface of the tread, in line with the face of the riser at the forward edge of the tread. This provision differs from the stairrail height requirements of Subpart X, which was never intended to apply to scaffold stairways.

- Scaffold stairways shall be provided with landing platforms that are at least 18 in. wide and at least 18 in. long at each level. This provision provides adequate protection for employees without impeding the use of most scaffold stairways now in use.

- Each scaffold stairway shall be at least 18 in. wide between stairrails.

- Treads and landings shall have slip-resistant surfaces.

- Scaffold stairways shall be installed between 40 degrees and 60 degrees from the horizontal. OSHA has determined that scaffold stairways installed in the range of 40 degrees to 60 degrees from the horizontal will provide safe employee access and will still be capable of fitting into the confines of the scaffold frames.

- Guardrails meeting the requirements of OSHA 1926.451(g)(4) shall be provided on the open sides and ends of each landing.

- Riser heights within each flight of scaffold stairs shall be uniform within 1/4 in.

Note: OSHA believes that a uniform riser height within 1/4 in. for all steps in each flight of stairs is necessary in order to minimize the possibility that employees will slip, trip and fall while they are on the stairs. OSHA recognizes that there are situations where the level of the ground or of the structure to which the stair tower is connected will cause the spacing of the top or bottom step of the stairway system to deviate from uniformity with the other steps by more than 1/4 in. OSHA has determined that such deviation will not compromise employee safety, so long as the stair tower otherwise complies with the requirements of OSHA 1926.451(e)(4).

- Tread depth shall be uniform, within 1/4 in., for each flight of stairs.

The following requirements shall be considered for ramps and walkways used to access scaffolds:

- Ramps and walkways 6 ft. or more above lower levels shall be provided with guardrail systems in accordance with the provisions of Part 1926, Subpart M - Fall Protection.

- Ramps and walkways shall not exceed a slope of one (1) vertical to three (3) horizontal (20 degrees above the horizontal).

- If the slope of a ramp or walkway is steeper than one (1) vertical in eight (8) horizontal, the ramp or walkway shall have cleats not more than fourteen (14) in. apart which are securely fastened to the planking to provide secure footing.

The following shall be considered for integral prefabricated scaffold access frames:

- Frames shall be specifically designed and constructed for use as ladder rungs.

- The frames shall have a rung length of at least 8 in.

- Rungs less than 11 1/2 in. in length shall be used for access only and not as work platforms unless fall protection, or a positioning device, is used.

- That integral prefabricated scaffold access frames be uniformly spaced within each frame section; provided with rest platforms at 35 ft. maximum vertical intervals on all supported scaffolds more than 35 ft. high; and have a maximum spacing between rungs of 16 3/4 in., respectively.

- That non-uniform rung spacing caused by joining end frames together is allowed, provided the resulting spacing does not exceed 16 3/4 in.

All steps and rungs of all ladder and stairway type access shall line up vertically with each other between rest platforms per OSHA 1926.451(e)(7).

Direct access to or from another surface shall be allowed only when the pertinent surfaces are not more than 14 in. apart horizontally and not more than 24 in. apart vertically per OSHA 1926.451(e)(8).

Access requirements for employees erecting or dismantling supported scaffolds are addressed in OSHA 1926.451(e)(9). The introductory language of OSHA 1926.451(e)(9), access requirements of employees erecting or dismantling supported scaffold, requires employers to comply with paragraphs (e)(9)(i) - (iv) as follows:

- That the means of access for erectors or dismantlers shall be determined by a competent person, based on specific site conditions and the type of scaffold being erected. Employers shall have the erection, dismantling or alteration of a scaffold conducted under the supervision and direction of a competent person who is qualified in the pertinent subject matter.

- That hook-on or attachable ladders be installed as soon as practical after the scaffold erection has progressed to the point permitting their installation and use. Sectional ladders can be used for access once adequate support is available.

  Note: This entire section applies to only erectors and dismantlers.

- That the end frames of tubular welded frame scaffolds that meet certain requirements can be safely used as a means of access for scaffold erectors and dismantlers.

• That cross bracing is not an acceptable means of access on tubular welded frame scaffolds, because cross braces are designed to provide diagonal stability to the scaffold and are not designed to withstand the forces that could be applied by employees climbing up and down on them.

USE
OSHA 1926.451(f)

Safe work practices for the use of scaffolds and the activities that take place on scaffolds are addressed per OSHA 1926.451(f).

Scaffolds and scaffold components shall not be loaded in excess of their maximum intended loads or rated capacities, whichever is less. Compliance with this rule ensures that the scaffold's capacity is not exceeded. OSHA believes it is appropriate to take into account the "expected" burden as well as the burden a scaffold "can" support without failure.

Safety Tip: All scaffolds and components shall be inspected by a competent person before each work shift

The use of shore or lean-to scaffolds. Such scaffolds are not properly designed nor properly constructed, and pose a serious threat to anyone working on them.

Scaffolds and scaffold components shall be inspected for visible defects by a competent person prior to each work shift and after any occurrence which could affect a scaffold's structural integrity. OSHA has determined that inspections conducted by a competent person before each shift and after any occurrence that would affect the scaffold's integrity will adequately protect employees working on scaffolds and ensure that defects are detected in a timely fashion.

A scaffold whose strength has been reduced to less than that required by OSHA 1926.451(a) shall be immediately repaired or replaced, braced to meet those provisions, where appropriate, or be removed from service until repaired. This paragraph applies whenever a scaffold component, for any reason, lacks the required strength. In particular, under this provision employers shall follow through to address problems identified pursuant to paragraph OSHA 1926.451(f)(3).

Scaffolds shall not be moved horizontally while employees are on them, unless they have been designed by a registered professional engineer specifically for such movement or, for mobile scaffolds, where provisions of OSHA 1926.452(w) are followed.

Safety Tip: Proper distance must be maintained, when erecting scaffolds near exposed energized lines.

Scaffolds near exposed and energized power lines. In particular, this paragraph requires employers to maintain clearance between power lines and scaffolds, including any conductive materials on the scaffold. The minimum clearance for all uninsulated lines and for insulated lines of more than 300 volts is 10 ft. The minimum clearance for insulated lines of less than 300 volts is 3 ft. **(See Figure 12-10)**

Exception: Scaffolds and materials may be closer to power lines than specified above only where necessary to do the work, and only after the utility company or electrical system operator has been notified of the need to work closer and the utility company or electrical system operator has deenergized the lines, relocated the lines or installed protective coverings to prevent accidental contact with the lines.

Scaffolds shall only be erected, moved, dismantled or altered under the supervision and direction of a competent person. It further provides that the listed activities shall be performed only by experienced and trained employees selected for such work by the competent person.

Employees shall use caution and are prohibited from working on scaffolds covered with snow, ice or other slippery material except as necessary for removal of such materials.

Figure 12-10. This illustration shows the minimum distances for scaffolds near exposed and energized power lines.

INSULATED LINES VOLTAGE	MINIMUM DISTANCE ALTERNATIVES	
LESS THAN 300 VOLTS	3 FT.	
300 VOLTS TO 50 kV	10 FT.	
MORE THAN 50 kV	10 FT. PLUS .4 IN. FOR EACH kV OVER 50 kV	2 TIMES THE LENGTH OF THE LINE INSULATOR, BUT NEVER LESS THAN 10 FT.

UNINSULATED LINES VOLTAGE	MINIMUM DISTANCE ALTERNATIVES	
LESS THAN 50 kV	10 FT.	
MORE THAN 50 kV	10 FT. PLUS .4 IN. FOR EACH kV OVER 50 kV	2 TIMES THE LENGTH OF THE LINE INSULATOR, BUT NEVER LESS THAN 10 FT.

USE
OSHA 1926.451(f)

Where swinging loads are being hoisted on, to or near scaffolds, such that the loads could contact the scaffold, tag lines or equivalent measures shall be utilized to stabilize the loads. This provision covers all hoisting operations in proximity to scaffolds, because a swinging load can pose a hazard regardless of its destination.

Suspension ropes used with adjustable suspension scaffolds shall have sufficient diameter for functioning of the brakes and the hoist mechanism.

Safety Tip: Precautions shall be taken when employees are working on scaffolds during storms or high winds.

Suspension ropes shall be shielded when a heat-producing process is performed. When acids or other corrosive substances are used on a scaffold, the ropes shall be shielded, treated to protect against the corrosive substances or shall be of a material that is not adversely affected by the substance being used.

Work on or from scaffolds during storms or high winds is prohibited unless a competent person has determined that it is safe for employees to be on the scaffold and these employees are protected by a personal fall arrest system or wind screens. Wind screens shall not be used unless the scaffold is secured against the forces imposed.

Debris shall not be allowed to accumulate on platforms, where it could pose a slip, trip or fall hazard to employees on or below the platform.

Makeshift devices, such as but not limited to boxes and barrels, shall not be used on top of scaffold platforms to increase the working level height of employees.

The use of ladders on scaffolds to increase the employee's working level is prohibited except when the employees are on large area scaffolds and the ladder is used in accordance with the applicable provisions discussed below:

When a ladder is placed against a structure which is not a part of the scaffold, the scaffold must be secured against the sideways thrust exerted by the ladder.

- Platform units shall be secured to the scaffold to prevent them from moving

- Ladder legs are all on the same platform unit unless other means have been provided to stabilize the ladder against platform unit deflection

- Ladder legs shall be secured to prevent them from slipping and being pushed off the platform unit

Note: OSHA believes that compliance with these provisions will prevent the tipping and instability hazards that led OSHA to propose a prohibition against the use of ladders on all scaffolds.

Platform units shall not deflect more than 1/60 of the span when loaded. This provision is intended to limit the amount platform units can deflect under load without becoming overstressed and without their ends being pulled from their supports.

Employers shall reduce the possibility of welding current arcing through suspension wire rope while employees are performing welding from suspended scaffolds by insulating the suspended platform and its rigging. This provision is intended to protect employees from the electrocution and platform collapse hazards posed by arcing welding current. In particular, OSHA requires that employer's rig affected scaffolds with the following:

• Insulated thimbles
• Insulated wire rope
• Insulated hoist mechanisms

See Figure 12-11 for a detailed illustration pertaining to suspended scaffold platform welding precautions.

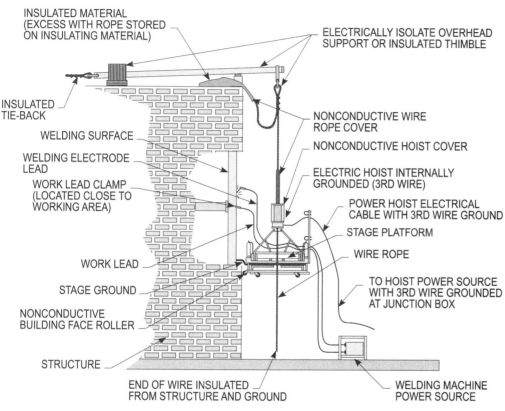

Figure 12-11. This illustration shows the requirements for suspended scaffold platform welding.

SUSPENDED SCAFFOLD PLATFORM WELDING PRECAUTIONS
OSHA 1926.451(f)(17)

This paragraph also specifies precautions for grounding the scaffold to the structure on which welding is being performed.

Compliance with these provisions taken together, will minimize the hazards of electric arcing during welding operations on suspended scaffolds.

FALL PROTECTION
OSHA 1926.451(g)

Fall protection requirements for employees working on scaffolds, including criteria for guardrail systems are addressed in OSHA 1926.451(g). Fall hazards account for a high percentage of the injuries and fatalities experienced by scaffold workers. OSHA has determined that compliance with this paragraph will effectively protect employees from those hazards.

Safety Tip: The threshold height where fall protection is required is 10 ft.

OSHA sets 10 ft. as the threshold height above which fall protection is required and indicates what fall protection measures are required for particular types of scaffolds, and in addition, addresses the fall protection requirements for employees erecting and dismantling supported scaffolds.

OSHA 1926.451(g)(1)(i) through (vii) specifies the types of fall protection to be used on particular types of scaffolds as follows:

- Personal fall arrest systems, not guardrails, are appropriate for use on boatswains' chairs, catenary scaffolds, float scaffolds, needle beam scaffolds and ladder jack scaffolds.

- Personal fall arrest systems and guardrail systems are required for all single-point adjustable suspension scaffolds (except boatswains' chairs), and for all two-point adjustable suspension scaffolds. The requirement to have guardrails and personal fall arrest systems on two-point scaffolds is based on the fact that a guardrail system alone does not provide adequate fall protection when a suspension rope fails and causes the scaffold to tip or hang from only one end. Personal fall arrest system protection is also necessary for single-point systems, because the fall hazard related to suspension rope failure is as serious as it is with the two-point scaffold. However, because personal fall arrest systems would be the primary means of fall protection on single-point and two-point systems, the provision allows a lower minimum strength guardrail system to be used.

- Each employee on a crawling board (chicken ladder) shall be protected by a personal fall arrest system, a guardrail system (with minimum 200 pound toprail capacity), or by a 3/4 in. diameter grabline or equivalent handhold securely fastened beside each crawling board.

- Employees on self-contained scaffolds shall be protected by both personal fall arrest systems and guardrail systems when the platform is supported by ropes (as when the scaffold is being raised or lowered on some systems) and by guardrail systems when the platform is supported directly by the scaffold frame.

- Guardrails shall be used along scaffold walkways and be located within 9 in. horizontally of at least one side of the walkway. The provision that guardrails need only to be provided along one side applies only when the platform is used solely as a means of access to get from one point on the scaffold to another. If work activities other than access are performed on or from the walkway, then the platform is not considered to be a walkway (see definition of "walkway"), and other provisions of paragraphs OSHA 1926.451(g)(1), as appropriate, would apply.

- Fall protection (i.e., a personal fall arrest system or guardrail) shall be provided on all open sides and ends of scaffolds from which employees are performing overhand bricklaying operations and/or related work, except those sides and ends next to the wall being laid.

- Employees performing overhand bricklaying operations from a supported scaffold shall be protected from falling from all open sides and ends of the scaffold (except at the side next to the wall being laid) by the use of a personal fall arrest system or guardrail system (with minimum 200 pound toprail capacity).

- Employees on scaffolds not addressed elsewhere in OSHA 1926.451(g)(1) shall be protected either by guardrails or personal fall arrest systems.

Note: OSHA 1926.451(g)(1) does not apply where there are no "open sides or ends" on the scaffold (see definition in OSHA 1926.451(b)). For the scaffold to be considered completely enclosed, no perimeter face of the scaffold may be more than 14 in. from a wall. The requirements for fall protection will apply at openings such as hoist ways, elevator shafts, stairwells or similar openings in the scaffold platform, or openings in the walls of the structure surrounding the platform.

Fall protection for employees erecting or dismantling supported scaffolds is addressed per OSHA 1926.451(g)(2). It requires that employers whose employees erect or dismantle supported scaffolds ensure that a competent person determines the feasibility and safety of providing fall protection for such employees. This paragraph further requires that affected employers provide fall protection for employees erecting or dismantling supported scaffolds where the installation and use of such protection is feasible and does not create a greater hazard.

Personal fall arrest systems shall comply with the pertinent provisions of OSHA 1926.502(d) and, in addition, shall be attached by lanyard to a vertical lifeline, horizontal lifeline or scaffold structural member. However, when overhead obstructions such as overhead protection or additional platform levels are part of a single-point or two-point adjustable suspension scaffold, then vertical lifelines shall not be used, because, in the event of a scaffold collapse, the overhead components would injure an employee who was tied off to a vertical lifeline.

Safety Tip: Fall arrest systems shall comply with 1926.502(d).

Vertical lifelines, when used, shall be fastened to a fixed safe point of anchorage, be independent of the scaffold and be protected from sharp edges and abrasion.

Horizontal lifelines, when used, shall be secured to two or more structural members of the scaffold, and shall not be attached only to the suspension ropes.

When lanyards are connected to horizontal lifelines or structural members on a single-point or two-point adjustable suspension scaffold, the scaffold shall be equipped with additional independent support lines and automatic locking devices capable of stopping the fall of the scaffold in the event one or more of the suspension ropes fail. The independent support lines shall be equal in number and strength to the suspension ropes. OSHA believes that in the event of a suspension rope failure, the additional support lines will keep the scaffold from falling.

Vertical lifelines, independent support lines and suspension ropes shall not be attached to each other, or be attached to or use the same point of anchorage or be attached to the same point on the scaffold or body belt/harness system.

Guardrail systems used to provide fall protection for employees working on scaffolds is addressed in OSHA 1926.451(g)(4).

Guardrail systems shall be installed along all open sides and ends of platforms. In the case of suspended scaffolds, guardrails shall be installed before any employee is allowed on a hoisted scaffold. In the case of supported scaffolds, installation shall occur before employees are permitted to work from the scaffold.

The top edge height of toprails or equivalent members on supported scaffolds manufactured or placed into service after January 1, 2000 shall be between 38 in. and 45 in. above the platform surface. The top edge height of guardrails on supported scaffolds manufactured and placed into service before January 1, 2000 and on all suspended scaffolds where both a guardrail and a personal fall arrest system are required shall be between 36 in. and 45 in. This paragraph also provides that toprail height may exceed 45 in. if the other criteria of paragraph OSHA 1926.451(g)(4) have been satisfied. **(See Figure 12-12)**

Figure 12-12. This illustration shows the guardrail requirements for scaffolds over 10 ft.

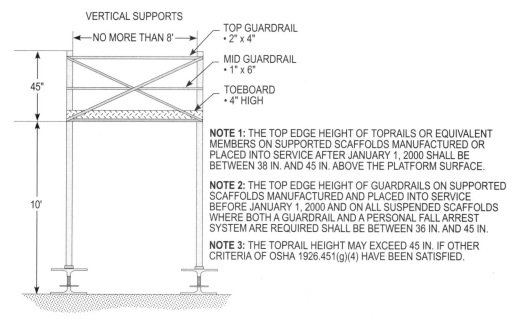

VERTICAL SUPPORTS

NO MORE THAN 8'

TOP GUARDRAIL
• 2" x 4"

MID GUARDRAIL
• 1" x 6"

TOEBOARD
• 4" HIGH

45"

10'

NOTE 1: THE TOP EDGE HEIGHT OF TOPRAILS OR EQUIVALENT MEMBERS ON SUPPORTED SCAFFOLDS MANUFACTURED OR PLACED INTO SERVICE AFTER JANUARY 1, 2000 SHALL BE BETWEEN 38 IN. AND 45 IN. ABOVE THE PLATFORM SURFACE.

NOTE 2: THE TOP EDGE HEIGHT OF GUARDRAILS ON SUPPORTED SCAFFOLDS MANUFACTURED AND PLACED INTO SERVICE BEFORE JANUARY 1, 2000 AND ON ALL SUSPENDED SCAFFOLDS WHERE BOTH A GUARDRAIL AND A PERSONAL FALL ARREST SYSTEM ARE REQUIRED SHALL BE BETWEEN 36 IN. AND 45 IN.

NOTE 3: THE TOPRAIL HEIGHT MAY EXCEED 45 IN. IF OTHER CRITERIA OF OSHA 1926.451(g)(4) HAVE BEEN SATISFIED.

GENERAL REQUIREMENTS
OSHA 1926.451(g)(4)

When midrails, screens, mesh, intermediate vertical members (such as balusters), solid panels or equivalent structural members are used, they shall be installed between the top edge of the guardrail system and the scaffold platform.

The criteria necessary to ensure that the midrails, screens, mesh and baluster type protection required by OSHA 1926.451(g)(4)(iii) will be properly placed and effective are as follows:

- Midrails, when used, shall be installed at a height midway between the top edge of the guardrail system and the platform surface.

- Screens and mesh, when used, shall extend from the top edge of the guardrail system to the scaffold platform, and along the entire opening between the supports.

- Intermediate vertical members (such as balusters or additional rails), when used, shall not be more than 19 in. apart.

Safety Tip: Toprails shall be capable of withstanding at least 100 pounds, without failure.

Toprails or equivalent members shall be capable of withstanding, without failure, a force applied in any downward or horizontal direction at any point along their top edge of at least 100 pounds for guardrail systems installed on single-point adjustable suspension scaffolds and on two-point adjustable suspension scaffolds, and at least 200 pounds for guardrail systems installed on all other scaffolds.

When the loads specified in OSHA 1926.451(g)(4)(vii) are applied in a downward direction, the top edge may not drop below the height above the platform surface prescribed in OSHA 1926.451(g)(4)(ii).

Midrails, screens, mesh, intermediate vertical members, solid panels and equivalent structural members shall be capable of withstanding, without failure, a force applied in any downward or horizontal direction at any point along the midrail or other member of at least 75 pounds for guardrail systems with a minimum 100 pound toprail capacity, and at least 150 pounds for guardrail systems with a minimum 200 pound toprail capacity.

A separate guardrail section is not required on the ends of suspension scaffolds when the scaffold's support system (stirrup) or hoist prevents passage of employees.

Guardrail systems shall be so surfaced as to prevent injury to an employee from punctures or lacerations, and to prevent the snagging of clothing.

Toprails and midrails shall not be so long as to constitute a hazard.

Steel banding and plastic banding as toprails or midrails is prohibited. Although such banding can often withstand a 200 pound load, it can tear easily if twisted. In addition, such banding often has sharp edges that can cut a hand if seized.

Guardrail systems using manila, plastic or synthetic rope as rails shall be inspected by a competent person as frequently as necessary to ensure that the guardrails comply with the performance criteria in OSHA 1926.451(g).

Cross bracing can be used in lieu of either a midrail or a toprail when certain criteria are met. Cross bracing would be accepted in lieu of a toprail when the crossing point is between 38 in. and 48 in. above the work surface. Also, cross bracing would be accepted in lieu of a midrail when the crossing point is between 20 in. and 30 in. above the work surface. In addition, the end points of each upright shall be no more than 48 in. apart, which will reduce the slope of the cross bracing and result in a surface that is similar to that of a standard guardrail.

FALLING OBJECT PROTECTION
OSHA 1926.451(h)

Employees working on scaffolds shall wear hardhats and be protected from falling hand tools, debris and other small objects through the installation of toeboards, screens or guardrail systems or through the erection of debris nets, catch platforms or canopy structures that deflect falling objects. In addition, when the falling objects to which employees on scaffolds may be exposed are too large, heavy or massive to be contained or deflected by any of the above-listed measures, the employer shall protect affected employees by placing any such potential falling objects away from the edge of a surface from which they might fall and shall secure those materials as necessary to prevent their falling.

Safety Tip: Employees working on scaffolds shall be protected from falling debris.

Employers shall protect employees working below from objects falling from scaffold.

Barricades can be used on lower levels to exclude employees from areas where falling objects might land. Compliance with this provision will enable employers to eliminate employee exposure to the hazard.

Employers shall provide toeboards along the edge of platforms more than 10 ft. above lower levels for a distance sufficient to protect workers below, except that on float (ship) scaffolds, an edging of 3/4 in. x 1 1/2 in. wood, or a material with equivalent strength, may be used in lieu of a toeboard.

As an alternative, for erection of paneling or screening in cases where tools or other materials are piled to a height higher than the top edge of a toeboard, the panel or screen shall extend from the toeboard (or platform) to the top of the guardrail and be erected for a distance sufficient to protect employees below. In addition, the panel or screen would need to be capable of withstanding, without failure, a force of at least 150 pounds, applied in any downward or outward direction at any point along the screen (to comply with OSHA 1926.451(g)(4)(ix)).

Employers are allowed to protect employees from falling objects through the installation of a guardrail system which complies with OSHA 1926.451(g)(4) and that has openings small enough to reject passage of potential falling objects.

Employers can protect employees working below scaffolds from falling objects through the installation of debris nets, catch platforms or canopies that have sufficient strength to withstand the impact forces of potential falling objects.

The following uses of canopies are addressed in OSHA 1926.451(h)(3):

- Canopies shall be installed between the falling object hazard and the employees.

- Use of additional independent support lines to support the scaffold in the event of suspension support rope failure is required, in cases where canopies are used for falling object protection on suspended scaffolds.

- Independent support lines and suspension ropes shall not be attached to the same point of anchorage.

This provision will prevent the loss of the backup safety systems in the event of suspension rope anchorage failure.

The following strength criteria for toeboards are addressed in OSHA 1926.451(h)(4).

- Toeboards shall be capable of withstanding, without failure, a force of at least 50 pounds applied in any downward or horizontal direction at any point along the toeboard.

- Toeboards shall be at least 3 1/2 in. high, fastened securely in place, and have not more than 1/4 in. clearance above the walking/working surface. In addition, toeboards shall be solid or have openings no greater than 1 in. in the greatest dimension

ADDITIONAL REQUIREMENTS APPLICABLE TO SPECIFIC TYPES OF SCAFFOLDS
OSHA 1926.452

OSHA 1926.452 contains requirements that supplement the requirements of OSHA 1926.451 with regard to particular types of scaffolds. The identified scaffolds have unique features, which require specific attention.

OSHA has determined that compliance with the performance-oriented provisions of OSHA 1926.451 and OSHA 1926.452, taken together, will provide adequate protection for employees working on scaffolds. Further, the Agency believes that the specification language suggested by the commenters would limit innovation and impose unreasonable burdens on employers.

POLE SCAFFOLDS
OSHA 1926.452(a)

The proper use of bearers, braces and runners on pole scaffolds are addressed in OSHA 1926.452(a). In addition, pole scaffolds over 60 ft. in height shall be designed by a registered professional engineer, and shall be constructed and loaded in accordance with that design. The provision also notes that non-mandatory Appendix A contains examples of criteria that will enable an employer to comply with design and loading requirements for pole scaffolds under 60 ft. in height. OSHA requirements for pole scaffolds are as follows:

- When platforms are being moved to the next level, the existing platform shall be left undisturbed until the new bearers have been set in place and braced, prior to receiving the new platforms.

- Cross bracing shall be installed between the inner and outer sets of poles on double pole scaffolds.

- Diagonal bracing in both directions shall be installed across the entire inside face of double-pole scaffolds used to support loads equivalent to a uniformly distributed load of 50 pounds or more per square foot.

- Diagonal bracing in both directions shall be installed across the entire outside face of all double- and single-pole scaffolds.

- Runners and bearers shall be installed on edge.

- Bearers shall extend a minimum of 3 in. over the outside edges of runners.

- Runners shall extend over a minimum of two poles, and shall be supported by bearing blocks securely attached to the poles.

- Braces, bearers and runners shall not be spliced between poles.

- Where wooden poles are spliced, the ends shall be squared and the upper section shall rest squarely on the lower section. Wood splice plates shall be provided on at least two adjacent sides, and shall extend at least 2 ft. on either side of the splice, overlap the abutted ends equally and have at least the same cross-sectional areas as the pole. Splice plates of other materials of equivalent strength may be used.

- Pole scaffolds over 60 ft. in height shall be designed by a registered professional engineer, and shall be constructed and loaded in accordance with that design. Non-mandatory Appendix A contains examples of criteria that will enable an employer to comply with design and loading requirements for pole scaffolds under 60 ft. in height.

TUBE AND COUPLER SCAFFOLDS
OSHA 1926.452(b)

The use of bearers, bracing, runners and couplers on tube and coupler scaffolds are addressed in OSHA 1926.452(b). In addition, OSHA provides that tube and coupler scaffolds over 125 ft. in height shall be designed by a registered professional engineer, and be constructed and loaded in accordance with such design. **(See Figure 12-13)**

Platforms shall not be moved until the next location has been properly prepared to support the platform being moved.

The installation of transverse bracing is required at the scaffold ends and, at least, at every third set of posts horizontally and every fourth post vertically. This paragraph provides for diagonal bracing from the outer or inner posts or runners upward to the next outer or inner posts or runners. In addition, building ties shall be installed at the bearer levels between the diagonal braces in conformance with OSHA 1926.451(c)(1).

The installation of longitudinal bracing across the inner and outer rows of posts for straight run scaffolds are addressed in OSHA 1926.452(b)(3). In particular, such bracing shall be installed diagonally in both directions and shall extend from the base of the end posts upward to the top of the scaffold at a 45 degree angle. Where scaffold length is greater than height, bracing shall be repeated at least at every fifth post. Where scaffold

Figure 12-13. This illustration shows the requirements for tube and coupler scaffold.

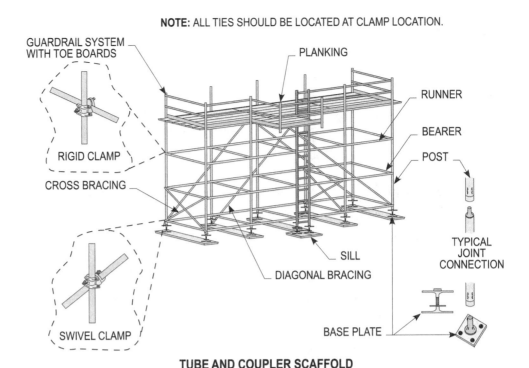

NOTE: ALL TIES SHOULD BE LOCATED AT CLAMP LOCATION.

GUARDRAIL SYSTEM WITH TOE BOARDS

PLANKING

RUNNER

BEARER

POST

RIGID CLAMP

CROSS BRACING

SILL

DIAGONAL BRACING

TYPICAL JOINT CONNECTION

SWIVEL CLAMP

BASE PLATE

TUBE AND COUPLER SCAFFOLD
OSHA 1926.452(b)

length is less than height, such bracing shall be installed from the base of the end posts upward to the opposite end posts and then in alternating directions until reaching the top of the scaffold. In addition, bracing shall be installed as close as possible to the intersection of the bearer and post or of the runner and post.

Safety Tip: Where conditions preclude attachment of bracing posts, bracing shall be attached to the runners as close to the post as possible.

Bracing shall be attached to the runners as close to the post as possible, where conditions preclude attachment of bracing to posts. OSHA recognizes that attachment to the post, while the most desirable option, is not always possible. In circumstances where such attachment is not possible, OSHA has determined that attachment to the runner, as close as possible to the post, will still maximize directional stability and provide the strength necessary to properly brace the scaffold.

Bearers shall be installed transversely between posts, and when coupled to the posts, shall have the inboard coupler bear directly on the runner coupler. When the bearers are coupled to the runners, the couplers shall be as close to the posts as possible. Bearers shall also extend beyond the posts and runners, and shall provide full contact with the coupler.

Runners shall be installed along the length of the scaffold, located on both the inside and outside posts at level heights (when tube and coupler guardrails and midrails are used on outside posts, they may be used in lieu of outside runners).

Runners shall be interlocked on straight runs to form continuous lengths, and shall be coupled to each post. The bottom runners and bearers shall be located as close to the base as possible.

Couplers shall be of a structural metal, such as drop-forged steel, malleable iron or structural grade aluminum. The use of gray cast iron is prohibited.

Tube and coupler scaffolds over 125 ft. in height shall be designed by a registered professional engineer, and shall be constructed and loaded in accordance with such design. Refer to non-mandatory Appendix A of Subpart L for examples of criteria that will help to comply with design and loading requirements for tube and coupler scaffolds less than 125 ft. in height.

FABRICATED FRAME SCAFFOLDS
OSHA 1926.452(c)

Additional requirements for fabricated frame scaffolds (tubular welded frame scaffolds) are addressed in OSHA 1926.452(c).

Platforms shall not be moved until the next location is properly prepared and ready to support the platform being moved.

OSHA requires the locking together of end frames. This requirement only applies where uplift forces are strong enough to displace the end frames or panels, such as when a hoist is being used that could snag the scaffold during a hoist operation.

The proper placement of platform support brackets are addressed in OSHA 1926.452(c)(5). Improper placement of such cantilever supports can significantly reduce their support capacity and thus endanger employees working on top of the platform.

Scaffolds over 125 ft. in height above their base plates shall be designed by a registered professional engineer, and shall be constructed and loaded in accordance with such design.

PLASTERERS', DECORATORS' AND LARGE AREA SCAFFOLDS
OSHA 1926.452(d)

Plasterers', decorators' and large area scaffolds shall be constructed in accordance with OSHA 1926.452(a), (b) or (c). OSHA 1926.452(d) references the provisions of OSHA 1926.452(a), (b) and (c) because plasterers', decorators' and large area scaffolds are almost always constructed using pole scaffolds, tube and coupler scaffolds or fabricated frame scaffolds.

BRICKLAYERS' SQUARE SCAFFOLDS (SQUARES)
OSHA 1926.452(e)

Scaffolds made of wood shall be reinforced with gussets on both sides of each corner; that diagonal braces shall be installed on all sides of each square; that diagonal braces shall be installed between squares on the rear and front sides of the scaffold, and extend from the bottom of each square to the top of the next square; and that scaffolds of this type shall not exceed three tiers in height, that they be constructed and arranged so that one square rests directly above the other, and that the upper tiers stand on a continuous row of planks laid across the next lower tier and be nailed down or otherwise secured to prevent displacement.

HORSE SCAFFOLDS
OSHA 1926.452(f)

Horse scaffolds shall not be constructed or arranged more than two tiers or 10 ft. in height, whichever is less; when arranged in tiers, that each horse shall be placed directly over the horse in the tier below; when arranged in tiers, the legs of each horse shall be nailed down or otherwise secured to prevent displacement; and that, when arranged in tiers, each tier shall be cross braced.

FORM SCAFFOLDS AND CARPENTERS' BRACKET SCAFFOLDS
OSHA 1926.452(g)

Additional rules for form scaffolds and carpenters' bracket scaffolds are addressed in OSHA 1926.452(g).

Each bracket, except those for wooden bracket-form scaffolds, shall be attached to the supporting form work or structure by means of one or more of the following:

- Nails

- Metal stud attachment device

- Welding; hooking over a secured structural supporting member, with the form wales either bolted to the form or secured by snap ties or tie bolts extending through the form and securely anchored

- For carpenters' bracket scaffolds only, by a bolt extending through to the opposite side of the structure's wall.

Wooden bracket-form scaffolds shall be an integral part of the form panel.

Folding type metal brackets, when extended for use, shall be either bolted or secured with a locking-type pin.

ROOF BRACKET SCAFFOLDS
OSHA 1926.452(h)

Scaffold brackets shall be constructed to fit the pitch of the roof and provide a level support for the platform; and that brackets be anchored in place by nails unless it is impractical to use nails. OSHA 1926.452(h)(2) further provides that brackets shall be held in place with first-grade manila rope of at least 3/4 in. diameter, or a rope with equivalent strength, when nails are not used. Reference OSHA 451(g)(1)(viii) for fall protection.

OUTRIGGER SCAFFOLDS
OSHA 1926.452(i)

OSHA 1926.452(i)(1) through (i)(4) set requirements for the proper positioning and securing of outrigger beams. OSHA 1926.452(i)(5) and (i)(6) require that the inboard ends of outrigger beams be securely anchored and that the entire supporting structure be securely braced.

Platform units shall be nailed, bolted or otherwise secured to outriggers, to prevent displacement.

Scaffolds and scaffold components shall be designed by a registered professional engineer and constructed and loaded in accordance with such design. This provision reflects OSHA's determination that the design of this type of scaffold involves calculations that required the skills of a registered professional engineer, and that the criteria in the proposed rule had such limited applicability as to be of virtually no help to employers in almost all situations. **(See Figure 12-14)**

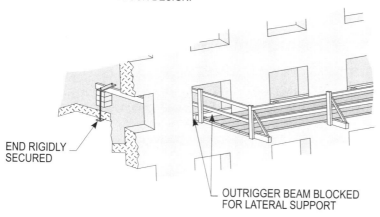

NOTE: SCAFFOLDS AND SCAFFOLD COMPONENTS SHALL BE DESIGNED BY A REGISTERED PROFESSION ENGINEER AND CONSTRUCTED AND LOADS IN ACCORDANCE WITH SUCH DESIGN.

END RIGIDLY SECURED

OUTRIGGER BEAM BLOCKED FOR LATERAL SUPPORT

**OUTRIGGER SCAFFOLDS
OSHA 1926.452(i)**

Figure 12-14. This illustration shows the requirements for outrigger scaffolds.

PUMP JACK SCAFFOLDS
OSHA 1926.452(j)

Pump jack brackets, braces and accessories shall be fabricated from metal plates and angles. In addition, each pump jack bracket shall have two positive gripping mechanisms to prevent any failure or slippage.

Poles shall be secured to the structure by rigid triangular bracing or equivalent, at the bottom, top and other points as necessary. In addition, that provision further requires that when the pump jack has to pass bracing that is already installed, an additional brace shall be installed approximately 4 ft. above the brace to be passed. That additional brace shall be left in place until the pump jack has been moved and the original brace reinstalled.

When guardrails are used for fall protection, a workbench may be used as the toprail only if the workbench complies with the requirements of OSHA 1926.451(g)(4)(ii), (vii), (viii) and (xiii).

Work benches shall not be used as scaffold platforms.

When poles are made of wood, the pole lumber shall be straight-grained, free of shakes, large loose or dead knots and other defects that might impair strength.

When wood poles are constructed of two continuous lengths, the lengths shall be joined together with the seam parallel to the bracket.

When two by fours (2 in. x 4 in.) are spliced to make a pole, that mending plates shall be installed at all splices to develop the full strength of the member.

LADDER JACK SCAFFOLDS
OSHA 1926.452(k)

Platforms shall not exceed a height of 20 ft.

All ladders used to support ladder jack scaffolds shall meet the requirements of Subpart X of 29 CFR Part 1926 - Stairways and Ladders, except that job-made ladders, which are permitted by Subpart X, are not permitted to be used to support ladder jack scaffolds.

The ladder jack shall be so designed and constructed that it will bear either on the side rails and ladder rungs or on the ladder rungs alone. This paragraph further requires that the bearing area for a ladder jack that bears only on the rungs shall be at least 10 in. on each rung to ensure adequate support.

Ladders used to support ladder jacks shall be placed, fastened or equipped with devices to prevent slipping.

Scaffold platforms shall not be bridged one to another. The provision would prohibit situations where, for example, four ladders are used to support three platforms. OSHA is prohibiting bridging because this practice often leads to overloading of the two ladders in the middle. This provision does not prohibit passage from one scaffold to another if the scaffolds are close enough for employees to walk (but not to jump or swing) from one scaffold to the other.

WINDOW JACK SCAFFOLDS
OSHA 1926.452(l)

This paragraph provides that window jack scaffolds shall be securely attached to the window opening, shall be used only for the purpose of working at the window opening through which the jack is placed and shall not be used to support planks placed between one window jack and another, or to support other elements of scaffolding. These requirements are necessary to ensure the safety of employees working from these platforms.

CRAWLING BOARDS (CHICKEN LADDERS)
OSHA 1926.452(m)

Additional requirements for crawling boards (chicken ladders) are addressed in OSHA 1926.452(m). It requires that crawling boards shall extend from the roof peak to the eaves when used in connection with roof construction, repair or maintenance, and that crawling boards shall be secured to the roof by ridge hooks or by means that satisfy equivalent criteria (e.g., strength and durability). These requirements are designed to ensure that crawling boards used by employees performing roof work are as secure as possible. Reference OSHA 451(g)(1)(iii) for fall protection.

STEP, PLATFORM AND TRESTLE LADDER SCAFFOLDS
OSHA 1926.452(n)

Scaffold platforms shall not be placed any higher than the second highest rung or step of the ladder supporting the platform. This provision is consistent with paragraphs 17.4 and 17.5 of ANSI A10.8-1988, and is intended to ensure the stability of this type of scaffold.

All ladders used in conjunction with step, platform and trestle ladder scaffolds shall meet the requirements of Subpart X of 29 CFR Part 1926 - Stairways and Ladders, except that job-made ladders shall not be used to support such scaffolds.

Ladders used to support step, platform and trestle ladder scaffolds shall be placed, fastened or equipped with devices to prevent slipping.

Scaffolds shall not be bridged one to another. Bridging, as discussed above under OSHA 1926.452(k)(5), occurs when four ladders are used to support three platforms. OSHA is prohibiting bridging because this practice often leads to overloading of the two ladders in the middle.

SINGLE-POINT ADJUSTABLE SCAFFOLDS
OSHA 1926.452(o)

This paragraph combines single-point adjustable suspension scaffolds, and boatswains' chairs, because boatswains' chairs are a form of single-point adjustable suspension scaffold.

When two single-point adjustable suspension scaffolds are combined to form a two-point adjustable suspension scaffold, the resulting scaffold shall meet the requirements for two-point adjustable suspension scaffolds in OSHA 1926.452(p).

The circumstances under which the supporting rope between a scaffold and a suspension device is permitted to deviate from a vertical position (i.e., at a 90 degree angle from level grade) requires that the supporting rope between the scaffold and the suspension device be kept vertical unless the following four conditions are met:

- Rigging shall have been designed by a qualified person

- Scaffold shall be accessible to rescuers

- Supporting rope shall be protected to ensure that it will not chafe at any point where a change in direction occurs

- Scaffold shall not be able to sway into another surface.

Whenever swaying of the scaffold could bring the scaffold into contact with another surface, the supporting rope shall be vertical, with no exceptions.

The tackle used with boatswains' chairs shall be ball bearing or bushed blocks containing safety hooks and properly "eye" spliced minimum five-eight (5/8) in. diameter first grade manila rope, or other rope that meets the performance criteria of the above-specified manila rope. OSHA recognizes that the use of an open hook could allow a chair to be dislodged if the rigging hung up on an obstruction. The corresponding ANSI standard, A10.8-1988, paragraph 6.14.5, provides for the use of a hook with a safety latch over the opening (safety hook) to prevent dislodging of the chair. The Agency agrees that it is appropriate to explicitly require that employers who have their employees use boatswains' chair rig their scaffolds with safety hooks. In addition, OSHA believes that locking safety hooks, such as are required for use with crane and derrick suspended personnel platforms (OSHA 1926.550(g)(4)(iv)(B)), would provide the most effective protection for affected employees. **(See Figure 12-15)**

Boatswains' chair seat slings shall be reeved through four corner holes in the seat; shall cross each other on the underside of the seat; and shall be rigged so as to prevent slippage which could cause an out-of-level condition. This paragraph is intended to prevent tipping of the chair.

OSHA 1926.452(o)(5) requires, except as provided in OSHA 1926.452(o)(6), that boatswains' chair seat slings shall be a minimum of five-eight (5/8) in. diameter fiber or synthetic rope or other rope that satisfies equivalent performance criteria.

Boatswains' chair seat slings shall be a minimum of three-eight (3/8) in. wire rope when a heat-producing process such as gas or arc welding is being conducted. This provision is necessary to ensure that the chair's sling is made of fire-resistant materials.

Non-cross-laminated wood boatswains' chairs shall be reinforced on their underside by cleats securely fastened to prevent the board from splitting.

Figure 12-15. This illustration shows the requirements for a boatswains chair.

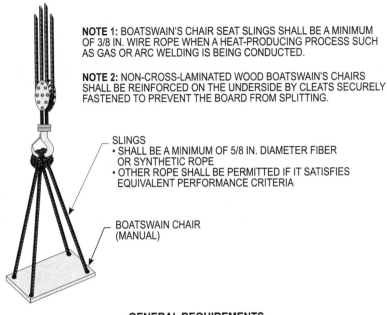

NOTE 1: BOATSWAIN'S CHAIR SEAT SLINGS SHALL BE A MINIMUM OF 3/8 IN. WIRE ROPE WHEN A HEAT-PRODUCING PROCESS SUCH AS GAS OR ARC WELDING IS BEING CONDUCTED.

NOTE 2: NON-CROSS-LAMINATED WOOD BOATSWAIN'S CHAIRS SHALL BE REINFORCED ON THE UNDERSIDE BY CLEATS SECURELY FASTENED TO PREVENT THE BOARD FROM SPLITTING.

SLINGS
• SHALL BE A MINIMUM OF 5/8 IN. DIAMETER FIBER OR SYNTHETIC ROPE
• OTHER ROPE SHALL BE PERMITTED IF IT SATISFIES EQUIVALENT PERFORMANCE CRITERIA

BOATSWAIN CHAIR (MANUAL)

GENERAL REQUIREMENTS
OSHA 1926.452(o)(3)

TWO-POINT ADJUSTABLE SUSPENSION SCAFFOLDS
OSHA 1926.452(p)

Platforms cannot be more than 36 in. wide unless designed by a qualified person to prevent unstable conditions.

Platforms shall be securely fastened to hangers (stirrups) by U-bolts or other means that satisfy OSHA 1926.451(a).

Blocks for fiber or synthetic ropes shall consist of at least one double and one single block, and the sheaves of all blocks shall fit the size of the rope used.

All platforms shall be of the ladder-type, plank-type, beam-type or light-metal type. Light metal-type platforms having a rated capacity of 750 pounds or less and platforms 40 ft. or less in length shall be tested and listed by a nationally recognized testing laboratory.

Safety Tip: Unless the bridge connectors are articulated and the hoists properly sized, two-point scaffolds cannot be bridged or otherwise connected on to another during raising and lowering operations.

Two-point scaffolds cannot be bridged or otherwise connected one to another during raising and lowering operations unless the bridge connections are articulated and the hoists properly sized. It is not intended to prohibit passage from one scaffold to another, but to prevent significant overloading of the hoist nearest the bridging device during operation of the hoist, or displacement of the bridge if the hoist is used to raise or lower one of the scaffolds. Many hoists are only sized to support one end of a two-point system. If one of two bridged scaffolds were to be raised by a hoist, a bridge laid between the scaffolds could be displaced unless the bridge is articulated (connected). This could also significantly increase the load on the hoist if it is not properly sized. The final rule addresses these two hazards by requiring bridge connections to be articulated and requiring that hoists be properly sized.

Passage from one platform to another is permitted only when the platforms are at the same height, when the platforms abut each other and when walk-through stirrups specifically designed for this purpose are used.

MULTI-POINT SUSPENSION SCAFFOLDS, STONE SETTERS' MULTI-POINT ADJUSTABLE SUSPENSION SCAFFOLDS AND MASONS' MULTI-POINT ADJUSTABLE SUSPENSION SCAFFOLDS
OSHA 1926.452(q)

When two or more scaffolds are used, they shall not be bridged one to another unless they are designed to be bridged, the bridge connections are articulated (connected) and the hoists are properly sized.

If bridges are not used, passage may be made from one platform to another only when the platforms are at the same height and are abutting.

Scaffolds shall be suspended from metal outriggers, brackets, wire rope slings, hooks or equivalent means.

CATENARY SCAFFOLDS
OSHA 1926.452(r)

Not more than one platform shall be placed between consecutive vertical pickups, and no more than two platforms shall be used on a catenary scaffold. These requirements are intended to prevent overloading of this type of scaffold.

Platforms supported by wire ropes are to have hook-shaped stops on each end of the platforms to prevent the platforms from slipping off the wire ropes. These hooks shall be so placed that they will prevent the platforms from falling if one of the horizontal wire ropes breaks.

Wire ropes shall not be tightened to the extent that the application of a scaffold load will overstress them.

Wire ropes shall be continuous and without splices between anchors. This is necessary to ensure that the rope has sufficient integrity to handle the load.

Safety Tip: To prevent platforms from slipping of the wire ropes, platforms supported by wire ropes are to have hook-shaped stops on each end of the platform.

FLOAT (SHIP) SCAFFOLDS
OSHA 1926.452(s)

Platforms are to be supported by a minimum of two bearers, each of which shall project a minimum of 6 in. beyond the platform on both sides. This will ensure that the platform will be fully supported. In addition, each bearer shall be securely fastened to the platform to prevent slippage.

Rope connections shall be such that the platform cannot shift or slip. Platform slippage is a significant factor in scaffold accidents.

When only two ropes are used with each float, those ropes shall be arranged so as to provide four ends that are securely fastened to overhead supports, and each supporting rope shall be hitched around one end of the bearer and pass under the platform to the other end of the bearer where it is hitched again, leaving sufficient rope at each end for the supporting ties.

INTERIOR HUNG SCAFFOLDS
OSHA 1926.452(t)

Scaffolds can be suspended only from the roof structure or other structural members such as ceiling beams. This requirement is necessary to ensure that suspended scaffolds are supported by structural members with adequate capacity for safe use.

Supporting members shall be inspected and checked for strength before the scaffold is erected. This requirement is necessary because such points of support cannot be assumed to be strong enough to support a scaffold since they may already be loaded to their capacity or they may have deteriorated over time.

Suspension ropes and cables shall be connected to the overhead supporting members by shackles, clips, thimbles or by other means that provide equivalent strength, security and durability.

NEEDLE BEAM SCAFFOLDS
OSHA 1926.452(u)

Scaffold support beams shall be installed on edge.

Ropes or hangers shall be used for supports, except that one end of a needle beam scaffold may be supported by a permanent structural member. This is necessary to ensure that these scaffolds are properly supported by rope or hangers that meet the strength criteria of OSHA 1926.451(a). The ropes shall be securely attached to the needle beams.

The support connections shall be arranged so as to prevent the needle beam from rolling or becoming displaced, which could result in tipping of the platform.

Platform units shall be securely attached to the needle beams by bolts or equivalent means. In addition, cleats and overhang are not considered to be adequate means of attachment.

MULTI-LEVELED SUSPENDED SCAFFOLDS
OSHA 1926.452(v)

Safety Tip: Independent support lines and suspension ropes shall not be attached to the same point of anchorage.

Multi-level suspended platform scaffolds shall be equipped with additional independent support lines, equal in number to the number of points supported and of equivalent strength to the suspension ropes, and be rigged to support the scaffold in the event the suspension rope(s) fail. These additional lines would support the scaffold, and prevent collapse in the event of primary support line failure.

Independent support lines and suspension ropes shall not be attached to the same points of anchorage. This provision reflects OSHA concern that the independent support lines would not protect workers from scaffold collapse if the independent lines and the suspension ropes were attached to the same anchorage point when the anchorage failed.

Supports for platforms shall be attached directly to the support stirrup and not to any other platform. This provision is intended to protect against platform overloading.

MOBILE SCAFFOLDS (DOES NOT INCLUDE SCISSOR LIFTS)
OSHA 1926.452(w)

This paragraph applies to all mobile scaffolds, not just to those which are manually propelled.

Scaffolds shall be braced by cross, horizontal or diagonal braces, or combination thereof, to prevent racking or collapse of the scaffold and to secure vertical members together laterally so as to automatically square and align the vertical members. In addition, scaffolds shall be plumb, level and squared. All brace connections shall be secured. This paragraph also provides that scaffolds constructed of tube and coupler components shall conform to the requirements of OSHA 1926.452(b), and that scaffolds constructed of fabricated frame components shall conform to the requirements of 1926.452(c). **(See Figure 12-16)**

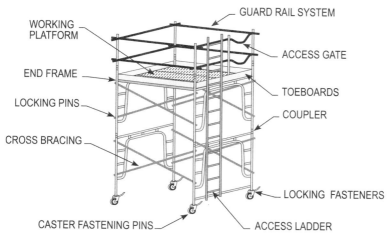

Figure 12-16. This illustration shows the requirements for mobile scaffolds.

WORKING PLATFORM
GUARD RAIL SYSTEM
ACCESS GATE
END FRAME
TOEBOARDS
LOCKING PINS
COUPLER
CROSS BRACING
LOCKING FASTENERS
CASTER FASTENING PINS
ACCESS LADDER

NOTE 1: SCAFFOLD CASTERS AND WHEELS SHALL BE LOCKED WITH POSITIVE WHEEL AND/OR WHEEL AND SWIVEL LOCKS, OR EQUIVALENT MEANS, TO PREVENT MOVEMENT OF THE SCAFFOLD WHILE THE SCAFFOLD IS USED IN A STATIONARY MANNER.

NOTE 2: MANUAL FORCE USED TO MOVE THE SCAFFOLD SHALL BE APPLIED AS CLOSE TO THE BASE AS PRACTICABLE, BUT NOT MORE THAN 5 FT. ABOVE THE SUPPORTING SURFACE.

NOTE 3: POWER SYSTEMS USED TO PROPEL MOBILE SCAFFOLDS SHALL BE DESIGNED FOR SUCH USE. IN ADDITION, FORKLIFTS, TRUCKS, SIMILAR MOTOR VEHICLES OR ADD-ON MOTORS SHALL NOT BE USED TO PROPEL SCAFFOLDS UNLESS THE SCAFFOLD IS DESIGNED FOR SUCH PROPULSION SYSTEMS.

**MOBILE SCAFFOLDS
OSHA 1926.452(w)**

Scaffold casters and wheels shall be locked with positive wheel and/or wheel and swivel locks, or equivalent means, to prevent movement of the scaffold while the scaffold is used in a stationary manner.

Manual force used to move the scaffold shall be applied as close to the base as practicable, but not more than 5 ft. above the supporting surface. OSHA limits the height at which the force can be applied to 5 ft. above the supporting surface, to minimize overturning forces.

Safety Tip: Manual force used to move scaffolds shall be applied as close to the base a practical, but not more than 5 ft.

Power systems used to propel mobile scaffolds shall be designed for such use. In addition, forklifts, trucks, similar motor vehicles or add-on motors shall not be used to propel scaffolds unless the scaffold is designed for such propulsion systems.

Scaffolds shall be stabilized to prevent tipping during movement.

Employees shall not be allowed to ride on scaffolds unless the following conditions exist:

• The surface on which the scaffold is being moved shall be within three degrees of level, and free of pits, holes and obstructions

• The height to base width ratio of the scaffold during movement shall be two to one or less, unless the scaffold is designed and constructed to meet or exceed nationally recognized stability test requirements

- Outrigger frames, when used, shall be installed on both sides of the scaffold per OSHA

- When power systems are used, the propelling force shall be applied directly to the wheels, and shall not produce a speed in excess of one foot per second (0.3 mps)

- No employee is on any part of the scaffold that extends outward beyond the wheels, casters or other supports

Platforms shall not extend outward beyond the base supports of the scaffold unless outrigger frames or equivalent devices are used to ensure stability. Compliance with this provision will prevent eccentric loading of the scaffold frame that could cause the scaffold to tip over.

Where leveling of the scaffold is necessary, screw jacks or equivalent means shall be used.

Caster stems and wheel stems shall be pinned or otherwise secured to scaffold legs or adjustment screws.

Before a scaffold is moved, employees on the scaffold shall be made aware of the move.

REPAIR BRACKET SCAFFOLDS
OSHA 1926.452(x)

OSHA has described such scaffolds as consisting of platforms supported by brackets that are secured in place by one or more wire ropes placed in an approximately horizontal plane around the circumference of the structure and tensioned by a turnbuckle.

Employers shall secure brackets in place with 1/4 in. diameter wire rope that extends around the circumference of the chimney.

Each bracket shall be attached to the securing wire rope (or ropes) by a positive locking device capable of preventing the unintentional detachment of the bracket from the rope, or by some other means that prevents unintentional detachment.

Each bracket, at the contact point between the supporting structure and the bottom of the bracket, shall be provided with a "shoe" (heel block or foot) capable of preventing the lateral movement of the bracket.

Platform units shall be secured to brackets in a manner that prevents the separation of platform units from brackets and prevents movement of platform units or brackets on a completed scaffold.

When a wire rope is placed around a structure to provide safe anchorage for personal fall arrest systems that are used by employees erecting or dismantling repair bracket scaffolds, the wire rope shall be at least 5/16 in. in diameter and shall, in all other respects, satisfy the requirements of Subpart M, OSHA's Fall Protection Standard.

Each wire rope used for securing brackets in place or as an anchorage for personal fall arrest systems shall be protected from damage due to contact with edges, corners, protrusions or other discontinuities of the supporting structure or scaffold components.

Tensioning of each wire rope used for securing brackets in place or as an anchorage for personal fall arrest systems shall be by means of a turnbuckle at least 1 in. in diameter, or by some other equivalent means. OSHA has allowed employers the flexibility to use means other than a single turnbuckle for tensioning wire ropes, where the alternative means provide equivalent tension, because the Agency wants to encourage innovation and provide flexibility. In addition, OSHA anticipates that there may be circumstances where more than one turnbuckle will be needed to tension the wire rope, depending on the diameter of the chimney.

Each turnbuckle shall be connected to the other end of its rope by use of a proper-size eye splice thimble.

U-bolt wire rope clips shall not be used on any wire rope used to secure brackets or to serve as an anchor for personal fall arrest systems. OSHA is concerned that the use of U-bolt wire rope clips as wire rope fasteners on the horizontal support ropes could result in damage to the dead end of the rope. Further, if a segment of damaged dead end later were to become part of the live end due to an increase in the circumference of the structure, the Agency was concerned that the wire rope would be unable to support the loads imposed on it.

Employers shall ensure that materials are not dropped to the outside of the supporting structure.

Erection of a repair bracket scaffold shall be performed in only one direction around the structure.

STILTS
OSHA 1926.452(y)

Employees shall not wear stilts on scaffolds except when the employees are on large area scaffolds.

When employees wearing stilts are on large area scaffolds where guardrail systems are being used, the dimensions of the guardrail system shall be increased to offset the height of the stilts.

All surfaces on which stilts are used shall be flat and free of pits, holes and obstructions, such as debris, as well as all other tripping and falling hazards.

Stilts shall be properly maintained and that any alterations of the original equipment shall be approved by the manufacturer.

AERIAL LIFTS
OSHA 1926.453

Unless otherwise provided in this section, aerial lifts acquired for use on or after January 22, 1973 shall be designed and constructed in conformance with the applicable requirements of the American National Standards for "Vehicle Mounted Elevating and Rotating Work Platforms," ANSI A92.2-1969, including appendix. Aerial lifts acquired before January 22, 1973 that do not meet the requirements of ANSI A92.2-1969 may not be used after January 1, 1976 unless they shall have been modified so as to conform with the applicable design and construction requirements of ANSI A92.2-1969. Aerial lifts include the following types of vehicle-mounted aerial devices used to elevate personnel to job-sites above ground:

- Extensible boom platforms
- Aerial ladders
- Articulating boom platforms
- Vertical towers

• A combination of any such devices. Aerial equipment may be made of metal, wood, fiberglass reinforced plastic (FRP) or other material; may be powered or manually operated; and are deemed to be aerial lifts whether or not they are capable of rotating about a substantially vertical axis.

Aerial lifts may be "field modified" for uses other than those intended by the manufacturer provided the modification has been certified in writing by the manufacturer or by any other equivalent entity, such as a nationally recognized testing laboratory, to be in conformity with all applicable provisions of ANSI A92.2-1969 and this section and to be at least as safe as the equipment was before modification. **(See Figure 12-17)**

Figure 12-17. This illustration shows two different types of aerial lifts.

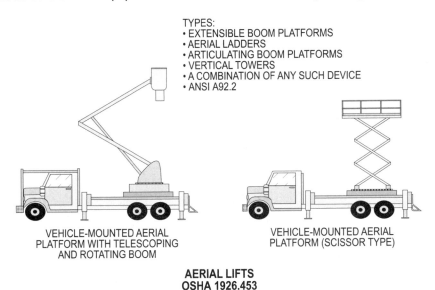

TYPES:
• EXTENSIBLE BOOM PLATFORMS
• AERIAL LADDERS
• ARTICULATING BOOM PLATFORMS
• VERTICAL TOWERS
• A COMBINATION OF ANY SUCH DEVICE
• ANSI A92.2

VEHICLE-MOUNTED AERIAL PLATFORM WITH TELESCOPING AND ROTATING BOOM

VEHICLE-MOUNTED AERIAL PLATFORM (SCISSOR TYPE)

AERIAL LIFTS
OSHA 1926.453

Safety Tip: Prior to each day's use, the lift controls of the aerial lift shall be tested by the lift operator to determine if the controls are in a safe working condition.

Aerial ladders shall be secured in the lower traveling position by the locking device on top of the truck cab, and the manually operated device at the base of the ladder before the truck is moved for highway travel.

Each day prior to use the operator of the aerial lift shall test the lift controls to determine that such controls are in safe working condition. Only authorized persons shall operate an aerial lift.

Never belt off to an adjacent pole, structure or equipment while working from an aerial lift. Employees shall always stand firmly on the floor of the basket, and shall not sit or climb on the edge of the basket or use planks, ladders or other devices for a work position. A body belt shall be worn and a lanyard attached to the boom or basket when working from an aerial lift. **(See Figure 12-18)**

Note: As of January 1, 1998, Subpart M (OSHA 1926.502(d)) provides that body belts are not acceptable as part of a personal fall arrest system.

Climbers shall not be worn while performing work from an aerial lift.

The brakes shall be set and when outriggers are used, they shall be positioned on pads or a solid surface. Wheel chocks shall be installed before using an aerial lift on an incline, provided they can be safely installed.

An aerial lift truck shall not be moved when the boom is elevated in a working position with men in the basket, except for equipment that is specifically designed for this type of operation.

Articulating boom and extensible boom platforms, primarily designed as personnel carriers, shall have both platform (upper) and lower controls. Upper controls shall be in

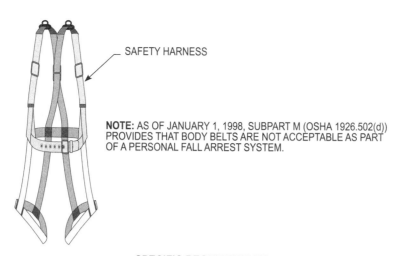

SAFETY HARNESS

NOTE: AS OF JANUARY 1, 1998, SUBPART M (OSHA 1926.502(d)) PROVIDES THAT BODY BELTS ARE NOT ACCEPTABLE AS PART OF A PERSONAL FALL ARREST SYSTEM.

SPECIFIC REQUIREMENTS
OSHA 1926.453(b)(1)

Figure 12-18. This illustration shows the requirements for fall protection while working from an aerial lift.

or beside the platform within easy reach of the operator. Lower controls shall provide for overriding the upper controls. Controls shall be plainly marked as to their function. Lower level controls shall not be operated unless permission has been obtained from the employee in the lift, except in case of emergency.

The insulated portion of an aerial lift shall not be altered in any manner that might reduce its insulating value.

Before moving an aerial lift for travel, the boom(s) shall be inspected to see that it is properly cradled and outriggers are in stowed position.

All electrical tests shall conform to the requirements of ANSI A92.2-1969, Section 5.

TRAINING REQUIREMENTS
OSHA 1926.454

Training requirements for employers who have employees working on scaffolds are addressed in OSHA 1926.454(a). It requires employers to ensure that each employee whose employment involves being on a scaffold is trained to recognize the hazards associated with the type of scaffold being used and to understand the procedures which must be followed to control or minimize those hazards.

OSHA 1926.454(a)(1) through (a)(5) address five areas in which training shall be provided, as applicable. These are as are as follows:

- Affected employees shall be trained in the nature of any electrical hazards, fall hazards and falling object hazards in the work area. Many employees have been killed or seriously injured because they were unaware of workplace hazards or did not understand the consequences of exposure to those hazards. This provision clearly indicates the hazards (i.e., electrocution, falls and falling objects) regarding which training must be provided.

- Affected employees shall be trained in the correct procedures for protection from electrical hazards and for erecting, maintaining and disassembling the required fall protection systems and falling object protection systems. Employees who are on scaffolds while working need to know how protective systems function, so that they know how to install, maintain or remove these systems, as necessary.

Safety Tip: Employees erecting and working on scaffolds shall be properly trained.

OSHA Required Training
1926.454(a)(1) thru (5) and (b)(1) thru (4) and (c)(1) thru (3)

For example: Where a scaffold has been erected without the protective measures necessary for work to be performed on or from the scaffold, the employees subsequently coming onto the scaffold would need to install them. Even where the scaffold erectors have installed the required protection for affected employees, the employees working on the scaffold need to know when and how to maintain that protection, so that a hazardous situation does not develop during scaffold use.

- Employees shall be trained in the proper use of the scaffold and in the proper handling of materials on the scaffold.

- Employees shall be trained in the maximum intended load and the load-carrying capacities of the scaffolds used.

- Employees shall be trained in the pertinent requirements of Subpart L.

Employers shall have each employee who erects, disassembles, moves, operates, repairs, maintains or inspects a scaffold trained by a competent person so that the employee can recognize any hazards related to such work duties. It is designed to differentiate clearly between the training needed by employees erecting and dismantling scaffolds and the training needed by employees who are on scaffolds in the course of their work. The employer shall ensure that each affected employee has been trained by a competent person in four areas, as follows:

- Affected employees shall be trained in the nature of scaffold hazards.

- Affected employees shall be trained in the correct procedures for erecting, disassembling, moving, operating, repairing, inspecting and maintaining the type of scaffold in question. Training provided to an employee to construct, repair or dismantle one type of scaffold will not necessarily enable that employee to repair another type.

- Affected employees shall be trained in the design criteria, maximum load-carrying capacity and intended use of the scaffold.

- Affected employees shall be trained in the pertinent requirements of Subpart L.

Employers shall retrain any employee when the employer has reason to believe that the employee does not have the understanding and skill required by OSHA 1926.454(a) or (b). Employees shall be retrained, as necessary, to restore the requisite scaffold-related proficiency. Circumstances where the provision requires retraining include, but are not limited to, the following situations:

- Whenever there is a change at the worksite that presents a hazard about which the employee has not been trained

- Where changes in the types of scaffolds, fall protection, falling object protection, or other equipment present a hazard about which the employee has not been trained

- Where inadequacies in an affected employee's work practices involving scaffolds indicate that the employee has not retained the requisite proficiency

APPENDIX A

Appendix A provides non-mandatory guidelines to assist employers in complying with the requirements of Subpart L (scaffolds). An employer may use these guidelines and tables as a starting point for designing scaffold systems. However, the guidelines do not provide all the information necessary to build a complete system, and the employer is still responsible for designing and assembling these components in such a way that the completed system will meet the requirements of OSHA 1926.451(a). Scaffold components that are not selected and loaded in accordance with this Appendix, and components for which no specific guidelines or tables are given in this Appendix (e.g., joints, ties, components for wood pole scaffolds more than 60 ft. in height, components for heavy-duty horse scaffolds, components made with other materials and components with other dimensions, etc.) shall be designed and constructed in accordance with the capacity requirements of OSHA 1926.451(a), and loaded in accordance with OSHA 1926.451(d)(1).

GENERAL GUIDELINES AND TABLES

The following tables, and the tables in Part 2 — Specific guidelines and tables, assume that all load-carrying timber members (except planks) of the scaffold are a minimum of 1500 lb-f/in(2) (stress grade) construction grade lumber. All dimensions are nominal sizes as provided in the American Softwood Lumber Standards, dated January 1970, except that, where rough sizes are noted, only rough or undressed lumber of the size specified will satisfy minimum requirements.

Solid sawn wood used as scaffold planks shall be selected for such use following the grading rules established by a recognized lumber grading association or by an independent lumber grading inspection agency. Such planks shall be identified by the grade stamp of such association or agency. The association or agency and the grading rules under which the wood is graded shall be certified by the Board of Review, American Lumber Standard Committee, as set forth in the American Softwood Lumber Standard of the U.S. Department of Commerce.

- Allowable spans shall be determined in compliance with the National Design Specification for Wood Construction published by the National Forest Products Association; paragraph 5 of ANSI A10.8-1988 Scaffolding-Safety Requirements published by the American National Standards Institute; or for 2 in. x 10 in. (nominal) or 2 in. x 9 in. (rough) solid sawn wood planks. **(See Figure 12-19)**

MAXIMUM INTENDED NOMINAL LOAD (LB/FT²)	MAXIMUM PERMISSIBLE SPAN USING FULL THICKNESS UNDRESSED LUMBER (FT)	MAXIMUM PERMISSIBLE SPAN USING NOMINAL THICKNESS LUMBER (FT)
25	10	8
50	8	6
75	6	-

Figure 12-19. This illustration shows the maximum permissible span using full thickness undressed lumber and nominal thickness lumber.

- The maximum permissible span for 1 1/4 in. x 9 in. or wider wood plank of full thickness with a maximum intended load of 50 lb/ft.(2) shall be 4 ft.

Fabricated planks and platforms may be used in lieu of solid sawn wood planks. Maximum spans for such units shall be as recommended by the manufacturer based on the maximum intended load being calculated. **(See Figure 12-20)**

RATED LOAD CAPACITY	INTENDED LOAD
LIGHT-DUTY	25 POUNDS PER SQUARE FOOT APPLIED UNIFORMLY OVER THE ENTIRE SPAN AREA
MEDIUM-DUTY	50 POUNDS PER SQUARE FOOT APPLIED UNIFORMLY OVER THE ENTIRE SPAN AREA
HEAVY-DUTY	75 POUNDS PER SQUARE FOOT APPLIED UNIFORMLY OVER THE ENTIRE SPAN AREA
ONE-PERSON	250 POUNDS PLACED AT THE CENTER OF THE SPAN (TOTAL 250 POUNDS)
TWO-PERSON	250 POUNDS PLACED 18 INCHES TO THE LEFT AND RIGHT OF THE CENTER OF THE SPAN (TOTAL 500 POUNDS)
THREE-PERSON	250 POUNDS PLACED AT THE CENTER OF THE SPAN AND 250 POUNDS PLACED 18 INCHES TO THE LEFT AND RIGHT OF THE CENTER OF THE SPAN (TOTAL 750 POUNDS)

Figure 12-20. This illustration shows the intended load per sq. ft. applied uniformly over the entire span area.

Note: Platform units used to make scaffold platforms intended for light-duty use shall be capable of supporting at least 25 pounds per sq. ft. applied uniformly over the entire unit-span area, or a 250-pound point load placed on the unit at the center of the span, whichever load produces the greater shear force.

Guardrails shall be as follows:

(a) Toprails shall be equivalent in strength to
  • 2 in. by 4 in. lumber
  • 1 1/4 in. x 1/8 in. structural angle iron
  • 1 in. x .070 in. wall steel tubing
  • 1.990 in. x .058 in. wall aluminum tubing.

(b) Midrails shall be equivalent in strength to
  • 1 in. by 6 in. lumber
  • 1 1/4 in. x 1 1/4 in. x 1/8 in. structural angle iron
  • 1 in. x .070 in. wall steel tubing
  • 1.990 in. x .058 in. wall aluminum tubing.

(c) Toeboards shall be equivalent in strength to
  • 1 in. by 4 in. lumber
  • 1 1/4 in. x 1 1/4 in. structural angle iron
  • 1 in. x .070 in. wall steel tubing
  • 1.990 in. x .058 in. wall aluminum tubing

(d) Posts shall be equivalent in strength to
  • 2 in. by 4 in. lumber
  • 1 1/4 in. x 1 1/4 in. x 1/8 structural angle iron
  • 1 in. x .070 in. wall steel tubing
  • 1.990 in. x .058 in. wall aluminum tubing

(e) Distance between posts shall not exceed 8 ft.

(f) Overhead protection shall consist of 2 in. nominal planking laid tight, or 3/4 in. plywood.

(g) Screen installed between toeboards and midrails or toprails shall consist of No. 18 gauge U.S. Standard wire 1 in. mesh.

SPECIFIC GUIDELINES AND TABLES

The following pole scaffolds shall meet specific guidelines:

  • Single wood pole scaffolds
  • Independent wood pole scaffolds

See Figures 12-21(a) and (b) for a table pertaining to single wood pole scaffolds and independent wood pole scaffolds.

Note: All members except planking are used on edge. All wood bearers shall be reinforced with 3/16 in. x 2 in. steel strip, or the equivalent, secured to the lower edges for the entire length of the bearer.

TUBE AND COUPLER SCAFFOLDS

Tube and coupler scaffolds shall meet specific guidelines. **(See Figures 12-22(a) and (b))**

Single Wood Pole Scaffolds				
	Light Duty up to 20 feet high	Light Duty up to 60 feet hight	Medium Duty up to 60 feet high	Heavy Duty up to 60 feet high
Maximum Intended Load (lbs/ft^2)	25	25	50	75
Poles or Uprights	2" x 4"	4" x 4"	4" x 4"	4" x 6"
Maximum Pole Spacing (Longitudinal)	6'	10'	8'	6'
Maximum Pole Spacing (Transverse)	5'	5'	5'	5'
Runners	1" x 4"	1 1/4" x 9"	2" x 10"	2" x 10"
Bearers and Maximum Spacing of Bearers				
3'	2" x 4"	2" x 4"	2" x 10" or 3" x 4"	2" x 10" or 3" x 5"
5'	2" x 6" or 3" x 4"	2" x 6" or 3" x 4" (rough)	2" x 10" or 3" x 4"	2" x 10" or 3" x 5"
6'	-	-	2" x 10" or 3" x 4"	2" x 10" or 3" x 5"
8'	-	-	2" x 10" or 3" x 4"	-
Planking	1 1/4" x 9"	2" x 10"	2" x 10"	2" x 10"
Maximum Vertical Spacing of Horizontal Members	7'	9'	7'	6' 6"
Bracing Horizontal	1" x 4"	1" x 4"	1" x 6" or 1 1/4" x 4"	2" x 4"
Tie-ins	1" x 4"	1" x 4"	1" x 4"	1" x 4"

Note: All members except planking are used on edge. All wood bearers shall be reinforced with 3/16" x 2" steel strip, or equivalent, secured to the lower edges for the entire length of the bearer.

Figure 12-21(a). This illustration shows a table pertaining to the requirements for single wood pole scaffolds.

Independent Wood Pole Scaffolds				
	6 feet	10 feet	8 feet	8 feet
Runners	1 1/4" x 4"	1 1/4" x 9"	2" x 10"	2" x 10"
Bearers and Maximum Spacing of Bearers				
3'	2" x 4"	2" x 4"	2" x 10"	2" x 10" (rough)
6'	2" x 6" or 3" x 4"	2" x 10" (rough) or 2" x 4"	2" x 10"	2" x 10" (rough)
8'	2" x 6" or 3" x 4"	2" x 10" (rough) 3" x 8"	2" x 10"	-
10'	2" x 6" or 3" x 4"	2" x 10" (rough) 3" x 3"	2" x 10"	-
Planking	1 1/4" x 9"	2" x 10"	2" x 10"	2" x 10"
Maximum Vertical Spacing of Horizontal Members	7'	7'	6'	6'
Bracing Horizontal	1" x 4"	1" x 4"	1" x 6" or 1 1/4" x 4"	2" x 4"
Bracing Diagonal	1" x 4"	1" x 4"	1" x 4"	2" x 4"
Tie-ins	1" x 4"	1" x 4"	1" x 4"	1" x 4"

Note: All members except planking are used on edge. All wood bearers shall be reinforced with 3/16" x 2" steel strip, secured to the lower edges for the entire length of the bearer.

Figure 12-21(b). This illustration shows a table pertaining to the requirements for independent wood pole scaffolds.

Minimum Size of Members			
	Light Duty	Medium Duty	Heavy Duty
Maximum Intended load	25 lbs/ft^2	50 lbs/ft^2	75 lbs/ft^2
Posts, Runner and braces	Nominal 2" (1.90") OD steel tube or pipe	Nominal 2" (1.90") OD Steel tube or pipe	Nominal 2" (1.90") OD steel tube or pipe
Bearers	Nominal 2" (1.90") OD Steel tube or pipe and a maximum post spacing of 4' x 10'	Nominal 2" (2.375") OD Steel tube or pipe and a maximum post spacing of 4' x 7' or Nominal 2 1/2" (1.90") OD Steel tube or pipe and a maximum post spacing of 6' x 8'*	Nominal 2 1/2" (2.375") OD Steel tube or pipe and a maximum post spacing of 6' x 6'
Maximum Runner Spacing Vertically	6' 6"	6' 6"	6' 6"

* Bearers shall be installed in the direction of the shorter dimension.
Note: Longitudinal diagonal bracing shall be installed at an angle of 45° (±5°)

Figure 12-22(a). This illustration shows the minimum size of members for tube and coupler scaffolds.

Maximum Number of Planked Levels				
Number of Working Levels	Maximum Number of Additional Planked Levels			Maximum Height of Scaffold (in feet)
	Light Duty	Medium Duty	Heavy Duty	
1	16	11	6	125
2	11	1	0	125
3	6	0	0	125
4	1	0	0	125

Figure 12-22(b). This illustration shows the maximum number of planked levels for tube and coupler scaffolds.

FABRICATED FRAME SCAFFOLDS

Because of their prefabricated nature, no additional guidelines or tables for these scaffolds are being adopted in this Appendix.

PLASTERERS', DECORATORS' AND LARGE AREA SCAFFOLDS

The guidelines for pole scaffolds or tube and coupler scaffolds (Pole Scaffolds – Tube and Coupler Scaffolds – Appendix A) may be applied.

BRICKLAYERS' SQUARE SCAFFOLDS

The following shall be considered for bricklayers' square scaffolds:

- Maximum intended load: 50 lb/ft.(2)(*)

Note: The squares shall be set not more than 8 ft. apart for light duty scaffolds and not more than 5 ft. apart for medium duty scaffolds.

- Maximum width: 5 ft.
- Maximum height: 5 ft.
- Gussets: 1 in. x 6 in.
- Braces: 1 in. x 8 in.
- Legs: 2 in. x 6 in.
- Bearers (horizontal members): 2 in. x 6 in.

HORSE SCAFFOLDS

The following shall be considered for horse scaffolds:

- Maximum intended load (light duty): 25 lb/ft.(2)

Note: Horses shall be spaced not more than 8 ft. apart for light duty loads, and not more than 5 ft. apart for medium duty loads.

- Maximum intended load (medium duty): 50 lb/ft.(2)

Note: Horses shall be spaced not more than 8 ft. apart for light duty loads, and not more than 5 ft. apart for medium duty loads.

The following shall be considered for horizontal members or bearers:

- Light duty: 2 in. x 4 in.
- Medium duty: 3 in. x 4 in.
- Legs: 2 in. x 4 in.
- Longitudinal brace between legs: 1 in. x 6 in.
- Gusset brace at top of legs: 1 in. x 8 in.
- Half diagonal braces: 2 in. x 4 in.

FORM SCAFFOLDS AND CARPENTERS' BRICK SCAFFOLDS

The following shall be considered for form scaffolds and carpenters' brick scaffolds:

- Brackets shall consist of a triangular-shaped frame made of wood with a cross-section not less than 2 in. by 3 in., or of 1 1/4 in. x 1 1/4 in. x 1/8 in. structural angle iron.

- Bolts used to attach brackets to structures shall not be less than 5/8 in. in diameter.

- Maximum bracket spacing shall be 8 ft. on centers.

- No more than two employees shall occupy any given 8 ft. of a bracket or form scaffold at any one time.

- Tools and materials shall not exceed 75 pounds in addition to the occupancy.

WOODEN FIGURE-FOUR SCAFFOLDS

The following shall be considered for wood figure-four scaffolds:

- Maximum intended load: 25 lb/ft.(2)
- Uprights: 2 in. x 4 in. or 2 in. x 6 in.
- Bearers (two): 1 in. x 6 in.
- Braces: 1 in. x 6 in.
- Maximum length of bearers (unsupported): 3 ft. 6 in.

Outrigger bearers shall consist of two pieces of 1 in. x 6 in. lumber nailed on opposite sides of the vertical support.

Bearers for wood figure-four brackets shall project not more than 3 ft. 6 in. from the outside of the form support, and shall be braced and secured to prevent tipping or turning. The knee or angle brace shall intersect the bearer at least 3 ft. from the form at an angle of approximately 45 degrees, and the lower end shall be nailed to a vertical support.

METAL BRACKET SUPPORTS

The following shall be considered for metal bracket supports:

- Maximum intended load: 25 lb/ft.(2)
- Uprights: 2 in. x 4 in.
- Bearers: As designed.
- Braces: As designed.

WOOD BRACKET SCAFFOLDS

The following shall be considered for wood bracket scaffolds:

- Maximum intended load: 25 lb/ft.(2)
- Uprights: 2 in. x 4 in or 2 in. x 6 in
- Bearers: 2 in. x 6 in
- Maximum scaffold width: 3 ft. 6 in.
- Braces: 1 in. x 6 in.

ROOF BRACKET SCAFFOLDS

No specific guidelines or tables are given.

OUTRIGGER SCAFFOLDS (SINGLE LEVEL)

No specific guidelines tables are given.

PUMP JACK SCAFFOLDS

Wood poles shall not exceed 30 ft. in height. Maximum intended load — 500 lbs between poles; applied at the center of the span. Not more than two employees shall be on a pump jack scaffold at one time between any two supports. When 2 ft. x 4 ft. lumber are spliced together to make a 4 in. x 4 in. wood pole, they shall be spliced with "10 penny" common nails no more than 12 in. center to center, staggered uniformly from the opposite outside edges.

LADDER JACK SCAFFOLDS

Maximum intended load — 25 lb/ft(2). However, not more than two employees shall occupy any platform at any one time. Maximum span between supports shall be 8 ft.

WINDOW JACK SCAFFOLDS

Not more than one employee shall occupy a window jack scaffold at any one time.

CRAWLING BOARDS (CHICKEN LADDERS)

Crawling boards shall be not less than 10 in. wide and 1 in. thick, with cleats having a minimum 1 in. x 1 1/2 in. cross-sectional area. The cleats shall be equal in length to the width of the board and spaced at equal intervals not to exceed 24 in.

STEP, PLATFORM AND TRESTLE LADDER SCAFFOLDS

No additional guidelines or tables are given.

SINGLE-POINT ADJUSTABLE SUSPENSION SCAFFOLDS

Maximum intended load — 250 lbs. Wood seats for boatswains' chairs shall be not less than 1 in. thick if made of non-laminated wood, or 5/8 in. thick if made of marine quality plywood.

TWO-POINT ADJUSTABLE SUSPENSION SCAFFOLDS

The following shall be considered for two-point adjustable suspension scaffolds:

- In addition to direct connections to buildings (except window cleaners' anchors) acceptable ways to prevent scaffold sway include angulated roping and static lines. Angulated roping is a system of platform suspension in which the upper wire rope sheaves or suspension points are closer to the plane of the building face than the corresponding attachment points on the platform, thus causing the platform to press against the face of the building. Static lines are separate ropes secured at their top and bottom ends closer to the plane of the building face than the outermost edge of the platform. By drawing the static line taut, the platform is drawn against the face of the building.

- On suspension scaffolds designed for a working load of 500 pounds, no more than two employees shall be permitted on the scaffold at one time. On suspension scaffolds with a working load of 750 pounds, no more than three employees shall be permitted on the scaffold at one time.

- Ladder-type platforms. The side stringer shall be of clear straight-grained spruce. The rungs shall be of straight-grained oak, ash, or hickory, at least 1 1/8 in. in diameter, with 7/8 in. tenons mortised into the side stringers at least 7/8 in. The stringers shall be tied together with tie rods not less than 1/4 in. in diameter, passing through the stringers and riveted up tight against washers on both ends. The flooring strips shall be spaced not more than 5/8 in. apart, except at the side rails where the space may be 1 in. **(See Figures 12-23(a) and (b))**

Schedule for Ladder Type Platforms			
Length of Platform	12'	14' and 16'	18' and 20'
Side Stringers, Minimum Cross Section (Finished Sizes):			
At Ends	1 3/4" x 2 3/4"	1 3/4" x 2 3/4"	1 3/4" x 3"
At Middle	1 3/4" x 3 3/4"	1 3/4" x 3 3/4"	1 3/4" x 4"
Reinforcing Strip (Minimum)	A 1/8" x 7/8" steel reinforcing strip attached to the side or underside, full length		
Rungs	Rungs shall be 1 1/8" minimum diameter with at least 7/8" in diameter tenons, and the maximum spacing shall be 12" to center		
Tie Rods:			
Number (Minimum)	3	4	4
Diameter (Minimum)	1/4"	1/4"	1/4"
Flooring, Minimum Finished Size	1/2" x 2 3/4"	1/2" x 2 3/4"	1/2" x 2 3/4"

Figure 12-23(a). This illustration shows a schedule for ladder-type platforms.

Schedule for Ladder Type Platforms		
Length of Platform	22' and 24'	28 and 30'
Side Stringers, Minimum Cross Section (Finished Sizes):		
At Ends	1 3/4" x 3"	1 3/4" x 3 1/2"
At Middle	1 3/4" x 4 1/4"	1 3/4" x 5"
Reinforcing Strip (Minimum)	A 1/8" x 7/8" steel reinforcing strip attached to the side or underside, full length	
Rungs	Rungs shall be 1 1/8" minimum diameter with at least 7/8" in diameter tenons, and the maximum spacing shall be 12" to center	
Tie Rods:		
Number (Minimum)	5	6
Diameter (Minimum)	1/4"	1/4"
Flooring, Minimum Finished Size	1/2" x 2 3/4"	1/2" x 2 3/4"

Figure 12-23(b). This illustration shows a schedule for ladder-type platforms.

- Plank-type platforms. Plank-type platforms shall be composed of not less than nominal 2 in. x 8 in. unspliced planks, connected together on the underside with cleats at intervals not exceeding 4 ft., starting 6 in. from each end. A bar or other effective means shall be securely fastened to the platform at each end to prevent the platform from slipping off the hanger. The span between hangers for plank-type platforms shall not exceed 10 ft.

- Beam type platforms. Beam platforms shall have side stringers of lumber not less than 2 x 6 inches set on edge. The span between hangers shall not exceed 12 feet when beam platforms are used. The flooring shall be supported on 2 x 6 inch cross beams, laid flat and set into the upper edge of the stringers with a snug fit, at intervals of not more than 4 feet, securely nailed to the cross beams. Floor-boards shall not be spaced more than 1/2 inch apart.

MULTI-POINT ADJUSTABLE SUSPENSION SCAFFOLDS AND STONESETTERS' MULTI-POINT ADJUSTABLE SUSPENSION SCAFFOLDS

No specific guidelines or tables are given for these scaffolds.

MASONS' MULTI-POINT ADJUSTABLE SUSPENSION SCAFFOLDS

Maximum intended load — 50 lb/ft(2). Each outrigger beam shall be at least a standard 7 in., 15.3 pound steel I-beam, at least 15 ft. long. Such beams shall not project more than 6 ft. 6 in. beyond the bearing point. Where the overhang exceeds 6 ft. 6 in., outrigger beams shall be composed of stronger beams or multiple beams.

CATENARY SCAFFOLDS

The following shall be considered for catenary scaffolds:

- Maximum intended load — 500 lbs.
- Not more than two employees shall be permitted on the scaffold at one time.
- Maximum capacity of come-along shall be 2000 lbs.
- Vertical pickups shall be spaced not more than 50 ft. apart.
- Ropes shall be equivalent in strength to at least 1/2 in. diameter improved plow steel wire rope.

FLOAT (SHIP) SCAFFOLDS

The following shall be considered for float (ship) scaffolds:

- Maximum intended load — 750 lbs.
- Platforms shall be made of 3/4 in. plywood, equivalent in rating to American Plywood Association Grade B-B, Group I, Exterior.
- Bearers shall be made from 2 in. x 4 in., or 1 in. x 10 in. rough lumber. They shall be free of knots and other flaws.
- Ropes shall be equivalent in strength to at least 1 in. diameter first grade manila rope.

INTERIOR HUNG SCAFFOLDS

The following shall be considered for interior hung scaffolds:

- Bearers (use on edge): 2 in. x 10 in.
- Maximum intended load: Maximum span
- 25 lb/ft.(2): 10 ft.
- 50 lb/ft.(2): 10 ft.
- 75 lb/ft.(2): 7 ft.

NEEDLE BEAM SCAFFOLDS

The following shall be considered for needle beam scaffolds:

- Maximum intended load: 25 lb/ft.(2)
- Beams: 4 in. x 6 in.
- Maximum platform span: 8 ft.
- Maximum beam span: 10 ft.

Ropes shall be attached to the needle beams by a scaffold hitch or an eye splice. The loose end of the rope shall be tied by a bowline knot or by a round turn and a half hitch.

Ropes shall be equivalent in strength to at least 1 in. diameter first grade manila rope.

MULTI-LEVEL SUSPENSION SCAFFOLDS

No additional guidelines or tables are being given for these scaffolds.

MOBILE SCAFFOLDS

Stability test as described in the ANSI A92 series documents, as appropriate for the type of scaffold, can be used to establish stability for the purpose of OSHA 1926.452(w)(6).

REPAIR BRACKET SCAFFOLDS

No additional guidelines or tables are being given for these scaffolds.

STILTS

No specific guidelines or tables are given.

TANK BUILDERS' SCAFFOLD

The following shall be considered for tank builders' scaffold:

- The maximum distance between brackets to which scaffolding and guardrail supports are attached shall be no more than 10 ft. 6 in.

- Not more than three employees shall occupy a 10 ft. 6 in. span of scaffold planking at any time.

- A taut wire or synthetic rope supported on the scaffold brackets shall be installed at the scaffold plank level between the innermost edge of the scaffold platform and the curved plate structure of the tank shell to serve as a safety line in lieu of an inner guardrail assembly where the space between the scaffold platform and the tank exceeds 12 in. In the event the open space on either side of the rope exceeds 12 in., a second wire or synthetic rope appropriately placed, or guardrails in accordance with OSHA 1926.451(e)(4), shall be installed in order to reduce that open space to less than 12 in.

- Scaffold planks of rough full-dimensioned 2 9n. x 12 in. Douglas Fir or Southern Yellow Pine of Select Structural Grade shall be used. Douglas Fir planks shall have a fiber stress of at least 1900 lb/in(2) (130,929 n/cm(2)) and a modulus of elasticity of at least 1,900,000 lb/in(2) (130,929,000 n/cm(2)), while Yellow Pine planks shall have a fiber stress of at least 2500 lb/in(2) (172,275 n/cm(2)) and a modulus of elasticity of at least 2,000,000 lb/in(2) (137,820,000 n/cm(2)).

- Guardrails shall be constructed of a taut wire or synthetic rope, and shall be supported by angle irons attached to brackets welded to the steel plates. These guardrails shall comply with OSHA 1926.451(e)(4). Guardrail supports shall be located at no greater than 10 ft. 6 in. intervals.

Scaffolds

Section	Answer
_____	T F
_____	T F
_____	T F
_____	T F
_____	T F
_____	T F
_____	T F
_____	T F
_____	T F
_____	T F
_____	T F
_____	T F
_____	T F
_____	T F

Scaffolds

1. Each scaffold and scaffold component shall be capable of supporting, without failure, its own weight and at least 3 times the maximum intended load applied or transmitted to it.

2. OSHA provides that the 4 to 1 factor for a component applies only to the load which is actually applied or transmitted to that component, and not to the total load placed on the scaffold.

3. Suspension rope, including its connecting hardware, used on non-adjustable suspension scaffolds shall be capable of supporting, without failure, at least 6 times the maximum intended load applied or transmitted to that rope.

4. OSHA believes that platforms used solely as walkways or solely by employees erecting or dismantling scaffolds should be at least 3 planks wide.

5. Each end of a platform unit, unless cleated or otherwise restrained by hooks or equivalent means, shall extend over the center line of its support at least 6 in.

6. Where platforms are overlapped to create a long platform, the overlap shall occur only over supports, and shall not be less than 10 in. unless the platforms are nailed together or otherwise restrained to prevent movement.

7. Scaffolds with a height to base width ratio of more than 3 to 1 (including outrigger supports, if used) shall be restrained from tipping by guying, tying, bracing, or equivalent means.

8. All suspension scaffold support devices, such as outrigger beams, cornice hooks, parapet clamps, and similar devices, shall rest on surfaces capable of supporting at least 4 times the loads imposed on them by the scaffold operating at the rated load of the hoist.

9. Swaged attachments or spliced eyes on wire suspension ropes shall not be used unless they are made by the wire rope manufacturer or a qualified person.

10. Scaffolds and scaffold components shall not be loaded in excess of their maximum intended loads or rated capacities, whichever is less.

11. The minimum clearance between power lines and scaffolds, including any conductive materials on the scaffold for all uninsulated lines and for insulated lines of more than 300 volts is 10 ft.

12. Platform units shall not deflect more than 1/50 of the span when loaded.

_____ T   F   **13.** OSHA sets 6 ft. as the threshold height above which fall protection is required and indicates what fall protection measures are required for particular types of scaffolds.

_____ T   F   **14.** The top edge height of toprails or equivalent members on supported scaffolds manufactured or placed into service after January 1, 2000 shall be between 38 in. and 45 in. above the platform surface.

_____ T   F   **15.** Toeboards shall be capable of withstanding, without failure, a force of at least 60 pounds applied in any downward or horizontal direction at any point along the toeboard.

_____ T   F   **16.** Pole scaffolds over 80 ft. in height shall be designed by a registered professional engineer.

_____ T   F   **17.** Tube and coupler scaffolds over 125 ft. in height shall be designed by a registered professional engineer.

_____ T   F   **18.** Horse scaffolds shall not be constructed or arranged more than two tiers or 20 ft. in height.

_____ T   F   **19.** The tackle used with boatswains' chairs shall be ball bearing or bushed blocks containing safety hooks and properly "eye" spliced minimum five-eight (5/8) in. diameter first grade manila rope.

_____ T   F   **20.** Two-point scaffolds cannot be bridged or otherwise connected one to another during raising and lowering operations unless the bridge connections are articulated and the hoists properly sized.

13

Fall Protection

In the construction industry in the U.S., falls are the leading cause of worker fatalities. Each year, on average, between 150 and 200 workers are killed and more than 100,000 are injured as a result of falls at construction sites. OSHA recognizes that accidents involving falls are generally complex events frequently involving a variety of factors. Consequently the standard for fall protection deals with both the human and equipment-related issues in protecting workers from fall hazards. For example, employers and employees need to do the following:

- Where protection is required, select fall protection systems appropriate for given situations

- Use proper construction and installation of safety systems

- Supervise employees properly

- Use safe work procedures

- Train workers in the proper selection, use, and maintenance of all protection systems

OSHA has revised its construction industry safety standards (29 Code of Federal Regulations, Subpart M, Fall Protection, 1926.500, 1926.501, 1926.502 and 1926.503) and developed systems and procedures designed to prevent employees from falling off, onto or through working levels and to protect employees from being struck by falling objects. The performance-oriented requirements make it easier for employers to provide the necessary protection.

The rule covers most construction workers except those inspecting, investigating or assessing workplace conditions prior to the actual start of work or after all work has been completed.

Safety Tip: Employees working at 6 ft. or more above a lower level shall be protected from fall hazards.

The rule identifies areas or activities where fall protection is needed. These include, but are not limited to, ramps, runways and other walkways; excavations; hoist areas; holes; formwork and reinforcing steel; leading edge work; unprotected sides and edges; overhand bricklaying and related work; roofing work; precast concrete erection; wall openings; residential construction; and other walking/working surfaces. The rule sets a uniform threshold height of 6 ft. (1.8 m), thereby providing consistent protection. This means that construction employers shall protect their employees from fall hazards and falling objects whenever an affected employee is 6 ft. (1.8 m) or more above a lower level. Protection shall also be provided for construction workers who are exposed to the hazard of falling into dangerous equipment.

Under the new standard, employers will be able to select fall protection measures compatible with the type of work being performed. Fall protection generally can be provided through the use of guardrail systems, safety net systems, personal fall arrest systems, positioning device systems and warning line systems, among others.

The OSHA rule clarifies what an employer shall do to provide fall protection for employees, such as identifying and evaluating fall hazards and providing specific training. Requirements to provide fall protection for workers on scaffolds and ladders and for workers engaged in steel erection of buildings are covered in other subparts of OSHA regulations.

The new standard prescribes the duty to provide fall protection, sets the criteria and practices for fall protection systems, and requires training. It covers hazard assessment and fall protection and safety monitoring systems. Also addressed are controlled access zones, safety nets, and guardrail, personal fall arrest, warning line and positioning device systems.

SCOPE, APPLICATION AND DEFINITIONS
OSHA 1926.500

Fall protection requirements of Subpart M shall apply to all construction workplaces, except where another subpart of Part 1926 specifies what fall protection systems shall be used and sets the criteria for those fall protection systems. OSHA notes there are some activities that will be classified as either general industry or construction depending on other activities occurring at the same time or same site.

For example: When surface preparation work and sandblasting work are being performed in connection with painting activities or other construction activities, then these two activities are considered construction work and employers engaged in these activities shall follow the requirements of Subpart M as it pertains to fall hazards associated with surface preparation and sandblasting. On the other hand, when these activities are conducted as part of general maintenance work, the fall protection requirements of the general industry standards (Part 1910) would apply.

The provisions of Subpart M shall not apply when the employer establishes that employees are only inspecting, investigating or assessing workplace conditions prior to the actual start of the work or after work has been completed. OSHA has set this exception because employees engaged in inspecting, investigating and assessing workplace conditions before the actual work begins or after work has been completed are exposed to fall hazards for very short durations, if at all, since they most likely would be able to accomplish their work without going near the danger zone. Also, the Agency's experience is that such individuals who are not continually or routinely exposed to fall hazards tend to be very focused on their footing, ever alert and aware of the hazards associated with falling. These practical considerations would make it unreasonable, the agency believes, to require the installation of fall protection systems either prior to the start of construction work or after such work has been completed. Such requirements would impose an unreasonable burden on employers without demonstrable benefits.

OSHA notes that operations are normally conducted in good weather, that the nature of such work normally exposes the employee to the fall hazard only for a short time, if at all, and that requiring the installation of fall protection systems under such circumstances would expose the employee who installs those systems to falling hazards for a longer time than the person performing an inspection or similar work. In addition, OSHA anticipates that employees who inspect, investigate or assess workplace conditions will be more aware of their proximity to an unprotected edge than, for example, a roofer who is moving backwards while operating a felt laying machine or a plumber whose attention is on overhead pipe and not on the floor edge.

Some subparts within Part 1926, aside from Subpart M, contain fall protection requirements. Those other provisions, however, are not comprehensive. Therefore, when an employee is exposed to a falling hazard, such as that of falling more than 6 ft. to a lower level, which is not specifically addressed in another subpart, OSHA intends that the general provisions of Subpart M apply.

For example: Subpart N contains requirements for fall protection when certain cranes are used; it does not address other equipment or workplace conditions otherwise covered by Subpart N that may also expose employees to a fall hazard. Also, OSHA 1926.500(a)(3) requires the use of specified fall protection systems, but does not set criteria to that those systems shall meet.

For example: Subpart L - Scaffolds, requires that employers provide guardrails and safety belts (body belts) when employees are working on scaffolds. Subpart L sets criteria for the use of guardrail systems on scaffolds, but does not set criteria for the use of body belts. Under those circumstances, body belts used by employees working on scaffolds shall satisfy the criteria in Subpart M, while guardrails would be required to meet the criteria in Subpart L.

Aside from Subpart L, the subparts in Part 1926 that address the subject of fall protection are Subparts N, R, S, V and X. A brief summary of the fall protection topics covered for each of the above-mentioned subparts is listed below.

- Subpart N - Cranes and Derricks. Requirements to have fall protection for employees working on certain cranes and derricks are contained in Subpart N.

- Subpart R - Steel Erection. Requirements to have fall protection for employees engaged in the construction of skeleton steel buildings are contained in Subpart R.

- Subpart S - Underground Construction, Caissons, Cofferdams and Compressed Air. Requirements to have fall protection on specified pieces of equipment used in underground construction operations are contained in Subpart S.

- Subpart V - Power Transmission and Distribution. Requirements to have fall protection for employees engaged in the construction of electric transmission and distribution lines and equipment are contained in Subpart V.

- Subpart X - Stairways and Ladders. Requirements to have fall protection for employees working on stairways and fixed ladders are contained in Subpart X. Subpart X already references the current fall protection criteria set out in Subpart M, and that reference will now incorporate the revised criteria in OSHA 1926.502 of this final rule.

DUTY TO HAVE FALL PROTECTION
OSHA 1926.501

This section specifies the areas and operations where fall protection systems are required. The criteria to be met by fall protection systems, and the training necessary to use the systems properly, are covered in OSHA 1926.502 and OSHA 1926.503, respectively.

OSHA 1926.501 sets forth the type of fall protection systems employers shall use in various areas and operations. In addition, it mandates that all fall protection systems required to be used by 1926.501 conform to the criteria and work practices set forth in OSHA 1926.502. OSHA notes that most of the provisions provide several choices for providing fall protection, but some provisions limit the choices.

For example: Only guardrail systems shall be used to protect employees on ramps and runways and other walkways. In these situations, OSHA believes guardrail systems offer the appropriate level of fall protection and the record supports this conclusion.

OSHA has consistently maintained that all construction employers are responsible for obtaining information about the workplace hazards to which their employees may be exposed and for taking appropriate action to protect affected employees from any such hazards. The Occupational Safety and Health Review Commission has held that an employer shall make a reasonable effort to anticipate the particular hazards to which its employees may be exposed in the course of their scheduled work. Specifically, an employer shall inspect the area to determine what hazards exist or may arise during the work before permitting employees to work in that area, and the employer shall then give specific and appropriate instructions to prevent exposure to unsafe conditions.

OSHA considers subcontractors to have a reciprocal responsibility to determine what protective measures the general contractors have identified as necessary and have implemented. Furthermore, the same considerations arise at a multi-employer worksite, because each contractor needs to know about any hazards that other contractors may confront or create so that contractors can take the appropriate precautions for employee protection.

Safety Tip: Employers shall protect employees from fall hazards.

OSHA requires employers to protect employees performing construction work from fall hazards, and sets criteria for the proper implementation of fall protection through the requirements in Subpart M and the specific standards referenced in OSHA 1926.500(a)(2) and (a)(3). However, there is much more to workplace safety than an employer arriving at a work site with a copy of the pertinent standards in hand. Employers have a duty to anticipate the need to work at heights and to plan their work activities accordingly. Careful planning and preparation (e.g., project design that incorporates fall protection and employee training) lay the necessary groundwork for an accident-free workplace.

OSHA is aware that many falls have occurred because employers have not taken fall protection into account when they plan and undertake construction even when it is known that the work involves employee exposure to fall hazards. In some cases, an employer has recognized the hazard and established appropriate fall protection procedures, but has failed to ensure that employees followed those procedures. In other cases, employers have misidentified the hazard, selected inappropriate measures or completely failed to address fall hazards. The foreseeable consequence is that, as discussed above in the Background section, falls from elevations account for a large percentage of construction-related injuries and are the leading cause of death on construction jobs.

Employers need information about the work they are to perform so that they can make fall protection an integral part of their projects. An employer's communication and coordination with customers, other contractors (particularly at multi-employer worksites) and suppliers are critical elements of that employer's ability to protect its employees and to avoid creating hazards for other employees. Initially, the employer needs to develop or obtain information regarding the work to be performed, so that all anticipated fall hazards are identified. The employer would then determine how to protect its employees from those hazards.

For example: Many employers are minimizing exposure to fall hazards by having anchorage points for personal fall arrest systems fabricated or designed into structural members and by installing perimeter lines on structural members before those members are lifted into position.

OSHA anticipates that the trend towards providing "100 percent fall protection" will spur even more effective efforts, from the design stage through to project completion, to increase employee protection. To this end, employers will need to reexamine their "traditional methods" and, when possible, update them by incorporating available fall protection technology and design concepts. OSHA believes that while there may initially be some increased costs and disruption associated with these efforts, subsequent productivity gains and reductions in the cost of workers' compensation will clearly make it highly cost effective in the long run to provide effective fall protection. (See Regulatory Impact and Regulatory Flexibility Analysis, Section V.)

An employer who controls its own construction projects will generally find it adequate to make its design and equipment decisions part of the project blueprints and workplan. Where employers are bidding a contract to perform specified construction work, making fall protection a bid item gives potential customers a clear idea of how prospective contractors plan to comply with Subpart M. This is a point at which effective communication between a contractor and a customer is critical. Prospective contractors shall obtain sufficient information from the customer to enable them to develop responsive bids, incorporating fall protection that complies with Subpart M. By doing so, the bidding contractor reassures the customer that it has taken into account the full cost of performing the work in question. Both parties need to recognize that employee protection is an integral part of every construction project. Employers will not be permitted to gain a competitive advantage by exposing their workers to fall hazards.

Employers shall ensure the structural integrity of walking/working surfaces before employees are permitted to be on those surfaces.

There are 15 requirements contained in OSHA 1926.501(b) that set forth the options from which employers may choose to protect employees exposed to fall hazards when on "walking/working surfaces." Employers shall choose and use a fall protection system (or combination of systems) as provided by OSHA 1926.501(b)(1) through (b)(15) which cover the fall protection needs of particular walking and working surfaces. The following 15 requirements are covered for employers to choose when protecting employees exposed to fall hazards when on walking/working surfaces:

- Unprotected sides and edges
- Leading edges
- Hoist areas
- Holes
- Formwork and reinforcing steel
- Excavations
- Dangerous equipment
- Overhand bricklaying and related work
- Roofing work on low-slope roofs
- Steep roofs
- Precast concrete erection
- Residential construction
- Wall openings
- Walking/working surfaces not otherwise addressed

UNPROTECTED SIDES AND EDGES
OSHA 1926.501(b)(1)

Employees shall be protected when they are exposed to falls from unprotected sides and edges of walking/working surfaces that are 6 ft. or more above lower levels. The options from which an employer can choose to provide this protection are guardrail systems, safety net systems and personal fall arrest systems. OSHA considers these three types of systems to be "conventional fall protection systems." **(See Figure 13-1)**

Figure 13-1. This illustration shows the requirements for unprotected sides and edges.

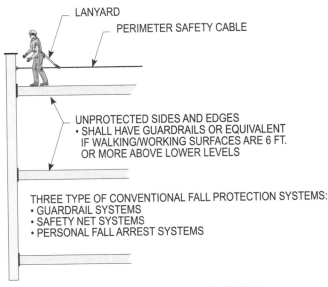

LANYARD

PERIMETER SAFETY CABLE

UNPROTECTED SIDES AND EDGES
• SHALL HAVE GUARDRAILS OR EQUIVALENT IF WALKING/WORKING SURFACES ARE 6 FT. OR MORE ABOVE LOWER LEVELS

THREE TYPE OF CONVENTIONAL FALL PROTECTION SYSTEMS:
• GUARDRAIL SYSTEMS
• SAFETY NET SYSTEMS
• PERSONAL FALL ARREST SYSTEMS

UNPROTECTED SIDES AND EDGES
OSHA 1926.501(b)(1)

LEADING EDGES
OSHA 1926.501(b)(2)

Employees who are exposed to fall hazards while constructing leading edges and employees who are working on the same level as a leading edge shall be protected. As defined in the final rule, a leading edge is the edge of a floor, roof or formwork that changes location as an additional floor, roof or formwork sections are placed, formed or constructed. Leading edges not actively and continuously under construction are considered to be "unprotected sides and edges," and are covered by OSHA 1926.501(b)(1).

Safety Tip: Employees constructing leading edges shall be protected from fall hazards by the use of guardrails, safety nets, or fall arrest systems.

Employers shall protect employees actively engaged in constructing leading edges from fall hazards through the use of guardrail systems, safety net systems, or personal fall arrest systems. In addition, if the employer can demonstrate that it is infeasible or would create a greater hazard to use any of these systems, the employers shall develop and implement a fall protection plan that meets the requirements of OSHA 1926.502(k). The fall protection plan, in turn, requires, among other criteria and conditions for use, that the employer designate all areas where conventional fall protection systems cannot be used as controlled access zones. Employers shall also implement a safety monitoring system in those zones if no other alternative measure has been implemented. Criteria for controlled zone systems and safety monitoring systems are found in OSHA 1926.502(g) and (h), respectively.

Employees on walking/working surfaces where leading edges are under construction, but who are not constructing the leading edge, shall be protected from fall hazards by guardrail systems, safety net systems or personal fall arrest system.

HOIST AREAS
OSHA 1926.501(b)(3)

Employees in hoist areas of walking and working surfaces that are 6 ft. or more above lower levels shall be protected. Employees shall be protected through the use of guardrail systems or personal fall arrest systems. If guardrails (or chains or gates if they are being used in lieu of guardrails at the hoist area) are removed to facilitate hoisting operations, then employees who lean through the access opening or out over the edge of the access opening to perform their duties shall be protected by the use of personal fall arrest systems.

HOLES
OSHA 1926.501(b)(4)

Employees shall be protected from hazards associated with holes. In particular, employees may be injured or killed if they step into holes, trip over holes, fall through holes or are hit by objects falling through holes. Some workplaces may present all of these hazards; while others may have only one of them. The proposed rule has been revised to indicate clearly which protective measures are applicable to a particular hole situation. Covers that comply with the criteria of OSHA 1926.502(i) will protect employees from all of the above-described hazards. **(See Figure 13-2)**

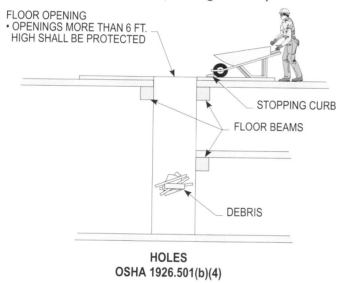

Figure 13-2. This illustration shows the requirements for floor hole openings that are to be protected.

Employees shall be protected from falling into or through holes (including skylight openings) 6 ft. or more above lower levels by covers over the hole, erecting a guardrail system around the hole or by the use of a personal fall arrest system. The Agency has revised the proposed rule to include personal fall arrest systems as an acceptable fall protection option because OSHA believes that a properly rigged system can protect an employee from falling though a hole. Employees shall be protected from tripping in or stepping into holes by covers; and employees shall be protected from objects falling through holes by covers. **(See Figure 13-3)**

Safety Tip: Employees shall be protected from falling through hole openings.

FORMWORK AND REINFORCING STEEL
OSHA 1926.501(b)(5)

Employees working on formwork and reinforcing steel 6 ft. or more above lower levels shall be protected by a personal fall arrest system, safety net system or positioning device system. **(See Figure 13-4)**

Figure 13-3. This illustration shows an improperly covered floor hole opening.

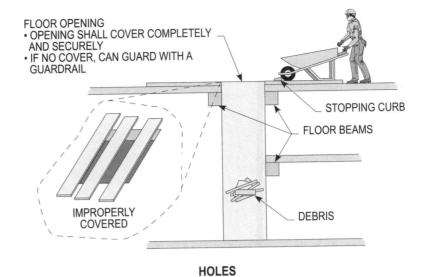

FLOOR OPENING
• OPENING SHALL COVER COMPLETELY AND SECURELY
• IF NO COVER, CAN GUARD WITH A GUARDRAIL

STOPPING CURB

FLOOR BEAMS

DEBRIS

IMPROPERLY COVERED

**HOLES
OSHA 1926.501(b)(4)**

Figure 13-4. This illustration shows the requirements for employees working on formwork and reinforcing steel.

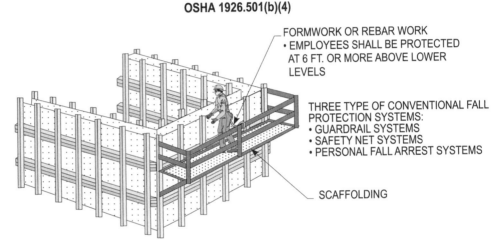

FORMWORK OR REBAR WORK
• EMPLOYEES SHALL BE PROTECTED AT 6 FT. OR MORE ABOVE LOWER LEVELS

THREE TYPE OF CONVENTIONAL FALL PROTECTION SYSTEMS:
• GUARDRAIL SYSTEMS
• SAFETY NET SYSTEMS
• PERSONAL FALL ARREST SYSTEMS

SCAFFOLDING

**FORMWORK AND REINFORCING STEEL
OSHA 1926.501(b)(5)**

RAMPS, RUNWAYS AND OTHER WALKWAYS
OSHA 1926.501(b)(6)

Ramps, runways and other walkways shall be equipped with guardrails.

EXCAVATIONS
OSHA 1926.501(b)(7)

The edges of excavations that are not readily seen (i.e., concealed from view by plant growth, etc.) shall be protected with guardrail systems, fences or barricades to prevent employees from falling into them if the excavation depth is 6 ft. or more. In addition, walls, pits, shafts and similar excavations with depths of 6 ft. or more shall be guarded to prevent employees from falling into them. **(See Figure 13-5)**

DANGEROUS EQUIPMENT
OSHA 1926.501(b)(8)

Employers shall protect employees from falling onto dangerous equipment. Where a floor, roof or other walking or working surface is less than 6 ft. above such hazards, employees shall be protected by guardrails or equipment guards that shield the hazard.

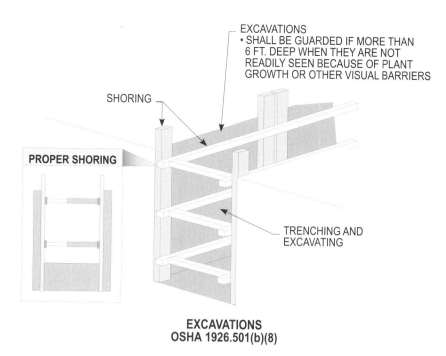

SHORING

EXCAVATIONS
• SHALL BE GUARDED IF MORE THAN
6 FT. DEEP WHEN THEY ARE NOT
READILY SEEN BECAUSE OF PLANT
GROWTH OR OTHER VISUAL BARRIERS

PROPER SHORING

TRENCHING AND
EXCAVATING

EXCAVATIONS
OSHA 1926.501(b)(8)

Figure 13-5. This illustration shows the requirements for excavations.

Employers shall protect employees on floors, roofs and other walking or working surfaces 6 ft. or more above dangerous equipment with guardrail systems, personal fall arrest systems or safety net systems. "Dangerous equipment" is defined in OSHA 1926.500(b) as equipment such as pickling or galvanizing tanks, degreasing units, machinery, electrical equipment and other units that, as a result of form or function, may be hazardous to employees who fall onto or into such equipment.

OVERHAND BRICKLAYING AND RELATED WORK
OSHA 1926.501(b)(9)

This Subpart addresses the fall protection requirements for employees engaged in overhand bricklaying operations and related work, except as set in OSHA 1926.451(g)(1)(vii). These employees are involved in the construction of masonry walls and must lean over the wall to complete the joint work. Related work, as used in this paragraph, means mason tending as well as electrical work that must be incorporated into the brick wall during the bricklaying process.

Employees performing overhand bricklaying and related work 6 ft. or more above lower levels shall be protected by guardrail systems, safety net systems, or personal fall arrest systems or they shall work in a controlled access zone. However, when these employees are reaching more than 10 in. below the level of the walking/working surfaces, only guardrail systems, safety net systems or personal fall arrest systems shall be used; controlled access zones are not acceptable in this situation.

ROOFING WORK ON LOW-SLOPE ROOFS
OSHA 1926.501(b)(10)

Paragraph (b)(10) applies to employees performing roofing operations on low-slope roofs with unprotected sides and edges 6 ft. or more above lower levels. Employers shall protect employees from fall hazards by one of the following systems:

• Guardrail systems, safety net systems or personal fall arrest systems

• Combinations of warning line systems and guardrail systems, warning line systems and safety net systems, warning line systems and personal fall arrest system, or warning line systems and safety monitoring system

Where the roof is 50 ft. or less in width, the employer may protect employees by the use of a safety monitoring system alone.

STEEP ROOFS
OSHA 1926.501(b)(11)

Employees on roofs with slopes greater than 4 in. x 12 in. (i.e., 4 inches vertical to 12 inches horizontal run) shall be protected from falling when the roof has unprotected sides or edges more than 6 ft. above lower levels by the use of guardrail systems with toeboards, personal fall arrest systems or safety net systems.

PRECAST CONCRETE ERECTION
OSHA 1926.501(b)(12)

Employees erecting precast concrete members 6 ft. or more above a lower level shall be protected from falling by guardrail systems, safety net systems or personal fall arrest systems, unless the employer can demonstrate that such systems would be infeasible or would create a greater hazard at the site where the affected employees are working.

An exception is also allowed if another provision allows an alternative fall protection measure, such as covers over holes. Those alternative measures are also acceptable and do not need to be documented in a fall protection plan in order to be used.

RESIDENTIAL CONSTRUCTION
OSHA 1926.501(b)(13)

Employers engaged in residential construction work shall protect employees from falls of 6 ft. or more to lower levels by the use of one of the three conventional fall protection systems unless such systems are infeasible or would create a greater hazard for affected employees. In those situations, OSHA requires the employer to develop and implement a fall protection plan that meets the criteria of OSHA 1926.502(k).

WALL OPENINGS
OSHA 1926.501(b)(14)

Employees shall be protected when exposed to the hazard of falling out or through wall openings. Wall openings (defined as openings 30 in. or more high and 18 in. or more wide, that have a bottom edge to lower level fall distance of 6 ft. or more on the side away from the employees, and a bottom edge to walking/working surface height of less than 39 in. on the side facing the employees) shall be equipped with a guardrail system, safety net system or personal fall arrest system. OSHA believes the most practical method of compliance is the guardrail system because it provides protection at all times and for all employees who may have exposure at the wall opening. However, OSHA recognizes that there may be cases where employers may desire to use safety net systems or personal fall arrest systems, which also will provide an appropriate level of protection. For that reason, the provision has been revised to permit the use of these other systems.

WALKING/WORKING SURFACES NOT OTHERWISE ADDRESSED
OSHA 1926.501(b)(15)

This subpart is a "catch all" provision intended to clarify the overall thrust of paragraph (b). It sets forth clearly that all employees exposed to falls of 6 ft. or more to lower levels shall be protected by a guardrail system, safety net system or personal fall arrest system except where otherwise provided by OSHA 1926.501(b) or by fall protection standards in other subparts of Part 1926.

PROTECTION FROM FALLING OBJECTS
OSHA 1926.501(c)

Employers shall protect employees from falling objects by either:

- Using toeboards, screens or guardrail systems

- Erecting a canopy structure and placing potential fall objects away from edges

- Barricading the area to which objects could fall, prohibiting employees from entering that area and placing potential fall objects away from the edges

FALL PROTECTION SYSTEMS CRITERIA AND PRACTICES
OSHA 1926.502

The following shall be considered for fall protection systems criteria and practices:
- General
- Guardrail systems
- Safety net systems
- Personal fall arrest systems
- Positioning device systems
- Warning line systems
- Controlled access zones
- Safety monitoring systems
- Covers
- Protection from falling objects
- Fall protection plan

GENERAL
OSHA 1926.502(a)

All fall protection shall conform to the criteria set in OSHA 1926.502(b) for the particular system being used, and all fall protection equipment shall be provided and installed before employees begin any other work on or from the surface on which they will be protected. To be fully effective, fall protection shall be in place at the earliest possible time.

Safety Tip: Guardrails shall be 42 in +/- 3 in. in height.

GUARDRAIL SYSTEMS
OSHA 1926.502(b)

The top edge of guardrail systems shall be 42 in., plus or minus 3 in., above the walking/working surface, except when conditions warrant, the height of the top edge of the top rail may exceed the 45 in. limit. Where employees are using stilts, the height of the top rail shall be increased a height equal to the height of the stilts, which in effect serve as the walking/working surface. **(See Figure 13-6)**

Midrails, screens, mesh, intermediate vertical members (i.e., balusters), solid panels or equivalent structural members shall be installed between the top edge of the system and the walking/working surface when there is no wall or parapet wall at least 21 in. high.

Figure 13-6. This illustration shows the requirements for guardrail systems.

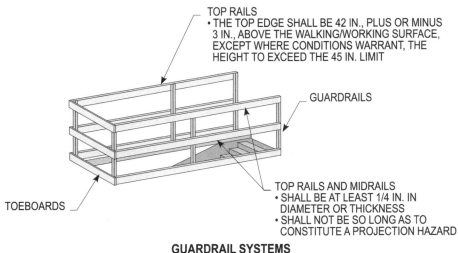

NOTE: GUARDRAIL SYSTEMS SHALL BE CAPABLE OF WITHSTANDING A 200 POUND FORCE APPLIED WITHIN 2 IN. OF THE TOP EDGE IN AN OUTWARD OR DOWNWARD DIRECTION.

TOP RAILS
• THE TOP EDGE SHALL BE 42 IN., PLUS OR MINUS 3 IN., ABOVE THE WALKING/WORKING SURFACE, EXCEPT WHERE CONDITIONS WARRANT, THE HEIGHT TO EXCEED THE 45 IN. LIMIT

GUARDRAILS

TOP RAILS AND MIDRAILS
• SHALL BE AT LEAST 1/4 IN. IN DIAMETER OR THICKNESS
• SHALL NOT BE SO LONG AS TO CONSTITUTE A PROJECTION HAZARD

TOEBOARDS

GUARDRAIL SYSTEMS
OSHA 1926.502(b)

When midrails are used, they shall be installed midway between the top edge of the guardrail system and the walking/working level.

Safety Tip: Guardrail systems shall be capable of withstanding 200 lbs of force applied within 2 in. of the top edge in an outward or downward direction.

The proper placement of screens, mesh, intermediate vertical members and other structural members shall be used in lieu of midrails in the guardrail system.

Guardrail systems shall be capable of withstanding a 200 pound force applied within 2 in. of the top edge in an outward or downward direction. When the 200 pound load is applied in a downward direction, the top edge of the guardrail shall not deflect to a height less than 39 in. above the walking/working level.

Midrails, screens, mesh, intermediate vertical members, solid panels and equivalent structural members shall be capable of withstanding, without failure, a force of at least 150 pounds, applied in any downward or outward direction at any point along the midrail or other member.

Guardrail systems shall be smooth surfaced to prevent employee injury due to lacerations or tripping caused by snagged clothing.

Top rails and midrails shall not be so long as to constitute a projection hazard.

The use of steel banding and plastic banding as top rails or midrails is prohibited. While such banding can often withstand a 200 pound load, it can tear easily if twisted. In addition, such banding often has sharp edges that can easily cut a hand if seized.

Top rails and midrails shall be at least 1/4 in. in nominal diameter or thickness. OSHA believes that the minimum thickness requirement is needed to prevent the use of rope that would cause cuts or lacerations. In addition, top rails constructed of wire rope shall be flagged at not more than 6 ft. intervals with high-visibility material. This requirement supplements the strength requirements for guardrails specified in OSHA 1926.502(b)(3), (b)(4) and (b)(5) of this section. The purpose of this requirement is to assure that rails made of high strength materials are not so thin that a worker grabbing a rail is injured, such as by cuts or lacerations, because of the small size of the rail.

OSHA 1926.502(b)(10) through (b)(13) address the use of guardrail systems.

Guardrail systems on ramps and runways shall be erected along each unprotected side or edge. It is OSHA's contention that the purpose of installing guardrails on ramps and runways is solely to keep employees from falling off the unprotected sides or edges of such ramps and runways when employees are exposed to falls of 6 ft. or more to a lower level. OSHA recognizes that there may be circumstances where the movement of materials or equipment across ramps or runways would be impeded by guardrails and situations where that interference is such that compliance with this provision would be infeasible (i.e., the work cannot be done) or would create a greater hazard. OSHA believes, in general, that preplanning of work will ensure that compliance is feasible and does not create a greater hazard.

Manila, plastic and synthetic rope used in guardrail systems shall be inspected as frequently as necessary to detect deterioration.

OSHA observes that non-mandatory Appendix B contains detailed specifications for minimum sizes of guardrail system components. These specifications are based on existing OSHA 1926.500(f)(1)(i), (ii) and (iii) and should provide useful information to help employers to design guardrail systems. The transfer of this guidance from existing regulatory text to a non-mandatory appendix does not reduce the level of safety achieved through compliance with the existing standard. The existing specific provisions are consistent with the performance-oriented requirements in the final rule. The promulgation of non-mandatory Appendix B removes redundant provisions from the standard.

SAFETY NET SYSTEMS
OSHA 1926.502(c)

The following provisions shall be complied with for safety net systems and their use:

- The installation of safety nets shall be as close as practicable under the walking/ working surface where employees need to be protected, but in no case more than 30 ft. below such level. **(See Figure 13-7)**

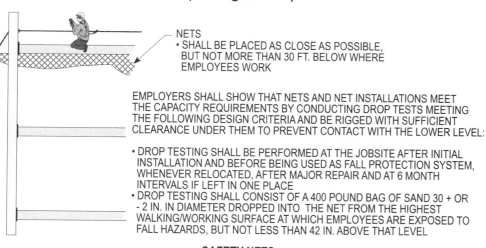

NETS
- SHALL BE PLACED AS CLOSE AS POSSIBLE, BUT NOT MORE THAN 30 FT. BELOW WHERE EMPLOYEES WORK

EMPLOYERS SHALL SHOW THAT NETS AND NET INSTALLATIONS MEET THE CAPACITY REQUIREMENTS BY CONDUCTING DROP TESTS MEETING THE FOLLOWING DESIGN CRITERIA AND BE RIGGED WITH SUFFICIENT CLEARANCE UNDER THEM TO PREVENT CONTACT WITH THE LOWER LEVEL:

- DROP TESTING SHALL BE PERFORMED AT THE JOBSITE AFTER INITIAL INSTALLATION AND BEFORE BEING USED AS FALL PROTECTION SYSTEM, WHENEVER RELOCATED, AFTER MAJOR REPAIR AND AT 6 MONTH INTERVALS IF LEFT IN ONE PLACE
- DROP TESTING SHALL CONSIST OF A 400 POUND BAG OF SAND 30 + OR - 2 IN. IN DIAMETER DROPPED INTO THE NET FROM THE HIGHEST WALKING/WORKING SURFACE AT WHICH EMPLOYEES ARE EXPOSED TO FALL HAZARDS, BUT NOT LESS THAN 42 IN. ABOVE THAT LEVEL

SAFETY NETS
OSHA 1926.502(c)

Figure 13-7. This illustration shows the requirements for safety nets.

- Where nets are used on bridges, there shall be an unobstructed fall to the net. In other words, nets shall not be used when a falling employee could hit an obstruction before reaching the net.

- Nets shall extend at least 8 ft. from the outer edge of the net from the edge of the working surface. **(See Figure 13-8)**

- Employers shall show that nets and net installations meet the capacity requirements by conducting drop tests meeting the following designed criteria and be rigged with sufficient clearance under them to prevent contact with the lower level:

Figure 13-8. This illustration shows a table pertaining to the distances from the outer edge of the net from the edge of the working surface.

VERTICAL DISTANCE FROM WORKING LEVEL TO HORIZONTAL PLANE OF NET	MINIMUM REQUIRED HORIZONTAL DISTANCE OF OUTER EDGE OF NET FROM THE EDGE OF THE WORKING SURFACE
UP TO 5 FT.	8 FT.
MORE THAN 5 FT. UP TO 10 FT.	10 FT.
MORE THAN 10 FT.	13 FT.

SAFETY NET SYSTEMS
OSHA 1926.502(c)(3)

(a) Drop testing shall be performed at the jobsite after initial installation and before being used as a fall protection system, whenever relocated, after major repair and at 6 month intervals if left in one place.

(b) Drop testing shall consist of a 400 pound bag of sand 30 + or − 2 in. in diameter dropped into the net from the highest walking/working surface at which employees are exposed to fall hazards, but not less than 42 in. above that level.

• Where an employer can demonstrate that drop testing is not feasible or practicable as required by OSHA 1926.502(c)(4)(i), the employer or a designated competent person shall certify the net and net installation is in compliance with OSHA 1926.502(c)(3) and (c)(4)(i) by preparing a certification record prior to the net being used as a fall protection system. The certification record must include the following:

(a) Identification of the net and net installation for which the certification record is being repaired
(b) The date that it was determined that the identified net and installations were in compliance with OSHA 1926.502(c)(3)
(c) The signature of the person making the determination and certification

• The most recent certification record for each net and net installation shall be available at the jobsite for inspection.

Note: OSHA considers two or more net panels joined together to be one net. Safety net installations that do not share the same net are considered to be separate systems. In addition, each time a safety net system is erected, it is considered to be a separate installation that shall be tested or certified.

• Defective nets shall not be used. Safety net systems shall be inspected at least once a week for wear, damage or other deterioration. Safety nets shall be inspected after any occurrence that could affect the integrity of the safety net system. Defective components shall be removed from service.

• Debris and tools shall be removed as soon as possible from the net, but not later than the start of the next work shift. Such materials pose safety hazards to anyone who falls into the net.

• The maximum size of each safety net opening shall be meet the following criteria:

(a) Opening shall not exceed 36 sq. in. nor be longer than 6 in. on any side
(b) Opening shall not be longer than 6 in. where measured center-to-center of mesh ropes or webbing
(c) All mesh crossings shall be secured to prevent enlargement of the mesh opening

• Each safety net shall have a minimum breaking strength of 5000 pounds for border ropes used for net webbing.

• Connections between net panels shall be as strong as integral components and to be spaced not more than 6 in. apart.

PERSONAL FALL ARREST SYSTEMS
OSHA 1926.502(d)

This paragraph replaces all of the existing provisions in OSHA 1926.104 — Safety Belts, Lifelines and Lanyards and relocates coverage of personal fall arrest systems to revised Subpart M. This is being done as part of the consolidation of fall protection requirements for construction. **(See Figure 13-9)**

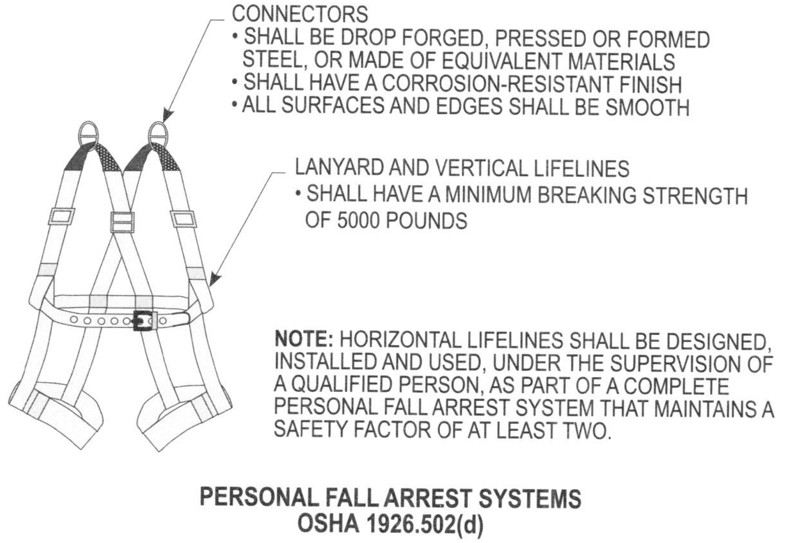

Figure 13-9. This illustration show a personal fall arrest system.

CONNECTORS
• SHALL BE DROP FORGED, PRESSED OR FORMED STEEL, OR MADE OF EQUIVALENT MATERIALS
• SHALL HAVE A CORROSION-RESISTANT FINISH
• ALL SURFACES AND EDGES SHALL BE SMOOTH

LANYARD AND VERTICAL LIFELINES
• SHALL HAVE A MINIMUM BREAKING STRENGTH OF 5000 POUNDS

NOTE: HORIZONTAL LIFELINES SHALL BE DESIGNED, INSTALLED AND USED, UNDER THE SUPERVISION OF A QUALIFIED PERSON, AS PART OF A COMPLETE PERSONAL FALL ARREST SYSTEM THAT MAINTAINS A SAFETY FACTOR OF AT LEAST TWO.

**PERSONAL FALL ARREST SYSTEMS
OSHA 1926.502(d)**

OSHA is phasing out, and then prohibiting, the use of body belts as a component of personal fall arrest systems. After December 31, 1997, body belts will no longer be permitted for use in a personal fall arrest system. They will, however, continue to be acceptable for use as part of a positioning device system [See OSHA 1926.502(e)] or as a part of a ladder safety device system required in Subpart X of Part 1926 since positioning device systems and ladder safety device systems are not used to arrest a fall.

Safety Tip: Body belts are not allowed to be used as a component of a personal fall arrest system.

Connectors shall be drop forged, pressed or formed steel, or made of equivalent materials.

Connectors shall have a corrosion-resistant finish, and all surfaces and edges shall be smooth to prevent damage to interfacing parts of the system.

Dee-rings and snaphooks shall have a minimum breaking strength of 5000 pounds.

Dee-rings and snaphooks shall be 100 percent proof-tested to a minimum tensile load of 3600 pounds without cracking, breaking or taking permanent deformation.

Employers shall either use snaphooks that are sized to be compatible with the members to which they are connected, or use locking type snaphooks that have been designed to prevent disengagement. OSHA considers a hook to be compatible in size where the diameter of the dee-ring to which the snaphook is attached is greater than the inside length of the snaphook measured from the bottom (hinged end) of the snaphook keeper to the inside curve of the top of the snaphook, so that no matter how the dee-ring is positioned or moved (rolls) with the snaphook attached, the dee-ring cannot touch the outside of the keeper so as to depress it open. The intent of this requirement is to prevent unintentional disengagement (roll out) of the snaphook. This provision also prohibits the use of nonlocking snaphooks after December 31, 1997.

The use of snaphooks is limited for certain connections unless the snaphook is a locking type, designed for those connections. Only locking snaphooks designed to be connected directly to webbing, rope or wire rope; to other snaphooks; to a dee-ring that already has another snaphook, or other connector attached; to a horizontal lifeline; or to any object that could depress the snaphook keeper because it is incompatibly sized or dimensioned in relation to the snaphook can be used for these connections. This provision reflects OSHA's determination that certain connections increase the likelihood of roll-out and that only locking snaphooks specifically designed for such connections are needed to provide adequate assurance of employee safety. Accordingly, even before outright prohibiting the use of nonlocking snaphooks, OSHA has limited the circumstances in which they can be used.

A device shall be used to connect to a horizontal lifeline that may become a vertical lifeline to be capable of locking in both directions on the lifeline. This provision applies only when horizontal lifelines are used on suspended scaffolds or similar work platforms, and the horizontal lifeline would become a vertical lifeline if the scaffold or platform were to fall.

Horizontal lifelines shall be designed, installed and used, under the supervision of a qualified person, as part of a complete personal fall arrest system that maintains a safety factor of at least two.

Lanyards and vertical lifelines shall have a minimum breaking strength of 5000 pounds.

Safety Tip: Not more than one employee can be attached to any one lifeline.

More than one employee shall not be attached to any one lifeline, except as provided in paragraph OSHA 1926.502(d)(10)(ii). The exception allows two employees to be attached to the same lifeline during construction of elevators, provided the employees are working atop a false car that is equipped with guardrails and the breaking strength of the lifeline has been increased to 10,000 pounds [5000 pounds per worker attached] and all other criteria of OSHA 1926.502(d) for personal fall arrest systems have been met. This exception recognizes the potential for a greater hazard (entanglement) in the elevator shaft with the additional lifeline.

Lifelines shall be protected against being cut or abraded.

Paragraph (d)(12) requires that, when in the fully extended position, self-retracting lifelines and lanyards that automatically limit free fall distance to 2 ft. or less be capable of sustaining a minimum tensile load of 3000 pounds (13.3 kN).

When in the fully extended position, self-retracting lifelines and lanyards that do not limit free fall to 2 ft. or less, as well as ripstitch, tearing and deforming lanyards, shall be capable of sustaining a minimum tensile load of 5000 pounds.

Ropes and straps (webbing) used in lanyards, lifelines and strength components of body belts and body harnesses shall not be made from natural fibers.

Anchorages used for the attachment of personal fall arrest equipment shall be capable of supporting at least 5000 pounds per employee attached, or the anchorage shall be designed, installed and used under the supervision of a qualified person and as part of a complete personal fall arrest system that maintains a safety factor of at least two.

Personal fall arrest systems shall comply with the following when stopping a fall:

- Limit the maximum arresting forces on an employee to 900 pounds when a body belt is used

- Limit the maximum arresting forces on an employee to 1800 pounds when a body harness is used

Note: As discussed in relation to the introductory text of OSHA 1926.502(d), above, OSHA has decided that body belts shall be phased out from use in personal fall arrest systems because employees wearing them have been seriously injured by the impact loads transmitted and by the pressures imposed while suspended after fall arrest. OSHA 1926.502(d)(16)(i) and (d)(16)(ii) reflect OSHA's determination that fall arrest systems that use body belts up to the time the prohibition takes effect shall minimize the related hazards by limiting the impact load to half that allowed when body harnesses are worn.

- Personal fall arrest system shall bring an employee to a complete stop and limit maximum deceleration distance an employee travels to 3.5 ft.

- Personal fall arrest system shall have sufficient strength to withstand twice the potential impact energy of an employee free falling a distance of 6 ft., or the free fall distance permitted by the system, whichever is less.

Personal fall arrest systems shall be worn so that the attachment point for body belts is located in the center of the wearer's back, and that the attachment point for body harnesses is located either in the center of the wearer's back near shoulder level, or above the wearer's head. Proper positioning of the lanyard or deceleration device is crucial for the prevention of injuries in a fall situation.

Safety Tip: Fall arrest systems shall be worn so that the attachment point is in the center of the wearer's back.

Body belts, harnesses and components shall be used only for employee fall protection or positioning. This means that those systems or components may not be used as material or equipment hoist slings, bundle ties or for other such purposes.

Personal fall arrest systems or components that are subject to impact loading (as distinguished from static load testing) shall be immediately removed from service, and prohibits subsequent use unless inspected by a competent person who determines the system or component to be undamaged and suitable for reuse.

Personal fall arrest systems shall be inspected prior to each use for damage and deterioration, and that defective components be removed from service.

Body belts shall be at least one and five-eighths (1 5/8) in.

The attachment of personal fall arrest systems to hoists or guardrail systems is prohibited, except where otherwise provided in Part 1926. Neither hoists nor guardrail systems are designed as anchorages for personal fall arrest systems since they are not built to withstand the impact forces generated by a fall. Therefore, in the interest of employee safety, OSHA is prohibiting the use of hoists and guardrails as attachment points.

Safety Tip: Personal fall arrest systems shall not be attached to a guard rail.

Personal fall arrest systems used at hoist areas shall be rigged to allow the movement of employees only as far as the edge of the walking/ working surface.

POSITIONING DEVICE SYSTEMS
OSHA 1926.502(e)

This paragraph sets the minimum performance criteria for "positioning devices," which are systems similar to personal fall arrest systems and which can be comprised of many of the same components. The significant difference is that personal fall arrest systems are used to arrest falls, whereas employees use positioning devices so they can maintain a leaning position without using their hands while working on vertical surfaces.

For example: These devices may be used during the placement of reinforcing bars in the vertical face of a wall under construction. The employees often stand on bars already in place and must lean backward, similar to a lineman on a telephone pole, to place additional bars.

The positioning device allows this to be done without the employees having to use their hands to maintain position. Positioning device systems and their use shall conform to the following provisions:

- Positioning device systems shall be rigged so that an employee cannot free fall more than 2 ft.

- Positioning devices shall be secured to an anchorage capable of supporting at least twice the potential impact load of an employee's fall or 3000 pounds, whichever is greater.

- Connectors shall be dropped forged, pressed or formed steel, or made of equivalent materials.

- Connectors shall be of a corrosion-resistant finish and that all surfaces and edges be smooth to prevent damage to interfacing parts of the systems.

- Connecting assemblies (which are dee-rings, snaphooks, lanyards and other components of the positioning device system) shall have a minimum breaking strength of 5000 pounds.

- Dee-rings and snaphooks shall be proof-tested to a minimum tensile load of 3600 pounds without cracking, breaking or taking permanent deformation.

- Snaphooks shall be sized to be compatible with the member to which they are connected or be of a locking type designed to prevent disengagement of the snaphook and, that after December 31, 1997, only locking type snaphooks can be used in positioning device systems.

- Unless the snaphook is a locking type and designed for the following conditions:

  (a) Snaphooks shall not be engaged directly to webbing, rope or wire rope.
  (b) Snaphooks shall not be engaged to each other.
  (c) Snaphooks shall not be engaged to a dee-ring to that another snaphook or other connector is attached
  (d) Snaphooks shall not be engaged to a horizontal lifeline
  (e) Snaphooks shall not be engaged to any object which is incompatibly shaped or dimensioned in relation to the snaphook such that unintentional disengagement could occur by the connected object being able to depress the snaphook keeper and release itself

- Positioning device systems shall be inspected prior to each use for damage and deterioration, and defective components shall be removed from service.

- Body belts, harnesses and components shall be used only for employee fall protection or positioning and not to hoist materials.

WARNING LINE SYSTEMS
OSHA 1926.502(f)

This paragraph provides the criteria for use of a warning line system. The warning line system is permitted when work conditions make it impossible to use conventional fall protection systems. Warning line systems shall and their use shall comply with the following criteria:

- Warning lines shall be erected around all sides of the roof work area and comply with the following:

  (a) When mechanical equipment is not being used, warning lines shall be erected not less than 6 ft. from the roof edge
  (b) When mechanical equipment is being used, warning lines shall be erected not less than 6 ft. from the roof edge that is parallel to the direction of mechanical equipment operation, and not less than 10 ft. from the roof edge that is perpendicular to the direction of mechanical equipment operation

- Points of access, materials handling areas, storage areas and hoisting areas shall be connected to the work area by an access path formed by two warning lines.

- When a path to a point of access is not in use, a rope, wire, chain or other barricade, equivalent in strength and height to the warning line, shall be placed across the path at the point where the path intersects the warning line erected around the work area, or the path offset such that a person cannot walk directly into the work area.

Warning lines shall consist of ropes, wires or chains, and supporting stanchions erected as follows:

- The rope, wire or chain shall be flagged at not more than 6 ft. intervals with high-visibility material.

- The rope, wire or chain shall be rigged and supported in such a way that its lowest point (including sag) is no less than 34 in. from the walking/working surface and its highest point is no more than 39 in. from the walking/working surface.

- After being erected, with the rope, wire or chain attached, stanchions shall be capable of resisting, without tripping over, a force of at least 16 pounds applied horizontally against the stanchion, 30 in. above the walking/working surface, perpendicular to the warning line, and in the direction of the floor, roof or platform edge.

- The rope, wire or chain shall have a minimum tensile strength of 500 pounds, and after being attached to the stanchions, shall be capable of supporting, without breaking, the loads applied to the stanchions as prescribed in OSHA 1926.502(f)(2)(iii).

- The line shall be attached at each stanchion in such a way that pulling on one section of the line between stanchions will not result in slack being taken up in adjacent section before the stanchion tips over.

No employee shall be allowed in the area between a roof edge and a warning line unless the employee is performing roofing work in that area.

Mechanical equipment on roofs shall be used or stored only in areas where employees are protected by a warning line system, guardrail system or personal fall arrest system.

CONTROLLED ACCESS ZONES
OSHA 1926.502(g)

This paragraph sets minimum performance criteria for controlled access zones (CAZ). The use of controlled access zones is a way to limit the number of workers who would be exposed to the hazard of falling from unprotected sides or edges at those locations where the use of conventional fall protection systems is infeasible or creates a greater hazard. The only work situation where use of a controlled access zone is specifically permitted instead of conventional fall protection systems is where overhand bricklaying operations are taking place. However, employers who develop a fall protection plan under 1926.501(b)(2), (b)(12) or (b)(13), will also be required to establish controlled access zones.

Employers, engaged in overhand bricklaying work, may use a controlled access zone as long as the employee does not have to reach more than 10 in. below the walking/working level to do the work. Employers engaged in leading edge work, precast concrete erection work or residential construction work who demonstrate infeasibility or greater hazard with the use of conventional fall protection systems will be required to develop and implement a fall protection plan that meets the requirements of OSHA 1926.502(k). OSHA 1926.502(k)(7) requires the employer to establish a controlled access zone that meets the requirements of this OSHA 1926.502(g).

In general, a controlled access zone is formed by erecting a line or lines - referred to as control lines - to restrict access to an area or to define the area in which employees will work without conventional fall protection. Sometimes only one line will be needed to define the area. The control line warns the employee that access to the controlled access zone is limited to authorized personnel. The line also designates the area where conventional fall protection systems are not in use.

Safety Tip: Control lines must not be closer than 6 ft. or farther than 25 ft. away from the leading edge.

Paragraph (g)(1) sets the distance from an unprotected side or edge that control zone lines are to be erected when leading edge operations or other activities are being performed and controlled access zones are permitted. When control lines are used, they shall be erected no closer than 6 ft. nor farther than 25 ft. away from the leading edge or unprotected edge. An exception is provided for the erection of precast concrete members, in which case the control line must be no closer than 6 feet nor farther than one-half the size of the precast member being erected, to a maximum of 60 ft. This exception is being made for precast concrete erection because it is sometimes necessary to "turn" a precast member, which may be as long as 120 feet. If the control lines are too close, they could become entangled or uprooted as the concrete member is being positioned.

The following provisions shall be conformed to for controlled access zones and their use:

- The control line shall be connected on each side to a guardrail system or to a wall. The following controlled access zones shall be defined by a control line or by any other means that restricts access when used to control access to areas where leading edge and other operations are taking place:

  (a) Except when erecting precast concrete members, control lines shall be erected not less than 6 ft. nor more than 25 ft. from the unprotected or leading edge.
  (b) Where precast members are erected, the control line shall be erected not less than 6 ft. nor more than 60 ft. or half the length of the member being erected, whichever is less, from the leading edge.
  (c) The control line shall extend along the entire length of the unprotected or leading edge and shall be approximately parallel to the unprotected or leading edge.
  (d) The control line shall be connected on each side to a guardrail system or wall.

Note: OSHA proposed this language to ensure that there was no gap between the coverage of the controlled access zone and that of the fall protection required for other areas of the pertinent work zone. All employees working outside the controlled access zone shall be provided fall protection as required by 1926.501(b)(1) if they may be exposed to fall hazards. As the controlled access zone changes [moves forward as the work progresses at the leading edge], it exposes unprotected sides and edges perpendicular to the leading edge. The employer shall ensure that any employees who may be exposed to falls of 6 ft. or more at those perimeters are provided with fall protection that complies with OSHA 1926.501(b)(1). Again, OSHA notes that this situation only occurs when two groups of workers are working on the same level and one group of workers is working in a controlled access zone and the other is being protected by conventional fall protection systems.

For example: Precast concrete workers may be connecting floor or roof members at the leading edge while other workers are engaged in "grouting" activities outside the controlled access zone. As each precast member is added, the controlled access zone moves forward and the control line moves forward, creating sections of unprotected sides and edges outside the controlled access zone from which workers engaged in grouting or other activities could fall. Those employees shall be afforded protection from falls of 6 ft. or more from the unprotected sides and edges of the floor, roof or other walking/ working surface as required by OSHA 1926.501(b)(1); or as required under fall protection plans where such plans are permitted.

The following shall be considered for control access to areas where overhand bricklaying and related work are taking place:

- Controlled access zones used during overhand bricklaying operations shall not be less than 10 ft. nor more than 15 ft. from the working edge where the overhand bricklaying operations are underway.

- The controlled access zone shall enclose all employees performing overhand bricklaying and related work at the working edge and shall be approximately parallel to the working edge.

- Additional control lines shall be erected at each end to enclose the controlled access zone.

- Only employees engaged in overhand bricklaying or related work shall be permitted in the controlled access zone.

Control lines shall be made of ropes, wires, tapes or other equivalent materials (i.e., material that can meet the requirements of OSHA 1926.502(g)(3)) and supported on stanchions as follows:

- Systems shall be flagged or otherwise clearly marked at 6 ft. intervals with high-visibility material.

- Control lines shall be rigged and supported in such a way that its lowest point (including sag) is not less than 39 in. from the walking/working surface and its highest point is not more than 45 in. (50 in. when overhand bricklaying operations are being performed) from the walking/working surface. Overhand bricklaying control zone height limits are higher than those for other work to allow the ready passage of materials underneath the line.

- Control lines shall have a minimum breaking strength of 200 pounds. This minimum strength is required to assure that the lines will not break if an inattentive worker walks into the line.

Where guardrail systems are not in place on floors and roofs prior to the beginning of overhand bricklaying operations, controlled access zones shall be enlarged, as necessary, to enclose all points of access, material handling areas and storage areas.

Where guardrail systems are in place on floors and roofs and need to be removed to allow overhand bricklaying work or leading edge work to take place, only that portion of the guardrail necessary shall be removed to accomplish that day's work.

SAFETY MONITORING SYSTEMS
OSHA 1926.502(h)

This paragraph contains the criteria that shall be followed when safety monitoring systems are being used. Safety monitoring systems may be used to protect employees engaged in roofing operations on low-slope roofs (See OSHA 1926.501(b)(10)) and employees engaged in leading edge operations, precast concrete or residential construction work through the use of safety monitoring systems as part of a fall protection plan (See OSHA 1926.502(k)). The following shall be complied with for safety monitoring systems:

Safety Tip: The safety monitor shall be a "competent person" and shall be able to recognize fall hazards.

• The employer shall designate a competent person as the safety monitor and ensure that the monitor meets certain requirements including being able to recognize fall hazards. The safety monitor is required to warn an employee who appears to be unaware of a fall hazard or is acting in an unsafe manner. The monitor shall also be on the same surface and within visual sighting distance of the monitored employee and close enough to communicate orally with the monitored employee. The monitor may have additional supervisory or non-supervisory responsibilities, provided that the monitor's other responsibilities do not interfere with the monitoring function.

• The use of mechanical equipment is prohibited where safety monitoring systems are being used to protect employees from falling off low-slope roofs.

• Employees shall not engage in roofing work on low-sloped roofs or employees covered by a fall protection plan from being in an area where other employees are protected by a safety monitoring system.

Note: OSHA believes that the presence of extraneous employees in these areas can interfere with work procedures necessary for the effective use of the safety monitoring system. OSHA notes that this provision is consistent with the provisions of OSHA 1926.502(k), which also prohibits employees from entering a controlled access zone because a safety monitoring system or other non-conventional fall protection system is in use in the controlled access zone.

• Each employee performing work in safety monitoring systems areas shall comply with directions from safety monitors to avoid fall hazards.

Note: The safety monitor shall be a "competent person"; that means that the monitor shall be capable of identifying workplace hazards and have the authority to take prompt corrective measures. Within the context of the safety monitoring system, the "corrective measures" are to have the affected employees move away from the unprotected side or edge or use other work procedures to avoid fall hazards.

COVERS
OSHA 1926.502(i)

The following requirements shall be complied with for covers for holes in floors, roofs and other walking/working surfaces:

• Covers in roadways and vehicular aisles shall be capable of supporting, without failure, at least twice the maximum axle load of the largest vehicle expected to cross over the cover.

- All other covers (those not addressed by paragraph OSHA 1926.502(i)(1)) shall be capable of supporting, without failure, at least twice the weight of any employee (including any equipment or material the employee may be carrying) who may be on the cover.

- Covers shall be secured when installed so as to prevent accidental displacement by wind, equipment and employees.

- Covers shall be color coded or the word "HOLE" or "COVER" shall appear on the cover to serve as a warning to employees of the hazard.

Note: OSHA does not intend for employers to color code or mark the permanent cast iron manhole covers or steel grates that cover street or roadways openings or similar kinds of covers that may be encountered on a construction worksite.

PROTECTION FROM FALLING OBJECTS
OSHA 1926.502(j)

The following shall be complied with for protection from falling objects:

- Toeboards, when used, shall be erected along the edge of overhead walking/working surfaces for a distance sufficient to protect employees working below.

- Toeboards shall be capable of withstanding, without failure, a force of at least 50 pounds. This is to ensure the ability of the toeboard to restrain falling objects.

- Toeboards shall be a minimum of 3 1/2 in. in vertical height from their top edge to the level of the walking/working surface.

- Toeboards shall have not more than 1/4 in. clearance above the walking/working surface.

- Toeboards shall be solid or have openings not over 1 in. in greatest dimension.

- Additional protection, such as paneling or screening erected from the working level or toeboard to the top of the top rail or midrail, shall be used where tools, equipment or materials are higher than the top of a toeboard.

- When guardrails are used to prevent objects from falling, the openings in the guardrail shall be small enough to retain the potential falling objects.

- Housekeeping provisions for overhand bricklaying operations shall be performed to prevent tripping and to prevent displacement of materials and equipment to areas below the walking/working surface.

- Materials and equipment shall not be stored within 6 ft. of a roof edge unless guardrails are erected at the edge. Materials that are piled, grouped or stacked near a roof edge shall be stable and self-supporting.

- When canopy structures are erected, they shall be strong enough to prevent collapse or penetration of falling objects.

FALL PROTECTION PLAN
OSHA 1926.502(k)

Safety Tip: the person developing the fall protection plan shall be a "competent person".

A written fall protection plan shall be prepared by a "qualified" person (as defined by OSHA 1926.32(m)), that the plan be developed specifically for the site where the work is being done and that the plan be maintained up to date. The definition in OSHA 1926.32(m) describes "qualified" as a person who "has successfully demonstrated his ability to solve or resolve problems relating to the subject matter, the work or the project." **(See Figure 13-10)**

Figure 13-10. This illustration shows the requirements for a fall protection plan.

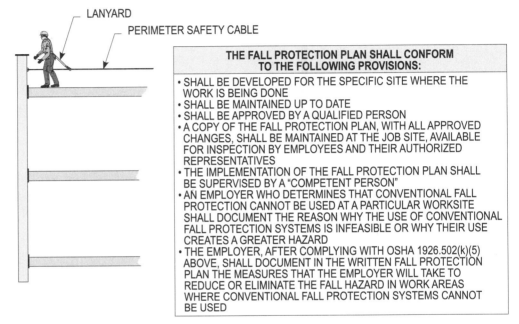

LANYARD

PERIMETER SAFETY CABLE

THE FALL PROTECTION PLAN SHALL CONFORM TO THE FOLLOWING PROVISIONS:

- SHALL BE DEVELOPED FOR THE SPECIFIC SITE WHERE THE WORK IS BEING DONE
- SHALL BE MAINTAINED UP TO DATE
- SHALL BE APPROVED BY A QUALIFIED PERSON
- A COPY OF THE FALL PROTECTION PLAN, WITH ALL APPROVED CHANGES, SHALL BE MAINTAINED AT THE JOB SITE, AVAILABLE FOR INSPECTION BY EMPLOYEES AND THEIR AUTHORIZED REPRESENTATIVES
- THE IMPLEMENTATION OF THE FALL PROTECTION PLAN SHALL BE SUPERVISED BY A "COMPETENT PERSON"
- AN EMPLOYER WHO DETERMINES THAT CONVENTIONAL FALL PROTECTION CANNOT BE USED AT A PARTICULAR WORKSITE SHALL DOCUMENT THE REASON WHY THE USE OF CONVENTIONAL FALL PROTECTION SYSTEMS IS INFEASIBLE OR WHY THEIR USE CREATES A GREATER HAZARD
- THE EMPLOYER, AFTER COMPLYING WITH OSHA 1926.502(k)(5) ABOVE, SHALL DOCUMENT IN THE WRITTEN FALL PROTECTION PLAN THE MEASURES THAT THE EMPLOYER WILL TAKE TO REDUCE OR ELIMINATE THE FALL HAZARD IN WORK AREAS WHERE CONVENTIONAL FALL PROTECTION SYSTEMS CANNOT BE USED

FALL PROTECTION PLAN
OSHA 1926.502(k)

Employers whose workplace situations satisfy the criteria of OSHA 1926.501(b)(2), (b)(12) or (b)(13) still must develop and implement alternative measures that protect affected employees from fall hazards. It is clear that the abilities of the personnel who develop those measures will strongly influence their effectiveness. OSHA has determined that the abilities of a qualified person are needed to ensure that the fall protection plan minimizes fall hazards. OSHA notes that an employer may use the services of more than one qualified person to comply with these requirements, as long as (1) those persons, collectively, are qualified to prepare the fall protection plan and approve any changes; and (2) the resulting plan complies with the applicable requirements of the standards. The fall protection plan shall conform to the following provisions:

- The fall protection plan shall be developed for the specific site where the work is being done.

    Note: OSHA notes that the designs, erection plans and circumstances for one site will, in many cases, differ significantly from those for another site. Accordingly, the fall protection plan for the one site may well be inapplicable to the other or may require substantial modification before it can be used.

- The fall protection plan shall be maintained up to date.

    Note: The employer must review the fall protection plan as necessary to determine if it still fits the workplace situation and must modify the plan as necessary to maintain its effectiveness, such as when elements of the plan have become inapplicable due to changes in the design, erection plan or other circumstances of a site.

- Changes in a fall protection plan shall be approved by a qualified person.

Note: The qualified person may either sign or initial and date the changed portion of the fall protection plan to indicate approval of the plan as modified. This provision, like OSHA 1926.502(k)(1), reflects OSHA's belief that the characteristics set out in OSHA 1926.32(m) are needed to assure that the person who sets the terms of the fall protection plan has the requisite ability and judgment.

- A copy of the fall protection plan, with all approved changes, shall be maintained at the job site, available for inspection by employees and their authorized representatives.

Note: In many cases, the opportunity to review the plan will provide the necessary reassurance to employees that the employer has taken appropriate measures to minimize exposure to fall hazards. In other cases, review of the plan will alert employees to deficiencies that need to be brought to the employer's attention for correction.

- The implementation of the fall protection plan shall be supervised by a "competent person".

Note: That term is defined in OSHA 1926.32(f) to mean "one who is capable of identifying existing and predictable hazards in the surroundings or working conditions which are unsanitary, hazardous or dangerous to employees, and who has authorization to take prompt corrective measures to eliminate them. 'The proper implementation of a fall protection plan will require unflagging vigilance and decisive action. Without the "built-in" protection of a guardrail, safety net or personal fall arrest system, employees must rely on safety monitors and other measures to warn them away from fall hazards.

A "competent person" who also has the abilities of a "qualified person" will be particularly well positioned to develop and implement solutions to fall protection problems. OSHA has not required that the supervisor be both "competent" and "qualified," because OSHA believes that such consolidation of function is not essential for employee protection. Also, the Agency recognizes that more than one person can be a "competent person" for the purposes of paragraph OSHA 1926.502(k)(6), as long as those persons, collectively, exercise the requisite oversight and authority.

- An employer who determines that conventional fall protection cannot be used at a particular worksite shall document the reason why the use of conventional fall protection systems (guardrail systems, safety net systems or personal fall arrest systems) is infeasible or why their use creates a greater hazard. Employers must explain in writing, before work begins, for each of the three conventional systems, why those systems cannot be used at the specific location where the fall hazard is present.

- The employer, after complying with OSHA 1926.502(k)(5) above, shall document in the written fall protection plan the measures that the employer will take to reduce or eliminate the fall hazard in work areas where conventional fall protection systems cannot be used.

For example: If safety monitoring systems and control zone systems are going to be used, the written plan shall so state.

The employer shall comply with all of the provisions described in the fall protection plan as alternative measures. It will not be acceptable for employers, under OSHA 1926.502(k)(6), to list "nothing" or "no measures to be taken" as the alternative measure. At the very minimum, the safety monitoring system (see OSHA 1926.502(k)(8)) shall be employed and all of the criteria in OSHA 1926.502(h) of this section followed. OSHA notes, at this point, that if a safety monitoring system is to be used, the designated monitor shall fulfill all of the criteria in OSHA 1926.502(h). If monitors are given other work assignments, such as those discussed under OSHA 1926.502(h) of this section, that render them unable to monitor other employees effectively, OSHA will view that situation as "not in compliance." Therefore, employers may need to designate more than one monitor so that a monitor is always available to fulfill the criteria of OSHA 1926.502(h).

In situations where conventional systems are not used, OSHA does not encourage employers to elect the safety monitoring system as a first choice. Rather, the Agency will permit it to be used in those circumstances when no other alternative, more protective measures can be implemented. Examples of such more protective measures include having employees work from scaffolds, ladders or vehicle mounted work platforms to provide a safer working surface and thereby reduce the hazard of falling. The written plan shall include a discussion of these other measures and the extent to which they can be used. The employer should also note where the use of those measures would not reduce exposure, would be unreasonable, infeasible or would create a greater hazard. The employer's failure to perform this evaluation as part of the plan will support an OSHA determination that the employer does not have a fall protection plan, and OSHA will consider the employer to be in violation of OSHA 1926.501(b)(2), (b)(12) or (b)(13). OSHA will also expect safe work practices to be elements of the alternative measures.

For example: Employees engaged in grouting operations would be expected to position themselves so their backs are not to the fall hazard. Employees on ladders would use a leg lock to position themselves more securely than they would otherwise be. In brief, employers need to preplan the work and plan the use of safe work practices that eliminate or reduce the possibility of a fall.

The choice of alternative fall protection systems will be particularly important when, pursuant to OSHA 1926.501(b)(2), (b)(12) or (b)(13), an employer establishes that it shall use alternatives to conventional fall protection. Accordingly, OSHA has determined that the employer must do what it can to minimize exposure to fall hazards, before turning to the use of safety monitoring systems (see OSHA 1926.502(h)) under a fall protection plan.

- Employers shall identify in the plan, each location where conventional fall protection cannot be used and to classify those locations as controlled access zones. Controlled access zones shall conform to the criteria in OSHA 1926.502(g). Compliance with this provision will provide a reference point to enable the employer to distinguish between those work areas where the fall protection plan applies and those where it does not.

  Note: OSHA has determined that, when it is impossible to perform the work with conventional fall protection, the work shall be performed in a controlled access zone. The controlled access zone prevents employees who are not engaged in the activities covered by the fall protection plan from being exposed to fall hazards in the areas where those activities are being conducted.

- Safety monitoring systems shall be implemented where no other alternative measures have been implemented. Safety monitoring systems shall comply with the criteria in OSHA 1926.502(h).

  Note: OSHA has added this requirement because it believes that employers shall, at a minimum, have a competent person assigned to monitor those employees who have not been provided conventional fall protection to warn the employees when they are acting in an unsafe manner or approaching an unprotected side or edge, among other activities when other, more protective measures, are not used.

• Fall protection plan shall identify, by name or other method, those employees who are authorized to work in controlled work zones. The paragraph further requires that only employees identified in the fall protection plan be allowed to enter controlled access zones.

Note: OSHA anticipates that compliance with this paragraph will enable an employer to maintain control over access to a controlled access zone, minimizing the number of employees exposed to fall hazards. This provision, like the rest of OSHA 1926.502(k), reflects the Agency's position that although there may be situations where fall protection cannot be used, any deviation from the general requirements for fall protection shall be construed as narrowly as possible.

• If an employee falls while performing work covered by a fall protection plan or there is other reason to believe that the substance or implementation of the plan is deficient (e.g., a near miss), the employer shall review the fall protection plan and make any changes in work practices, training, erection procedures or construction practices needed to correct any deficiencies in the plan.

Note: Given the immediacy of the hazards to which employees covered by a fall protection plan may be exposed, it is essential that contractors promptly revise their plans to incorporate what they learn through experience.

TRAINING REQUIREMENTS
OSHA 1926.503

This section supplements and clarifies the requirements of OSHA 1926.21 regarding the hazards addressed in Subpart M.

Employers shall provide a training program for each employee exposed to fall hazards so that each employee can recognize fall hazards and know how to avoid them. This section identifies components of the requisite training, but does not specify the details of the training program.

Employers need not retrain employees who were trained by a previous employer or were trained prior to the effective data of the standard, as long as the employee demonstrates an understanding of the subjects covered by OSHA 1926.503(a).

OSHA 1926.503(a) also states the subject areas to be addressed in the required training programs. The list of subjects reflects OSHA's determination that fall protection equipment and systems are only effective when they are properly designed, built, located, maintained and used. Employers are required to ensure that each employee is trained, as necessary, by a competent person qualified in the following areas:

• Nature of the fall hazards in the work area
• Correct procedures for erecting, maintaining, disassembling, using and inspecting the fall protection systems to be used
• Role of employees in the safety monitoring systems when used
• Role of employees in fall protection plans
• Standards contained in Subpart M.

Employees covered by a fall protection plan will, for example, need training to understand and to work effectively within the constraints of a controlled access zone. Affected

Safety Tip: Employers shall provide a training program for employees so that they can recognize fall hazards and know how to avoid them.

OSHA Required Training:
1926.503(a)(1) and (2)(ii) thru (vii)

employees will also need training on how to work with a safety monitor if a monitor is in use, to ensure that they respond appropriately when they hear a warning. OSHA recognizes that much of the information covered by training will be site-specific, so the Agency is framing this provision in performance-oriented terms.

This approach to training provides flexibility for the employer in designing the training program.

Employers shall verify that employees have been trained as required by OSHA 1926.503(a). In particular, final rule OSHA 1926.503(b)(1) requires employers to prepare a written certification record. The written certification record shall contain the name or other identity of the employee trained, the date(s) of the training and the signature of the person who conducted the training or the signature of the employer.

Note: OSHA does not require retraining provided the employee can demonstrate the ability to recognize the hazards of falling and the procedures to be followed to minimize fall hazards as required by OSHA 1926.503(a). OSHA recognizes that in many cases an employer will be unable to identify the date on which the previous training was provided. Accordingly, when employers relying on previous training prepare their certification records, they shall indicate the date the employer determined the prior training was adequate rather than the date of actual training.

The certification record can be prepared in any format an employer chooses, including preprinted forms, computer generated lists or 3 x 5 cards.

OSHA recognizes that many employers have already been providing affected employees with training that complies with OSHA 1926.503(a) and that requiring those employers to repeat the pertinent training would be unreasonably burdensome.

The latest certification record shall be maintained.

Fall protection training shall be repeated when changes in workplace conditions or changes in the types of fall protection systems or equipment to be used render previous training obsolete, and when inadequacies in an affected employee's knowledge or use of fall protection systems or equipment indicate that the employee has not retained the understanding or skill required by OSHA 1926.503(a).

Fall Protection

Section	Answer
_____	T F
_____	T F
_____	T F
_____	T F
_____	T F
_____	T F
_____	T F
_____	T F
_____	T F
_____	T F
_____	T F
_____	T F
_____	T F

Fall Protection

1. Employees shall be protected when they are exposed to falls from unprotected sides and edges of walking/working surfaces that are 6 ft. or more above lower levels.

2. Employees working on formwork and reinforcing steel 4 ft. or more above lower levels shall be protected by a personal fall arrest system, safety net system or positioning device system.

3. Where a roof is 50 ft. or less in width, the employer may protect employees by the use of a safety monitoring system alone.

4. Employees on roofs with slopes greater than 4 in. x 12 in. (i.e., 4 in. vertical to 12 in. horizontal run) shall be protected from falling when the roof has unprotected sides or edges more than 6 ft. above lower levels by the use of guardrail systems with toeboards, personal fall arrest systems or safety net systems.

5. The top edge of guardrail systems shall be 42 in., plus or minus 3 in., above the walking/working.

6. Guardrail systems shall be capable of withstanding a 300 pound force applied within 2 in. of the top edge in an outward or downward direction.

7. The installation of safety nets shall be as close as practicable under the walking/working surface where employees need to be protected, but in no case more than 35 ft. below such level.

8. Nets shall extend at least 8 ft. from the outer edge of the net from the edge of the working surface.

9. As of December 31, 1997, body belts are no longer permitted for use in a personal fall arrest system.

10. Lanyards and vertical lifelines shall have a minimum breaking strength of 8000 pounds.

11. When in the fully extended position, self-retracting lifelines and lanyards that do not limit free fall to 2 ft. or less, as well as ripstitch, tearing and deforming lanyards, shall be capable of sustaining a minimum tensile load of 5000 pounds.

12. Personal fall arrest system shall bring an employee to a complete stop and limit maximum deceleration distance an employee travels to 3.5 ft.

13. Positioning device systems shall be rigged so that an employee cannot free fall more than 4 ft.

Section	Answer	
_____	T	F

14. Employers engaged in overhand bricklaying work may use a controlled access zone as long as the employee does not have to reach more than 10 in. below the walking/working level to do the work.

15. When control lines are used, they shall be erected no closer than 6 ft. nor farther than 25 ft. away from the leading edge or unprotected edge.

16. Covers in roadways and vehicular aisles shall be capable of supporting, without failure, at least 3 times the maximum axle load of the largest vehicle expected to cross over the cover.

17. Toeboards shall be capable of withstanding, without failure, a force of at least 50 pounds. This is to ensure the ability of the toeboard to restrain falling objects.

18. Employers shall provide a training program for each employee exposed to fall hazards so that each employee can recognize fall hazards and know how to avoid them.

19. An employer who determines that conventional fall protection cannot be used at a particular worksite must explain to the job foreman the reason why the use of conventional fall protection systems (guardrail systems, safety net systems or personal fall arrest systems) is infeasible or why their use creates a greater hazard.

20. Toeboards shall be a minimum of 3 1/2 in. in vertical height from their top edge to the level of the walking/working surface.

14

Cranes, Derricks, Hoists, Elevators and Conveyors

The employer has the responsibility to comply with manufacturer's specifications and limitations applicable to the operation of any and all cranes and derricks. The employer shall designate a competent person who shall inspect all machinery and equipment prior to each use, and during use, to make sure it is in safe operating conditon.

Helicopter cranes shall comply with any applicable regulations of the Federal Aviation Administration. The helicopter operator shall be responsible for size, weight and manner in which loads are connected to the helicopter.

The employer has the responsibility to comply with manufacturer's specifications and limitations applicable to the operation of all hoists and elevators.

CRANES AND DERRICKS
OSHA 1926.550

The following types of cranes and derricks shall be considered:

- General requirements
- Crawler, locomotive and truck cranes
- Hammerhead tower cranes
- Overhead and gantry cranes
- Derricks
- Floating cranes and derricks
- Crane or derrick suspended personnel platforms

GENERAL REQUIREMENTS
OSHA 1926.550(a)

Safety Tip: Employers shall comply with manufacturer's specifications and limitations applicable to the operation of oil cranes.

The employer shall comply with the manufacturer's specifications and limitations applicable to the operation of any and all cranes and derricks. Where manufacturer's specifications are not available, the limitations assigned to the equipment shall be based on the determinations of a qualified engineer competent in this field and such determinations will be appropriately documented and recorded. Attachments used with cranes shall not exceed the capacity, rating or scope recommended by the manufacturer. Cranes and Derricks shall also meet the applicable requirements for design, inspection, construction, testing, maintenance and operation as prescribed in the ANSI B30 Standard for cranes.

Rated load capacities, and recommended operating speeds, special hazard warnings or instruction, shall be conspicuously posted on all equipment. Instructions or warnings shall be visible to operators while they are at their control stations.

OSHA Required Training:
1926.550(a)(1); (5) and (6)

Hand signals to crane and derrick operators shall be those prescribed by the applicable ANSI standard for the type of crane in use. An illustration of the signals shall be posted at the job site.

Safety Tip: A competent person shall inspect all machinery and equipment prior to each use.

The employer shall designate a competent person who shall inspect all machinery and equipment prior to each use, and during use, to make sure it is in safe operating condition. Any deficiencies shall be repaired, or defective parts replaced, before continued use. Inspection shall be both frequent and periodic.

All new and altered cranes or hoists shall have an initial inspection by a qualified person. Thereafter, crane and hoists shall have regular inspections classified as frequent or periodic. Inspections of these classifications shall be made at intervals dependent upon the critical components of the crane or hoist and the degree of exposure to wear, deprecation or malfunctions. **(See Figure 14-1)**

Figure 14-1. This illustration shows that all new and altered cranes or hoists shall have an initial inspection by a qualified person.

OPERATIONAL TESTING
• HOISTING AND LOWERING
• TROLLEY TRAVEL
• BRIDGE TRAVEL
• SWING MOTION
• BRAKES AND CLUTCHES
• LIMIT-LOCKING AND SAFETY DEVICES
• ASSEMBLY, FOUNDATION AND ERECTION
• CONTROL OPERATIONS
• RUNNING GEAR AND DRIVES
• ALL OTHER MECHANISMS PERTINENT TO THE SAFE OPERATION OF THE EQUIPMENT

NOTE: FREQUENT OR PERIODIC INSPECTIONS SHALL BE MADE AT INTERVALS DEPENDENT UPON THE CRITICAL COMPONENTS OF THE CRANE OR HOIST AND THE DEGREE OF EXPOSURE TO WEAR, DEPREDATION OR MALFUNCTIONS.

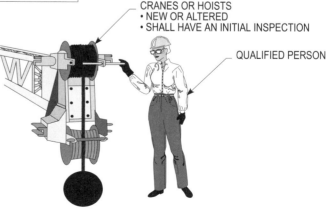

CRANES OR HOISTS
• NEW OR ALTERED
• SHALL HAVE AN INITIAL INSPECTION

QUALIFIED PERSON

**GENERAL REQUIREMENTS
OSHA 1926.550(a)**

Before initial use, all new, altered, reinstalled, excessively repaired or modified cranes and hoists shall be operationally tested to include (but not be limited to) the following as applicable to the specific crane or hoist:

- Hoisting and lowering,
- Trolley travel,
- Bridge Travel,
- Swing motion,
- Brakes and clutches,
- Limit-locking and safety devices,
- Assembly, foundation, and erection,
- Control operations,
- Running gear and drives and
- All other mechanisms pertinent to the safe operation of the equipment.

Before initial use, all new, altered, reinstalled and excessively repaired or modified cranes or hoists shall be rate-load tested according to the weight percentage specified by the manufacturer or the appropriate ASME/ ANSI standard.

Safety Tip: Frequent inspection (daily to monthly) shall be performed by a designated person.

Frequent inspections shall be daily to monthly (dependent upon use and condition) and shall be performed by a designated person. The inspections include (but are not limited to) checking control mechanisms for maladjustment and excessive wear, checking safety devices for proper function and checking hydraulic hoses and systems, hooks and latches, ropes and rope reeving, and electrical and mobile parts as applicable and as recommended by the manufacturer.

Periodic inspections of cranes and hoists shall be made by a qualified person at intervals of 1 to 12 months depending on the severity of equipment use and the environment. Inspections shall be in-depth; all structures and mechanisms shall be inspected for wear, deterioration, leaks, damage and operation. Any deficiency shall be examined, and it shall be determined whether or not it constitutes a hazard.

Cranes and hoists not in regular use for periods of more than one month and not more than one year shall be thoroughly inspected before use.

Dated inspection reports, especially those of periodic inspections, shall be maintained and be readily available for review.

A thorough, annual inspection of the hoisting machinery shall be made by a competent person, or by a government or private agency recognized by the U.S. Department of Labor. The employer shall maintain a record of the dates and results of inspections for each hoisting machine and piece of equipment. **(See Figure 14-2)**

ADJUSTMENT, MAINTENANCE AND REPAIR

A preventive maintenance program shall be established based on the crane or hoist manufacturer's recommendations, and a maintenance record shall be maintained.

Safety Tip: A preventive maintenance program shall be established.

Replacement parts shall be at least equivalent to the original manufacturer's specifications.

Before any major adjustments or repairs are made to a crane or hoist, every precaution shall be taken to ensure that the crane is in a safe location, has stops applied, has power in the "OFF" position, is properly tagged "Out of Order" and is locked out as applicable to ensure safe worker conditions.

Figure 14-2. This illustration shows that a thorough, annual inspection of the hoisting machinery shall be made by a competent person, or by a government agency recognized by the U.S. Department of Labor.

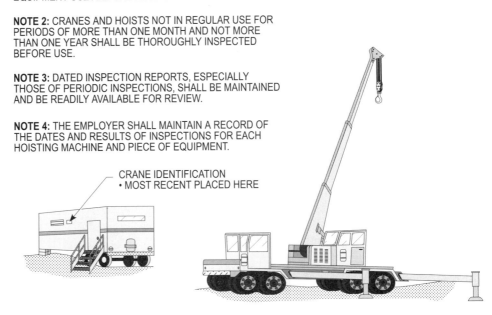

NOTE 1: PERIODIC INSPECTIONS OF CRANES AND HOISTS SHALL BE MADE BY A QUALIFIED PERSON AT INTERVALS OF 1 TO 12 MONTHS DEPENDING ON THE SEVERITY OF EQUIPMENT USE AND THE ENVIRONMENT.

NOTE 2: CRANES AND HOISTS NOT IN REGULAR USE FOR PERIODS OF MORE THAN ONE MONTH AND NOT MORE THAN ONE YEAR SHALL BE THOROUGHLY INSPECTED BEFORE USE.

NOTE 3: DATED INSPECTION REPORTS, ESPECIALLY THOSE OF PERIODIC INSPECTIONS, SHALL BE MAINTAINED AND BE READILY AVAILABLE FOR REVIEW.

NOTE 4: THE EMPLOYER SHALL MAINTAIN A RECORD OF THE DATES AND RESULTS OF INSPECTIONS FOR EACH HOISTING MACHINE AND PIECE OF EQUIPMENT.

CRANE IDENTIFICATION
• MOST RECENT PLACED HERE

GENERAL REQUIREMENTS
OSHA 1926.550(a0

Following adjustments, repairs or maintenance work, the crane or hoist shall not be returned to service until all guards are replaced, safety devices are reactivated, tools and work equipment are removed and warning tags or signs are removed.

Any defects found in inspections shall be repaired before the crane or hoist is used, except when a qualified person certifies that it may be operated without undue hazard. All repairs shall be made by a designated person.

Equipment shall be maintained and adjusted to ensure correct functioning of components, as applicable to the crane type or hoist mechanism.

Moving parts for which lubrication is specified by the manufacturer should be lubricated regularly. A check should be made for proper delivery of lubricant. The equipment shall be stopped during a worker-performed lubrication process.

WIRE ROPE INSPECTION AND MAINTENANCE

Safety Tip: All wire rpes (running rope) that are in service shall be inspected daily.

All running ropes (wire ropes) that are in continuous service should be visually inspected once each day. Wire rope shall be taken out of service when any of the following conditions exist:

- In running ropes, six randomly distributed broken wires in one lay or three broken wires in one strand in one lay;

- Wear of one-third the original diameter of outside individual wires. Kinking, crushing, bird caging or any other damage resulting in distortion of the rope structure;

- Evidence of any heat damage from any cause;

- Reductions from nominal diameter of more than one-sixty-fourth inch for diameters up to and including five-sixteenths inch, one-thirty-second inch for diameters three-eighths inch to and including one-half inch, three-sixty-fourths inch for diameters nine-sixteenths inch to and including three-fourths inch, one-sixteenth inch for diameters seven-eighths inch to 1 in. inclusive, three-thirty-seconds inch for diameters 1 to 1 inches inclusive;

- In standing ropes, more than two broken wires in one lay in sections beyond end connections or more than one broken wire at an end connection.

Wire rope safety factors shall be in accordance with American National Standards Institute B30.5-1968 or SAE J959-1966.

See Figure 14-3 for a detailed illustration pertaining to wire rope inspection.

NOTE: ALL RUNNING ROPES (WIRE ROPES) THAT ARE IN CONTINUOUS SERVICE SHOULD BE VISUALLY INSPECTED ONCE EACH DAY.

Figure 14-3. This illustration shows the requirements for wire rope inspection.

WIRE ROPE SHALL BE TAKEN OUT OF SERVICE WHEN ANY OF THE FOLLOWING CONDITIONS EXIST:

- IN RUNNING ROPES, SIX RANDOMLY DISTRUBTED BROKEN WIRES IN ONE LAY OR THREE BROKEN WIRES IN ONE STRAND IN ONE LAY

- WEAR OF ONE-THIRD THE ORIGINAL DIAMETER OF OUTSIDE INDIVIDUAL WIRES, KINKING, CRUSHING, BIRD CAGING OR ANY OTHER DAMAGE RESULTING IN DISTORTION OF THE ROPE STRUCTURE

- EVIDENCE OF ANY HEAT DAMAGE FROM ANY CAUSE

- REDUCTIONS FROM NOMINAL DIAMETER OF MORE THAN ONE-SIXTY-FOURTH INCH FOR DIAMETER THREE-EIGHTHS INCH TO AND INCLUDING THREE-FOURTHS INCH, ONE-SIXTEENTH INCH FOR DIAMETERS SEVEN-EIGHTHS INCH TO 1 INCH INCLUSIVE, THIRTY-THREE-SECONDS INCH FOR DIAMETERS 1 TO 1 INCHES INCLUSIVE

- IN STRANDING ROPES, MORE THAN TWO BROKEN WIRES IN ONE LAY IN SECTIONS BEYOND END CONNECTIONS OR MORE THAN 1 BROKEN WIRE AT AN END CONNECTION

WIRE ROPE INSPECTION AND MAINTENANCE

Replacement rope shall have the strength rating of the original ropes. Any change in rope size, grade or construction shall be specified by the rope manufacturer, crane or hoist manufacturer or qualified person.

Rope shall be stored in a manner to avoid damage and shall be unreeled or uncoiled with care to avoid kinking or twisting.

Rope shall be maintained in a well-lubricated condition with a lubricant equal to that of the original.

At least two wraps of rope shall remain on the drum. The end attachment shall be by a clamp securely attached to the drum or a wedge/socket arrangement that is approved by the equipment or rope manufacturers. **(See Figure 14-4)**

Nonrotating (rotation-resistant) rope and fiber core rope shall not be used for boom-hoist reeving on cranes. When nonrotating rope is used for other purposes, special care shall be taken during installation to avoid damage.

Eye splices shall be made as recommended by the rope manufacturer, and rope thimbles should be used in the eye splice. **(See Figure 14-4)**

Figure 14-4. This illustration shows the requirements for wire rope on the drum, eye splices and wire rope clips.

NOTE 1: NONROTATING (ROTATION-RESISTANT) ROPE AND FIBER CORE ROPE SHALL NOT BE USED FOR BOOM-HOIST REEVING ON CRANES. WHEN NONROTATING ROPE IS USED FOR OTHER PURPOSES, SPECIAL CARE SHALL BE TAKEN DURING INSTALLATION TO AVOID DAMAGE.

NOTE 2: ROPES SHALL BE MAINTAINED IN A WELL-LUBRICATED CONDITION WITH A LUBRICANT EQUAL TO THAT OF THE ORIGINAL.

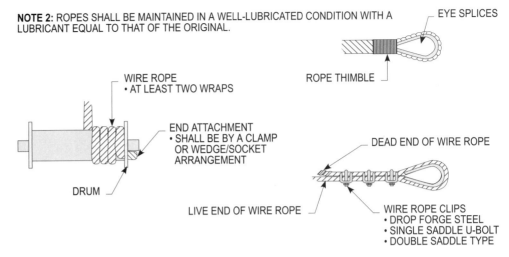

WIRE ROPE INSPECTION AND MAINTENANCE

Wire rope clips shall be drop-forged steel of the single-saddle U-bolt or double-saddle type. Spacing, number of clips, and torque shall be determined according to manufacturer's recommendations. Wire rope clips shall have the U-bolt over the dead end and the saddle over the live end of the wire rope.

Rope with an independent wire-rope, wire strand core or other temperature-resistant core shall be used when wire is exposed to temperatures in excess of 180°F at the rope.

OPERATOR QUALIFICATIONS AND PRACTICES

Cranes and hoists shall be operated by designated persons, trainees under the direct supervision of designated persons, maintenance and test personnel in duty performance and crane and hoist inspectors.

Operators of cranes and hoists shall pass a written or oral practical operating examination or furnish satisfactory evidence of qualification and experience. Examination shall be limited to the type of equipment for which the operator is qualified. Operators and operator trainees of cranes or hoists shall meet the following physical qualifications:

- Snellen vision test 20/30 in one eye and 20/50 in the other (corrected or uncorrected), normal depth perception, field of vision and color vision (if required by the operation);

- Sufficient strength, endurance, agility, coordination and responsiveness to meet the demands of the equipment operation;

- Adequate hearing (with or without hearing aid) for specific operations;

- No physical defects or emotional disorders that could cause hazard to the operator or others: and no evidence of seizures or loss of physical control.

Note: Medical or special clinical judgements may be required in determining physical or emotional conditions.

Operators shall not engage in any practices that might interfere with operating the crane or hoist, and they shall be responsible for those operations under their control.

Operators shall respond to signals from an appointed signal person but shall respond to a stop signal no matter who may give the signal.

Before starting a crane or hoist, the operator shall see that all controls are in "OFF" or "NEUTRAL" position and that all personnel are clear of machinery.

If power fails during the operation, the operator shall set all brakes and locking devices, move all clutches or other power controls to "OFF" or "NEUTRAL" position, and, if practical, land the suspended load under mechanical brake control.

Before leaving a crane or hoist unattended, the operator shall:

- Land the load,
- Set brakes or locking devices,
- Bring hook to highest position,
- Lower the boom to rest or cradle position,
- Put controls in "OFF" or "NEUTRAL" position,
- Secure the equipment from any accidental movement by setting brakes or other locking devices, and
- Shut off power.

Exceptions include those cranes and hoists that are securely blocked, dogged pawled, rocket locked or secured by other equivalent means. Operators shall follow any specific equipment's manufacturer's instruction for shutdown procedures. **(See Figure 4-5)**

BEFORE LEAVING A CRANE OR HOIST UNATTENDED, THE OPERATOR SHALL PERFORM THE FOLLOWING:
• LAND THE LOAD • SET BRAKES OR LOCKING DEVICES • BRING HOOK TO HIGHEST POSITION • LOWER THE BOOM TO REST OR CRADLE POSITION • PUT CONTROLS IN "OFF" OR "NEUTRAL" POSITION • SECURE THE EQUIPMENT FROM ANY ACCIDENTAL MOVEMENT BY SETTING BRAKES OR OTHER LOCKING DEVICES • SHUT OFF POWER

NOTE 1: OPERATORS SHALL RESPOND TO SIGNALS FROM AN APPOINTED SIGNAL PERSON BUT SHALL RESPOND TO A STOP SIGNAL NO MATTER WHO MAY GIVE THE SIGNAL.

NOTE 2: BEFORE STARTING A CRANE OR HOIST, THE OPERATOR SHALL SEE THAT ALL CONTROLS ARE IN "OFF" OR "NEUTRAL" POSITION AND THAT ALL PERSONNEL ARE CLEAR OF MACHINERY.

NOTE 3: IF POWER FAILS DURING THE OPERATION, THE OPERATOR SHALL SET ALL BRAKES AND LOCKING DEVICES, MOVE ALL CLUTCHES OR OTHER POWER CONTROLS TO "OFF" OR "NEUTRAL" POSITION, AND, IF PRACTICAL, LAND THE SUSPENDED LOAD UNDER MECHANICAL BRAKE CONTROL.

Figure 14-5. This illustration shows the requirements that shall be followed before a operator may leave a crane or hoist unattended.

Safety Tip: No crane or hoist shall be loaded beyond the specifications of the load-rating chart.

OPERATOR QUALIFICATIONS AND PRACTICES

No crane or hoist shall be loaded beyond the specifications of the load-rating chart, except for test purposes. The load to be lifted is to be within the rated capacity of the existing configuration. When the load is not accurately known, the person responsible for the lift shall determine that the weight of the load does not exceed the capacity of the crane at the configuration at which the load is to be lifted.

Prior to initiating a lift, the hook shall be positioned over the load in such a manner as to prevent the load from swinging. **(See Figure 14-6)**

The hoisting rope shall not be wrapped around the load. The load shall be attached to the hook with a sling or other device of sufficient capacity.

The crane shall not be operated if a worker is on the load or hook.

The operator shall not move a load over an area where workers or other persons are located. **(See Figure 14-7)**

Figure 14-6. This illustration shows that a hook shall be positioned over the load to prevent the load from swinging.

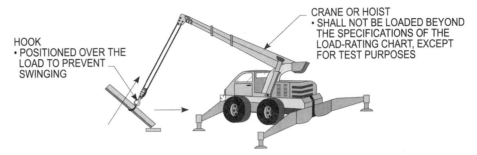

NOTE: WHEN THE LOAD IS NOT ACCURATELY KNOWN, THE PERSON RESPONSIBLE FOR THE LIFT SHALL DETERMINE THAT THE WEIGHT OF THE LOAD DOES NOT EXCEED THE CAPACITY OF THE CRANE AT THE CONFIGURATION AT WHICH THE LOAD IS TO BE LIFTED.

HOOK
• POSITIONED OVER THE LOAD TO PREVENT SWINGING

CRANE OR HOIST
• SHALL NOT BE LOADED BEYOND THE SPECIFICATIONS OF THE LOAD-RATING CHART, EXCEPT FOR TEST PURPOSES

OPERATOR REQUIREMENTS AND PRACTICES

Figure 14-7. This illustration shows that a operator shall not move a load over an area where workers or other persons are located.

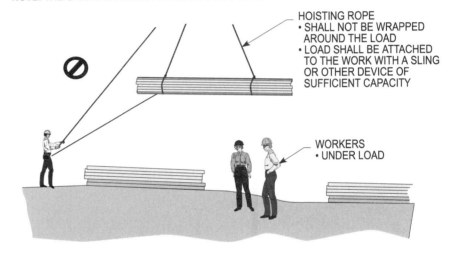

NOTE: THE CRANE SHALL NOT BE OPERATED IF A WORKER IS ON THE LOAD OR HOOK.

HOISTING ROPE
• SHALL NOT BE WRAPPED AROUND THE LOAD
• LOAD SHALL BE ATTACHED TO THE WORK WITH A SLING OR OTHER DEVICE OF SUFFICIENT CAPACITY

WORKERS
• UNDER LOAD

OPERATOR QUALIFICATIONS AND PRACTICES

Safety Tip: While a load is suspended, the operator shall not leave the cab or operating controls.

The operator shall not leave the cab or operating controls while a load is suspended, except under circumstances in which a load is to be suspended for a period of time exceeding normal lift operations. The operator may then leave the controls provided that prior to that time, the appointed individual and operator shall establish the requirements for retraining the boom hoist, telescoping, load, swing and outrigger functions, and provide notice, barricades or whatever other precautions may be necessary. **(See Figure 14-8)**

When it is difficult for the crane operator to judge any clearance distance, a designated worker shall observe the clearance distance and give timely warnings or signals to the operator.

OPERATOR QUALIFICATIONS (FLOOR AND REMOTE)

Operators shall be required to pass a practical examination for cranes and hoists that have remotes or floor controls. Operator examinations shall be limited to the specific equipment.

THE OPERATOR MAY LEAVE THE CAB OR OPERATING CONTROLS UNDER THE FOLLOWING CONDITIONS:

• WHERE A LOAD IS TO BE SUSPENDED FOR A PERIOD OF TIME EXCEEDING NORMAL LIFT OPERATIONS

• THE OPERATOR MAY LEAVE THE CONTROLS PROVIDED THAT PRIOR TO THAT TIME, THE APPOINTED INDIVIDUAL AND OPERATOR SHALL ESTABLISH THE REQUIREMENTS FOR RETRAINING THE BOOM HOIST, TELESCOPING, LOAD, SWING AND OUTRIGGER FUNCTIONS AND PROVIDE NOTICE, BARRICADES OR WHATEVER PRECAUTIONS MAY BE NECESSARY

SUSPENDED LOAD
• THE OPERATOR SHALL NOT LEAVE THE CAB OR OPERATING CONTROLS

OPERATOR QUALIFICATIONS AND PRACTICES

Figure 14-8. This illustration shows the requirements for a operator having a suspended load.

OTHER EQUIPMENT, CONDITIONS AND OPERATIONS

Warning devices or audible alarms shall be provided and routinely tested for the specific crane or hoist. The workers and operators shall understand the purpose and use of the alarm, warning and directional signals.

Standard crane operational signals shall be used when communicating with the equipment operator. Special operations may require special signals that shall be agreed upon and understood by the operator and signal person before their use.

The load hoisting unit of a crane or hoist shall be equipped with at least one braking means capable of holding a full-rated load (100 percent to 125 percent of the rated load depending on the crane or hoist type and design) at the point where the brake is applied; a braking means capable of providing a controlled lowering speed in the event of loss of power or pressure; an automatic means to stop and hold the rated load in the event of brake actuating power.

Refer to the appropriate ANSI B30 crane or hoist standard for any load hoist holding exceptions or additional requirements.

Hooks shall not be overloaded and shall meet the manufacturer's recommendations. Swiveling hooks shall rotate freely. All hooks shall have a safety latch and a functional spring except when application makes the latch impractical or unnecessary. **(See Figure 14-9)**

Load hooks and hook blocks shall be weighted to overhaul the line for the highest hook position. All hook ball assemblies and load blocks shall be permanently labeled with their rated capacity and weight.

Sheave grooves shall be smooth and free from surface defects that could damage the rope.

Belts, gears, shafts, pulleys, sprockets, spindles, drums, fly wheels, chains or other reciprocating, rotating moving parts or equipment shall be guarded if such parts are exposed to contact by employees, or otherwise create a hazard. Guarding shall meet

Safety Tip: Standard crane operational signals shall be used when communicating with the operator.

the requirements of the American National Standards Institute B15.1-1958 Rev., Safety Code for Mechanical Power Transmission Apparatus. **(See Figure 14-10)**

Accessible areas within the swing radius of the rear of the rotating superstructure of the crane, either permanently or temporarily mounted, shall be barricaded in such a manner as to prevent an employee from being struck or crushed by the crane. **(See Figure 14-11)**

All exhaust pipes shall be guarded or insulated in areas where contact by employees is possible in the performance of normal duties.

Whenever internal combustion engine powered equipment exhausts in enclosed spaces, tests shall be made and recorded to see that employees are not exposed to unsafe concentrations of toxic gases or oxygen deficient atmospheres.

All windows in cabs shall be of safety glass, or equivalent, that introduces no visible distortion that will interfere with the safe operation of the machine.

Where necessary for rigging or service requirements, a ladder, or steps, shall be provided to give access to a cab roof. Guardrails, handholds and steps shall be provided on cranes for easy access to the car and cab, conforming to American National Standards Institute B30.5. Platforms and walkways shall have anti-skid surfaces. **(See Figure 14-12)**

Fuel tank filler pipe shall be located in such a position, or protected in such manner, as to not allow spill or overflow to run onto the engine, exhaust or electrical equipment of any machine being fueled. Portable containers used in refueling shall be of an approved safety-can type. During the refueling process, the engine must be turned off, and no smoking or open flame shall be allowed.

An accessible fire extinguisher of 5BC rating, or higher, shall be available at all operator stations or cabs of equipment.

All fuels shall be transported, stored and handled to meet the rules of Subpart F, Fire Protection and Prevention. When fuel is transported by vehicles on public highways, Department of Transportation rules contained in 49 CFR Parts 177 and 393 concerning such vehicular transportation are considered applicable.

Figure 14-9. This illustration shows the requirements for hooks.

NOTE 1: HOOKS SHALL NOT BE OVERLOADED AND SHALL MEET THE MANUFACTURER'S RECOMMENDATIONS.

NOTE 2: LOAD HOOKS AND HOOK BLOCKS SHALL BE WEIGHTED TO OVERHAUL THE LINE FOR THE HIGHEST HOOK POSITION.

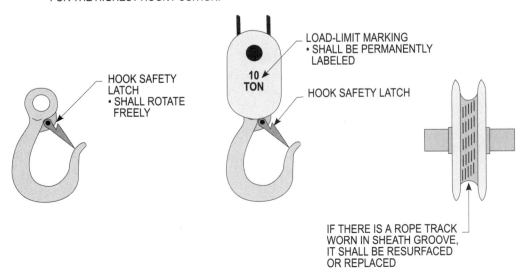

HOOK SAFETY LATCH
• SHALL ROTATE FREELY

LOAD-LIMIT MARKING
• SHALL BE PERMANENTLY LABELED

10 TON

HOOK SAFETY LATCH

IF THERE IS A ROPE TRACK WORN IN SHEATH GROOVE, IT SHALL BE RESURFACED OR REPLACED

OTHER EQUIPMENT, CONDITIONS AND OPERATIONS

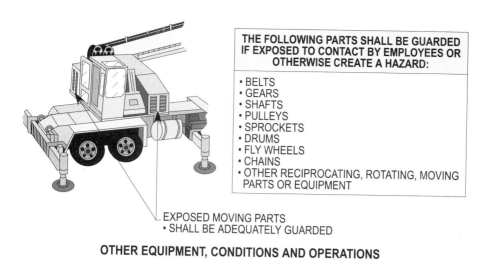

THE FOLLOWING PARTS SHALL BE GUARDED IF EXPOSED TO CONTACT BY EMPLOYEES OR OTHERWISE CREATE A HAZARD:

- BELTS
- GEARS
- SHAFTS
- PULLEYS
- SPROCKETS
- DRUMS
- FLY WHEELS
- CHAINS
- OTHER RECIPROCATING, ROTATING, MOVING PARTS OR EQUIPMENT

EXPOSED MOVING PARTS
• SHALL BE ADEQUATELY GUARDED

OTHER EQUIPMENT, CONDITIONS AND OPERATIONS

Figure 14-10. This illustration shows that parts shall be grounded if exposed to contact by employees, or otherwise create a hazard.

NOTE: ALL EXHAUST PIPES SHALL BE GUARDED OR INSULATED IN AREAS WHERE CONTACT BY EMPLOYEES IS POSSIBLE IN THE PERFORMANCE OF NORMAL DUTIES.

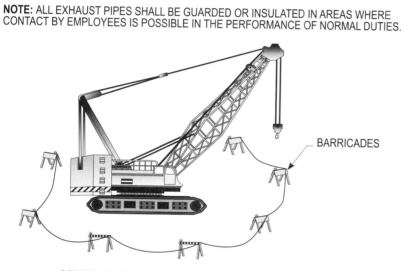

BARRICADES

OTHER EQUIPMENT, CONDITIONS AND OPERATIONS

Figure 14-11. This illustration shows that barricades shall be used to prevent an employee from being struck or crushed by the crane.

PLATFORMS AND WALKWAYS
• ANTI-SKID SURFACES

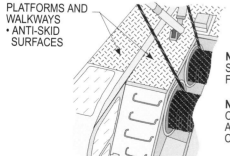

NOTE 1: GUARDRAILS, HANDHOLDS AND STEPS SHALL BE PROVIDED ON CRANES FOR EASY ACCESS TO THE CAR OR CAB.

NOTE 2: AN ACCESSIBLE FIRE EXTINGUISHER OF 5BC RATING, OR HIGHER, SHALL BE AVAILABLE AT ALL OPERATOR STATIONS OR CABS OF EQUIPMENT.

OTHER EQUIPMENT, CONDITIONS AND OPERATIONS

Figure 14-12. This illustration shows that platforms and walkways shall have anti-skid surfaces.

WORK NEAR ELECTRICAL POWER LINES

Except where electrical distribution and transmission lines have been deenergized and visibly grounded at point of work or where insulating barriers, not a part of or an attachment to the equipment or machinery, have been erected to prevent physical contact with the lines, equipment or machines shall be operated proximate to power lines only in accordance with the following:

- For lines rated 50 kV or below, minimum clearance between the lines and any part of the crane or load shall be 10 ft.;

- For lines rated over 50 kV minimum clearance between the lines and any part of the crane or load shall be 10 ft. plus 0.4 in. for each 1 kV over 50 kV, or twice the length of the line insulator, but never less than 10 ft.;

- In transit with no load and boom lowered, the equipment clearance shall be a minimum of 4 ft. for voltages less than 50 kV, and 10 ft. for voltages over 50 kV, up to and including 345 kV, and 16 ft. for voltages up to and including 750 kV.

- A person shall be designated to observe clearance of the equipment and give timely warning for all operations where it is difficult for the operator to maintain the desired clearance by visual means;

- Cage-type boom guards, insulating links or proximity warning devices may be used on cranes, but the use of such devices shall not alter the requirements of any other regulation of this part even if such device is required by law or regulation;

- Any overhead wire shall be considered to be an energized line unless and until the person owning such line or the electrical utility authorities indicate that it is not an energized line and it has been visibly grounded;

- Prior to work near transmitter towers where an electrical charge can be induced in the equipment or materials being handled, the transmitter shall be deenergized or tests shall be made to determine if electrical charge is induced on the crane.

- The following precautions shall be taken when necessary to dissipate induced voltages:

  (a) The equipment shall be provided with an electrical ground directly to the upper rotating structure supporting the boom; and

  (b) Ground jumper cables shall be attached to materials being handled by boom equipment when electrical charge is induced while working near energized transmitters. Crews shall be provided with nonconductive poles having large alligator clips or other similar protection to attach the ground cable to the load.

Combustible and flammable materials shall be removed from the immediate area prior to operations. **(See Figure 14-13)**

No modifications or additions that affect the capacity or safe operation of the equipment shall be made by the employer without the manufacturer's written approval. If such modifications or changes are made, the capacity, operation, and maintenance instruction plates, tags, or decals shall be changed accordingly. In no case shall the original safety factor of the equipment be reduced.

DISSIPATING INDUCED VOLTAGES

- THE EQUIPMENT SHALL BE PROVIDED WITH AN ELECTRICAL GROUND DIRECTLY TO THE UPPER ROTATING STRUCTURE SUPPORTING THE BOOM

- GROUND JUMPER CABLES SHALL BE ATTACHED TO MATERIALS BEING HANDLED BY BOOM EQUIPMENT WHEN ELECTRICAL CHARGE IS INDUCED WHILE WORKING NEAR ENERGIZED TRANSMITTERS

- CABLES SHALL BE PROVIDED WITH NONCONDUCTIVE POLES HAVING LARGE ALLIGATOR CLIPS OR OTHER SIMILAR PROTECTION TO ATTACH THE GROUND CABLE TO THE LOAD

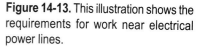

Figure 14-13. This illustration shows the requirements for work near electrical power lines.

CLEARANCES
- 50 kV OR BELOW - AT LEAST 10 FT.

- OVER 50 kV - AT LEAST 10 FT. PLUS 0.4 IN. FOR EACH 1 kV OVER 50 kV, OR TWICE THE LENGTH OF THE LINE INSULATOR, BUT NEVER LESS THAN 10 FT.

- IN TRANSIT WITH NO LOAD AND BOOM LOWERED, THE CLEARANCE SHALL BE AT LEAST 4 FT. FOR VOLTAGES LESS THAN 50 kV, 10 FT. FOR VOLTAGES OVER 50 kV, UP TO AND INCLUDING 345 kV AND 16 FT. FOR VOLTAGES UP TO AND INCLUDING 750 kV

DANGER ZONE

WORK NEAR ELECTRICAL POWER LINES

The employer shall comply with Power Crane and Shovel Association Mobile Hydraulic Crane Standard No. 2.

Sideboom cranes mounted on wheel or crawler tractors shall meet the requirements of SAE J743a-1964.

CRAWLER, LOCOMOTIVE AND TRUCK CRANES
OSHA 1926.550(b)

All crawler, truck, or locomotive cranes in use shall meet the applicable requirements for design, inspection, construction, testing, maintenance and operation as prescribed in the ANSI B30.5-1968, Safety Code for Crawler, Locomotive and Truck Cranes. However, the written, dated and signed inspection reports and records of the monthly inspection of critical items prescribed in section 5-2.1.5 of the ANSI B30.5-1968 standard are not required. Instead, the employer shall prepare a certification record that includes the date the crane items were inspected; the signature of the person who inspected the crane items; and a serial number, or other identifier, for the crane inspected. The most recent certification record shall be maintained on file until a new one is prepared. **(See Figure 14-14)**

All jibs shall have positive stops to prevent their movement of more than 5° above the straight line of the jib and boom on conventional type crane booms. The use of cable type belly slings does not constitute compliance with this rule. Boom stops shall be provided to prevent the boom from falling backward and should be one of the following types:

- A fixed or telescoping bumper,
- A shock absorbing bumper, or
- Hydraulic boom elevation cylinder(s).

Figure 14-14. This illustration shows the requirements for crawler, locomotive and truck cranes.

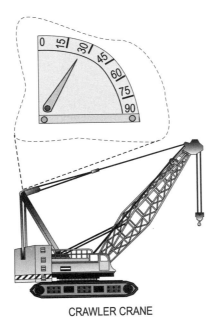

BOOM STOPS SHALL BE PROVIDED TO PREVENT THE BOOM FROM FALLING BACKWARD AND SHOULD BE ONE OF THE FOLLOWING:

• A FIXED OR TELESCOPING BUMPER
• A SHOCK ABSORBING BUMPER
• HYDRAULIC BOOM ELEVATION CYLINDER(S)

NOTE: A BOOM HOIST DISCONNECT, SHUTOFF OR HYDRAULIC RELIEF SHALL BE PROVIDED TO AUTOMATICALLY STOP THE BOOM HOIST WHEN THE BOOM REACHES A PREDETERMINED HEIGHT.

CRAWLER CRANE

TRUCK CRANE

CRAWLER, LOCOMOTIVE AND TRUCK CRANES

Jibs shall be restrained from backward overturning.

A boom angle indicator readable from the operator's station shall be provided. A boom hoist disconnect, shutoff or hydraulic relief shall be provided to automatically stop the boom hoist when the boom reaches a predetermined high angle.

A boom-length indicator, readable from the operator's station, shall be provided for all telescoping booms.

A means shall be provided for the operator to visually determine from the operator's seat the levelness of the crane. (The accuracy of the boom-level indicator shall be checked during periodic inspections.)

A means shall be provided to hold outriggers in the retracted position during travel and in the extended position when in use. A means shall be provided for fastening outrigger floats to the outriggers when in use. **(See Figure 14-15)**

When in use, the crane shall be supported by a firm surface, level within 1 percent grade. Substantial timbers, cribbing or other structural members may be used for this purpose.

Figure 14-15. This illustration shows the requirements for outriggers during times of travel.

NOTE 1: A MEANS SHALL BE PROVIDED FOR FASTENING OUTRIGGER FLOATS TO THE OUTRIGGER WHEN IN USE.

NOTE 2: WHEN IN USE, THE CRANE SHALL BE SUPPORTED BY A FIRM SURFACE, LEVEL WITHIN 1% GRADE. SUBSTANTIAL TIMBERS, CRIBBING OR OTHER STRUCTURAL MEMBERS MAY BE USED FOR THIS PURPOSE.

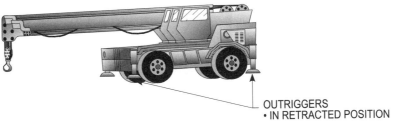

OUTRIGGERS
• IN RETRACTED POSITION

CRAWLER, LOCOMOTIVE AND TRUCK CRANES

A device or boom support shall be provided to prevent the boom and superstructure from rotating when the crane is in transit; this device or boom support shall be constructed to minimize inadvertent engagement or disengagement.

Travel brakes on crawler, locomotive and wheelmounted cranes shall be capable of holding the crane in position during working cycles on level ground and capable of stopping the crane when it is descending the maximum grade recommended for travel. Commercial truck, vehicle-mounted cranes shall meet the braking requirements of the Department of Transportation (DOT).

Telescopic boom cranes shall have either an anti-two-block device or a two-block damage prevention feature at all points of two-blocking. Those cranes manufactured before the effective date of ANSI B30.5 should have those features retrofitted or equipped with a two-blocking warning device.

HAMMERHEAD TOWER CRANES
OSHA 1926.550(c)

Adequate clearance shall be maintained between moving and rotating structures of the crane and fixed objects to allow the passage of employees without harm. **(See Figure 14-16)**

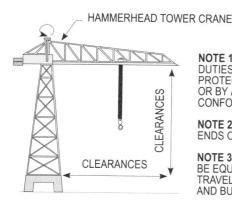

HAMMERHEAD TOWER CRANE

NOTE 1: EACH EMPLOYEE REQUIRED TO PERFORM DUTIES ON THE HORIZONTAL BOOM SHALL BE PROTECTED AGAINST FALLING BY GUARDRAILS OR BY A PERSONAL FALL ARREST SYSTEM IN CONFORMANCE WITH SUBPART M, FALL PROTECTION.

NOTE 2: BUFFERS SHALL BE PROVIDED AT BOTH ENDS OF TRAVEL OF THE TROLLEY.

NOTE 3: CRANES MOUNTED ON RAIL TRACKS SHALL BE EQUIPPED WITH LIMIT SWITCHES LIMITING THE TRAVEL OF THE CRANE ON THE TRACK AND STOPS AND BUFFERS AT EACH END OF THE TRACKS.

HAMMERHEAD TOWER CRANES
OSHA 1926.550(c)

Figure 14-16. This illustration shows that adequate clearance shall be maintained to allow the passage of employees without harm.

Each employee required to perform duties on the horizontal boom of hammerhead tower cranes shall be protected against falling by guardrails or by a personal fall arrest system in conformance with Subpart M, Fall Protection.

Buffers shall be provided at both ends of travel of the trolley.

Cranes mounted on rail tracks shall be equipped with limit switches limiting the travel of the crane on the track and stops or buffers at each end of the tracks.

All hammerhead tower cranes in use shall meet the applicable requirements for design, construction, installation, testing, maintenance, inspection, and operation as prescribed by the manufacturer.

OVERHEAD AND GANTRY CRANES
OSHA 1926.550(d)

The rated load of the crane shall be plainly marked on each side of the crane, and if the crane has more than one hoisting unit, each hoist shall have its rated load marked on it or its load block, and this marking shall be clearly legible from the ground or floor. **(See Figure 14-17)**

Figure 14-17. This illustration shows the requirements for overhead and gantry cranes.

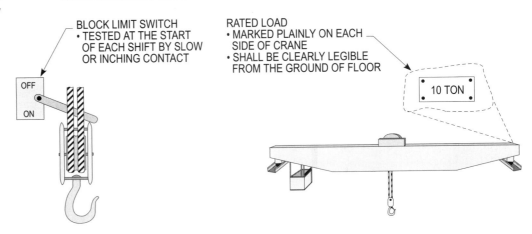

NOTE: EACH HOIST SHALL HAVE ITS RATED LOAD MARKED ON IT OR ITS LOAD BLOCK.

BLOCK LIMIT SWITCH
• TESTED AT THE START OF EACH SHIFT BY SLOW OR INCHING CONTACT

RATED LOAD
• MARKED PLAINLY ON EACH SIDE OF CRANE
• SHALL BE CLEARLY LEGIBLE FROM THE GROUND OF FLOOR

OFF
ON

10 TON

OVERHEAD AND GANTRY CRANES
OSHA 1926.550(d)

Bridge trucks shall be equipped with sweeps that extend below the top of the rail and project in front of the truck wheels.

Except for floor-operated cranes, a gong or other effective audible warning signal shall be provided for each crane equipped with a power traveling mechanism.

Each hoisting unit shall be provided with an upper movement limit switch that shall be tested at the start of each shift by slow or inching contact.

All overhead and gantry cranes in use shall meet the applicable requirements for design, construction, installation, testing, maintenance, inspection and operation as prescribed in the ANSI B30.2.0-1967, Safety Code for Overhead and Gantry Cranes.

DERRICKS
OSHA 1926.550(e)

All derricks in use shall meet the applicable requirements for design, construction, installation, inspection, testing, maintenance and operation as prescribed in American National Standards Institute B30.6-1969, Safety Code for Derricks.

FLOATING CRANES AND DERRICKS
OSHA 1926.550(f)

The following shall be considered for floating cranes and derricks:

- Mobile cranes mounted on barges
- Permanently mounted floating cranes and derricks
- Protection of employees working on barges

MOBILE CRANES MOUNTED ON BARGES

When a mobile crane is mounted on a barge, the rated load of the crane shall not exceed the original capacity specified by the manufacturer. **(See Figure 14-18)**

NOTE 1: A LOAD RATING CHART WITH CLEARLY LEGIBLE LETTERS AND FIGURES SHALL BE PROVIDED WITH EACH CRANE AND SECURELY FIXED AT A LOCATION EASILY VISIBLE TO THE OPERATOR.

NOTE 2: WHEN LOAD RATINGS ARE REDUCED TO STAY WITHIN THE LIMITS FOR OF THE BARGE WITH A CRANE MOUNTED ON IT, A NEW LOAD RATING CHART SHALL BE PROVIDED.

Figure 14-18. This illustration shows the requirements for mobile cranes mounted on barges.

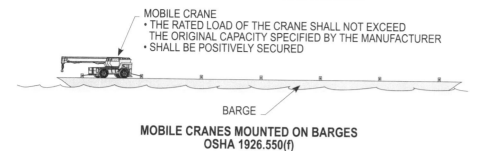

MOBILE CRANE
• THE RATED LOAD OF THE CRANE SHALL NOT EXCEED THE ORIGINAL CAPACITY SPECIFIED BY THE MANUFACTURER
• SHALL BE POSITIVELY SECURED

BARGE

MOBILE CRANES MOUNTED ON BARGES
OSHA 1926.550(f)

A load rating chart, with clearly legible letters and figures, shall be provided with each crane, and securely fixed at a location easily visible to the operator.

When load ratings are reduced to stay within the limits for list of the barge with a crane mounted on it, a new load rating chart shall be provided.

Mobile cranes on barges shall be positively secured.

PERMANENTLY MOUNTED FLOATING CRANES AND DERRICKS

When cranes and derricks are permanently installed on a barge, the capacity and limitations of use shall be based on competent design criteria.

A load rating chart with clearly legible letters and figures shall be provided and securely fixed at a location easily visible to the operator.

Floating cranes and floating derricks in use shall meet the applicable requirements for design, construction, installation, testing, maintenance and operation as prescribed by the manufacturer.

PROTECTION OF EMPLOYEES WORKING ON BARGES

The employer shall comply with the applicable requirements for protection of employees working onboard marine vessels specified in OSHA 1926.605.

CRANE OR DERRICK SUSPENDED PERSONNEL PLATFORMS
OSHA1926.550(g)

The following shall be considered for crane or derrick suspended personnel platforms:

- Hoisting a personnel platform
- Design and use of personnel platforms
- Trail lifts
- Hoisting workers

HOISTING A PERSONNEL PLATFORM

Use of cranes or derricks to hoist workers on personnel platforms is prohibited except when it is not possible (or is more hazardous) to erect a conventional means to reach the worksite. **(See Figure 14-19)**

Figure 14-19. This illustration shows the requirements for hoisting a personnel platform.

OSHA Required Training:
1926.550(g)(4)(i)(A)
1926.550(g)(5)(iv)

NOTE 1: THE PERSONNEL PLATFORM SHALL BE HOISTED IN A SLOW, CONTROLLED, CAUTIOUS MANNER.

NOTE 2: LOAD- AND BEAM-HOIST DRUM BRAKES, SWING BRAKES AND LOCKING DEVICES, SUCH AS PAWLS OR DOGS, SHALL BE ENGAGED WHEN AN OCCUPIED PERSONNEL PLATFORM IS IN A STATIONARY WORKING POSITION.

NOTE 3: THE CRANE SHALL BE UNIFORMLY LEVEL TO WITHIN 1 PERCENT OR GRADE AND LOCATED ON A FIRM FOOTING. ALL OUTRIGGERS SHALL BE FULLY EXTENDED.

NOTE 4: THE TOTAL WEIGHT OF THE LOADED PERSONNEL PLATFORM AND RELATED RIGGING SHALL NOT EXCEED 50 PERCENT OF THE RATED CAPACITY FOR THE RADIUS AND CONFIGURATION OF THE CRANE OR DERRICK.

LOAD LINES
- SHALL BE CAPABLE OF SUPPORTING AT LEAST SEVEN TIMES THE MAXIMUM INTENDED LOAD
- SHALL BE CAPABLE OF SUPPORTING AT LEAST TEN TIMES THE MAXIMUM INTENDED LOAD WHEN ROTATION RESISTANT ROPE IS USED

SUSPENDED PERSONNEL PLATFORM

HOISTING A PERSONNEL PLATFORM
OSHA 1926.550(g)

The personnel platform shall be hoisted in a slow, controlled, cautious manner. Load lines shall be capable of supporting at least seven times the maximum intended load; when rotation resistant rope is used, the line shall be capable of supporting ten times the maximum intended load.

Load- and boom-hoist drum brakes, swing brakes and locking devices, such as pawls or dogs, shall be engaged when an occupied personnel platform is in a stationary working position.

The crane shall be uniformly level to within 1 percent of grade and located on a firm footing. All outriggers shall be fully extended.

The total weight of the loaded personnel platform and related rigging shall not exceed 50percent of the rated capacity for the radius and configuration of the crane or derrick.

Machines equipped with booms in which lowering is controlled only by a brake are prohibited.

Cranes and derricks with variable-angle booms shall be equipped with a boom-angle indicator that is visible to the operator.

Cranes with telescoping booms shall be equipped with a device to indicate at all times the boom's extended length, or the load radius to be used during the lift shall be determined prior to hoisting personnel.

An anti-two-blocking device or means to deactivate the hoist shall be used to prevent contact between the load block or overhaul ball and the boom tip.

The loading hoist drum shall have a governor (or equivalent device other than the load hoist brake) on the power train to regulate the lowering rate of speed of the hoist mechanism. Free fall of the hoist is prohibited.

DESIGN AND USE OF PERSONNEL PLATFORMS

Each personnel platform shall be equipped with a guardrail system and shall be enclosed from the toeboard to the midrail with solid or expanded metal that has openings no greater than 1/2 in.

The personnel platform and suspension system shall be designed by a qualified engineer or a qualified person competent in structural design. The personnel platform suspension system shall be designed to minimize tipping caused by workers moving on the platform.

Personnel platforms shall be capable of supporting their own weight and at least five times the maximum intended load.

A grab rail shall be installed inside the entire perimeter of the personnel platform.

Access gates, including those that slide or fold, shall be equipped with restraining devices to prevent them from being opened accidentally.

Sufficient space shall be provided so that workers can stand upright on the personnel platform.
The personnel platform shall have overhead protection in situations where workers might be exposed to falling objects.

A qualified welder shall perform all welding on the personnel platform and its components.

The number of workers occupying the personnel platform shall not be more than required for the work being performed.

Personnel platforms shall be used only for workers, their tools and the materials necessary to do their work and shall not be used to hoist only materials and tools.

The personnel platform's weight and load rating or maximum intended load shall be conspicuously posted on a plate or other permanent marking on the platform. **(See Figure 14-20)**

Materials and tools hoisted during a personnel lift shall be secured to prevent displacement. When a wire-rope bridle is used to connect the personnel platform to the load line, each bridle leg shall be connected to a master link or shackle to ensure that the load is evenly divided among the bridle legs. **(See Figure 14-21)**

Hooks on overhaul ball assemblies, lower load blocks or other attachment assemblies shall be of a type that can be closed and locked. An alloy anchor shackle with a bolt, nut and retaining pin may also be used. Bridles and associated rigging used for personnel lifting shall not be used for any other purpose.

TRIAL LIFTS

Before the trial lift, a meeting shall be held at each new location to review all appropriate regulations with the crane or derrick operator, signal person, workers to be lifted and the person responsible for the task to be performed. Meetings shall be repeated for workers who are assigned to the operation after the trial lift takes place.

Figure 14-20. This illustration shows that the load tag shall be conspicuously posted on a plate or other permanent marking on the platform.

NOTE 1: THE NUMBER OF WORKERS OCCUPYING THE PERSONNEL PLATFORM SHALL NOT BE MORE THAN REQUIRED FOR THE WORK BEING PERFORMED.

NOTE 2: PERSONNEL PLATFORMS SHALL BE USED ONLY FOR WORKERS, THEIR TOOLS AND THE MATERIALS NECESSARY TO DO THEIR WORK AND SHALL NOT BE USED TO HOIST ONLY MATERIALS AND TOOLS.

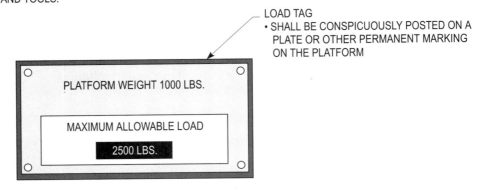

HOISTING A PERSONNEL PLATFORM
OSHA 1926.550(g)

Figure 14-21. This illustration shows a wire-rope bridle master link for hoisting materials and equipment.

NOTE 1: MATERIALS AND TOOLS HOISTED DURING A PERSONNEL LIFT SHALL BE SECURED TO PREVENT DISPLACEMENT.

NOTE 2: WHEN A WIRE-ROPE BRIDLE IS USED TO CONNECT THE PERSONNEL PLATFORM TO THE LOAD LINE, EACH BRIDLE LEG SHALL BE CONNECTED TO A MASTER LINK OR SHACKLE TO ENSURE THAT THE LOAD IS EVENLY DIVIDED AMONG THE BRIDLE LEGS.

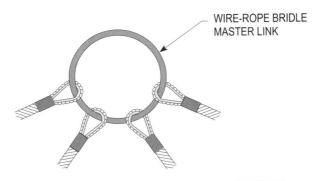

DESIGN AND USE OF PERSONNEL PLATFORMS
OSHA 1926.550(g)

A trial lift shall be made from ground level, or any other location where workers will enter the platform, to each location where the personnel platform is to be hoisted and positioned. The personnel platform shall be unoccupied and loaded to the anticipated lift weight.

The trial lift shall be repeated before hoisting workers or whenever the crane or derrick is moved and set up in a new location.

Between the trial lift and the time workers are first hoisted, the platform shall be hoisted a few inches above the ground and inspected to ensure that it is secure and properly balanced.

Workers shall not be hoisted unless the hoist ropes are free of kinks, multiple-part lines are not twisted around each other, the primary attachment is centered over the platform and the hoisting system is inspected to ensure that all ropes are properly set on drums and in sheaves.

Platforms shall be proof tested at 125 percent of the rated capacity at each job site prior to hoisting personnel or after any repair or modification.

A visual inspection of the crane or derrick, rigging, personnel platform and the crane or derrick base support or ground shall be conducted by a competent person immediately after the trial lift or proof test to determine if the test has exposed any defect or adversely affected any component or structure.

HOISTING WORKERS

Any defects found during inspections that create a safety hazard shall be corrected before workers are hoisted.

Safety Tip: Workers shall keep all parts of their bodies inside the platform.

Workers shall be instructed to keep all parts of the body inside the platform during raising, lowering and positioning. Before workers exit or enter a hoisted personnel platform that has not landed, the platform shall come to a complete stop and be secured to the structure on which the work is to be performed unless such securing creates an unsafe condition.

Tag lines shall be used unless they create an unsafe condition. **(See Figure 14-22)**

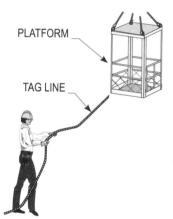

PLATFORM

TAG LINE

NOTE 1: WORKERS SHALL BE INSTRUCTED TO KEEP ALL PARTS OF THE BODY INSIDE THE PLATFORM DURING RAISING, LOWERING AND POSITIONING.

NOTE 2: BEFORE WORKERS EXIT OR ENTER A HOISTED PERSONNEL PLATFORM THAT HAS NOT LANDED, THE PLATFORM SHALL COME TO A COMPLETE STOP AND BE SECURED TO THE STRUCTURE ON WHICH THE WORK IS TO BE PERFORMED UNLESS SUCH SECURING CREATES AN UNSAFE CONDITION.

HOISTING WORKERS
OSHA 1926.550(g)

Figure 14-22. This illustration shows that tag lines shall be used unless they create an unsafe condition.

The crane or derrick operator shall remain at the controls at all times if the crane engine is running and if the platform is occupied.

Safety Tip: Each worker on the platform must use fall protection with a lanyard attached to the lower load block or overhaul ball or a strustural member that can support an impact from a fall.

When dangerous weather conditions or any other danger is imminent, hoisting of workers shall be stopped.

Workers being hoisted shall remain in continuous sight of, and in direct communication with, the hoist operator or signal person. When direct visual contact with the operator is not feasible, direct communication alone may be used.

Each worker on the personnel platform shall use a body belt or safety harness with a lanyard attached to the lower load block or overhaul ball or to a structural member on the platform that can support an impact from a fall. **(See Figure 14-23)**

Lifts shall not be made with one of the cranes or derrick's remaining load lines if workers are suspended on a platform.

The employer shall implement the following procedures for all circumstances in which a crane might travel while it is hoisting workers:

Figure 14-23. This illustration shows that each worker on the personnel platform shall use a body belt or safety harness with a lanyard.

NOTE 1: WHEN DANGEROUS WEATHER CONDITIONS OR ANY OTHER DANGER IS IMMINENT, HOISTING OF WORKERS SHALL BE STOPPED.

NOTE 2: EACH WORKER ON THE PERSONNEL PLATFORM SHALL USE A BODY BELT OR SAFETY HARNESS WITH A LANYARD ATTACHED TO THE LOWER LOAD BLOCK OR OVERHAUL BALL OR TO A STRUCTURAL MEMBER ON THE PLATFORM THAT CAN SUPPORT AN IMPACT FROM A FALL.

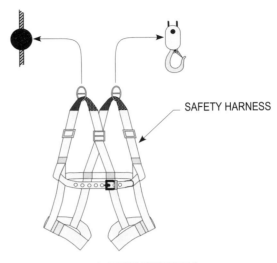

SAFETY HARNESS

HOISTING WORKERS
OSHA 1926.550(g)

- Crane travel shall be restricted to a fixed track or runway.

- Travel shall be limited to the load radius of the boom used during the lift.

- The boom shall be parallel to the direction of travel.

- A complete trial run shall be performed to test the route of travel before workers are allowed to occupy the platform.

- If travel is done with a rubber-tired carrier, the overall condition and air pressure of the fires shall be checked.

HELICOPTERS
OSHA1926.551

The following shall be considered for helicopters:

- Helicopter regulations
- Briefing
- Slings and tag lines
- Cargo hooks
- Personal protective equipment
- Loose gear and objects
- Housekeeping
- Operator responsibility
- Hooking and unhooking loads
- Static charge
- Weight limitation
- Ground lines
- Visibility
- Signal systems

- Approach distance
- Approaching helicopter
- Personnel
- Communications
- Fires

HELICOPTER REGULATIONS
OSHA 1926.551(a)

Helicopter cranes shall be expected to comply with any applicable regulations of the Federal Aviation Administration.

BRIEFING
OSHA 1926.551(b)

Prior to each day's operation a briefing shall be conducted. This briefing shall set forth the plan of operation for the pilot and ground personnel.

SLINGS AND TAG LINES
OSHA 1926.551(c)

Load shall be properly slung. Tag lines shall be of a length that will not permit their being drawn up into rotors. Pressed sleeve, swedged eyes or equivalent means shall be used for all freely suspended loads to prevent hand splices from spinning open or cable clamps from loosening.

CARGO HOOKS
OSHA 1926.551(d)

All electrically operated cargo hooks shall have the electrical activating device so designed and installed as to prevent inadvertent operation. In addition, these cargo hooks shall be equipped with an emergency mechanical control for releasing the load. The hooks shall be tested prior to each day's operation to determine that the release functions properly, both electrically and mechanically. **(See Figure 14-24)**

PERSONAL PROTECTIVE EQUIPMENT
OSHA 1926.551(e)

Personal protective equipment for employees receiving the load shall consist of complete eye protection and hard hats secured by chin straps. Loose-fitting clothing likely to flap in the downwash, and thus be snagged on hoist line, shall not be worn.

LOOSE GEAR AND OBJECTS
OSHA 1926.551(f)

Every practical precaution shall be taken to provide for the protection of the employees from flying objects in the rotor downwash. All loose gear within 100 ft. of the place of lifting the load, depositing the load, and all other areas susceptible to rotor downwash shall be secured or removed.

Figure 14-24. This illustration shows the requirements for helicopters.

NOTE 1: PERSONAL PROTECTIVE EQUIPMENT FOR EMPLOYEES RECEIVING THE LOAD SHALL CONSIST OF COMPLETE EYE PROTECTION AND HARD HATS SECURED BY CHIN STRAPS.

NOTE 2: LOOSE-FITTING CLOTHING LIKELY TO FLAP IN THE DOWNWASH SHALL NOT BE WORN.

NOTE 3: ALL LOOSE GEAR WITHIN 100 FT. OF THE PLACE OF LIFTING THE LOAD, DEPOSITING THE LOAD AND ALL OTHER AREAS SUSCEPTIBLE TO ROTOR DOWNWASH SHALL BE SECURED OR REMOVED.

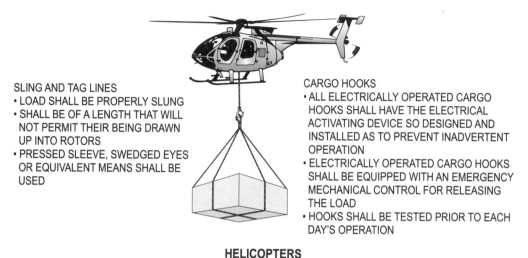

SLING AND TAG LINES
• LOAD SHALL BE PROPERLY SLUNG
• SHALL BE OF A LENGTH THAT WILL NOT PERMIT THEIR BEING DRAWN UP INTO ROTORS
• PRESSED SLEEVE, SWEDGED EYES OR EQUIVALENT MEANS SHALL BE USED

CARGO HOOKS
• ALL ELECTRICALLY OPERATED CARGO HOOKS SHALL HAVE THE ELECTRICAL ACTIVATING DEVICE SO DESIGNED AND INSTALLED AS TO PREVENT INADVERTENT OPERATION
• ELECTRICALLY OPERATED CARGO HOOKS SHALL BE EQUIPPED WITH AN EMERGENCY MECHANICAL CONTROL FOR RELEASING THE LOAD
• HOOKS SHALL BE TESTED PRIOR TO EACH DAY'S OPERATION

HELICOPTERS
OSHA 1926.551

HOUSEKEEPING
OSHA 1926.551(g)

Good housekeeping shall be maintained in all helicopter loading and unloading areas.

OPERATOR RESPONSIBILITY
OSHA 1926.551(h)

The helicopter operator shall be responsible for size, weight and manner in which loads are connected to the helicopter. If, for any reason, the helicopter operator believes the lift cannot be made safely, the lift shall not be made.

HOOKING AND UNHOOKING LOADS
OSHA 1926.551(i)

When employees are required to perform work under hovering craft, a safe means of access shall be provided for employees to reach the hoist line hook and engage or disengage cargo slings. Employees shall not perform work under hovering craft except when necessary to hook or unhook loads.

STATIC CHARGE
OSHA 1926.551(j)

Static charge on the suspended load shall be dissipated with a grounding device before ground personnel touch the suspended load, or protective rubber gloves shall be worn by all ground personnel touching the suspended load.

WEIGHT LIMITATION
OSHA 1926.551(k)

The weight of an external load shall not exceed the manufacturer's rating.

GROUND LINES
OSHA 1926.551(l)

Hoist wires or other gear, except for pulling lines or conductors that are allowed to "pay out" from a container or roll off a reel, shall not be attached to any fixed ground structure, or allowed to foul on any fixed structure.

VISIBILITY
OSHA 1926.551(m)

When visibility is reduced by dust or other conditions, ground personnel shall exercise special caution to keep clear of main and stabilizing rotors. Precautions shall also be taken by the employer to eliminate as far as practical reduced visibility.

SIGNAL SYSTEMS
OSHA 1926.551(n)

Signal systems between aircrew and ground personnel shall be understood and checked in advance of hoisting the load. This applies to either radio or hand signal systems. Hand signals shall be as shown in Figure N-1 in 1926.551(n).

APPROACH DISTANCE
OSHA 1926.551(o)

No unauthorized person shall be allowed to approach within 50 ft. of the helicopter when the rotor blades are turning.

APPROACHING HELICOPTER
OSHA 1926.551(p)

Whenever approaching or leaving a helicopter with blades rotating, all employees shall remain in full view of the pilot and keep in a crouched position. Employees shall avoid the area from the cockpit or cabin rearward unless authorized by the helicopter operator to work there.

PERSONNEL
OSHA 1926.551(q)

Sufficient ground personnel shall be provided when required for safe helicopter loading and unloading operations.

COMMUNICATIONS
OSHA 1926.551(r)

There shall be constant reliable communication between the pilot, and a designated employee of the ground crew who acts as a signalman during the period of loading and unloading. This signalman shall be distinctly recognizable from other ground personnel.

FIRES
OSHA 1926.551(s)

Open fires shall not be permitted in an area that could result in such fires being spread by the rotor downwash.

MATERIAL HOISTS, PERSONNEL HOISTS AND ELEVATORS
OSHA 1926.552

The following shall be considered for material hoists, personnel hoists and elevators:

- General requirements
- Material hoists
- Personnel hoists

GENERAL REQUIREMENTS
OSHA 1926.552(a)

OSHA Required Training:
1926.552(a)(1)
1926.552(c)(15) and (17)(i)

The employer shall comply with the manufacturer's specifications and limitations applicable to the operation of all hoists and elevators. Where manufacturer's specifications are not available, the limitations assigned to the equipment shall be based on the determinations of a professional engineer competent in the field.

Rated load capacities, recommended operating speeds and special hazard warnings or instructions shall be posted on cars and platforms.

Wire rope shall be removed from service when any of the following conditions exists:

- In hoisting ropes, six randomly distributed broken wires in one rope lay or three broken wires in one strand in one rope lay;

- Abrasion, scrubbing, flattening or peening, causing loss of more than one-third of the original diameter of the outside wires;

- Evidence of any heat damage resulting from a torch or any damage caused by contact with electrical wires;

- Reduction from nominal diameter of more than three sixty-fourths inch for diameters up to and including three-fourths inch; one-sixteenth inch for diameters seven-eights to 1 inch; and three thirty-seconds inch for diameters 1 to 1 inches.

- Hoisting ropes shall be installed in accordance with the wire rope manufacturer's recommendations.
- The installation of live booms on hoists is prohibited.

- The use of endless belt-type manlifts on construction shall be prohibited.

MATERIAL HOISTS
OSHA 1926.552(b)

Operating rules shall be established and posted at the operator's station of the hoist. Such rules shall include signal system and allowable line speed for various loads. Rules and notices shall be posted on the car frame or crosshead in a conspicuous location, including the statement "No Riders Allowed."

No person shall be allowed to ride on material hoists except for the purposes of inspection and maintenance.

All entrances of the hoistways shall be protected by substantial gates or bars, which shall guard the full width of the landing entrance. All hoistway entrance bars and gates shall be painted with diagonal contrasting colors, such as black and yellow stripes.

Bars shall be not less than 2 in. by 4 in. wooden bars or the equivalent, located 2 ft. from the hoistway line. Bars shall be located not less than 36 in. nor more than 42 in. above the floor.

Gates or bars protecting the entrances to hoistways shall be equipped with a latching device.

Overhead protective covering of 2 in. planking, 2 in. plywood or other solid material of equivalent strength shall be provided on the top of every material hoist cage or platform.

The operator's station of a hoisting machine shall be provided with overhead protection equivalent to tight planking not less than 2 in. thick. The support for the overhead protection shall be of equal strength.

Hoist towers may be used with or without an enclosure on all sides. However, whichever alternative is chosen, the following applicable conditions shall be met:

- When a hoist tower is enclosed, it shall be enclosed on all sides for its entire height with a screen enclosure of 2 in. mesh, No. 18 U.S. gauge wire or equivalent, except for landing access.

- When a hoist tower is not enclosed, the hoist platform or car shall be totally enclosed (caged) on all sides for the full height between the floor and the overhead protective covering with 2 in. mesh of No. 14 U.S. gauge wire or equivalent. The hoist platform enclosure shall include the required gates for loading and unloading. A 6 ft. high enclosure shall be provided on the unused sides of the hoist tower at ground level.

- Car arresting devices shall be installed to function in case of rope failure.

- All material hoist towers shall be designed by a licensed professional engineer. All material hoists shall conform to the requirements of ANSI A10.5-1969, Safety Requirements for Material Hoists.

PERSONNEL HOISTS
OSHA 1926.552(c)

Hoist towers outside the structure shall be enclosed for the full height on the side or sides used for entrance and exit to the structure. At the lowest landing, the enclosure on the sides not used for exit or entrance to the structure shall be enclosed to a height of at least 10 ft. Other sides of the tower adjacent to floors or scaffold platforms shall be enclosed to a height of 10 ft. above the level of such floors or scaffolds.

Towers inside of structures shall be enclosed on all four sides throughout the full height.

Towers shall be anchored to the structure at intervals not exceeding 25 ft. In addition to tie-ins, a series of guys shall be installed. Where tie-ins are not practical the tower shall be anchored by means of guys made of wire rope at least one-half inch in diameter, securely fastened to anchorage to ensure stability.

Hoistway doors or gates shall be not less than 6 ft. 6 in. high and shall be provided with mechanical locks that cannot be operated from the landing side, and shall be accessible only to persons on the car.

Cars shall be permanently enclosed on all sides and the top, except sides used for entrance and exit that have car gates or doors. **(See Figure 14-25)**

Figure 14-25. This illustration shows that cars shall be permanently enclosed on all sides and the top.

NOTE 1: HOISTWAY DOORS OR GATES SHALL NOT BE LESS THAN 6 FT. 6 IN. HIGH AND SHALL BE PROVIDED WITH MECHANICAL LOCKS THAT CANNOT BE OPERATED FROM THE LANDING SITE, AND SHALL BE ACCESSIBLE ONLY TO PERSONS ON THE CAR.

NOTE 2: DOOR OR GATE SHALL BE PROVIDED AT EACH ENTRANCE TO THE CAR, WHICH SHALL PROTECT THE FULL WIDTH AND HEIGHT OF THE CAR ENTRANCE OPENING.

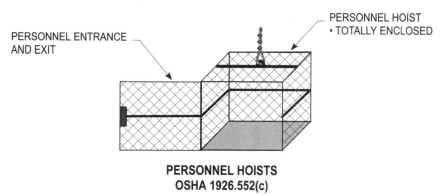

PERSONNEL ENTRANCE AND EXIT

PERSONNEL HOIST
• TOTALLY ENCLOSED

PERSONNEL HOISTS
OSHA 1926.552(c)

A door or gate shall be provided at each entrance to the car, which shall protect the full width and height of the car entrance opening.

Overhead protective covering of 2 in. planking, 2 in. plywood or other solid material or equivalent strength shall be provided on the top of every personnel hoist.

Doors or gates shall be provided with electric contacts that do not allow movement of the hoist when door or gate is open. **(See Figure 14-26)**

Safeties shall be capable of stopping and holding the car and rated load when traveling at governor tripping speed.

Cars shall be provided with a capacity and data plate secured in a conspicuous place on the car or crosshead.

Internal combustion engines shall not be permitted for direct drive.

NOTE 1: SAFETIES SHALL BE CAPABLE OF STOPPING AND HOLDING THE CAR AND RATED LOAD WHEN TRAVELING AT GOVERNOR TRIPPING SPEED.

NOTE 2: CARS SHALL BE PROVIDED WITH A CAPACITY AND DATA PLATE SECURED IN A CONSPICUOUS PLACE ON THE CAR OR CROSSHEAD.

NOTE 3: AN EMERGENCY STOP SWITCH SHALL BE PROVIDED IN THE CAR AND MARKED "STOP."

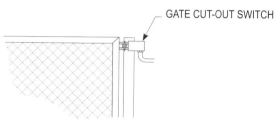

Figure 14-26. This illustration shows that doors and gates shall be provided with electrical contacts.

PERSONNEL HOISTS
OSHA 1926.552(c)

Normal and final terminal stopping devices shall be provided.

An emergency stop switch shall be provided in the car and marked "Stop."

The minimum number of hoisting ropes used shall be three for traction hoists and two for drum-type hoists. The minimum diameter of hoisting and counterweight wire ropes shall be 2 in. Minimum factors of safety for suspension wire ropes are shown in OSHA 1926.552(c)(14)(iii).

Following assembly and erection of hoists, and before being put in service, an inspection and test of all functions and safety devices shall be made under the supervision of a competent person. A similar inspection and test is required following major alteration of an existing installation. All hoists shall be inspected and tested at not more than 3-month intervals. The employer shall prepare a certification record that includes the date the inspection and test of all functions and safety devices was performed; the signature of the person who performed the inspection and test; and a serial number, or other identifier, for the hoist that was inspected and tested. The most recent certification record shall be maintained on file.

Safety Tip: Inspections and tests shall be made under the supervision of a competent person.

All personnel hoists used by employees shall be constructed of materials and components that meet the specifications for materials, construction, safety devices, assembly and structural integrity as stated in the American National Standard A10.4-1963, Safety Requirements for Workmen's Hoists. The requirements of this paragraph 1926.552(c)(16) do not apply to cantilever type personnel hoists.

Personnel hoists used in bridge tower construction shall be approved by a registered professional engineer and erected under the supervision of a qualified engineer competent in this field.

When a hoist tower is not enclosed, the hoist platform or car shall be totally enclosed (caged) on all sides for the full height between the floor and the overhead protective covering with 2 in. mesh of No. 14 U.S. gauge wire or equivalent. The hoist platform enclosure shall include the required gates for loading and unloading.

These hoists shall be inspected and maintained on a weekly basis. Whenever the hoisting equipment is exposed to winds exceeding 35 miles per hour it shall be inspected and put in operable condition before reuse.

Wire rope shall be taken out of service when any of the following conditions exist:

- In running ropes, six randomly distributed broken wires in one lay or three broken wires in one strand in one lay;

- Wear of one-third the original diameter of outside individual wires. Kinking, crushing, bird caging or any other damage resulting in distortion of the rope structure;

- Evidence of any heat damage from any cause;

- Reductions from nominal diameter of more than three-sixty-fourths inch for diameters to and including three-fourths inch, one-sixteenth inch for diameters seven-eights inch to 1 in. inclusive, three-thirty-seconds inch for diameters 1 to 1 inches inclusive;

- In standing ropes, more than two broken wires in one lay in sections beyond end connections or more than one broken wire at an end connection.

Permanent elevators under the care and custody of the employer and used by employees for work covered by this Act shall comply with the requirements of American National Standards Institute A17.1-1965 with addenda A17.1a-1967, A17.1b-1968, A17.1c-1969, A17.1d-1970, and inspected in accordance with A17.2-1960 with addenda A17.2a-1965, A17.2b-1967.

BASE-MOUNTED DRUM HOISTS
OSHA 1926.553

Exposed moving parts such as gears, projecting screws, set-screws, chain, cables, chain sprockets and reciprocating or rotating parts, which constitute a hazard, shall be guarded.

All controls used during the normal operation cycle shall be located within easy reach of the operator's station. Electric motor operated hoists shall be provided with:

- A device to disconnect all motors from the line upon power failure and not permit any motor to be restarted until the controller handle is brought to the "off" position;

- Where applicable, an overspeed preventive device;

- A means whereby remotely operated hoists stop when any control is ineffective.

All base-mounted drum hoists in use shall meet the applicable requirements for design, construction, installation, testing, inspection, maintenance and operations, as prescribed by the manufacturer. **(See Figure 14-27)**

OVERHEAD HOISTS
OSHA 1926.554

The safe working load of the overhead hoist, as determined by the manufacturer, shall be indicated on the hoist, and this safe working load shall not be exceeded.

The supporting structure to which the hoist is attached shall have a safe working load equal to that of the hoist.

The support shall be arranged so as to provide for free movement of the hoist and shall not restrict the hoist from lining itself up with the load.

The hoist shall be installed only in locations that will permit the operator to stand clear of the load at all times.

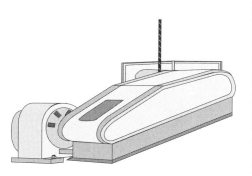

THE FOLLOWING EXPOSED WIRING PARTS SHALL BE GUARDED:
• GEARS • PROJECTING SCREWS • SET SCREWS • CHAIN • CABLES • CHAIN SPROCKETS • RECIPROCATING OR ROTATING PARTS

NOTE: ALL CONTROLS USED DURING THE NORMAL OPERATION CYCLE SHALL BE LOCATED WITHIN EASY REACH OF THE OPERATOR'S STATION.

BASE-MOUNTED DRUM HOISTS
OSHA 1926.553

Figure 14-27. This illustration shows that all exposed moving parts shall be guarded.

Air hoists shall be connected to an air supply of sufficient capacity and pressure to safely operate the hoist. All air hoses supplying air shall be positively connected to prevent their becoming disconnected during use.

All overhead hoists in use shall meet the applicable requirements for construction, design, installation, testing, inspection, maintenance and operation, as prescribed by the manufacturer.

CONVEYORS
OSHA1926.555

Means for stopping the motor or engine shall be provided at the operator's station. Conveyor systems shall be equipped with an audible warning signal to be sounded immediately before starting up the conveyor.

If the operator's station is at a remote point, similar provisions for stopping the motor or engine shall be provided at the motor or engine location.

Emergency stop switches shall be arranged so that the conveyor cannot be started again until the actuating stop switch has been reset to running or "on" position. **(See Figure 14-28)**

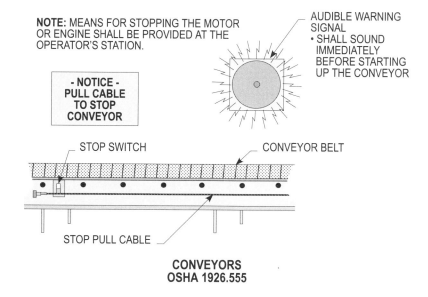

NOTE: MEANS FOR STOPPING THE MOTOR OR ENGINE SHALL BE PROVIDED AT THE OPERATOR'S STATION.

- NOTICE -
PULL CABLE TO STOP CONVEYOR

AUDIBLE WARNING SIGNAL
• SHALL SOUND IMMEDIATELY BEFORE STARTING UP THE CONVEYOR

STOP SWITCH

CONVEYOR BELT

STOP PULL CABLE

CONVEYORS
OSHA 1926.555

Figure 14-28. This illustration shows the general requirements for conveyors.

Screw conveyors shall be guarded to prevent employee contact with turning flights.

Where a conveyor passes over work areas, aisles or thoroughfares, suitable guards shall be provided to protect employees required to work below the conveyors.

All crossovers, aisles and passageways shall be conspicuously marked by suitable signs, as required by Subpart G, Signs, Signals and Barricades. **(See Figure 14-29)**

Conveyors shall be locked out or otherwise rendered inoperable, and tagged out with a "Do Not Operate" tag during repairs and when operation is hazardous to employees performing maintenance work.

All conveyors in use shall meet the applicable requirements for design, construction, inspection, testing, maintenance and operation, as prescribed in the ANSI B20.1-1957, Safety Code for Conveyors, Cableways, and Related Equipment

Figure 14-29. This illustration shows that all crossovers, aisles and passageways shall be conspicuously marked by suitable signs.

NOTE: CONVEYORS SHALL BE LOCKED OUT OR OTHERWISE RENDERED INOPERABLE, AND TAGGED OUT WITH A "DO NOT OPERATE" TAG DURING REPAIRS AND WHEN OPERATION IS HAZARDOUS TO EMPLOYEES PERFORMING MAINTENANCE WORK.

**CONVEYORS
OSHA 1926.555**

Cranes, Derricks, Hoists, Elevators and Conveyors

Section	Answer	
_____	T	F
_____	T	F
_____	T	F
_____	T	F
_____	T	F
_____	T	F
_____	T	F
_____	T	F
_____	T	F
_____	T	F
_____	T	F
_____	T	F
_____	T	F
_____	T	F
_____	T	F

Cranes, Derricks, Hoists, Elevators and Conveyors

1. The employer shall comply with the manufacturer's specifications and limitations applicable to the operation of any and all cranes and derricks.

2. Rated load capacities, and recommended operating speeds, special hazard warnings or instruction, shall be conspicuously posted in the shop.

3. Periodic inspections of cranes and hoists shall be made by a qualified person at intervals of 1 to 12 months depending on the severity of equipment use and the environment.

4. A thorough, annual inspection of the hoisting machinery shall be made by the supervisor.

5. Any defects found in inspections shall be repaired by the end of the shift.

6. All running ropes (wire ropes) that are in continuous service should be visually inspected once each day.

7. Wire rope shall be taken out of service if there are six randomly distributed broken wires in one lay or three broken wires in one strand in one lay.

8. Nonrotating (rotation-resistant) rope and fiber core rope shall not be used for boom-hoist reeving on cranes.

9. Rope with an independent wire-rope, wire strand core or other temperature-resistant core shall be used when wire is exposed to temperatures in excess of 150°F at the rope.

10. No crane or hoist shall be loaded beyond the specifications of the load-rating chart, except for test purposes.

11. The operator shall not leave the cab or operating controls while a load is suspended, except during scheduled breaks.

12. When it is difficult for the crane operator to judge any clearance distance, a designated worker shall observe the clearance distance and give timely warnings or signals to the operator.

13. An accessible fire extinguisher of 5BC rating, or higher, shall be available at all operator stations or cabs of equipment.

14. For lines rated 50 kV or below, minimum clearance between the lines and any part of the crane or load shall be 15 ft.

15. A boom-length indicator, readable from the operator's station, shall be provided for all telescoping booms.

_____ T F **16.** Telescopic boom cranes shall have either an anti-two-block device or a two-block damage prevention feature at all points of two-blocking.

_____ T F **17.** Each personnel platform shall be equipped with a guardrail system and shall be enclosed from the toeboard to the midrail with solid or expanded metal that has openings no greater than 3/4 in.

_____ T F **18.** Use of cranes or derricks to hoist workers on personnel platforms is prohibited except when it is not possible (or is more hazardous) to erect a conventional means to reach the worksite.

_____ T F **19.** Cranes or derricks used to hoist workers on personnel platforms shall be uniformly level to within 5 percent of grade and located on a firm footing. All outriggers shall be fully extended.

_____ T F **20.** Any defects found during inspections that create a safety hazard shall be corrected before workers are hoisted.

_____ T F **21.** No person shall be allowed to ride on material hoists except for the purposes of inspection and maintenance.

_____ T F **22.** Overhead protective covering of 2 in. planking, 3/4 in. plywood or other solid material of equivalent strength shall be provided on the top of every material hoist cage or platform.

_____ T F **23.** Personnel hoist towers outside the structure shall be enclosed for the full height on the side or sides used for entrance and exit to the structure. At the lowest landing, the enclosure on the sides not used for exit or entrance to the structure shall be enclosed to a height of at least 15 ft.

_____ T F **24.** Hoistway doors or gates shall be not less than 6 ft. 6 in. high and shall be provided with mechanical locks that cannot be operated from the landing side, and shall be accessible only to persons on the car.

_____ T F **25.** Personnel hoists shall be inspected and maintained on a weekly basis. Whenever the hoisting equipment is exposed to winds exceeding 35 miles per hour it shall be inspected and put in operable condition before reuse.

Motor Vehicles, Mechanical Equipment and Marine Operations

This chapter covers general requirements; coverage of vehicles; material handling equipment; pile-driving equipment; access to barges; work surface on barges; and life saving/diving for motor vehicles, mechanical equipment and marine operations.

GENERAL REQUIREMENTS
OSHA 1926.600(a)

All equipment left near a highway or active construction site at night shall be set off with appropriate warning devices. See Chapter 7 of this guide, "Signs, Signals, and Barricades."

Tires mounted on split or locking-ring rims shall be installed, removed or inflated in a safety tire rack, cage or area that offers equivalent protection. **(See Figure 15-1)**

Workers may not work under or between heavy machinery equipment or parts of machinery or equipment suspended from slings, hoists or jacks until the material is blocked or otherwise, supported.

For battery service and charging see Chapter 11 "Battery Servicing" in this guide and the appropriate CFR below. 29 CFR 1926.441; .600(a)(4)

Equipment cabs shall have safety glass or equivalent glass that does not distort the driver's vision.

Bulldozer and scraper blades, loader buckets, dump bodies and similar equipment shall be fully lowered or blocked during repair or when not in use. Except when otherwise required, controls shall be in neutral; motors shall be turned off; and brakes shall be set on all machinery.

When equipment is parked, the parking brake shall be set. On an incline, the wheels also shall be blocked. **(See Figure 15-2)**

Figure 15-1. This illustration shows the requirements for tires mounted on split or locking-ring rims.

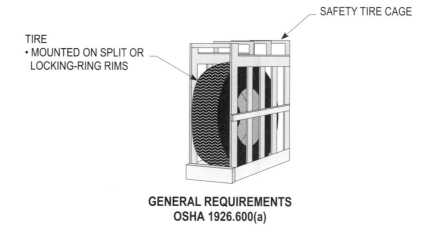

TIRE
• MOUNTED ON SPLIT OR LOCKING-RING RIMS

SAFETY TIRE CAGE

GENERAL REQUIREMENTS
OSHA 1926.600(a)

Figure 15-2. This illustration shows the requirements for bulldozers and scraper blades, loader buckets, dump bodies and similar equipment.

NOTE: WORKERS MAY NOT WORK UNDER OR BETWEEN HEAVY MACHINERY EQUIPMENT OR PARTS OF MACHINERY OR EQUIPMENT SUSPENDED FROM SLINGS, HOISTS OR JACKS UNTIL THE MATERIAL IS BLOCKED OR OTHERWISE, SUPPORTED.

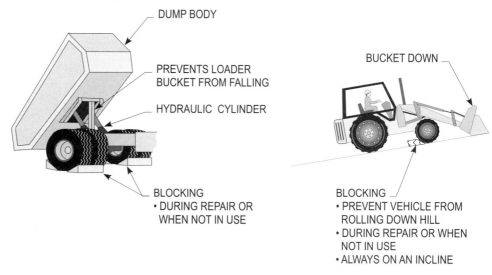

DUMP BODY

PREVENTS LOADER BUCKET FROM FALLING

HYDRAULIC CYLINDER

BLOCKING
• DURING REPAIR OR WHEN NOT IN USE

BUCKET DOWN

BLOCKING
• PREVENT VEHICLE FROM ROLLING DOWN HILL
• DURING REPAIR OR WHEN NOT IN USE
• ALWAYS ON AN INCLINE

GENERAL REQUIREMENTS
OSHA 1926.600(a)

All equipment that is operated near overhead power lines must comply with the clearance requirements of OSHA 1926.550(a)(15) and Chapter 22 of this guide.

Motor vehicles and bidirectional machines in which the rear view is obstructed shall have a reverse signal alarm audible above surrounding background noise or shall be backed up only when an observer signals backing is safe. **(See Figure 15-3)**

Only trained and authorized workers shall be allowed to operate powered industrial trucks.

Equipment used for lifting and handling (other than cranes, hoists, elevators, derricks and conveyors) shall meet applicable requirements of design, construction, stability, testing, maintenance, inspection and operation, as specified in American National Standard Institute (ANSI) B56.1, "Safety Standards for Powered Industrial Trucks."

For information on cranes, derricks and hoists, see Chapter 14 of this guide.

Trucks hauling loads over public roads with material that can be freed by wind conditions shall be securely covered with a tarp or other means to prevent the load material from flying free and causing damage to other vehicles or a hazard to people. All hauled loads, regardless of load material, shall be appropriately secured. (See 49 CFR 393.100 for additional details.)

Vehicles and equipment designed to move slowly (25 mph or less) on public roads shall display a slow-moving traffic identification symbol. **(See Figure 15-4)**

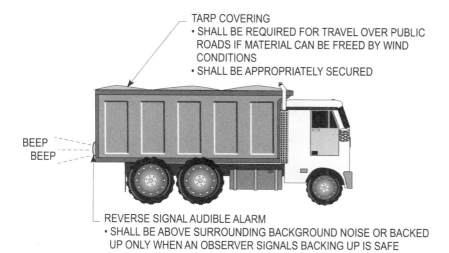

TARP COVERING
• SHALL BE REQUIRED FOR TRAVEL OVER PUBLIC ROADS IF MATERIAL CAN BE FREED BY WIND CONDITIONS
• SHALL BE APPROPRIATELY SECURED

BEEP
BEEP

REVERSE SIGNAL AUDIBLE ALARM
• SHALL BE ABOVE SURROUNDING BACKGROUND NOISE OR BACKED UP ONLY WHEN AN OBSERVER SIGNALS BACKING UP IS SAFE

GENERAL REQUIREMENTS
OSHA 1926.600(a)

Figure 15-3. This illustration shows that a reverse signal alarm is required for motor vehicles and bi-directional machines when rear view is obstructed.

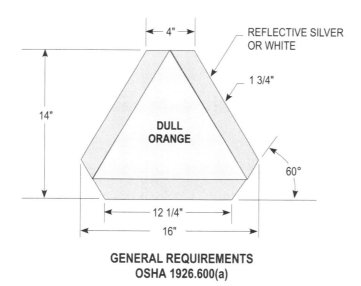

4"

REFLECTIVE SILVER OR WHITE

1 3/4"

14"

DULL ORANGE

60°

12 1/4"

16"

GENERAL REQUIREMENTS
OSHA 1926.600(a)

Figure 15-4. This illustration shows a slow-moving traffic identification symbol for vehicles and equipment designed to move 25 mph or less on public roads.

MOTOR VEHICLES
OSHA 1926.601

The requirements below apply only to motor vehicles operating inside an off-highway construction site.

COVERAGE
OSHA 1926.601(a)

All vehicles shall have a service, emergency and parking brake system; these systems may use the same components. All braking systems shall be in good operating condition.

When visibility conditions warrant additional light, vehicles shall be equipped with at least two headlights and two taillights. Vehicles shall have operable brake lights at all times.

Tools and materials that are carried in the same compartment with workers shall be secured.

Vehicles with cabs shall have windshields and powered wipers. Cracked and broken windshield glass shall be replaced. Vehicles that operate in fog or frost shall have defogging and defrosting devices.

Cracked windshields shall be replaced.

Vehicles whose payload is loaded by a crane, power shovel, loader or similar equipment shall have a cab shield and/or canopy adequate to protect the operator. **(See Figure 15-5)**

Figure 15-5. This illustration shows a cab shield and/or canopy shall be used to protect the operator when payload is loaded by a crane, power shovel, loader or similar equipment.

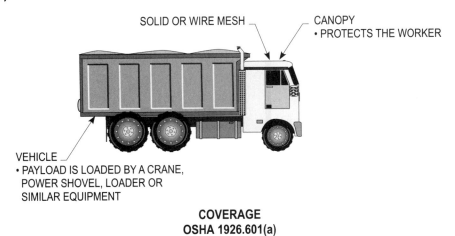

SOLID OR WIRE MESH

CANOPY
• PROTECTS THE WORKER

VEHICLE
• PAYLOAD IS LOADED BY A CRANE, POWER SHOVEL, LOADER OR SIMILAR EQUIPMENT

COVERAGE
OSHA 1926.601(a)

Motor vehicles used to carry personnel shall be equipped with seat belts for all personnel being carried.

Over-the-road vehicles shall have appropriate DOT approved flares, triangles or other warning devices available in the vehicle for appropriate warning if vehicle is disabled.

MATERIAL HANDLING EQUIPMENT
OSHA 1926.602

OSHA Required Training:
1926.602(c)(1)(iv)

Earth-moving equipment includes scrapers, loaders, crawlers or wheel tractors, bulldozers, off-highway trucks, graders, agricultural and industrial tractors and similar equipment.

Seat belts shall be provided on all earth-moving equipment. These belts shall be used by all operators of such equipment.

Seat belts are not required in equipment that is designed for stand up operation only and/orlacks a rollover protection structure (ROPS) or adequate canopy protection.

Construction equipment or vehicles shall not be moved on access roads or grades unless the road or grade has been constructed and maintained to safely accommodate such equipment.

When necessary, an emergency access ramp or berm shall be constructed to restrain runaway vehicles. **(See Figure 15-6)**

NOTE: CONSTRUCTION EQUIPMENT OR VEHICLES SHALL NOT BE MOVED ON ACCESS ROADS OR GRADES UNLESS THE ROAD OR GRADE HAS BEEN CONSTRUCTED AND MAINTAINED TO SAFELY ACCOMODATE SUCH EQUIPMENT.

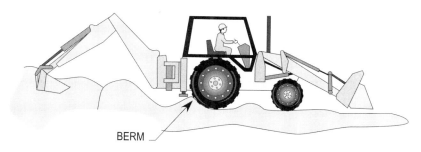

EARTH-MOVING EQUIPMENT
OSHA 1926.602(a)

Figure 15-6. This illustration shows a berm to restrain runaway vehicles.

Brakes on all earth-moving equipment detailed under 29 CFR 1926.602(a)(1) shall have service braking systems capable of stopping and holding the equipment while fully loaded as specified in the Society of Automotive Engineers (SAE) Recommended Practices.

PILE-DRIVING EQUIPMENT
OSHA 1926.603

Overhead protection shall be provided when workers are exposed to falling objects. Protection shall be 2 in. planking or the equivalent and shall not block the view of the operator.

A stop block shall be provided to prevent the hammer from being raised against the head block. A blocking device capable of supporting the hammer weight shall be placed in the leads under the hammer during the entire period that workers are under the hammer.

Cable guards shall be provided across the top of the head block to prevent cables from jumping out of the sheaves. **(See Figure 15-7)**

NOTE 1: A STOP BLOCK SHALL BE PROVIDED TO PREVENT THE HAMMER FROM BEING RAISED AGAINST THE HEAD BLOCK.

NOTE 2: A BLOCKING DEVICE CAPABLE OF SUPPORTING THE HAMMER WEIGHT SHALL BE PLACED IN THE LEADS UNDER THE HAMMER DURING THE ENTIRE PERIOD THAT WORKERS ARE UNDER THE HAMMER.

Figure 15-7. This illustration shows that cable guards shall be provided across the top of the head block.

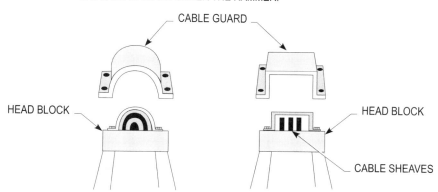

PILE-DRIVING EQUIPMENT
OSHA 1926.603

When a baiter pile is driven on an incline, the leads shall be stabilized.

Ladders will have adequate rings or similar attachments through which fixed leads can pass to attach to the loft worker's safety belt harness.

Pile-driver workers shall wear hearing protection when operations exceed the noise-time weighted-average exposure from high-impact noise. See Chapter 4 of this guide for details on hearing protection.

Steam hoses leading to a hammer or jet pipe shall be securely attached to the hammer with a tiedown that consists of at least a 1/4 in. chain or cable of adequate length. Safety chains or the equivalent shall be provided across each hose connection. **(See Figure 15-8)**

Figure 15-8. This illustration shows that steam hoses leading to a hammer or jet pipe shall be securely attached to the hammer with a tiedown.

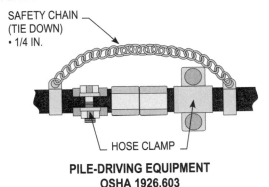

NOTE 1: STEAM HOSES LEADING TO A HAMMER OR SET PIPE SHALL BE SECURELY ATTACHED TO THE HAMMER WITH A TIE DOWN THAT CONSIST OF AT LEAST 1/4 IN. CHAIN OR CABLE OF ADEQUATE LENGTH.

NOTE 2: SAFETY CHAINS OR THE EQUIPMENT SHALL BE PROVIDED ACROSS EACH HOSE CONNECTION.

NOTE 3: STEAM-LINE CONTROLS SHALL BE EQUIPPED WITH TWO SHUTOFF VALVES, ONE OF WHICH SHALL BE A QUICK ACTING LEVER TYPE, THAT IS WITHIN REACH OF THE HAMMER OPERATOR.

SAFETY CHAIN (TIE DOWN) • 1/4 IN.

HOSE CLAMP

PILE-DRIVING EQUIPMENT
OSHA 1926.603

Steam-line controls shall be equipped with two shutoff valves, one of which shall be a quick-acting lever type, which is within easy reach of the hammer operator.

Guys, outriggers, thrustouts or counterbalances shall be provided as necessary to stabilize a pile-driver rig.

Engineers and winchers shall accept signals from a designated signaler only.

When piles are driven in an excavated pit, the pit walls shall be sloped to the angle of repose or sheet-piled and braced.

When steel tube piles are being "blown out," workers shall be out of range of falling material.

Pile-driving operations shall be suspended when driven piles are cut off except in a case where the cutting operations are located at least twice the length of the longest pile from the driver.

While installing piles with jacking equipment under existing structures, all access pits shall be provided with ladders and bulkheaded curbs to prevent material from falling into the pit.

MARINE OPERATIONS AND EQUIPMENT
OSHA 1926.605

The following shall be considered for marine operations and equipment:

- Access to barges
- Work surfaces on barges
- Life saving/diving

ACCESS TO BARGES

Vehicle ramps to or between barges shall be of adequate strength, have side boards, be well maintained, and be properly secured. **(See Figure 15-9)**

NOTE 1: A SAFE MEANS, WALKWAY OR RAMP SHALL BE PROVIDED FOR WORKERS SO THEY CAN SAFELY STEP TO OR FROM A WHARF, FLOAT, OR RIVER TOWBOAT.

NOTE 2: WHEN THE UPPER END OF THE ACCESS TO THE BARGE RESTS ON OR IS FLUSH WITH THE TOP OF A BULKWALK, SUBSTANTIAL STEPS EQUIPPED WITH A HANDRAIL AT LEAST 33 IN. HIGH SHALL BE INSTALLED BETWEEN THE TOP OF THE BULKWALK AND THE DECK.

Figure 15-9. This illustration shows the requirements for access to barges.

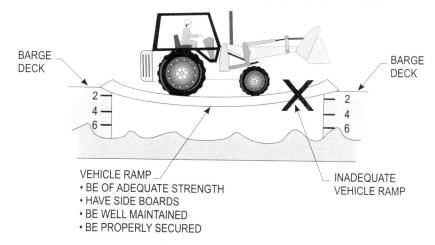

BARGE DECK

BARGE DECK

VEHICLE RAMP
- BE OF ADEQUATE STRENGTH
- HAVE SIDE BOARDS
- BE WELL MAINTAINED
- BE PROPERLY SECURED

INADEQUATE VEHICLE RAMP

ACCESS TO BARGES
OSHA 1926.605

A safe means, walkway or ramp shall be provided for workers so they can safely step to or from a wharf, float, barge or river towboat.

A Jacob's ladder shall be of the double rung or flat tread type, be well maintained and secured, and hang without slack from its lashings or be pulled up entirely.

When the upper end of the access to the barge rests on or is flush with the top of a bulwark, substantial steps equipped with a handrail at least 33 in. high shall be installed between the top of the bulwark and the deck.

Gangways shall be unobstructed and access shall be adequately lit.

Access to a barge shall be located so that loads do not pass over workers.

WORK SURFACES ON BARGES

Workers shall not walk along the sides of covered lighters or barges with coamings more than 5 ft. high unless a walkway 3 ft. wide, a grab rail or a taut handline is provided.

Decks and work surfaces shall be kept in safe condition.

Workers shall not pass fore and aft over or around deckloads unless there is a safe passageway.

If it is necessary to stand at the outboard or inboard edge of the deckload where there is less than 2 ft. of bulwark, rail or coamings, workers shall be protected from falling off the deckload.

LIFE SAVING/DIVING

Each barge shall have at least one life preserver (buoy or ring), approved by the U.S. Coast Guard, with at least 90 ft. of line attached. Each barge also shall have at least one portable ladder or permanent ladder that will reach from the top of the apron to the surface of the water. If this equipment is not available at the pier, the employer shall furnish it.

Workers walking or working on unguarded decks of barges shall wear life jackets approved by the US Coast Guard.

Commercial diving shall be done in accordance with Subpart T of 29 CFR 1910.

Motor Vehicles, Mechanical Equipment and Marine Operations

Section	Answer	
_____	T	F
_____	T	F
_____	T	F
_____	T	F
_____	T	F
_____	T	F
_____	T	F
_____	T	F
_____	T	F
_____	T	F

Motor Vehicles, Mechanical Equipment and Marine Operations

1. All equipment left near a highway or active construction site at night shall be set off with appropriate warning devices.

2. Bulldozer and scraper blades, loader buckets, dump bodies and similar equipment shall be fully lowered or blocked during repair or when not in use. Except when otherwise required, controls shall be in neutral; motors shall be turned off; and brakes shall be set on all machinery.

3. Seat belts are optional on all earth-moving equipment.

4. When equipment is parked, the parking brake shall be set. On an incline, the wheels also shall be blocked.

5. Only trained and authorized workers shall be allowed to operate powered industrial trucks.

6. Vehicles and equipment designed to move slowly (50 mph or less) on public roads shall display a slow-moving traffic identification symbol.

7. Motor vehicles used to carry personnel shall be equipped with seat belts for at least 2 of the occupants.

8. Overhead protection shall be provided when pile-driving equipment workers are exposed to falling objects. Protection shall be 2 in. planking or the equivalent and shall not block the view of the operator.

9. Steam hoses leading to a hammer or jet pipe shall be securely attached to the hammer with a tiedown that consists of at least a 3/4 in. chain or cable of adequate length. Safety chains or the equivalent shall be provided across each hose connection.

10. When the upper end of the access to the barge rests on or is flush with the top of a bulwark, substantial steps equipped with a handrail at least 33 in. high shall be installed between the top of the bulwark and the deck.

16

Excavations

Trenching and excavation procedures are performed thousands of times a day across the United States. Unfortunately, many workers are killed in trenching accidents each year. Contractors and construction laborers should understand the laws and regulations applicable to trenching and excavation occupations. These statutes are in effect for the express purpose of protecting those who work in trenching and excavation situations.

SOIL MECHANICS

In trenching and excavation practices, "soil" is defined as any material removed from the ground to form a hole, trench or cavity for the purpose of working below the earth's surface. This material is most often weathered rock and humus known as clays, silts and loams, but also can be gravel, sand and rock. It is necessary to know the characteristics of the soil at the particular job site. Soils information is used by contractors and engineers who are trained to identify the proper safety protective devices or procedures needed for each situation. OSHA requires a "competent person" to be in charge of all excavation and trenching activities at a job site.

Soil is an extremely heavy material, and may weigh more than 100 pounds per cubic foot (pcf). A cubic yard of soil (3 ft x 3 ft x 3 ft), which contains 27 cubic feet of material, may weigh more than 2700 pounds (lbs). That is nearly one and a half tons (the equivalent weight of a car) in a space less than the size of the average office desk. Furthermore, wet soil, rocky soil or rock is usually heavier. The human body cannot support such heavy loads without being injured. **(See Figure 16-1)**

From a soil mechanics point of view, one can visualize the soil as a series of multiple columns of soil blocks, with the blocks piled one on top of the other. In the soil column shown in Figure 16-1, each soil block measures one foot square, weighs approximately 100 lbs, and supports the weight of all of the blocks above. This means that a block sitting at a five-foot depth supports its own weight and the combined weight of the four blocks resting on it. The combined weight of this column is 500 lbs spread over a one-square-foot area; 500 pounds per square foot (psf). This five- block column constitutes a 500 pound force exerted vertically on whatever lies below.

A column of soil exerts not only a vertical force, but also a horizontal force in all outward directions. The outward force is equal to one-half the vertical force.

For example, the five-block column illustrated in Figure 16-1 has a downward vertical force of 500 lbs at the base of soil block number five. The horizontal force pushing out from the base of that same block is half of 500 lbs, or 250 lbs, in all outward directions. As the weight of the column increases, the soil blocks at the bottom of the column theoretically have a tendency to compress and spread outward. In undisturbed soil conditions, this process is stopped by the presence of the surrounding columns pushing back with equal pressure. These hypothetical columns press against each other, maintaining an equilibrium. Therefore, the horizontal pressures of all the columns are balanced, producing a stable relationship.

Figure 16-1. This illustration shows that it is necessary to know the characteristics of the soil at the particular job site.

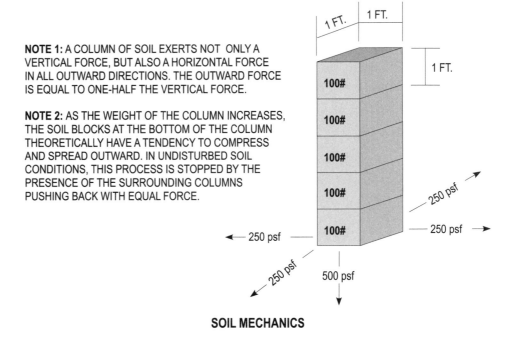

NOTE 1: A COLUMN OF SOIL EXERTS NOT ONLY A VERTICAL FORCE, BUT ALSO A HORIZONTAL FORCE IN ALL OUTWARD DIRECTIONS. THE OUTWARD FORCE IS EQUAL TO ONE-HALF THE VERTICAL FORCE.

NOTE 2: AS THE WEIGHT OF THE COLUMN INCREASES, THE SOIL BLOCKS AT THE BOTTOM OF THE COLUMN THEORETICALLY HAVE A TENDENCY TO COMPRESS AND SPREAD OUTWARD. IN UNDISTURBED SOIL CONDITIONS, THIS PROCESS IS STOPPED BY THE PRESENCE OF THE SURROUNDING COLUMNS PUSHING BACK WITH EQUAL FORCE.

SOIL MECHANICS

TRENCH FAILURE

When a trench is excavated, the stable relationship described in the previous section no longer exists. The horizontal pressure on the soil blocks along the trench wall is no longer in equilibrium, and a block may not be able to support its weight and the weight of any blocks above. At the point where the soil can no longer withstand the pressure, the wall will shear and break away from its stable position. The first failure occurs as the bottom of the wall moves into the trench. This movement creates an undercut area at the base of the trench as soil material along the wall falls into the trench. Often there is a second movement in which more of the wall material erodes. Finally, the erosion at the base of the trench leaves the upper part of the column supported only by cohesion to the columns around it, and more soil from the column will soon fall into the excavation. Many rescue attempts are unsuccessful because rescuers attempt to save victims before the second and third failures take place, often trapping the would-be rescuers along with the first victims.

INTRODUCTION

The Occupational Safety and Health Administration (OSHA) issued its first Excavation and Trenching Standard in 1971 to protect workers from excavation hazards. Since then, OSHA has amended the standard several times to increase worker protection and to reduce the frequency and severity of excavation accidents and injuries. Despite these efforts, excavation-related accidents resulting in injuries and fatalities continue to occur.

To better assist excavation firms and contractors, OSHA completely updated the existing standard to simplify many of the existing provisions, add and clarify definitions, eliminate duplicate provisions and ambiguous language, and give employers added flexibility in providing protection for employees. The standard was effective as of March 5, 1990.

In addition, the standard provides several new appendices. One appendix provides a consistent method of soil classification. Others provide sloping and benching requirements, pictorial examples of shoring and shielding devices, timber tables, hydraulic shoring tables and selection charts that provide a graphic summary of the requirements contained in the standard.

This discussion highlights the requirements in the updated standard for excavation and trenching operations, provides methods for protecting employees against cave-ins and describes safe work practices for employees.

SCOPE AND APPLICATION
OSHA 1926.650

OSHA's revised rule applies to all open excavations made in the earth's surface, which includes trenches.

According to the OSHA construction safety and health standards, a trench is referred to as a narrow excavation made below the surface of the ground in which the depth is greater than the width-the width not exceeding 15 ft. An excavation is any man-made cut, cavity, trench or depression in the earth's surface formed by earth removal. This can include excavations for anything from cellars to highways.

SPECIFIC EXCAVATION REQUIREMENTS
OSHA 1926.651

The following shall be considered for specific excavation requirements:

- Surface encumbrances
- Underground installations
- Access and egress
- Exposure to vehicular traffic
- Exposure to falling loads
- Hazardous atmospheres
- Water accumulation
- Stability of adjacent structures
- Inspections
- Fall protection

SURFACE ENCUMBRANCES
OSHA 1926.651(a)

All surface encumbrances that are located so as to create a hazard to employees shall be removed or supported, as necessary, to protect employees.

UNDERGROUND INSTALLATIONS
OSHA 1926.651(b)

Safety Tip: Employers shall determine the location of utility installations before an excavation can begin.

Before any excavation actually begins, the standard requires the employer to determine the estimated location of utility installations—sewer, telephone, fuel, electric, water lines or any other underground installations—that may be encountered during digging. Also, before starting the excavation, the contractor shall contact the utility companies or owners involved and inform them, within established or customary local response times, of the proposed work. The contractor shall also ask the utility companies or owners to find the exact location of the underground installations. If they cannot respond within 24 hours (unless the period required by state or local law is longer), or if they cannot find the exact location of the utility installations, the contractor may proceed with caution. To find the exact location of underground installations, workers shall use safe and acceptable means. If underground installations are exposed, OSHA regulations also require that they be removed, protected or properly supported.

When all the necessary specific information about the job site is assembled, the contractor is ready to determine the amount, kind and cost of the safety equipment needed. A careful inventory of the safety items on hand should be made before deciding what additional safety material shall be acquired. No matter how many trenching, shoring and backfilling jobs have been done in the past, each job should be approached with the utmost care and preparation.

ACCESS AND EGRESS
OSHA 1926.651(c)

Safety Tip: Employers shall provide safe access and egress to all excavations.

Under the standard, the employer shall provide safe access and egress to all excavations. According to OSHA regulations, when employees are required to be in trench excavations 4 ft. deep or more, adequate means of exit, such as ladders, steps, ramps or other safe means of egress, shall be provided and be within 25 ft. of lateral travel. If structural ramps are used as a means of access or egress, they shall be designed by a competent person if used for employee access or egress, or a competent person qualified in structural design if used by vehicles. Also, structural members used for ramps or runways must be uniform in thickness and joined in a manner to prevent tripping or displacement.

OSHA Required Training: 1926.651(c)(1)(i)

EXPOSURE TO VEHICULAR TRAFFIC
OSHA 1926.651(d)

When employees are exposed to vehicular traffic they shall be provided with and shall wear warning vests.

EXPOSURE TO FALLING LOADS
OSHA 1926.651(e), (j)

In addition to cave-in hazards and secondary hazards related to cave-ins, there are other hazards from which workers shall be protected during excavation-related work. These hazards include exposure to falls, falling loads and mobile equipment. To protect employees from these hazards, OSHA requires the employer to take the following precautions:

- Keep materials or equipment that might fall or roll into an excavation at least 2 ft. from the edge of excavations, or have retaining devices or both.

- Provide warning systems such as mobile equipment, barricades, hand or mechanical signals or stop logs, to alert operators of the edge of an excavation. If possible, keep the grade away from the excavation.

- Provide scaling to remove loose rock or soil or install protective barricades and other equivalent protection to protect employees against falling rock, soil or materials.

- Prohibit employees from working on faces of sloped or benched excavations at levels above other employees unless employees at lower levels are adequately protected from the hazard of falling, rolling or sliding material or equipment.

- Prohibit employees under loads that are handled by lifting or digging equipment. To avoid being struck by any spillage or falling materials, require employees to stand away from vehicles being loaded or unloaded. If cabs of vehicles provide adequate protection from falling loads during loading and unloading operations, the operators may remain in them.

HAZARDOUS ATMOSPHERES
OSHA 1926.651(g)

Under this provision, a competent person shall test excavations greater than 4 ft. in depth as well as ones where oxygen deficiency or a hazardous atmosphere exists or could reasonably be expected to exist, before an employee enters the excavation. If hazardous conditions exist, controls such as proper respiratory protection or ventilation shall be provided. Also, controls used to reduce atmospheric contaminants to acceptable levels shall be tested regularly.

Safety Tip: A competent person shall verify if an oxygen deficient or hazardous atmosphere exists.

Where adverse atmospheric conditions may exist or develop in an excavation, the employer also shall provide and ensure that emergency rescue equipment, (e.g., breathing apparatus, a safety harness and line, basket stretcher, etc.) is readily available. This equipment shall be attended when used.

When an employee enters bell-bottom pier holes and similar deep and confined footing excavations, the employee shall wear a harness with a lifeline. The lifeline shall be securely attached to the harness and shall be separate from any line used to handle materials. Also, while the employee wearing the lifeline is in the excavation, an observer shall be present to ensure that the lifeline is working properly and to maintain communication with the employee.

WATER ACCUMULATION
OSHA 1926.651(h)

The standard prohibits employees from working in excavations where water has accumulated or is accumulating unless adequate protection has been taken. If water removal equipment is used to control or prevent water from accumulating, the equipment and operations of the equipment shall be monitored by a competent person to ensure proper use.

OSHA Required Training:
1926.503(a)(1) and (2)(ii) thru (vii)

OSHA standards also require that diversion ditches, dikes or other suitable means be used to prevent surface water from entering an excavation and to provide adequate drainage of the area adjacent to the excavation. Also, a competent person shall inspect excavations subject to runoffs from heavy rains.

STABILITY OF ADJACENT STRUCTURES
OSHA 1926.651(i)

OSHA Required Training:
1926.651(i)(1)
1926.651(i)(2)(iii)
1926.651(i)(2)(iv)

The standard requires the employer to provide support systems such as shoring, bracing or underpinning to ensure the stability of adjacent structures such as buildings, walls, sidewalks or pavements.

The standard prohibits excavation below the level of the base or footing of any foundation or retaining wall unless (1) a support system such as underpinning is provided, (2) the excavation is in stable rock, or (3) a registered professional engineer determines that the structure is sufficiently removed from the excavation and that excavation will not pose a hazard to employees.

Excavations under sidewalks and pavements are also prohibited unless an appropriately designed support system is provided or another effective method is used.

INSPECTIONS
OSHA 1926.651(k)

Safety Tip: A competent person shall inspect excavations on a daily basis.

The standard requires that a competent person inspect, on a daily basis, excavations and the adjacent areas for possible cave-ins, failures of protective systems and equipment, hazardous atmospheres or other hazardous conditions. If these conditions are encountered, exposed employees shall be removed from the hazardous area until the necessary safety precautions have been taken. Inspections are also required after natural (e.g., heavy rains) or man-made events such as blasting that may increase the potential for hazards.

OSHA Required Training:
1926.651(k)(1) and (2)

Larger and more complex operations should have a full-time safety official who makes recommendations to improve the implementation of the safety plan. In a smaller operation, the safety official may be part-time and usually will be a supervisor.

Supervisors are the contractor's representatives on the job. Supervisors should conduct inspections, investigate accidents and anticipate hazards. They should ensure that employees receive on-the-job safety and health training. They should also review and strengthen overall safety and health precautions to guard against potential hazards, get the necessary worker cooperation in safety matters and make frequent reports to the contractor.

It is important that managers and supervisors set the example for safety at the job site. It is essential that when visiting the job site, all managers, regardless of status, wear the prescribed personal protective equipment such as safety shoes, safety glasses, hard hats and other necessary gear (see CFR 1926.100 and 102 and also Chapter 5 of this guide).

Employees shall also take an active role in job safety. The contractor and supervisor should make certain that workers have been properly trained in the use and fit of the prescribed protective gear and equipment, that they are wearing and using the equipment correctly, and that they are using safe work practices.

FALL PROTECTION
OSHA 1926.651(l)

If it is necessary for employees or equipment to pass over areas that have been excavated, walkways or bridges with guard rails shall be provided.

REQUIREMENTS FOR PROTECTIVE SYSTEMS
OSHA 1926.652

The following requirements shall be considered for protective systems:

- Protection for employees in excavations
- Design of sloping and benching systems

PROTECTION FOR EMPLOYEES IN EXCAVATIONS
OSHA 1926.652(a)

Excavation workers are exposed to many hazards, but the chief hazard is danger of cave-ins. OSHA requires that in all excavations employees exposed to potential cave-ins shall be protected by sloping, or benching the sides of the excavation; supporting the sides of the excavation or placing a shield between the side of the excavation and the work area.

DESIGN OF SLOPING AND BENCHING SYSTEMS
OSHA 1926.652(b)

Designing a protective system can be complex because of the number of factors involved - soil classification, depth of cut, water content of soil, changes due to weather and climate or other operations in the vicinity. The standard, however, provides several different methods and approaches (four for sloping and four for shoring, including the use of shields) for designing protective systems that can be used to provide the required level of protection against cave-ins. One method of ensuring the safety and health of workers in an excavation is to slope the sides to an angle not steeper than one and one-half horizontal to one vertical (34 degrees measured from the horizontal). These slopes shall be excavated to form configurations that are in accordance with those for Type C soil found in Appendix B of the standard. A slope of this gradation or less is considered safe for any type of soil.

EXCAVATION MADE IN TYPE C SOIL

All simple slope excavations 20 ft. or less in depth shall have a maximum allowable slope of 1 1/2:1. A second design method, which can be applied for both sloping and shoring, involves using tabulated data, such as tables and charts, approved by a registered professional engineer. This data shall be in writing and shall include sufficient explanatory information to enable the user to make a selection, including the criteria for determining the selection and the limits on the use of the data. **(See Figure 16-2)**

At least one copy of the information, including the identity of the registered professional engineer who approved the data, shall be kept at the worksite during construction of the protective system. Upon completion of the system, the data may be stored away from the job site, but a copy shall be made available, upon request, to the Assistant Secretary of Labor for OSHA. Contractors also may use a trench box or shield that is either designed or approved by a registered professional engineer or is based on tabulated data prepared or approved by a registered professional engineer. Timber, aluminum or other suitable materials may also be used. OSHA standards permit the use of a trench shield (also known as a welder's hut) as long as the protection it provides is equal to or greater than the protection that would be provided by the appropriate shoring system. **(See Figure 16-3)**

Figure 16-2. This illustration shows that all simple slope excavations 20 ft. or less in depth shall have a maximum allowable slope of 1 1/2:1.

NOTE: A SECOND DESIGN METHOD, WHICH CAN BE APPLIED FOR BOTH SLOPING AND SHORING INVOLVES USING TABULATED DATA, SUCH AS TABLES AND CHARTS, APPROVED BY A REGISTERED PROFESSIONAL ENGINEER.

20' MAX.

1

1 1/2

DESIGN OF SLOPING AND BENCHING SYSTEMS
OSHA 1926.652(b)

Figure 16-3. This illustration shows that a trench shield shall be permitted to be used as long as it provides protection equal to or greater than the protection that would be provided by the appropriate shoring system.

NOTE: OSHA STANDARDS PERMIT THE USE OF A TRENCH SHIELD AS LONG AS THE PROTECTION IT PROVIDES IS EQUAL TO OR GREATER THAN THE PROTECTION THAT WOULD BE PROVIDED BY THE APPROPRIATE SHORING SYSTEM.

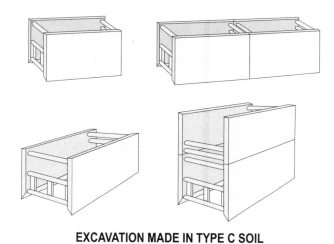

EXCAVATION MADE IN TYPE C SOIL
OSHA 1926.652(b)

TRENCH SHIELDS

The employer is free to choose the most practical design approach for any particular circumstance. Once an approach has been selected, however, the required performance criteria shall be met by that system. The standard does not require the installation and use of a protective system when an excavation (1) is made entirely in stable rock or (2) is less than 5 ft. deep and a competent person has examined the ground and found no indication of a potential cave-in.

INSTALLATION AND REMOVAL OF PROTECTIVE SYSTEMS

The standard requires the following procedures for the protection of employees when installing support systems:

• Securely connect members of support systems,

• Safely install support systems,

• Never overload members of support systems, and

• Install other structural members to carry loads imposed on the support system when temporary removal of individual members is necessary.

In addition, the standard permits excavation of 2 ft. or less below the bottom of the members of a support or shield system of a trench if (1) the system is designed to resist the forces calculated for the full depth of the trench and (2) there are no indications, while the trench is open, of a possible cave-in below the bottom of the support system. Also, the installation of support systems shall be closely coordinated with the excavation of trenches.

As soon as work is completed, the excavation should be back-filled as the protective system is dismantled. After the excavation has been cleared, workers should slowly remove the protective system from the bottom up, taking care to release members slowly.

MATERIALS AND EQUIPMENT

The employer is responsible for the safe condition of materials and equipment used for protective systems. Defective and damaged materials and equipment can result in the failure of a protective system and cause excavation hazards.

To avoid possible failure of a protective system, the employer shall ensure that (1) materials and equipment are free from damage or defects, (2) manufactured materials and equipment are used and maintained in a manner consistent with the recommendations of the manufacturer and in a way that will prevent employee exposure to hazards and (3) while in operation, damaged materials and equipment are examined by a competent person to determine if they are suitable for continued use. If materials and equipment are not safe for use, they shall be removed from service. These materials cannot be returned to service without the evaluation and approval of a registered professional engineer.

SUMMARY

Trenching and excavation work presents serious risks to all workers involved. The greatest risk, and one of primary concern, is that of a cave-in. Furthermore, when cave-in accidents occur, they are much more likely to result in worker fatalities than other excavation-related accidents. Strict compliance, however, with all sections of the standard will prevent or greatly reduce the risk of cave-ins as well as other excavation-related accidents.

Excavations

Section	Answer	
_____	T	F
_____	T	F
_____	T	F
_____	T	F
_____	T	F
_____	T	F
_____	T	F
_____	T	F
_____	T	F
_____	T	F

Excavations

1. Soil is an extremely heavy material, and may weigh more than 100 pounds per cubic foot (pcf).

2. According to the OSHA construction safety and health standards, a <u>trench</u> is referred to as a narrow excavation made below the surface of the ground in which the depth is greater than the width-the width not exceeding 10 ft.

3. Under the OSHA standards, the employer shall provide safe access and egress to all excavations. According to OSHA regulations, when employees are required to be in trench excavations 4 ft. deep or more, adequate means of exit, such as ladders, steps, ramps or other safe means of egress, shall be provided and be within 20 ft. of lateral travel.

4. Materials or equipment that might fall or roll into an excavation shall be kept at least 3 ft. from the edge of excavations, or have retaining devices, or both.

5. A competent person shall test excavations greater than 4 ft. in depth as well as ones where oxygen deficiency or a hazardous atmosphere exists or could reasonably be expected to exist, before an employee enters the excavation.

6. A competent person shall inspect, on a daily basis, excavations and the adjacent areas for possible cave-ins, failures of protective systems and equipment, hazardous atmospheres or other hazardous conditions.

7. One method of ensuring the safety and health of workers in an excavation is to slope the sides to an angle not steeper than one and one-half horizontal to one vertical (34 degrees measured from the horizontal).

8. OSHA prohibits excavation below the level of the base or footing of any foundation or retaining wall unless (1) a support system such as underpinning is provided, (2) the excavation is in stable rock or (3) a registered excavation engineer determines that the structure is sufficiently removed from the excavation and that excavation will not pose a hazard to employees.

9. A competent person shall inspect, on a weekly basis, excavations and the adjacent areas for possible cave-ins, failures of protective systems and equipment, hazardous atmospheres or other hazardous conditions.

10. OSHA requires that in all excavations employees exposed to potential cave-ins shall be protected by sloping, or benching the sides of the excavation; supporting the sides of the excavation or placing a shield between the side of the excavation and the work area.

_____   T   F

_____   T   F

_____   T   F

_____   T   F

_____   T   F

11. Tabulated data that can be applied for both sloping and shoring involves using tables and charts approved by a registered professional engineer. At least one copy of the information, including the identity of the registered professional engineer who approved the data, shall be kept at the main office during construction of the protective system.

12. OSHA standards permit the use of a trench shield (also known as a welder's hut) as long as the protection it provides is equal to or greater than the protection that would be provided by the appropriate shoring system.

13. The standard does not require the installation and use of a protective system when an excavation (1) is made entirely in stable rock or (2) is less than 6 ft. deep and a competent person has examined the ground and found no indication of a potential cave-in.

14. OSHA permits excavation of 2 ft. or less below the bottom of the members of a support or shield system of a trench if (1) the system is designed to resist the forces calculated for the full depth of the trench and (2) there are no indications, while the trench is open, of a possible cave-in below the bottom of the support system.

15. If materials and equipment used for protective systems are not safe for use, they shall be removed from service. These materials cannot be returned to service without the evaluation and approval of the shop mechanic.

Concrete and Masonry Construction

The Occupational Safety and Health Administration's standard, Subpart Q, Concrete and Masonry Construction, Title 29 of the Code of Federal Regulations (CFR), Part 1926.700 through 706, sets forth requirements with which construction employers shall comply to protect construction workers from accidents and injuries resulting from the premature removal of formwork, the failure to brace masonry walls, the failure to support precast panels, the inadvertent operation of equipment and the failure to guard reinforcing steel.

SCOPE AND APPLICATION
OSHA 1926.700

The standard, Subpart Q, prescribes performance-oriented requirements designed to help protect all construction workers from the hazards associated with concrete and masonry construction operations at construction, demolition, alteration or repair worksites. Other relevant provisions in both general industry and construction standards (29 CFR Parts 1910 and 1926) also apply to these operations.

GENERAL REQUIREMENTS
OSHA 1926.701

The following general requirements shall be considered for concrete and masonry construction:

- Construction loads
- Reinforcing steel
- Post-tensioning operations
- Concrete buckets
- Working under loads
- Personal protective equipment

CONSTRUCTION LOADS
OSHA 1926.701(a)

Employers shall not place construction loads on a concrete structure or portion of a concrete structure unless the employer determines, based on information received from a person who is qualified in structural design, that the structure or portion of the structure is capable of supporting the intended loads.

REINFORCING STEEL
OSHA 1926.701(b)

All protruding reinforcing steel, onto and into which employees could fall, shall be guarded to eliminate the hazard of impalement.

POST-TENSIONING OPERATIONS
OSHA 1926.701(c)

Employees (except those essential to the post-tensioning operations) shall not be permitted to be behind the jack during tensioning operations.

Signs and barriers shall be erected to limit employee access to the post-tensioning area during tensioning operations.

CONCRETE BUCKETS
OSHA 1926.701(d)

Employees shall not be permitted to ride concrete buckets.

WORKING UNDER LOADS
OSHA 1926.701(e)

Employees shall not be permitted to work under concrete buckets while the buckets are being elevated or lowered into position.

To the extent practicable, elevated concrete buckets shall be routed so that no employee or the fewest employees possible are exposed to the hazards associated with falling concrete buckets.

PERSONAL PROTECTIVE EQUIPMENT
OSHA 1926.701(f)

Employees shall not be permitted to apply a cement, sand and water mixture through a pneumatic hose unless they are wearing protective head and face equipment.

Employees shall not be permitted to place or tie reinforcing steel more than 6 ft. above any adjacent working surfaces unless they are protected by the use of a safety belt or equivalent fall protection meeting the criteria in OSHA standards on Personal Protective and Life Saving Equipment (29 CFR 1926 Subpart E).

EQUIPMENT AND TOOLS
OSHA 1926.702

The standard also includes requirements for the following equipment and operations:

· Bulk cement storage,
· Concrete mixers,
· Power concrete trowels,
· Concrete buggies,
· Concrete pumping systems,
· Concrete buckets,
· Tremies,
· Bull floats,
· Masonry saws and
· Lockout/tagout procedures.

CAST-IN-PLACE CONCRETE
OSHA 1926.703

The following shall be considered for cast-in-place concrete:

· General requirements for formwork
· Shoring and reshoring
· Vertical slip forms
· Reinforcing steel
· Removal of formwork

GENERAL REQUIREMENTS FOR FORMWORK
OSHA 1926.703(a)

Formwork shall be designed, fabricated, erected, supported, braced and maintained so that it will be capable of supporting without failure all vertical and lateral loads that might be applied to the formwork. As indicated in the Appendix to the standard, formwork that is designed, fabricated, erected, supported, braced and maintained in conformance with Sections 6 and 7 of the American National Standard for Construction and Demolition Operations - Concrete and Masonry Work (ANSI A10.9-1983) also meets the requirements of this paragraph. **(See Figure 17-1)**

DRAWINGS OR PLANS

Drawings and plans, including all revisions for the jack layout, formwork (including shoring equipment), working decks and scaffolds, shall be available at the jobsite.

SHORING AND RESHORING
OSHA 1926.703(b)

All shoring equipment (including equipment used in reshoring operations) shall be inspected prior to erection to determine that the equipment meets the requirements specified in the formwork drawings.

Damaged shoring equipment shall not be used for shoring. Erected shoring equipment shall be inspected immediately prior to, during and immediately after concrete placement.

OSHA Required Training:
1926.701(b)(8)(i)

OSHA Required Training:
1926.701(a)

Shoring equipment that is found to be damaged or weakened after erection shall be immediately reinforced.

If single-post shores are used one on top of another (tiered), then additional shoring requirements shall be met. The shores must be as follows:

· Designed by a qualified designer, and the erected shoring shall be inspected by an engineer qualified in structural design,

· Vertically aligned,

· Spliced to prevent misalignment and

· Adequately braced in two mutually perpendicular directions at the splice level. Each tier also must be diagonally braced in the same two directions.

Adjustment of single-post shores to raise formwork must not be made after the placement of concrete.

Reshoring shall be erected, as the original forms and shores are removed, whenever the concrete is required to support loads in excess of its capacity.

Figure 17-1. This illustration shows that formwork shall be designed, fabricated, erected, supported, braced and maintained.

NOTE: DRAWINGS AND PLANS, INCLUDING ALL REVISIONS FOR THE JACK LAYOUT, FORMWORK (INCLUDING SHORING EQUIPMENT), WORKING DECKS AND SCAFFOLDS, SHALL BE AVAILABLE AT THE JOBSITE.

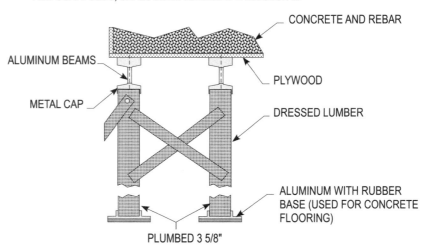

GENERAL REQUIREMENTS FOR FORMWORK
OSHA 1926.703(a)

VERTICAL SLIP FORMS
OSHA 1926.703(c)

The steel rods or pipes on which jacks climb or by which the forms are lifted shall be (1) specifically designed for that purpose and (2) adequately braced where not encased in concrete. Forms shall be designed to prevent excessive distortion of the structure during the jacking operation. Jacks and vertical supports shall be positioned in such a manner that the loads do not exceed the rated capacity of the jacks.

The jacks or other lifting devices shall be provided with mechanical dogs or other automatic holding devices to support the slip forms whenever failure of the power supply or lifting mechanisms occurs.

The form structure shall be maintained within all design tolerances specified for plumbness during the jacking operation.

The predetermined safe rate of lift shall not be exceeded.

All vertical slip forms shall be provided with scaffolds or work platforms where employees are required to work or pass. **(See Figure 17-2)**

NOTE 1: JACKS AND VERTICAL SUPPORTS SHALL BE POSITIONED IN SUCH A MANNER THAT THE LOADS DO NOT EXCEED THE RATED CAPACITY OF THE JACKS.

NOTE 2: A FORM STRUCTURE SHALL BE MAINTAINED WITHIN ALL DESIGN TOLERANCES SPECIFIED FOR PLUMBNESS DURING THE JACKING OPERATION.

Figure 17-2. This illustration shows that all vertical slip forms shall be provided with scaffolds or work platforms where employees are required to work or pass.

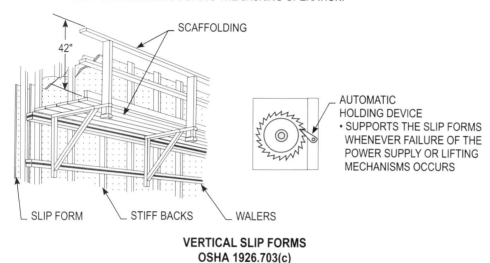

VERTICAL SLIP FORMS
OSHA 1926.703(c)

REINFORCING STEEL
OSHA 1926.703(d)

Reinforcing steel for walls, piers, columns and similar vertical structures shall be adequately supported to prevent overturning and collapse.

Employers shall take measures to prevent unrolled wire mesh from recoiling. Such measures may include, but are not limited to, securing each end of the roll or turning over the roll.

REMOVAL OF FORMWORK
OSHA 1926.703(e)

Forms and shores (except those used for slabs on grade and slip forms) shall not be removed until the employer determines that the concrete has gained sufficient strength to support its weight and superimposed loads. Such determination shall be based on compliance with one of the following:

· The plans and specifications stipulate conditions for removal of forms and shores, and such conditions have been followed or

· The concrete has been properly tested with an appropriate American Society for Testing and Materials (ASTM) standard test method designed to indicate the concrete compressive strength, and the test results indicate that the concrete has gained sufficient strength to support its weight and superimposed loads.

Reshoring shall not be removed until the concrete being supported has attained adequate strength to support its weight and all loads in place upon it.

PRECAST CONCRETE
OSHA 1926.704

Precast concrete wall units, structural framing and tilt-up wall panels shall be adequately supported to prevent overturning and to prevent collapse until permanent connections are completed. **(See Figure 17-3)**

Figure 17-3. This illustration shows that precast concrete wall units, structural framing and tilt-up wall panels shall be adequately supported.

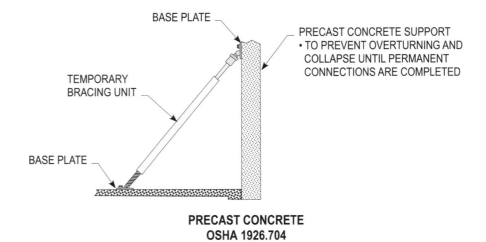

PRECAST CONCRETE
OSHA 1926.704

Lifting inserts that are embedded or otherwise attached to tilt-up wall panels shall be capable of supporting at least two times the maximum intended load applied or transmitted to them; lifting inserts for other precast members shall be capable of supporting four times the load. **(See Figure 17-4)**

Figure 17-4. This illustration shows that lifting inserts shall be embedded or otherwise attached to tilt-up wall panels.

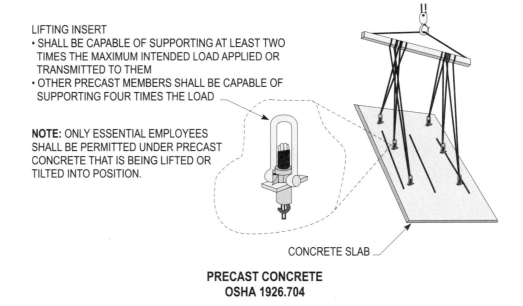

PRECAST CONCRETE
OSHA 1926.704

Only essential employees shall be permitted under precast concrete that is being lifted or tilted into position.

LIFT-SLAB OPERATIONS
OSHA 1926.705

Lift-slab operations shall be designed and planned by a registered professional engineer who has experience in lift-slab construction. Such plans and designs shall be implemented by the employer and shall include detailed instructions and sketches indicating the prescribed method of erection. The plans and designs shall also include provisions for ensuring lateral stability of the building/structure during construction.

Jacking equipment shall be capable of supporting at least two and one-half times the load being lifted during jacking operations, and the equipment shall not be overloaded. For the purpose of this provision, jacking equipment includes any load bearing component that is used to carry out the lifting operation(s). Such equipment includes, but is not limited to, the following: threaded rods, lifting attachments, lifting nuts, hook-up collars, T-caps, shearheads, columns and footings. No employee, except those essential to the jacking operation, shall be permitted in the building/structure while any jacking operation is taking place unless the building/structure has been reinforced sufficiently to ensure its integrity during erection. The phrase "reinforced sufficiently to ensure its integrity" used in this paragraph means that a registered professional engineer, independent of the engineer who designed and planned the lifting operation, has determined from the plans that if there is a loss of support at any jack location, that loss will be confined to that location and the structure as a whole will remain stable.

Under no circumstances shall any employee who is not essential to the jacking operation be permitted immediately beneath a slab while it is being lifted.

MASONRY CONSTRUCTION
OSHA 1926.706

Whenever a masonry wall is being constructed, employers shall establish a limited access zone prior to the start of construction. The limited access zone shall be as follows:

- Equal to the height of the wall to be constructed plus 4 ft., and shall run the entire length of the wall;

- On the side of the wall that will be unscaffolded;

- Restricted to entry only by employees actively engaged in constructing the wall; and

- Kept in place until the wall is adequately supported to prevent overturning and collapse unless the height of wall is more than 8 ft. and unsupported; in which case, it must be braced. The bracing shall remain in place until permanent supporting elements of the structure are in place. **(See Figure 17-5)**

Figure 17-5. This illustration shows the requirements for a masonry wall that is being constructed.

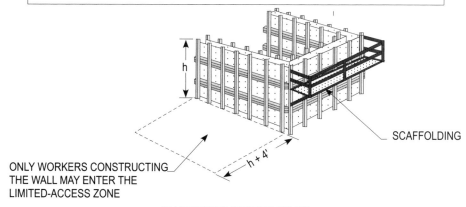

LIMITED ACCESS ZONE
• EQUAL TO THE HEIGHT OF THE WALL TO BE CONSTRUCTED PLUS 4 FT., AND SHALL RUN THE THE ENTIRE LENGTH OF THE WALL
• ON THE SIDE OF THE WALL THAT WILL BE UNSCAFFOLDED
• RESTRICTED TO ENTRY ONLY BY EMPLOYEES ACTIVELY ENGAGED IN CONSTRUCTING THE WALL
• KEPT IN PLACE UNTIL THE WALL IS ADEQUATELY SUPPORTED TO PREVENT OVERTURNING AND COLLAPSE UNLESS THE HEIGHT OF WALL IS MORE THAN 8 FT. AND UNSUPPORTED; IN WHICH CASE, IT SHALL BE BRACED. THE BRACING SHALL REMAIN IN PLACE UNTIL PERMANENT SUPPORTING ELEMENTS OF THE STRUCTURE ARE IN PLACE

h

SCAFFOLDING

h + 4'

ONLY WORKERS CONSTRUCTING THE WALL MAY ENTER THE LIMITED-ACCESS ZONE

MASONRY CONSTRUCTION
OSHA 1926.706

Concrete and Masonry Construction

Section	Answer
_____	T F
_____	T F
_____	T F
_____	T F
_____	T F
_____	T F
_____	T F
_____	T F
_____	T F
_____	T F
_____	T F
_____	T F

Concrete and Masonry Construction

1. Employers shall not place construction loads on a concrete structure or portion of a concrete structure unless the employer determines, based on information received from a person who is qualified in structural design, that the structure or portion of the structure is capable of supporting the intended loads.

2. All protruding reinforcing steel, onto and into which employees could fall, shall be capped to eliminate the hazard of impalement.

3. Employees are permitted to ride concrete buckets.

4. To the extent practicable, elevated concrete buckets shall be routed so that no employee or the fewest employees possible are exposed to the hazards associated with falling concrete buckets.

5. Formwork shall be designed, fabricated, erected, supported, braced and maintained so that it will be capable of supporting without failure all vertical and lateral loads that might be applied to the formwork.

6. Formwork shall be designed, fabricated, erected, supported, braced and maintained so that it will be capable of supporting without failure all vertical and lateral loads that might be applied to the formwork.

7. The steel rods or pipes on which jacks climb or by which **vertical slip** forms are lifted shall be (1) specifically designed for that purpose and (2) adequately braced where not encased in concrete.

8. Forms and shores (except those used for slabs on grade and slip forms) shall not be removed until the foreman determines that the concrete has gained sufficient strength to support its weight and superimposed loads.

9. Precast concrete wall units, structural framing and tilt-up wall panels shall be adequately supported to prevent overturning and to prevent collapse until all employees are clear.

10. Lift-slab operations shall be designed and planned by a registered concrete technician who has experience in lift-slab construction.

11. Jacking equipment shall be capable of supporting at least one and one-half times the load being lifted during jacking operations, and the equipment shall not be overloaded.

12. No employee, except those essential to the jacking operation, shall be permitted in the building/structure while any jacking operation is taking place unless the building/structure has been reinforced sufficiently to ensure its integrity during erection.

_____   T   F

13. Whenever a masonry wall is being constructed, employers shall establish a limited access zone prior to the start of construction.

_____   T   F

14. A limited access zone shall be equal to the height of the wall to be constructed plus 3 ft., and shall run the entire length of the wall.

_____   T   F

15. A limited access zone shalll be restricted to entry only by employees actively engaged in constructing the wall.

Steel Erection

On January 18, 2001, OSHA published a revised safety rule that addresses the major causes of fatalities and injuries in steel erection.

The new rule, which became effective January 18, 2002, phases in some provisions affecting the design of building components. Concepts addressed by the standard include:

- Site layout and construction sequence
- Site-specific erection plan
- Hoisting and rigging
- Structural steel assembly
- Column anchorage
- Beams and columns
- Open web steel joists
- Systems-engineered metal buildings
- Falling object protection
- Fall protection
- Training

SCOPE
OSHA 1926.750

OSHA 1926.750(a) provides that Subpart R applies to employers engaged in steel erection activities involved in the construction, alteration and/or repair of any type of building or structure — single and multi-story buildings, bridges and other structures — where steel erection occurs. The paragraph makes clear that differences in coverage under the previous standards between single and multi-story (or tiered) buildings, as well as buildings and other types of steel structures, are no longer relevant. All the provisions of revised Subpart R now apply irrespective of such distinctions. Subpart R does not cover electrical transmission towers, communication and broadcast towers or tanks.

Steel erection activities include the following:

- Hoisting, laying out, placing, connecting, welding, burning, guying, bracing, bolting, plumbing and rigging structural steel, steel joists and metal buildings

- Installing metal deck and siding systems, miscellaneous metals, ornamental iron and similar materials

- Moving point-to-point while performing these activities

OSHA 1926.750(b) lists a number of activities that are covered by Subpart R when they occur during and are a part of the steel erection activities. The standard explicitly states that coverage depends on whether an activity occurs during and is a part of steel erection. **(See Figure 8-1)**

Figure 8-1. This illustration shows a steel erection decision tree to use when determining if activities are part of the steel erection process.

STEEL ERECTION DECISION TREE

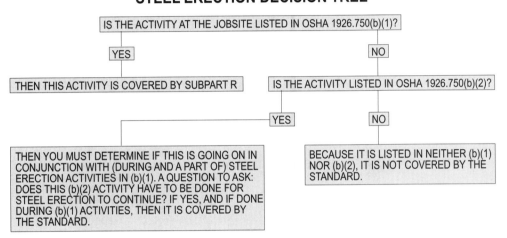

For example: There are standing seam metal roofing systems that incorporate a layer of insulation under the metal roof. In the installation process, a row of insulation is installed, which is then covered by a row of metal roofing. Once that row of roofing is attached, the process is repeated, row-by-row, until the roof is completed. The installation of the row of insulation is a part of the installation of the metal roofing (which is steel erection), and is covered by Subpart R.

OSHA 1926.750(c) provides that the duties of controlling contractors under this rule include, but are not limited to, the duties specified in OSHA 1926.752(a) (approval to begin steel erection), OSHA 1926.752(c) (site layout), OSHA 1926.755(b)(2) (notification of repair, replacement or modification of anchor bolts), OSHA 1926.759(b) (protection from falling objects) and OSHA 1926.760(a)(2)(i) (perimeter safety cables).

SITE LAYOUT, SITE-SPECIFIC ERECTION PLAN AND CONSTRUCTION SEQUENCE
OSHA 1926.752

This section sets forth OSHA's requirements for proper communication between the controlling contractor and the steel erector prior to the beginning of the steel erection operation and proper pre-planning by the steel erector to minimize overhead exposure during hoisting operations. Appendix A, which is referred to in this section, also provides guidelines for employers who elect to develop a site-specific erection plan. OSHA's current standard does not contain provisions similar to those being adopted in this section. The following shall be considered for site layout, site-specific erection plan and construction sequence:

- Approval to begin steel erection
- Commencement of steel erection
- Site layout
- Pre-planning of overhead hoisting operations
- Site-specific erection plan

APPROVAL TO BEGIN STEEL ERECTION
OSHA 1926.752(a)

The controlling contractor shall ensure that the following written notifications are provided to the steel erector:

- The concrete in the footings, piers and walls and the mortar in the masonry piers and walls have cured to a level that will provide the proper strength to support any forces imposed on the concrete during steel erection

- Any repairs, replacements and modifications made to anchor bolts meet the requirements of OSHA 1926.755(b).

  Note: The criteria for adequate strength for concrete footings depend on the results of required American Society for Testing and Materials (ASTM) standard test methods.

COMMENCEMENT OF STEEL ERECTION
OSHA 1926.752(b)

The commencement of steel erection shall not be done by a steel erection contractor unless it has received written notification that the concrete in the footings, piers and walls or the mortar in the masonry piers and walls has attained, on the basis of an appropriate ASTM standard test method of field-cured samples, either 75 percent of the intended minimum compressive design strength of sufficient strength to support the loads imposed during steel erection.

SITE LAYOUT
OSHA 1926.752(c)

The following shall be provided and maintained by the controlling contractor:

- Controlling contractor shall provide and maintain adequate access roads into and through the site and means and methods for pedestrian and vehicular control. This requirement does not apply to road outside of the construction site.

These conditions enable the steel erector to move around the site and perform necessary operations in a safe manner.

- Controlling contractor shall provide and maintain a firm, properly graded, drained area, readily accessible to the work and with adequate space for the safe storage of materials and the safe operation of the erector's equipment.

PRE-PLANNING OF OVERHEAD HOISTING OPERATIONS
OSHA 1926.752(d)

Safety Tip: All hoisting operations shall be pre-planned.

All hoisting operations in steel erection shall be pre-planned to ensure that they comply with the requirements of OSHA 1926.753(d), the paragraph regulating "working under loads."

The purpose of OSHA 1926.752(d), is to address the hazards associated with overhead loads. Specifically, these hazards include failure of the lifting device, which would create a crushing hazard, and items falling from the load, which creates a struck-by and crushing hazard, among others. Given the nature of the loads used in steel erection, either of these events could result in serious injury or death.

SITE-SPECIFIC ERECTION PLAN
OSHA 1926.752(e)

Site-specific erection plans shall be developed by a qualified person and be available at the worksite. The standard does not require such plans for all steel erection worksites; the three following specific provisions of this rule allow them as alternatives to specific provisions of the standard:

- An employer wishes to provide "equivalent protection", rather than deactivating or making safety latches on hoisting hooks inoperable per OSHA 1926.753(c)(5)

- An employer provides an alternative erection method for setting certain steel joists detailed per OSHA 1926.757(a)(4)

- An employer places decking bundles on steel joists and, under certain circumstances, shall document in an erection plan that the structure can support the load per OSHA 1926.757(e)(4)(i)

  Note: OSHA has provided Appendix A as a guideline for establishing the components of a site-specific erection plan.

HOISTING AND RIGGING
OSHA 1926.753

Rigging and hoisting of steel members and materials are essential activities in the steel erection process. This section sets safety requirements to address the hazards associated with these activities.

OSHA 1926.753(a) provides that all provisions of OSHA 1926.550, the general construction requirements for cranes and derricks, apply to hoisting and rigging operations in steel erection except for OSHA 1926.550(g)(2), the general requirements for crane or derrick suspended personnel platforms. Provisions for the use of suspended platforms in steel erection are in OSHA 1926.753(c)(4) of this section. **(See Figure 18-2)**

OSHA 1926.753(b) provides that, in addition to the OSHA 1926.550 provisions, the requirements in OSHA 1926.753(c) through (e) of this section apply as well. Subparts (a) and (b) to OSHA 1926.753 were added because hoisting safety is critical in steel erection operations and the OSHA 1926.550 provisions are, in many respects, outdated.

NOTE: OSHA 1926.753(a) PROVIDES THAT ALL PROVISIONS OF OSHA 1926.550, THE GENERAL CONSTRUCTION REQUIREMENTS FOR CRANES AND DERRICKS, APPLY TO HOISTING AND RIGGING OPERATIONS IN STEEL ERECTION EXCEPT FOR OSHA 1926.550(g)(2), THE GENERAL REQUIREMENTS FOR CRANE OR DERRICK SUSPENDED PERSONNEL PLATFORMS.

Figure 18-2. This illustration shows the activities covered by OSHA 1926.753(a) when they occur during and are a part of steel erection activity.

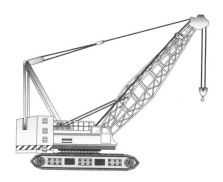

**HOISTING AND RIGGING
OSHA 1926.753**

GENERAL
OSHA 1926.753(c)

OSHA 1926.753(c) contains the requirements for pre-shift inspections of cranes and rigging used in steel erection.

A competent person shall perform a pre-shift visual inspection of the cranes to be used for steel erection. The inspection shall meet the requirements of OSHA 1926.550 along with the supplemental requirements listed in paragraph (c) of this section. These requirements are needed to ensure that safe equipment and procedures are used to perform the specialized and potentially hazardous types of hoisting operations in steel erection. These include the use of cranes to hoist employees on personnel platforms per OSHA 1926.753(c)(4); to suspend loads over certain employees per OSHA 1926.753(d); and to perform multiple lifts per OSHA 1926.753(e). In addition, more frequent inspections are needed for cranes being used for steel erection. An inspection prior to each shift is needed to provide an added measure of protection for the specialized and potentially hazardous hoisting operations. **(See Figure 18-3)**

Figure 18-3. This illustration shows a pre-shift visual inspection, by a competent person, of cranes to be used for steel erection .

INSPECTION SHALL INCLUDE THE FOLLOWING ITEMS:

- ALL CONTROL MECHANISMS FOR MALADJUSTMENT
- CONTROL AND DRIVE MECHANISMS FOR EXCESSIVE WEAR OF COMPONENTS AND CONTAMINATION BY LUBRICANTS, WATER OR OTHER FOREIGN MATTER
- SAFETY DEVICES, INCLUDING, BUT NOT LIMITED TO, BOOM ANGLE INDICATORS, BOOM STOPS, BOOM KICK-OUT DEVICES AND LOAD MOVEMENT INDICATORS WHERE REQUIRED
- AIR, HYDRAULIC AND OTHER PRESSURE LINES FOR DETERIORATION OR LEAKAGE, PARTICULARLY THOSE THAT FLEX IN NORMAL OPERATION
- HOOKS AND LATCHES FOR DEFORMATION, CHEMICAL DAMAGE, CRACKS OR WEAR
- WIRE ROPE REEVING FOR COMPLIANCE WITH HOISTING EQUIPMENT MANUFACTURER'S SPECIFICATIONS
- ELECTRICAL APPARATUS FOR MALFUNCTIONING, SIGNS OF EXCESSIVE DETERIORATION, DIRT OR MOISTURE ACCUMULATION
- HYDRAULIC SYSTEM FOR PROPER FLUID LEVEL
- TIRES FOR PROPER INFLATION AND CONDITION
- GROUND CONDITIONS AROUND THE HOISTING EQUIPMENT FOR PROPER SUPPORT, INCLUDING GROUND SETTING UNDER AND AROUND OUTRIGGERS, GROUND WATER ACCUMULATION OR OTHER SIMILAR CONDITIONS
- THE HOISTING EQUIPMENT FOR LEVEL PROTECTION
- THE HOISTING EQUIPMENT FOR LEVEL PROTECTION AFTER EACH MOVE AND SETUP DURING THE SHIFT

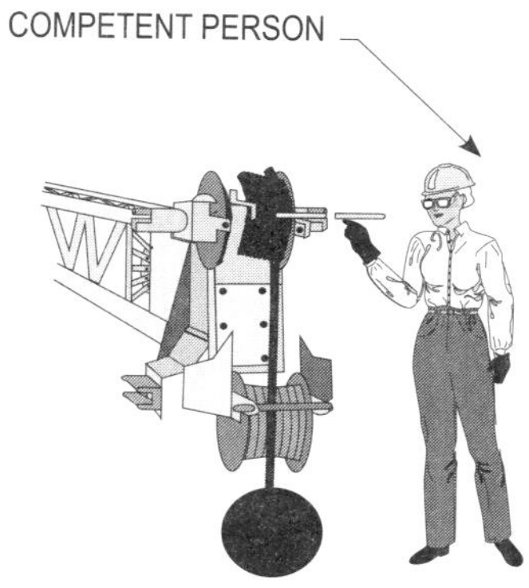

COMPETENT PERSON

**HOISTING AND RIGGING
OSHA 1926.753(c)**

Safety Tip: A competent person shall inspect all cranes and rigging before each shift.

A competent person shall perform a complete visual inspection before each shift. This person might be the operator or oiler of the hoisting equipment being used or, on a large project, the master mechanic who checks each crane. The pre-shift visual inspection shall also include "observation for deficiencies during operation" and is anticipated to take between 10 and 20 minutes. At a minimum, the inspection shall include the following items:

- All control mechanisms for maladjustment

- Control and drive mechanisms for excessive wear of components and contamination by lubricants, water or other foreign matter

- Safety devices, including, but not limited to, boom angle indicators, boom stops, boom kick-out devices, anti-two block devices and load moment indicators where required

- Air, hydraulic and other pressure lines for deterioration or leakage, particularly those which flex in normal operation

- Hooks and latches for deformation, chemical damage, cracks or wear

- Wire rope reeving for compliance with hoisting equipment manufacturer's specifications

- Electrical apparatus for malfunctioning, signs of excessive deterioration, dirt or moisture accumulation

- Hydraulic system for proper fluid level

- Tires for proper inflation and condition

- Ground conditions around the hoisting equipment for proper support, including ground settling under and around outriggers, ground water accumulation or other similar conditions

- The hoisting equipment for level position

- The hoisting equipment for level position after each move and setup during the shift.

If the inspection identifies a deficiency, a competent person shall immediately determine whether the deficiency constitutes a hazard per OSHA 1926.753(c)(1)(iii).

If a deficiency is determined to constitute a hazard, the hoisting equipment shall be removed from service until the deficiency is corrected.

The operator shall be responsible for operations under his/her direct control and gives the operator the authority to refuse any load that he/she deems unsafe.

Control of a heavy-lifting operation solely under the direction of a supervisor, or any other person who may be less qualified than he, is not prudent. The crane operator has instrumentation in the crane to base his action upon, and should be the ultimate person to make decisions about the capacity and safety of both the machine and lifting operations.

Unlike a qualified crane operator, who has the training and experience to make informed decisions about handling a crane load, a supervisor may not have the qualifications and experience necessary for safe crane operation.

A qualified rigger shall inspect the rigging prior to each shift in accordance with OSHA 1926.251.

Note: A qualified rigger is defined as a "qualified person" who is performing the inspection of the rigging equipment. Based on the definition of a "qualified person," a qualified rigger shall have demonstrated successfully the ability to solve or resolve rigging problems. Since there are no degree or certification programs for "riggers," they shall have extensive experience to support this demonstration.

The use of the headache ball, hook or load shall not be used to transport personnel except as provided in OSHA 1926.753(c)(4) of this section. These practices are widely recognized as unsafe because of the risk of falling off the ball, hook or load (or, in a case where the load falls, falling with the load).

Employers engaged in steel erection work do not have to comply with the requirements of OSHA 1926.550(g)(2) — Crane or Derrick Suspended Personnel Platforms, if they hoist employees on a personnel platform. OSHA 1926.550(g)(2) requires an employer to demonstrate that the use of conventional methods to access the workstation "would be more hazardous or is not possible because of structural design or workday conditions" if the employer wants to hoist employees on a personnel platform.

Safety latches on hooks shall not be deactivated or made inoperable except when a qualified rigger has determined that the hoisting and placing of purlins and single joists can be performed more safely by doing so, or when equivalent protection is provided in a site-specific erection plan.

There are some activities in steel erection in which it is safer to hoist lighter members with a deactivated safety latch. One example is when deactivating the latch eliminates the need for a worker to climb up or onto an unstable structural member, such as a single bar joist, to unhook the member. OSHA 1926.753(c)(5)(i) requires all latched hooks to be latched in the absence of a determination by the qualified rigger that using the latch is unsafe. OSHA 1926.753(c)(5)(ii) states that if the latch is deactivated without such a determination by a qualified rigger, the employer shall have some form of equivalent protection in its site-specific erection plan.

Safety Tip: Safety latches on hooks shall not be deactivated unless a qualified rigger has determined it to be safe to do so.

WORKING UNDER LOADS
OSHA 1926.753(d)

Routes for suspended loads shall be pre-planned and prohibit employees from working under a hoisted load except for workers engaged in initial connection activities or employees who are necessary for unhooking the load. The following criteria shall be met when working under suspended loads:

- Materials being hoisted shall be rigged to prevent unintentional displacement

- Hooks with self-closing safety latches (or their equivalent) shall be used to prevent components from slipping out of the hook

- A qualified rigger shall rig all loads

See Figure 18-4 for a detailed illustration pertaining to working under loads.

MULTIPLE LIFT RIGGING PROCEDURE
OSHA 1926.753(e)

Multiple lift rigging, when executed as prescribed in this subpart, is a safe and effective method for decreasing the number of total crane swings and employee exposure on the steel while connecting. In the past, OSHA has not looked favorably upon "Christmas Treeing" because, when performed incorrectly, it can present significant hazards to workers.

Figure 18-4. This illustration shows employees who work under loads.

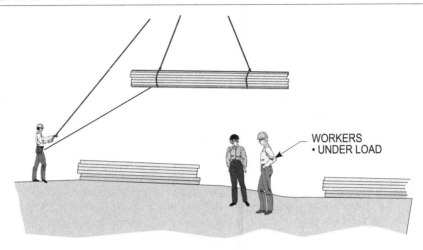

THE FOLLOWING CRITERIA SHALL BE MET WHEN WORKING UNDER SUSPENDED LOADS:
• MATERIALS BEING HOISTED SHALL BE RIGGED TO PREVENT UNINTENTIONAL DISPLACEMENT
• HOOKS WITH SELF-CLOSING SAFETY LATCHES (OR THEIR EQUIVALENT) SHALL BE USED TO PREVENT COMPONENTS FROM SLIPPING OUT OF THE HOOK
• A QUALIFIED RIGGER SHALL RIG ALL LOADS

WORKERS
• UNDER LOAD

WORKING UNDER LOADS
OSHA 1926.753(d)

This paragraph applies when a steel erector chooses to lift multiple pieces of steel at one time as an alternative to hoisting individual structural members. It limits the use of this procedure to the lifting of beams and similar structural members and requires specific equipment and work practices to be used. The following criteria shall be met when a multiple lift is performed:

• A multiple lift rigging assembly (defined in the definition section) is used.

Note: By definition, the assembly shall have been manufactured by a wire rope rigging supplier. Since this is a specialized type of lift, the rigging assembly shall have been designed specifically for the particular use in a multiple lift and meet each aspect of the definition.

• A multiple lift may not involve hoisting more than five members during the lift.

Note: Limiting the number of members hoisted is essential to safety. This limit takes into account the need to control both the load and the empty rigging. It also accounts for the fact that a typical bay, which consists of up to five members, can be filled with a single lift. Too many members in a lift may create a string that is too awkward to control or allow too much empty rigging to dangle loose, creating a hazard to employees.

• Only beams and similar structural members (like solid web beams and certain open web steel joists) are lifted.

Note: Other items, such as bundles of decking, meet the definition of structural members but do not lend themselves to the MLRP. A typical multiple lift member would be a wide flange beam section between 10 ft. and 30 ft. long, typically weighing less than 1800 pounds.

• All employees engaged in a multiple lift operation shall be trained in these procedures in accordance with OSHA 1926.761(c)(1), which contains specific training requirements for employees engaged in multiple lifts.

Note: Due to the specialized nature of multiple lifts and the knowledge necessary to perform them safely, this training requirement is necessary to ensure that employees are properly trained in all aspects of multiple lift procedures.

• No crane is to be used for a multiple lift where such use is contrary to the manufacturer's instructions.

Note: The use of a crane is prohibited in a multiple lift if the crane manufacturer recommends that the crane not be used for that purpose. Crane manufacturers often recommend that employers do not execute multiple lifting with their cranes. It has been argued that there are too many variables associated with attempting Christmas treeing and any miscalculations of those component variables (such as the weights and center of gravity of the beams, crane capacity, the stability of the load under lift conditions and inconsistent rigging techniques) could contribute to an accident.

OSHA remains consistent in requiring employers to follow the manufacturer's recommendations and specifications for its product. If the manufacturer of a crane prohibits the use of its crane in multiple lifts and an employer uses that crane to perform a multiple lift, that employer is in violation of both OSHA 1926.550(a) and OSHA 1926.760(e)(1)(v).

Employers shall perform multiple lifts using multiple lift rigging assembly components assembled and designed for a specified capacity. The employer shall ensure that each multiple lift rigging assembly is designed and assembled with a maximum capacity for both the total assembly and for each individual attachment point. This capacity, which shall be certified by the manufacturer or qualified rigger, shall be based on the manufacturer's specifications and must have a 5 to 1 safety factor for all components. The rigging shall be certified by the qualified rigger who assembles it or the manufacturer who provides the entire assembly to ensure that the assembly can support the whole load, and that each hook is capable of supporting the individual members. The appropriate rigging assembly to be used is the lightest one that will support the load. Typically, one assembly is manufactured and certified for the heaviest anticipated multiple lift on the job, and this rigging is then used for all the MLRPs. Therefore, the total load shall not exceed the following:

• The rated capacity of the hoisting equipment specified in the hoisting equipment load charts
• The rigging capacity specified in the rigging rating chart

The multiple lift rigging assembly shall be rigged with the members:

• Attached at their center of gravity and be kept reasonably level
• Be rigged from the top down
• Have a distance of at least 7 ft. between the members.

See Figure 18-5 for a detailed illustration pertaining to multiple lifting operations.

In practice, these procedures mean that the choker attached to the last structural member of the group to be connected is the one attached on the rigging assembly closest to the headache ball. The next-to-last member to be connected is attached to the next lower hook on the rigging assembly, and so on. As each member is attached, it is lifted approximately 2 ft. off the ground to verify the location of the center of gravity and to allow the choker to be checked for proper connection. Adjustments to choker location are made during this trial lift procedure. The choker length is then selected to ensure that the vertical distance between the bottom flange of the higher beam and the top flange of the next lower beam is never less than 7 ft. Thus, when the connector has made the initial end connections of the lower beam and moves to the center of each beam to remove the choker, there will be sufficient clearance to prevent the connector from contacting the upper suspended beam.

Figure 18-5. This illustration shows employees performing multiple lifting operations.

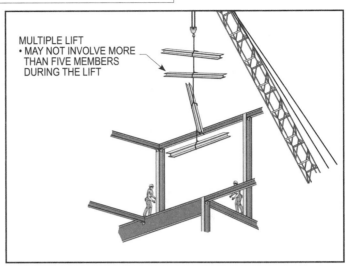

MULTIPLE LIFT RIGGING ASSEMBLY SHALL BE RIGGED WITH MEMBERS:	TOTAL LOAD SHALL NOT EXCEED THE FOLLOWING:
• ATTACHED AT THE CENTER OF GRAVITY AND BE KEPT REASONABLY LEVEL • BE RIGGED FROM THE TOP DOWN • HAVE A DISTANCE OF AT LEAST 7 FT. BETWEEN THE MEMBERS	• THE RATED CAPACITY OF THE HOISTING EQUIPMENT SPECIFIED IN THE HOISTING EQUIPMENT LOAD CHARTS • THE RIGGING CAPACITY SPECIFIED IN THE RIGGING RATING CHART

MULTIPLE LIFT
• MAY NOT INVOLVE MORE THAN FIVE MEMBERS DURING THE LIFT

MULTIPLE LIFT RIGGING PROCEDURE
OSHA 1926.753(e)

The members on the multiple lift rigging assembly shall be set from the bottom up. This is the only practical way that the members can be set, and OSHA is including this requirement for clarity and completeness.

Controlled load lowering (through the use of a controlled load lowering device) shall be used whenever the load is over the connectors. This means that the cranes in a multiple lift shall use controlled load lowering when lowering loads into position for the connectors to set the members. The record shows that control load lowering is essential to prevent accidents that could result from the crane operator's foot slipping off the brake, brake failure or from the load slipping through the brake. It assures that the operator has maximum control over the load.

All employees engaged in the multiple lift shall be trained in these procedures in accordance with section OSHA 1926.761(c)(1).

The standard requires that only the employees engaged in the multiple lift shall be trained in the requirements of this paragraph in accordance with OSHA 1926.761(c)(1), not all employees affected by the lift, as the comment seems to indicate.

STRUCTURAL STEEL ASSEMBLY
OSHA 1926.754

This section sets forth the requirements for the assembly of structural steel. It requires that the structural stability be maintained at all times during the erection process. This is a general requirement for any type of steel structure, including single story, multi-story and other structures. Since structural stability is essential to the successful erection of steel structures, this section is intended to prevent collapse due to lack of stability, a major cause of fatalities in this industry. **(See Figure 18-6)**

NOTE: THE STRUCTURAL STABILITY SHALL BE MAINTAINED AT ALL TIMES DURING THE ERECTION PROCESS. THIS IS A GENERAL REQUIREMENT FOR ANY TYPE OF STEEL STRUCTURE, INCLUDING SINGLE STORY, MULTISTORY AND OTHER STRUCTURES.

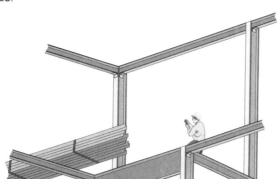

STRUCTURAL STEEL ASSEMBLY
OSHA 1926.754

Figure 18-6. This illustration shows that structural stability shall be maintained at all times during the erection process.

Permanent floors shall be installed as the erection of structural members progresses, and there shall be not more than eight stories between the erection floor and the uppermost permanent floor, except where the structural integrity is maintained as a result of the design.

At no time shall there be more than 4 floors or 48 ft., whichever is less, of unfinished bolting or welding above the foundation or uppermost permanently secured floor, except where the structural integrity is maintained as a result of the design. **(See Figure 18-7)**

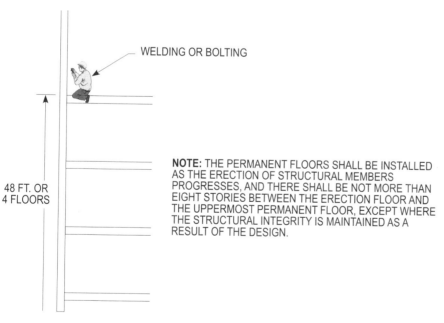

WELDING OR BOLTING

48 FT. OR 4 FLOORS

NOTE: THE PERMANENT FLOORS SHALL BE INSTALLED AS THE ERECTION OF STRUCTURAL MEMBERS PROGRESSES, AND THERE SHALL BE NOT MORE THAN EIGHT STORIES BETWEEN THE ERECTION FLOOR AND THE UPPERMOST PERMANENT FLOOR, EXCEPT WHERE THE STRUCTURAL INTEGRITY IS MAINTAINED AS A RESULT OF THE DESIGN.

PERMANENT FLOORING - SKELETON STEEL
CONSTRUCTION MULTI-TIERED BUILDINGS
OSHA 1926.754(a)

Figure 18-7. This illustration shows the requirements for permanent flooring of skeleton steel construction in tiered buildings.

A fully planked or decked floor or nets shall be maintained within 2 stories or 30 ft., whichever is less, directly under any erection work being performed. This provision serves many purposes: limits falls of employees to 30 ft., provides falling object protection and can be used as a staging area for emergency rescue. **(See Figure 18-8)**

Figure 18-8. This illustration shows the requirements for temporary flooring of skeleton steel construction in tiered buildings.

NOTE 1: WHEN SKELETON STEEL ERECTION IS BEING DONE, A TIGHTLY PLANKED AND SUBSTANTIAL FLOOR SHALL BE MAINTAINED WITHIN TWO STORIES OR 30 FT., WHICHEVER IS LESS, BELOW AND DIRECTLY UNDER THAT PORTION OF EACH TIER OF BEAMS ON WHICH ANY WORK IS BEING PERFORMED.

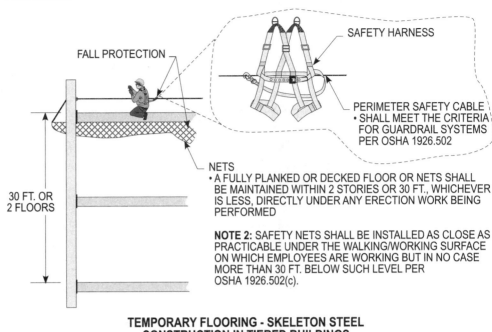

FALL PROTECTION

SAFETY HARNESS

PERIMETER SAFETY CABLE
• SHALL MEET THE CRITERIA FOR GUARDRAIL SYSTEMS PER OSHA 1926.502

30 FT. OR 2 FLOORS

NETS
• A FULLY PLANKED OR DECKED FLOOR OR NETS SHALL BE MAINTAINED WITHIN 2 STORIES OR 30 FT., WHICHEVER IS LESS, DIRECTLY UNDER ANY ERECTION WORK BEING PERFORMED

NOTE 2: SAFETY NETS SHALL BE INSTALLED AS CLOSE AS PRACTICABLE UNDER THE WALKING/WORKING SURFACE ON WHICH EMPLOYEES ARE WORKING BUT IN NO CASE MORE THAN 30 FT. BELOW SUCH LEVEL PER OSHA 1926.502(c).

**TEMPORARY FLOORING - SKELETON STEEL CONSTRUCTION IN TIERED BUILDINGS
OSHA 1926.754(b)**

WALKING/WORKING SURFACES
OSHA 1926.754(c)

OSHA 1926.754(c) sets forth requirements that address slipping/tripping hazards encountered when working on steel structures.

SHEAR CONNECTORS AND OTHER SIMILAR DEVICES
OSHA 1926.754(c)(1)

The attachment of shear connectors (such as headed steel studs, steel bars or steel lugs), reinforcing bars, deformed anchors or threaded studs shall not be attached to the top flanges of beams, joists or beam attachments so that they project vertically from or horizontally across the top flange of the member until after the decking or other walking/ working surface, has been installed. When shear connectors are used in the construction of composite floor, roofs and bridge decks, employees shall lay out and install the shear connectors after the decking has been installed, using the deck as a working platform. Shear connectors shall not be installed from within a controlled decking zone (CDZ), as specified in OSHA 1926.760(c)(8).

SLIP RESISTANCE OF SKELETAL STRUCTURAL STEEL
OSHA 1926.754(c)(3)

To reduce the risk of steel erection workers slipping on coated steel members, workers shall not be permitted to walk the top surface of any structural steel member installed after July 18, 2006, three years after the effective date of this standard. At that time, employees shall not be permitted to walk on the top surface of any structural steel member that has been coated with paint or similar material, unless the coating has achieved a minimum average slip-resistance of 0.50 when wet on an English XL tribometer, or the equivalent measurement on another device. This subpart does not require that the particular coated member be tested. Rather, it requires the test to be done on a sample of the paint formulation produced by the paint manufacturer. The testing laboratory shall use an acceptable ASTM method and an English XL tribometer or equivalent tester shall be used on a wetted surface, and the laboratory shall be capable of employing this method. The test results shall be available at the site and to the steel erector. Appendix B lists two appropriate ASTM standard test methods that may be used to comply with the paragraph. If other ASTM methods are approved, they too are allowed under this provision.

PLUMBING-UP
OSHA 1926.743(d)(1)

When deemed necessary by a competent person, plumbing-up equipment shall be installed in conjunction with the steel erection process to ensure the stability of the structure.

The approval of a competent person shall be made before plumbing-up equipment is removed.

METAL DECKING
OSHA 1926.754(e)

OSHA 1926.754(e) addresses specific requirements to protect employees during the installation of metal decking. Metal decking as defined in OSHA 1926.751 means a commercially manufactured, structural grade, cold rolled metal panel formed into a series of parallel ribs; for this subpart, this includes metal floor and roof decks, standing seam metal roofs, other metal roof systems and other products such as bar gratings, checker plate, expanded metal panels and similar products. After installation and proper fastening, these decking materials serve a combination of functions including, but not limited to:

- A structural element designed in combination with the rest of the structure to resist, distribute and transfer loads, stiffen the structure and provide a diaphragm action
- A walking/working surface
- A form for concrete slabs
- A support for roofing systems
- A finished floor or roof

The following are some of the common hazards associated with hoisting, landing and placing of deck bundles:

- Employers shall ensure that the packaging and strapping on the deck bundle are specifically designed for hoisting purposes.

  Note: Bundle straps usually are applied at the factory and are intended to keep the bundle together until it is placed for erection and the sheets are ready to be spread. Decking is bundled differently; some manufacturers design the strapping to be used as a lifting device. However, hoisting a bundle by straps that are not designed for lifting is extremely dangerous. The bundle straps can break apart or loosen, creating a falling object hazard or, if a structural member is hit by the bundle or its contents, it could cause the structure to collapse. OSHA believes that compliance with this requirement will prevent these hazards. There were no comments received regarding this requirement.

• Employers shall secure loose items such as dunnage, flashing or other materials placed on the top of deck bundles before a bundle is hoisted.

Note: Sometimes, to expedite unloading and hoisting, items such as dunnage or flashing are placed on the decking bundle to save time. Dunnage, for example, will be sent up with the bundle to help support it on the structure and to protect the decking that has already been installed. This requirement does not allow hoisting loose items or "piggy backing" unless the items are secured to prevent them from falling off the bundle in the event that it catches on the structure and tilts.

• Employers shall land bundles of decking on joists in accordance with OSHA 1926.757(e)(4), which sets out the six conditions that shall be met by employers before a bundle of decking is placed on steel joists where all bridging has not been installed and anchored.

Note: First, a qualified person shall determine, and document in the site-specific erection plan, that the structure or portion of the structure is capable of supporting the load. The bundle of decking shall be placed on a minimum of three steel joists and the joists supporting the bundle shall be attached at both ends. At least one row of bridging shall be installed and anchored, and the edge of the bundle shall be placed within 1 ft. of the bearing surface of the joist end. The total weight of the bundle of decking may not exceed 4000 pounds. The 4000 pound weight limit for decking bundles applies only if the employer has determined that all six conditions can be met prior to landing a bundle of decking on steel joists where all bridging has not been installed and anchored. At this time, the employer may negotiate with the manufacturer to restrict a specific bundle weight to 4000 pounds, or the employer may also opt to install and anchor all bridging in order to continue with the erection process without delay.

• Employers shall land bundles on framing members in such a manner that the decking can be unbanded without losing the support of the structure. If the blocking were to move while the bundle is being unbanded, the bundle would need to have enough support to prevent it from tilting and falling.

Note: OSHA considers hazards associated with cutting banding straps to be widely recognized throughout construction and general industries. In addition to falling straps and dunnage, cutting banding straps poses serious hazards to eyes as well as cuts, abrasions, as well as bruises, strains or other injuries while attempting to hold or secure the contents of the bundle. Training in the establishment, access, proper installation techniques and work practices required by OSHA 1926.754(e) would be covered by OSHA 1926.21(b)(2), OSHA's general training requirements for construction work. In addition, special training programs per OSHA 1926.761(c) that supplements OSHA 1926.21 specifically address employees who work in a controlled decking zone. All recognized hazards, including those associated with cutting banding straps, would be part of the work practices training to ensure that employees recognize unsafe conditions in the work environment and know the measures to control or eliminate hazards.

• Employers shall secure decking against displacement after the end of the shift or when environmental or job site conditions warrant.

Note: Decking may become dislodged from the structure or bundle because of conditions such as high winds. Wind can also move a sheet of loose decking and create a hazard where an employee inadvertently steps onto a sheet of loose piece of decking, believing it to be secured.

ROOF AND FLOOR HOLES AND OPENINGS
OSHA 1926.754(e)(2)

The following requirements shall be followed for installing metal decking to minimize the risks of falling through holes and openings in decking:

- Employers shall ensure that all framed metal deck openings have structural members turned down to allow continuous deck installation, except in cases where structural design constraints and constructibility do not allow this.

Note: Requiring framed deck openings to be turned down allows continuous decking to be performed without having to cut the deck around the opening. This procedure would apply to smaller openings rather than larger openings, such as elevator or mechanical shaft openings. Whereas smaller openings may be cut at a later time, it may not be appropriate to delay larger openings.

- Roof and floor openings shall be decked over. Where large size, configuration or other structural design does not allow for covering of the roof and floor holes and openings, they shall be protected in accordance with OSHA 1926.760(a)(1).

- Employers shall delay cutting decking holes and openings until immediately before they are permanently filled with the equipment or structure needed or intended to fulfill their specific use. That equipment or structure shall either meet the strength requirements of paragraph OSHA 1926.754(e)(3), or be immediately covered.

COVERING ROOF AND FLOOR OPENINGS
OSHA 1926.754(e)(3)

The following proper coverings for roof and floor openings, required by OSHA 1926.754(e)(2)(iii), will protect employees from falling into or through openings in roofs and floors:

- Cover shall be strong enough to withstand the weight of employees, equipment and materials by requiring that covers support twice that combined weight.

- All covers shall be secured when installed so as to prevent accidental displacement by the wind, equipment or employees.

Note: This provision eliminates a fall hazard. OSHA 1926.754(e)(3)(iii) requires that all covers be painted with high visibility paint or be marked with the word "HOLE" or "COVER" to warn of the hazard and to prevent an employee from inadvertently removing the cover. These provisions are consistent with the requirements in Subpart M.

- Installed smoke domes and skylight fixtures are not to be considered covers for the purposes of this section unless the strength requirement of paragraph OSHA 1926.754(e)(3)(i) is met.

Note: If these structures are not capable of supporting the load, they may give way, causing a fall. Unless they have adequate strength, these structures cannot be relied upon to protect employees from falls. Employees commonly lean or sit on skylights or smoke domes and these structures need to be capable of supporting the load without failure.

DECKING GAPS AROUND COLUMNS
OSHA 1926.754(e)(4)

Wire mesh, exterior plywood or equivalent shall be installed around columns where planks or metal decking do not fit tightly, thus leaving a gap. The materials used shall be of sufficient strength to provide fall protection for personnel and prevent objects from falling through.

INSTALLATION OF METAL DECKING
OSHA 1926.754(e)(5)

Safety Tip: Metal decking shall be laid tightly and immediately secured upon adjustment to prevent accidental movement or displacement.

OSHA 1926.754(e)(5) requires metal decking to be laid tightly and immediately secured upon adjustment to prevent accidental movement or displacement, except as provided in OSHA 1926.760(c). OSHA 1926.760(c) provides for a "controlled decking zone" (CDZ), which allows up to 3000 square feet of decking to be unsecured until adjustment, when safety attachment is then required (see discussion on "safety deck attachment" in OSHA 1926.760(c)).

DERRICK FLOORS
OSHA 1926.754(e)(6)

A derrick floor shall be fully decked and/or planked, and the steel member connections be completed to ensure that the floor will support the intended load. Temporary loads on a derrick floor shall be distributed over the underlying support members in order to prevent spot overloading.

COLUMN ANCHORAGE
OSHA 1926.755

This section addresses the hazards associated with column stability and, specifically, the proper use of anchor rods (anchor bolts) to ensure column stability. This section specifies the criteria for column anchorage.

GENERAL REQUIREMENTS FOR ERECTION STABILITY
OSHA 1926.755(a)

The following general requirements shall be applied for erection stability:

- All columns shall be anchored by a minimum of 4 anchor rods/bolts

- Each column anchor rod/bolt assembly, including the column-to-base plate weld and the column foundation, shall be designed to resist a minimum eccentric gravity load of 300 pounds located 18 in. from the extreme outer face of the column in each direction at the top of the column shaft

- Columns shall be set on level finished floors, pre-grouted leveling plates, leveling nuts or shim packs that are adequate to transfer the construction loads.

- All columns shall be evaluated by a competent person to determine whether guying or bracing is needed and, if needed, be installed

See Figure 18-9 for a detailed illustration pertaining to anchorage of columns.

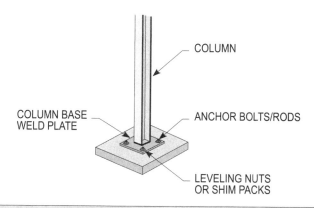

COLUMN

COLUMN BASE
WELD PLATE

ANCHOR BOLTS/RODS

LEVELING NUTS
OR SHIM PACKS

THE FOLLOWING SHALL BE APPLIED FOR ERECTION STABILITY

- ALL COLUMNS SHALL BE ANCHORED BY A MINIMUM OF 4 ANCHOR RODS/BOLTS
- EACH COLUMN ANCHOR ROD/BOLT ASSEMBLY, INCLUDING THE COLUMN-TO-BASE WELD AND THE COLUMN FOUNDATION, SHALL BE DESIGNED TO RESIST A MINIMUM ECCENTRIC GRAVITY LOAD OF 300 POUNDS LOCATED 18 IN. FROM THE EXTREME OUTER FACE OF THE COLUMN IN EACH DIRECTION AT THE TOP OF THE COLUMN SHAFT
- COLUMNS SHALL BE SET ON LEVEL FINISHED FLOORS, PRE- GROUTED LEVELING PLATES, LEVELING NUTS OR SHIM PACKS THAT ARE ADEQUATE TO TRANSFER THE CONSTRUCTION LOADS
- ALL COLUMNS SHALL BE EVALUATED BY A COMPETENT PERSON TO DETERMINE WHETHER GUYING OR BRACING IS NEEDED AND, IF NEEDED, BE INSTALLED

COLUMN ANCHORAGE
OSHA 1926.755

REPAIR, REPLACEMENT OR FIELD MODIFICATION OF ANCHOR RODS (ANCHOR BOLTS)
OSHA 1926.755(b)

OSHA 1926.755(b) addresses the situation where the steel erector encounters an anchor bolt that has been repaired, replaced or modified. The steel erector often cannot visually tell when an anchor bolt has been repaired and thus will not be aware of the repair unless notified that a repair has been made. If an anchor bolt has been improperly repaired, replaced or modified, it could lead to a collapse.

The repair, replacement or field modification of anchor rods (anchor bolts) is prohibited without the approval of the project structural engineer of record.

Prior to the erection of a column, the controlling contractor shall provide written notification to the steel erector if there has been any repair, replacement or modification of the anchor bolts for that column. This requirement, working in conjunction with OSHA 1926.752(a)(2), completes a crucial communication loop. The steel erector generally does not have contact with the project structural engineer of record. The steel erector cannot rely on the controlling contractor at present to convey the approval of the project structural engineer of record for repair, replacement or modification of anchor bolts because it is not required.

Safety Tip: The structural engineer of record shall approve the repair, replacement, or field modification of anchor rods.

BEAMS AND COLUMNS
OSHA 1926.756

OSHA 1926.756 sets forth requirements for connections of beams and columns to minimize the hazard of structural collapse during the early stages of the steel erection process. Recognizing inappropriate or inadequate connections of beams and columns is hazardous and can lead to collapses and worker fatalities, OSHA in this section establishes performance and specification requirements to address these hazards.

GENERAL
OSHA 1926.756(a)

During the final placing of solid web structural members, the load shall not be released from the hoisting line until the members are secured with at least two bolts per connection, of the same size and strength as shown in the construction documents. The members shall be drawn up snug tight or secured by an equivalent connection as specified by the project structural engineer of record. **(See Figure 18-10)**

Figure 18-10. This illustration shows the requirements for structural steel assemblies.

NOTE: SOLID WEB STRUCTURAL MEMBERS USED AS DIAGONAL BRACING SHALL BE SECURED BY AT LEAST ONE BOLT PER CONNECTION DRAWN SNUG TIGHT OR SECURED BY AN EQUIVALENT CONNECTION AS SPECIFIED BY THE PROJECT STRUCTURAL ENGINEER OF RECORD.

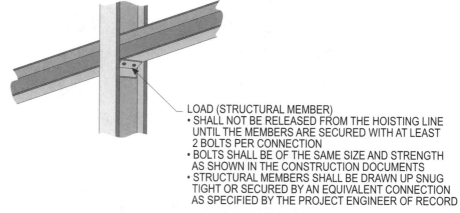

LOAD (STRUCTURAL MEMBER)
- SHALL NOT BE RELEASED FROM THE HOISTING LINE UNTIL THE MEMBERS ARE SECURED WITH AT LEAST 2 BOLTS PER CONNECTION
- BOLTS SHALL BE OF THE SAME SIZE AND STRENGTH AS SHOWN IN THE CONSTRUCTION DOCUMENTS
- STRUCTURAL MEMBERS SHALL BE DRAWN UP SNUG TIGHT OR SECURED BY AN EQUIVALENT CONNECTION AS SPECIFIED BY THE PROJECT ENGINEER OF RECORD

BEAMS AND COLUMNS
OSHA 1926.756

In addition, only bolts of the same strength and size as shown in the erection drawings to be used in securing the member until the final connections can be made. This will prevent collapses caused by the use of lesser strength/size bolts.

DIAGONAL BRACING
OSHA 1926.756(b)

Safety Tip: Solid web structural member used as diagonal bracing shall be secured by at least one bolt per connection drawn snug tight.

Solid web structural members used as diagonal bracing shall be secured by at least one bolt per connection drawn snug tight or secured by an equivalent connection as specified by the project structural engineer of record. In many cases, solid web structural members, such as channels or beams, are used as diagonal bracing or wind bracing. When used for this purpose, a one-bolt connection is sufficient. These members play a different role in erection stability than members used for other purposes since these members are designed to provide stability for the final completed structure and are not used as walking/working surfaces. Compliance with this provision will provide safe connections for these members. No comments were received addressing this paragraph, and the final rule is issued as proposed.

DOUBLE CONNECTIONS AT COLUMNS AND/OR AT BEAM WEBS OVER A COLUMN
OSHA 1926.756(c)

A double connection is a type of attachment in which the ends of two steel members join to opposite sides of a central (carrying) member — such as a beam, girder or column web — using the same bolts. The erection process is as follows:

• The first member is bolted to a beam, girder or column web. Later, a second member is added to the opposite side of the existing connection. This second member is attached using the same bolts (going through the same holes) that are being used to attach the first member. To attach the second member, the nuts on the first beam's bolts have to be removed and the bolts backed most of the way out; the ends of the bolts have to be flush with the surface of the central member so that the second member can be lined up with the existing holes. Only fractions of an inch of the ends of the bolts are now preventing the first beam from falling. Once the holes in the connection plate of the second member are lined up with the first beam's bolts, the bolts are pushed back through all the holes and the nuts are put back on the bolts and tightened to secure the three pieces of steel together. **(See Figure 18-11)**

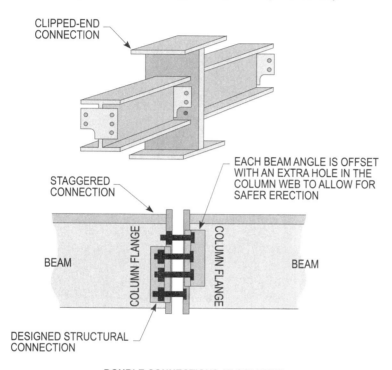

**DOUBLE CONNECTIONS AT COLUMNS
AND/OR AT BEAM WEBS OVER A COLUMN
OSHA 1926.756(c)**

Figure 18-11. This illustration shows a clipped end connection and staggered end connection.

Safety Tip: Double connections at columns and/or beam webs over a column is an extremely dangerous task, and must be performed with care.

This maneuver is extremely dangerous. The process often takes place with a worker sitting on the first beam. If the first beam collapses, the worker falls. The risk of collapse is high because of the tenuous grip of the loosened bolts and the possibility that the connector's spud wrench, which is used to align the second (incoming) member, may slip. If at any time the carrying member (the central member to which the first and second members are being attached) reacts to residual stresses developed through welding and/or misaligned connections at lower elevations, the carrying member can move suddenly, causing the bolts or the spud wrench to become dislodged. The second (incoming) member can also cause problems if it bumps up against the fitting or wrench end. Additionally, crane operators, wind, structural movements and the connector straining to make a tough connection impose stresses that can lead to disengagement of the connection.

When making a double connection, the first member shall remain connected to a supporting member by at least one connection bolt at all times unless a connection seat (see definition) or equivalent connection device is supplied with the members to secure the first member and prevent the column from being displaced. At a minimum, one bolt shall remain wrench tight in order to keep the first member from separating from the supporting member when the nuts are removed from the bolts that are to be shared with the second member.

Note: Appendix H provides examples of equivalent connection devices. They include "clipped end" and "staggered bolt" connections.

OSHA 1926.756(c)(2) does not permit such a practice. This subpart requires the erector to secure a seat (designed to support the load) to both the supporting and first members while the double connection is being made. The function of the seat is to provide support to the members until the double connection can be safely connected. Connecting the first member to the supporting member with the seat is a crucial step in making these double connections safely, since one of the dangers is that either the supporting member or the first member will be bumped or will pull away during the double connection process. The connection seat is only intended to facilitate that particular double connection.

This subpart also explicitly requires that seats or equivalent devices shall be designed to support the load during the double connection process. If these devices are to be used, they have to be capable of supporting the weight of the members involved; and that weight may vary significantly from job to job. The erector may not know the magnitude of the loads in time to have devices engineered and fabricated for the job. It is more efficient to incorporate this engineering determination into the design of the members and connections.

COLUMN SPLICES
OSHA 1926.756(d)

Each column splice shall be designed to resist a minimum eccentric gravity load of 300 pounds located 18 in. from the extreme outer face of the column in each direction at the top of the column shaft. This requirement, along with the requirements in OSHA 1926.755(a)(1) and (a)(2) for anchor rods/ bolts, will help to stabilize columns that employees have to climb during the erection process. By specifying requirements for certain key building elements, such as anchor bolts, column splices and double connections, the standard will prevent structural collapses. This section specifies a minimum force that a column splice shall withstand without failure before an employee is allowed to climb it.

PERIMETER COLUMNS
OSHA 1926.756(e)

The erection of perimeter columns is prohibited unless the column extends a minimum of 48 in. above the finished floor to permit installation of perimeter safety cables prior to the erection of the next tier, except where constructibility does not allow. OSHA 1926.760(a)(2) requires that the perimeter safety cables be installed at the final interior and exterior perimeters of the structure's finished floors of multi-story structures as soon as the decking has been installed. When the safety cables must be attached to the perimeter columns, the columns shall be at least 48 in. above the finished floor in order for the perimeter cable system to comply with the requirements of Subpart M. OSHA 1926.760(d) requires that perimeter safety cable systems conform to the criteria for guardrail systems in OSHA 1926.502.

The perimeter columns have holes or other devices in or attached to them at 42 to 45 in. above the finished floor and the midpoint between the finished floor and the top hole to permit the installation of perimeter cables, except where constructibility does not allow. This allows the erector to install the cables promptly when the columns have been erected.

OPEN WEB STEEL JOISTS
OSHA 1926.757

Some of the most serious risks facing the ironworker are encountered during the erection of open web steel joists, particularly landing loads on unbridged joists and improperly placing loads on joists. Based on an analysis of ironworker fatalities from January 1984 to December 1990 OSHA determined that of the approximately 40 fatalities caused by collapse, more than half were related to the erection of steel joists.

GENERAL
OSHA 1926.757(a)

Where steel joists are utilized, and columns are not framed in at least two directions with solid web structural steel members, a steel joist (commonly referred to as the "OSHA joist") shall be field-bolted at the column except as provided in OSHA 1926.757(a)(2), which addresses these joists installed near the column. **(See Figure 18-12)**

Figure 18-12. This illustration shows the requirements for structural steel assemblies.

> **THE FOLLOWING SHALL BE CONSIDERED FOR THE INSTALLATION OF STEEL JOISTS:**
>
> • A VERTICAL STABILIZER PLATE A MINIMUM OF 6 IN. x 6 IN. SHALL BE PROVIDED AT EACH COLUMN FOR STEEL JOISTS AND EXTEND AT LEAST 3 IN. BELOW THE BOTTOM CHORD OF THE JOIST, WITH A 13/16 IN. HOLE TO PROVIDE AN ATTACHMENT POINT FOR GUYING OR PLUMBING CABLES
>
> • TO PREVENT ROTATION DURING ERECTION, THE BOTTOM CHORDS OF STEEL JOISTS AT COLUMNS SHALL BE STABILIZED
>
> • HOISTING CABLES SHALL NOT BE RELEASED UNTIL THE SEAT AT EACH END OF THE STEEL JOIST IS FIELD-BOLTED, AND EACH END OF THE BOTTOM CHORD IS RESTRAINED BY THE COLUMN STABILIZER PLATE

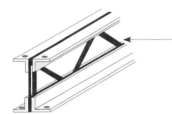

OPEN WEB STRUCTURAL MEMBER
• WHERE STEEL JOISTS ARE USED AND COLUMNS ARE NOT FRAMED IN AT LEAST 2 DIRECTIONS WITH SOLID WEB STRUCTURAL STEEL MEMBERS, A STEEL JOIST SHALL BE FIELD-BOLTED AT THE COLUMN TO PROVIDE STABILITY TO THE COLUMN DURING ERECTION

OPEN WEB STEEL JOISTS
OSHA 1926.757

The following shall be considered for the installation of steel joists:

- A vertical stabilizer plate a minimum of 6 in. by 6 in. shall be provided at each column for steel joists and extend at least 3 in. below the bottom chord of the joist, with a 13/16 in. hole to provide an attachment point for guying or plumbing cables

- To prevent rotation during erection, the bottom chords of steel joists at columns shall be stabilized

- Hoisting cables shall not be released until the seat at each end of the steel joist is field-bolted, and each end of the bottom chord is restrained by the column stabilizer plate

Where constructibility does not allow the steel joist to be installed at the column, an alternate means of stabilizing joists shall be installed on both sides near the column. Such alternate means shall provide stability equivalent to OSHA joists attached at the column; be designed by a qualified person; be shop installed; and be included in the erection drawings. OSHA believes that, even though OSHA joists are attached to the column the overwhelming majority of the time, workers need to receive the same protection from collapse when the OSHA joist is attached near the column. Thus, the alternate means of stabilization shall be considered and planned in the early stages of design and material preparation.

A steel joist (OSHA joist) at or near the column that spans 60 ft. or less shall be designed with sufficient lateral stiffness that the joist does not need erection bridging to maintain its stability when an employee goes out onto it to release the hoisting cable. Since the joist at the column is the OSHA joist and is either the first joist in place or the joist that boxes the bay, there is no other joist in place nearby for the erector to attach erection bridging. Therefore, without this provision, compliance with the bridging requirements would be infeasible for an OSHA joist. Consequently, the OSHA joist itself shall possess sufficient lateral stiffness to allow the erection process to progress safely.

Steel joists located at or near the column that spans more than 60 ft. shall be set in tandem, i.e., two steel joists shall be attached together, usually with all bridging installed (both bolted diagonal erection and horizontal bridging). These larger OSHA joists are commonly used in open structures such as warehouses, gymnasiums and arenas. This provision also allows the use of alternate means of erection of such long span steel joists, provided that the alternative is designed by a qualified person to ensure equivalent stability and is included in a site-specific erection plan.

Compliance with these provisions should help to satisfy the stability requirements of OSHA 1926.757(a)(5) of this section. This section prohibits the placement of steel joists or steel joist girders on any support structure unless it has been stabilized.

When steel joist(s) are landed on a structure, a hazard arises when a single steel joist or a bundle of joists are placed on the structure and then left unattended and unattached. An example of this might involve lighter steel joists, under 40 ft. in length, that would not require erection bridging under this section. A common practice in erecting these lighter joists, which can be set in place by hand, is to have a crane set the columns, steel joist girders or solid web primary members and bolted joists at the columns as required by OSHA 1926.757(a)(1) of this section, thus boxing the bays. The crane would then place a bundle of filler joists at an end or, more likely, at the center of the bay for installation by hand, and then move on to the next bay. Because cranes are among the more costly pieces of equipment on a steel erection job, minimizing crane time at the site is cost effective. This provision requires that, when steel joists are landed on structures, they be secured to prevent unintentional displacement, i.e., the bundles shall remain intact prior to installation until the time comes for them to be set. This paragraph also prevents those ironworkers who are shaking out the filler joists from getting too far ahead of those workers welding the joists, a practice that leaves many joists placed but unattached.

OSHA 1926.757(a)(7) addresses the potential for failure that can occur when a steel joist or joist girder is modified from its original manufactured state. OSHA believes modifications to joists can have disastrous consequences if performed by jobsite personnel without taking into account the design characteristics of the joist or joist girder. This provision prohibits modification without the prior approval of the project structural engineer of record.

FIELD-BOLTED JOISTS
OSHA 1926.757(a)(8)

Except for steel joists that have been pre-assembled into panels (panelized), connections of individual steel joists to steel structures in bays of 40 ft. or more shall not be made unless they have been fabricated to allow for field bolting during erection. This means that both the joists and the supporting member shall be fabricated with holes to allow the joists to be bolted to the supporting structure; otherwise they are prohibited from being erected. Unless constructibility does not allow, these connections shall be by field bolting.

The use of steel joists and steel joist girders shall not be used as anchorage points for a fall arrest system unless written direction allowing such use is obtained from a qualified person. Although performance criteria and manufacturer's specifications are not currently available regarding the adequacy of steel joists and steel joist girders as anchorages for fall protection systems, this provision recognizes that some joists and girders may be strong enough to meet the load requirements for anchorages in OSHA 1926.760.

OSHA 1926.757(a)(10) addresses the hazard posed by bridging joists without establishing an adequate terminus point for the bridging. Bridging is not effective until a terminus point is created. The following definitions are important to know for bridging joists:

- "Bridging," an operation integral to steel joist construction, refers to the steel elements that are attached between the joists (from joist to joist) to provide stability.

- "Erection bridging" is defined as "the bolted diagonal bridging that is required to be installed prior to releasing the hoisting cables from the steel joists."

- "Horizontal bridging," usually angle iron, is attached between steel joists, to the top and bottom chords of each joist, by welding.

See Figures 18-13(a) through (d) for a detailed illustration pertaining to bridging terminus points.

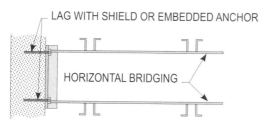

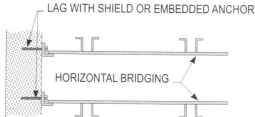

HORIZONTAL BRIDGING TERMINUS AT WALL
OSHA 1926.757(a)(10)

Figure 18-13(a). This illustration shows horizontal bridging terminus at wall.

Figure 18-13(b). This illustration shows horizontal bridging at structural shape.

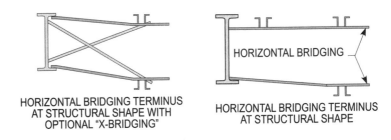

HORIZONTAL BRIDGING TERMINUS
AT STRUCTURAL SHAPE WITH
OPTIONAL "X-BRIDGING"

HORIZONTAL BRIDGING TERMINUS
AT STRUCTURAL SHAPE

HORIZONTAL BRIDGING TERMINUS AT STRUCTURAL SHAPE
OSHA 1926.757(a)(10)

Figure 18-13(c). This illustration shows bolted diagonal bridging terminus at wall.

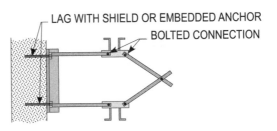

BOLTED DIAGONAL BRIDGING TERMINUS AT WALL

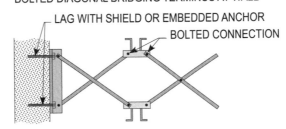

BOLTED DIAGONAL BRIDGING TERMINUS AT WALL

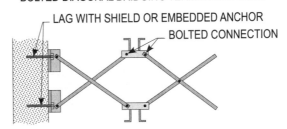

BOLTED DIAGONAL BRIDGING TERMINUS AT WALL

BOLTED DIAGONAL BRIDGING TERMINUS AT WALL
OSHA 1926.757(a)(10)

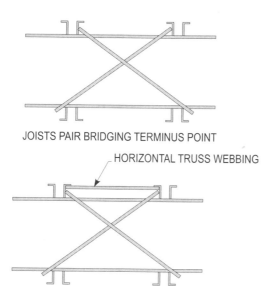

JOISTS PAIR BRIDGING TERMINUS POINT

HORIZONTAL TRUSS WEBBING

JOISTS PAIR BRIDGING TERMINUS POINT
WITH HORIZONTAL TRUSS

**JOISTS PAIR BRIDGING TERMINUS POINT
OSHA 1926.757(a)(10)**

Figure 18-13(d). This illustration shows joists pair bridging terminus point.

There are several provisions in this section that require bridging to be anchored. This means, by definition, that the steel joist bridging shall be connected to a bridging terminus point. The term, "bridging terminus point," is defined as follows:

> • Bridging terminus point means a wall, beam, tandem joists (with all bridging installed and a horizontal truss in the plane of the top chord) or other element at an end or intermediate point(s) of a line of bridging that provides an anchor point for the steel joist bridging.

Safety Tip: In order for the bridging to be anchored, a terminus point shall be established prior to installing bridging.

A terminus point shall be established prior to installing the bridging in order for the bridging to be anchored. OSHA is aware that steel erection is a progressive process that requires one piece to be erected before the subsequent piece can be attached to it. This provision requires pre- planning to determine the particular location of the terminus point for the attachment of bridging. To assist in developing or determining terminus points, OSHA is providing illustrative drawings of examples of bridging terminus points in non-mandatory Appendix C. In addition, OSHA 1926.757(c)(5), discussed below, deals with the situation in an erection sequence where the permanent bridging terminus points are not yet in existence at the time the joists and bridging are erected.

ATTACHMENT OF STEEL JOISTS AND STEEL JOIST GIRDERS
OSHA 1926.757(b)

The following three types of joists identified as being used in the steel erection industry:

- The K-Series open web steel joists, having joist depths from 8 in. through 30 in., are primarily used to provide structural support for floors and roofs of buildings. Although light in weight, they possess a high strength to weight ratio.
- The LH-Series steel joists span up to and including 96 ft. These joists are used for the direct support of floor or roof slabs or decks between walls, beams and main structural members, and their depths range from 18 in. to 48 in.

- The "Deep Longspan," or DLH-Series joists can run up to 144 ft. and have depths from 52 in. through 72 in.

The attachment of all three series of joists is addressed in OSHA 1926.757(b). The hazard addressed in this subpart is the adequacy of the attachment of joists that could affect the stability of the joist and thus the safety of the employee erecting the joist. OSHA 1926.757(b)(1) and (b)(2) specify the minimum attachment specifications for the lighter and the heavier joists, respectively. At a minimum, the K-Series shall be attached with either 2-1/8 in. fillet welds 1 in. long, or with 2-1/2 in. bolts. In addition, the provision provides alternative performance language "or the equivalent" to allow for attachment by any another means that provides at least equivalent connection strength. Similarly, at a minimum, the LH-Series and DLH-Series shall be attached with either 2-1/4 in. fillet welds 2 in. long, or with 2-3/4 in. bolts. Again, OSHA is providing performance language, "or the equivalent," for the reasons discussed above.

OSHA 1926.757(b)(3) addresses the hazards associated with the following improper erection sequence:

- Landing joists on the support structure
- Spreading them out unattached to their final position
- Attaching them

This procedure creates the potential for worker injury because joists handled in this manner may fall or the structure may collapse. To eliminate these hazards, this paragraph requires, with one exception discussed in OSHA 1926.757(b)(4) below, that each steel joist be attached, at least at one end on both sides of the seat, immediately upon placement in its final erection position, before any additional joists are placed.

OSHA 1926.757(b)(4) is an exception to the OSHA 1926.757(b)(3) "attachment upon final placement" requirement. It addresses the situation where steel joists have been pre-assembled into panels prior to placement on the support structure.

Pre-assembly of panels usually involves the installation of diagonal and horizontal bridging to form a platform at ground level, which eliminates fall hazards associated with attaching bridging at elevated work stations. Placing joists on the support structure in this manner eliminates the single joist instability concerns. Furthermore, because of the inherent stability of these pre-assembled panels, this paragraph requires only that the four corners of the panel be attached to the support structure before releasing the hoisting cables. The attachment can be either bolted or welded.

An additional benefit of panelizing joists is that, following installation on the primary support structure, in all likelihood, the panel will immediately provide anchorage points for fall protection systems.

Additionally, the pre-assembly allows for alternative joist erection methods such as a hybrid form of steel erection involving steel/wood-panelized roof structures, where wooden decking (dimensional wood and plywood) is attached to a single steel joist and the resulting panels are set on the support structure. Again, by placing joists on the support structure in this manner, the instability concerns and other hazards associated with attaching single joists are avoided.

ERECTION OF STEEL JOISTS
OSHA 1926.757(c)

For joists that require bridging as provided in Tables A and B, at least one end of each steel joist shall be attached on both sides of the seat to the support structure before the hoisting cables can be released. This will allow smaller lighter joists (that do not require bridging and can be landed in bundles) to be placed on the structure and spread out by hand. Once the joists have been placed in their final position, however, they shall be attached in accordance with OSHA 1926.757(b)(3) of this section.

See Figure 18-14(a) for an illustration pertaining to erection bridging for short span joists.

ERECTION BRIDGING FOR SHORT SPAN JOISTS		ERECTION BRIDGING FOR SHORT SPAN JOISTS		ERECTION BRIDGING FOR SHORT SPAN JOISTS		ERECTION BRIDGING FOR SHORT SPAN JOISTS	
JOIST	SPAN	JOIST	SPAN	JOIST	SPAN	JOIST	SPAN
8L1	NM	22K5	35-0	30K11	52-0	28KCS2	40-0
10K1	NM	22K6	36-0	30K12	54-0	28KCS3	45-0
12K1	23-0	22K7	40-0	10KCS1	NM	28KCS4	53-0
12K3	NM	22K9	40-0	10KCS2	NM	28KCS5	53-0
12K5	NM	22K10	40-0	10KCS3	NM	30KCS3	45-0
14K1	27-0	22K11	40-0	12KCS1	NM	30KCS4	54-0
14K3	NM	24K4	36-0	12KCS2	NM	30KCS5	54-0
14K4	NM	24K5	38-0	12KCS3	NM		
14K6	NM	24K6	39-0	14KCS1	NM		
16K2	29-0	24K7	43-0	14KCS2	NM		
16K3	30-0	24K8	43-0	14KCS3	NM		
16K4	32-0	24K9	44-0	16KCS2	NM		
16K5	32-0	24K10	NM	16KCS3	NM		
16K6	NM	24K12	NM	16KCS4	NM		
16K7	NM	26K5	38-0	16KCS5	NM		
16K9	NM	26K6	39-0	18KCS2	35-0		
18K3	31-0	26K7	43-0	18KCS3	NM		
18K4	32-0	26K8	44-0	18KCS4	NM		
18K5	33-0	26K9	45-0	18KCS5	NM		
18K6	35-0	26K10	49-0	20KCS2	36-0		
18K7	NM	26K12	NM	20KCS3	39-0		
18K9	NM	28K6	40-0	20KCS4	NM		
18K10	NM	28K7	43-0	20KCS5	NM		
20K3	32-0	28K8	44-0	22KCS2	36-0		
20K4	34-0	28K9	45-0	22KCS3	40-0		
20K5	34-0	28K10	49-0	22KCS4	NM		
20K6	36-0	28K12	53-0	22KCS5	NM		
20K7	39-0	30K7	44-0	26KCS2	39-0		
20K9	39-0	30K8	44-0	26KCS3	44-0		
20K10	NM	30K9	45-0	26KCS4	NM		
22K4	34-0	30K10	50-0	26KCS5	NM		

Figure 18-14(a). This table pertains to the erection bridging for short span joists.

See Figure 18-14(b) for an illustration pertaining to erection bridging for long span joists.

Based on the recognition of the inherent danger of employees working on unstable joists, no employee shall be allowed on steel joists where the span is equal to or greater than the span shown in Figures 18-14(a) and (b), unless the requirements of OSHA 1926.757(d) of this section are met.

OSHA 1926.757(c)(5) addresses the situation where the erection sequence calls for joists to be erected before the permanent bridging terminus points have been established. This situation commonly occurs in a single story structure that has masonry or architectural precast walls installed after the steel is partially or fully erected. Complying with this subpart will involve pre-planning and the addition of temporary bridging terminus points to provide stability and prevent structure collapse in this situation. **(See Figure 18-15)**

Note: Examples of bridging terminus points can be found in Appendix C.

Figure 18-14(b). This table pertains to the erection bridging for long span joists.

ERECTION BRIDGING FOR LONG SPAN JOISTS		ERECTION BRIDGING FOR LONG SPAN JOISTS	
JOIST	**SPAN**	**JOIST**	**SPAN**
18LH02	33-0	28LH11	NM
18LH03	NM	28LH12	NM
18LH04	NM	28LH13	NM
18LH05	NM	32LH06	47 - 60
18LH06	NM	32LH07	47 - 60
18LH07	27-0	32LH08	55 - 60
18LH08	NM	32LH09	NM - 60
18LH09	NM	32LH10	NM - 60
20LH02	33-0	32LH11	NM - 60
20LH03	38-0	32LH12	NM - 60
20LH04	NM	32LH13	NM - 60
20LH05	NM	32LH14	NM - 60
20LH06	NM	32LH15	NM - 60
20LH07	NM	36LH07	47 - 60
20LH08	NM	36LH08	47 - 60
20LH09	NM	36LH09	57- 60
20LH10	NM	36LH10	NM - 60
24LH03	35-0	36LH11	NM - 60
24LH04	39-0	36LH12	NM - 60
24LH05	40-0	36LH13	NM - 60
24LH06	45-0	36LH14	NM - 60
24LH07	NM	36LH15	NM - 60
24LH08	NM		
24LH10	NM		
24LH11	NM		
28LH05	42-0		
28LH06	46-0		
28LH07	NM		
28LH08	NM		
28LH09	NM		
28LH10	NM		

Figure 18-15. This illustration shows horizontal bridging terminus point secured by temporary guy cables.

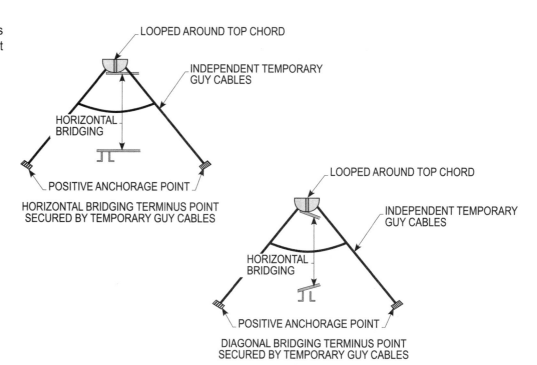

BRIDGING TERMINUS POINT SECURED BY TEMPORARY GUY CABLES
OSHA 1926.757(c)(5)

ERECTION BRIDGING
OSHA 1926.757(d)

The following shall apply where the span of the steel joist is equal to or greater than the span shown in the tables in Figures 18-14(a) and (b):

- A row of bolted diagonal erection bridging shall be installed near the midspan of the joist.

- Hoisting cables shall not be released until the bolted diagonal erection bridging is installed and anchored.

- No more than one employee shall be allowed on the joist until all the bridging is installed and anchored.

  Note: This provision will require that all bridging that is required for the joist (both bolted diagonal and horizontal bridging) be installed before additional employees are allowed on the joist.

The following shall apply for steel joists over 60 ft. through 100 ft.:

- All rows of bridging for these spans shall be bolted diagonal bridging.

- Two rows of bolted diagonal erection bridging shall be installed at the third points of the steel joists.

- Hoisting cables shall not be released until these two rows of erection bridging are installed and anchored.

- No more than two employees shall be allowed on a span until all other bridging is installed and anchored.

The following shall be applied to steel joists where the span is between 100 ft. through 144 ft.:

- All rows of bridging shall be bolted diagonal bridging.

- Hoisting cables shall not be released until all bridging is installed and anchored.

- No more than two employees shall be allowed on these spans until all bridging is installed and anchored.

For steel members spanning over 144 ft., the erection methods of these members shall be in accordance with OSHA 1926.756.

OSHA 1926.757(d)(5) requires that where any steel joist in paragraphs (c)(2) and (d)(1), (d)(2) and (d)(3) of this section is a bottom chord bearing joist, a row of bolted diagonal bridging shall be provided near the support(s). This bridging shall be installed and anchored before the hoisting cable(s) is released.

OSHA 1926.757(d)(6) specifies that when bolted diagonal erection bridging is required by this section, the erection drawings shall indicate the bridging and the erection drawings shall be the exclusive indicator of the proper bridging placement. This is to eliminate any confusion that might arise where bridging placement is specified through other means; reliance is to be placed only on the erection drawings for this information. In addition, shop-installed bridging clips or functional equivalents shall be provided where bridging bolts to the steel joists. When a common bolt and nut attach two pieces of bridging to a steel joist, the nut that secures the first piece of bridging may not be removed from the bolt for the attachment of the second piece. In addition, when bolted diagonal erection bridging is required, bridging attachments may not protrude above the top chord of the steel joist.

LANDING AND PLACING LOADS
OSHA 1926.757(e)

The work practice provisions found in OSHA 1926.754(e) regarding the hoisting, landing and placing of deck bundles, in general, have already been discussed above. The hazards of landing and placing loads on steel joists are also addressed. The following apply to all activities associated with metal decking that is used as a support element for either a floor or roof system.

> • The employer placing a load on steel joists, during the construction period, shall ensure that the load is distributed so as not to exceed the carrying capacity of any steel joists.

> **Note:** This paragraph requires that the load be adequately distributed so that the carrying capacity of any steel joist is not exceeded. After this general requirement is met, the employer shall meet the specific conditions set forth in the remainder of OSHA 1926.757(e).

> • No construction loads are allowed on steel joists until all bridging is installed and anchored and all joist-bearing ends are attached.

> **Note:** A construction load means any load other than the weight of the employee(s), the joists and the bridging bundle. Although bundles of decking constitute a construction load under this definition, under certain conditions decking can be placed safely on the steel joists before all the bridging is installed and anchored. These conditions form the basis for the exceptions in OSHA 1926.757(e)(4) of this section.

> • A bridging bundle is not considered a "construction load." The weight of the bridging bundle is limited to 1000 pounds because bridging will be placed on the joists before they have been fully stabilized. To ensure safe placement, this paragraph requires that the bundle of joist bridging be placed over a minimum of three steel joists that are secured at one end. Also, to ensure stability of the load, this provision requires that the edge of the bridging bundle be positioned within 1 ft. of the secured end (some clearance is necessary for material handling purposes and to provide employee access to the steel joist's attachment point).

> • Special conditions shall be met before an employer is permitted to place a bundle of decking on steel joists that do not yet have all bridging installed. This subpart applies only to bundles of decking and not to other construction loads. The following six conditions shall be met before the exception to the provisions of OSHA 1926.757(e)(2) applies:

> > **(a)** The employer shall determine, based on information from a qualified person, that the structure or portion of the structure is capable of safely supporting the load of decking. This determination shall be documented in a site-specific erection plan that is made available at the construction site.

> > **(b)** The bundle of metal decking is placed over a minimum of three joists to distribute the load.

> > **(c)** The three steel joists supporting the bundle of metal decking have both ends attached to the support structure. The attachments shall meet the requirements prescribed in OSHA 1926.757(b).

> > **(d)** At least one row of bridging shall be attached and anchored to the three joists specified in OSHA 1926.757(e)(4)(iii). The qualified person determines the type of bridging, erection bridging or horizontal bridging, needed to satisfy this requirement.

(e) The total weight of the bundle of metal decking does not exceed 4000 pounds.

(f) The edge of the bundle of metal decking shall be placed within 1 ft. of the bearing surface of the joist.

(g) The location for safe placement of all construction loads, not just metal decking, the edge of the construction load shall be positioned within 1 ft. of the secured end of the steel joists in order to enhance the stability of the load (some clearance is necessary for material handling purposes and for access to the steel joist's attachment point to the support structure).

See Figure 18-16 for a detailed illustration pertaining to landing and placing loads.

Figure 18-16. This illustration shows the requirements for landing and placing loads.

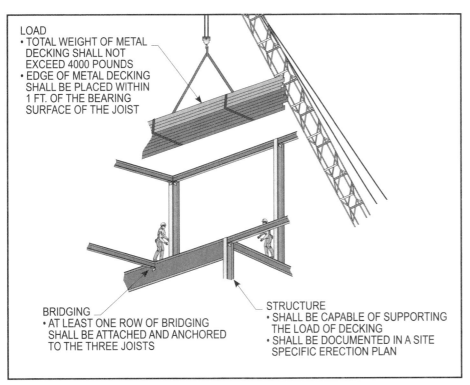

LOAD
• TOTAL WEIGHT OF METAL DECKING SHALL NOT EXCEED 4000 POUNDS
• EDGE OF METAL DECKING SHALL BE PLACED WITHIN 1 FT. OF THE BEARING SURFACE OF THE JOIST

BRIDGING
• AT LEAST ONE ROW OF BRIDGING SHALL BE ATTACHED AND ANCHORED TO THE THREE JOISTS

STRUCTURE
• SHALL BE CAPABLE OF SUPPORTING THE LOAD OF DECKING
• SHALL BE DOCUMENTED IN A SITE SPECIFIC ERECTION PLAN

LANDING AND PLACING LOADS
OSHA 1926.757(e)

SYSTEMS-ENGINEERED METAL BUILDINGS
OSHA 1926.758

This section sets forth requirements to erect systems-engineered metal buildings safely. Systems-engineered metal buildings are defined in the definition section of this proposal. Systems-engineered metal buildings include structures ranging from small sheds to larger structures such as warehouses, gymnasiums, churches, airplane hangars and arenas.

Systems-engineered metal buildings use different types of steel members and a different erection process than typical steel erection. Many contractors erect systems-engineered metal buildings exclusively. An overwhelming majority of these erectors are small employers. The erection of systems-engineered metal structures presents certain unique hazards. Although some of the hazards are similar to general steel erection, other hazards, such as those associated with anchor bolts, construction loads and double connections, are different.

Most of the requirements in this section are similar to those in other sections of the steel erection standard. Where a conflict arises between a provision in the systems-engineered metal building section and that of another section of Subpart R, to the extent that the work being performed is systems-engineered metal building work, the more specific systems-engineered metal building section would apply. This section, however, must not be interpreted to mean that (apart from OSHA 1926.755 and 1926.757), the other provisions of Subpart R do not apply to systems-engineered metal buildings where appropriate.

The requirements contained in Subpart R apply to systems-engineered metal buildings except for OSHA 1926.755 (Column Anchorage) and OSHA 1926.757 (Open Web Steel Joists).

All structural columns shall be anchored by at least four anchor bolts.

OSHA 1926.758(c) is unique to the erection of systems-engineered metal buildings because rigid frames are found only in this type of structure. Rigid frames shall have 50 percent of their bolts or the number of bolts specified by the manufacturer (whichever is greater) installed and tightened on both sides of the web adjacent to each flange before the hoisting equipment is released. Like final OSHA 1926.756(a), this provision requires an adequate number of bolts to ensure stability before the hoist line is released. Rigid frames are fully continuous frames that provide the main structural support for a systems-engineered metal building. They provide the support that is typically provided by columns and beams in conventional steel erection. Due to design and load requirements, connections in rigid frames occupy a greater area and require more than two bolts upon initial connection. The remaining bolts are used to attach other members to the structure and provide stability against wind loading. To allow these connections to be bolted only with two bolts would not be adequate in many cases to prevent a collapse hazard. No comments were received on this paragraph, and it is promulgated as proposed.

Safety Tip: Load shall not be placed on any structural steel framework unless the framework has been safely bolted.

Construction loads shall not be placed on any structural steel framework unless such framework has been safely bolted, welded or otherwise adequately secured. Without proper bolting or welding to provide stability, a construction load could cause a collapse of the structure.

When girts or eave struts share common connection holes, a double connection hazard exists. As with OSHA 1926.756(c), a seat or similar connection will prevent one member from becoming displaced during the double connection activity. In girt and eave strut to frame connections where girts or eave struts share common connection holes, at least one bolt with its wrench-tight nut should remain in place for the connection of the first member unless a field-attached seat or similar connection device is present to secure the first member so that the girt or eave strut is always secured against displacement. In addition, the seat or similar connection device shall be provided by the manufacturer of the girt or eave strut so that it is designed properly for the intended use. Because this form of double connection is unique to systems-engineered metal building construction and might not be considered a double connection under a literal reading of OSHA 1926.756(c), this provision specifically addresses girt and eave strut to frame connections.

Both ends of all steel joists or cold-formed joists shall be fully bolted and/or welded to the support structure before releasing the hoisting cables, allowing an employee on the joists, or allowing any construction loads on the joists.

The use of purlins and girts shall not be used as anchorage points for a fall arrest system unless written approval to do so is obtained from a qualified person. Generally, purlins and girts are lightweight members designed to support the final structure. They may not have been designed to resist the force of a fall arrest system. If, however, a qualified person determines that the purlin or girt is of sufficient strength to support a fall arrest system, it may be used for that purpose. The qualified person would be required to provide written documentation of this determination.

Purlins may only be used as a walking/working surface when installing safety systems, after all permanent bridging has been installed and fall protection is provided. Purlins are "Z" or "C" shaped lightweight members, generally less than 1/8 in. thick, 2 to 4 in. wide on the top and up to 40 ft. long. Purlins are not to be walked on and, because of their shape, are likely to roll over when used as a walking/working surface if not properly braced. OSHA has not included cold-formed joists in this paragraph because they provide greater stability than do purlins that are not to be used as walking/working surfaces without the addition of specific safety precautions.

Construction loads may be placed within a zone that is not more than 8 ft. from the centerline of the primary support member. Unlike conventional decking, systems-engineered metal building decking bundles are lighter, and the sheets in the bundle are staggered. The staggering of these bundles shall be set so that the end of one bundle overlaps another bundle since the lengths of the sheets vary. The zone needs to be big enough to allow for the lapping while still having the support of the structure. An 8 ft. zone allows enough room to meet these objectives.

FALLING OBJECT PROTECTION
OSHA 1926.759

This section sets forth the requirements for providing employees with protection from falling objects. A real, everyday hazard posed to steel erection employees is loose items that have been placed aloft that can fall and strike employees working below.

Safety Tip: Materials, tools, and equipment not in use, shall be stored.

All materials, equipment, and tools that are not in use while aloft shall be secured against accidental displacement.

When it is necessary to have work performed below on-going steel erection activities (other than hoisting), effective overhead protection shall be provided to those workers to prevent injuries from falling objects. If this protection is not provided, work by other trades is not to be permitted below steel erection work. One way controlling contractors can reduce the hazards associated with falling objects is by scheduling work in such a way that employees are not exposed.

OSHA 1926.759(b) states that, "The controlling contractor shall ensure that no other construction processes take place below steel erection unless adequate overhead protection for the employees below is provided." The use of the word "ensure" in this standard does not make the controlling contractor liable if it institutes reasonable measures to comply with the requirement. All defenses normally available to employers are equally available where a requirement is phrased using the term "ensure."

For a different reason, however, the Agency has rephrased the provision to read that the controlling contractor will "bar" other construction processes below steel erection. This change was made to more directly state that the employer shall institute measures to keep employees out of the area below the steel erection activities.

FALL PROTECTION
OSHA 1926.760

The following shall be considered for fall protection:

- General requirements
- Connectors
- Controlled decking zone
- Criteria for fall protection equipment
- Custody for fall protection

GENERAL REQUIREMENTS
OSHA 1926.760(a)

Safety Tip: Employees on a walking/ working surface with an unprotected side more than 15 ft. above a lower level shall be protected from falling.

Each employee engaged in a steel erection activity who is on a walking/working surface with an unprotected side or edge more than 15 ft. above a lower level shall be protected by conventional fall protection (systems/devices that either physically prevent a worker from falling or arrest a worker's fall). The following are exceptions to this rule:

- Exception 1 allows connectors to not use their personal fall protection to avoid hazards while working at heights between 15 ft. and 30 ft.

- Exception 2 allows workers engaged in decking in a controlled decking zone to work without conventional fall protection at heights between 15 ft. and 30 ft.

Prior to the revision to the steel erection standard, the fall protection requirements for steel erection were in three separate provisions. Depending on the structure and the type of fall exposure, one of the following applied:

- OSHA 1926.750(b)(1)(ii) and 1926.750(b)(2)(i) (both are in Subpart R)
- OSHA 1926.105(a) (Subpart E, Personal Protective and Life Saving Equipment)

These provisions were the subject of considerable litigation, the product of which was the following:

- In single story structures, OSHA 1926.105(a) applied, that required fall protection at and above 25 ft. for both fall hazards to the interior and exterior of the structure

- In multi-tiered buildings, OSHA 1926.750 applied to fall hazards to the interior of the building. Several courts held that, under that standard, fall protection was required at and above 30 ft.

- In multi-tiered buildings, OSHA 1926.105(a) applied to fall hazards to the exterior of the building, that required fall protection at and above 25 ft.

With the exception of OSHA 1926.754(b)(3), the revised standard eliminates distinctions between interior and exterior fall hazards and tiered versus untiered buildings for the fall protection trigger heights.

The fall protection rules for steel erection differ from the general fall protection rules in Subpart M, which set 6 ft. as the trigger height for fall protection. OSHA has determined that steel erection activities are different from most other construction activities. The different trigger height reflects these differences. OSHA also believes that the former fall protection rules relating to steel erection are insufficiently protective and need to be strengthened.

Steel erection differs from general construction in three major respects — the narrowness of the working surface, its location above, rather than below, the rest of the structure and a minimum distance of approximately 15 ft. to the next lower level. OSHA describes the steel erection process in the proposal as follows:

- Initially, vertical members, referred to as columns, are anchored to the foundation. The columns are then connected with solid web beams or steel joists and joist girders to form an open bay. In a multi-story building, the columns are usually two stories high. These structural members are set by connectors in conjunction with a hoisting device (typically a crane). When the two-story columns are set in place, the connector installs the header beams at the first level, which forms the first bay. Each floor is typically 12.5 ft. to 15 ft. in height. After an exterior bay is formed ("boxing the bay"), the filler beams or joists are placed in the bay. The connector then ascends the column to the next level, where the exterior members are connected to form a bay, and so on. The floor or roof decking process basically consists of hoisting and landing of deck bundles and the placement and securing of the metal decking panels.

In short, a new, very narrow working surface is constantly being created as skeletal steel is erected at various heights. For many steel erectors, especially connectors, the work starts at the top level of the structure.

The special circumstances of steel erection can make conventional fall protection very difficult to deploy below 15 ft. For many steel erectors, especially connectors, the work starts at the top level of the structure. This means that anchor points above foot level are often limited or unavailable. Because of the nature of the structure, the available fall arrest distance is usually about 15 ft.

The location of anchor points, in conjunction with a number of other factors, will affect the fall arrest distance — the distance a worker will fall before the fall arrest system stops the fall. The fall arrest distance is the sum of the distance the worker falls before the fall arrest system begins to stop the fall, plus the additional distance that it takes for the system to slow and then finally stop the fall completely. Other factors that affect the fall arrest distance include the type of fall protection system used, the type of components and how the system is configured and anchored. The degree of mobility needed for the worker, location of available anchor points and the need to limit the arresting forces on the worker's body also affect the choice of system and its installation.

Personal fall arrest systems commonly used by workers in full body harnesses often have one of the following:

- Shock absorbing lanyard
- Self-retracting lifeline
- Rope grab with vertical lifeline
- Shock absorbing lanyard with rope grab and vertical lifeline

Fall arrest distances can vary with different types and lengths of lanyards. The distances can also vary in systems that permit the user to adjust the amount of slack.

The three common types of anchorage systems include:

- Horizontally mobile and vertically rigid (such as a trolley connected to a flange of a structural beam)

- Horizontally fixed and vertically rigid (such as an eyebolt, choker or clamp connected to a structural beam, column or truss)

- Horizontally mobile and vertically flexible (such as a horizontal lifeline suspended between two structural columns or between stanchions, which are attached to a structural beam and designed to support the lifeline)

The total fall distance can differ significantly depending on how the system is configured. A system using an anchorage connector, harness and shock absorbing lanyard will have a total fall distance between 3 ft. and 23 ft., while the total fall distance for a system using an anchorage connector, harness and self-retracting lifeline will measure between 4 ft. and 10.5 ft. In 1995, one fall protection manufacturer indicated that the lowest point of the ironworker's body should be at least 12.5 ft. above the nearest obstacle in the potential fall path when using a properly rigged, rigidly anchored, personal fall arrest system of the shock absorbing lanyard type or self-retracting lifeline type. In view of the types of equipment available, potential locations of anchor points, and typical distance between work surfaces and the next lower level, OSHA determined that 15 ft. was an appropriate threshold for requiring fall protection, subject to the two exceptions mentioned above.

Perimeter safety cables shall be installed at the final interior and exterior perimeters of multi-story structures as soon as the decking has been installed. These cables shall be installed regardless of other fall protection systems in use. They shall meet the criteria for guardrail systems in Subpart M (OSHA 1926.502(b)). **(See Figure 18-17)**

Figure 18-17. This illustration shows the requirements for perimeter safety cables.

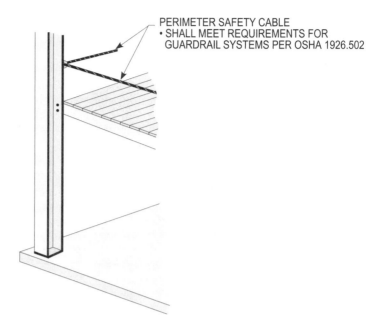

PERIMETER SAFETY CABLE
• SHALL MEET REQUIREMENTS FOR
  GUARDRAIL SYSTEMS PER OSHA 1926.502

PERIMETER SAFETY CABLES
OSHA 1926.760(2)

Connectors and employees working in controlled decking zones shall be protected from fall hazards as provided in OSHA 1926.760(b) and (c), respectively.

CONNECTORS
OSHA 1926.760(b)

Safety Tip: Connectors shall be protected from fall hazards of more than 2 stories or 30 ft.

Special rules shall be applied for employers of connectors:

- Each connector shall be protected from fall hazards of more than 2 stories or 30 ft. above a lower level, whichever is less.

- Connector training shall be accordance with OSHA 1926.761. Such training shall be specific to connecting and cover the recognition of hazards, and the establishment, access, safe connecting techniques and work practices required by OSHA 1926.756(c) and OSHA 1926.760(b).

Connectors shall be provided, at heights over 15 ft. and up to 30 ft. above a lower level, with a personal fall arrest system, positioning device system or fall restraint system and wear the equipment necessary to be tied off, or be provided with other means of protection from fall hazards in accordance with OSHA 1926.760(a)(1) (or, for protection against perimeter falls, OSHA 1926.760(a)(2)). **(See Figure 18-18)**

CONTROLLED DECKING ZONE (CDZ)
OSHA 1926.760(c)

The controlled decking zone is an alternative to fall protection for leading edge decking workers between 15 ft. and 30 ft. above a lower level. If an employer establishes a controlled decking zone that conforms to OSHA 1926.760(c), employees authorized to be in that zone who are trained pursuant to OSHA 1926.761, do not have to be provided with or use a fall protection system.

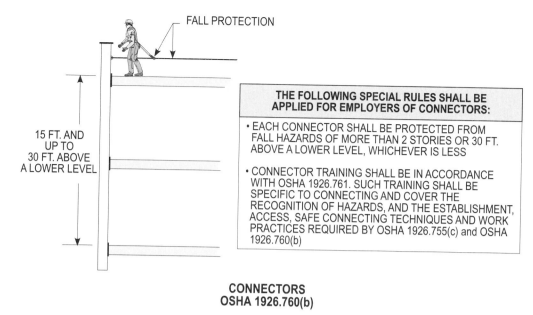

FALL PROTECTION

15 FT. AND
UP TO
30 FT. ABOVE
A LOWER LEVEL

THE FOLLOWING SPECIAL RULES SHALL BE APPLIED FOR EMPLOYERS OF CONNECTORS:

• EACH CONNECTOR SHALL BE PROTECTED FROM FALL HAZARDS OF MORE THAN 2 STORIES OR 30 FT. ABOVE A LOWER LEVEL, WHICHEVER IS LESS

• CONNECTOR TRAINING SHALL BE IN ACCORDANCE WITH OSHA 1926.761. SUCH TRAINING SHALL BE SPECIFIC TO CONNECTING AND COVER THE RECOGNITION OF HAZARDS, AND THE ESTABLISHMENT, ACCESS, SAFE CONNECTING TECHNIQUES AND WORK PRACTICES REQUIRED BY OSHA 1926.755(c) AND OSHA 1926.760(b)

**CONNECTORS
OSHA 1926.760(b)**

Figure 18-18. This illustration shows the requirements for connectors at heights over 15 ft. and up to 30 ft. above a lower level.

Each employee performing leading edge work in a controlled decking shall be protected from fall hazards of more than 2 stories or 30 ft., whichever is less. Controlled decking zones are inappropriate for decking operations at and above these heights. **(See Figure 18-19)**

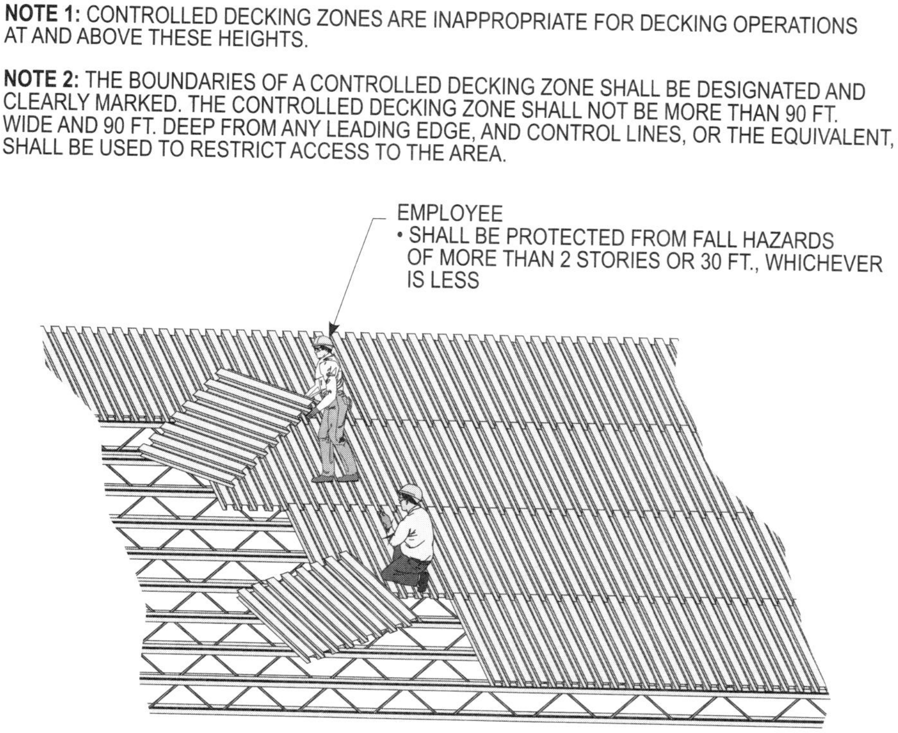

NOTE 1: CONTROLLED DECKING ZONES ARE INAPPROPRIATE FOR DECKING OPERATIONS AT AND ABOVE THESE HEIGHTS.

NOTE 2: THE BOUNDARIES OF A CONTROLLED DECKING ZONE SHALL BE DESIGNATED AND CLEARLY MARKED. THE CONTROLLED DECKING ZONE SHALL NOT BE MORE THAN 90 FT. WIDE AND 90 FT. DEEP FROM ANY LEADING EDGE, AND CONTROL LINES, OR THE EQUIVALENT, SHALL BE USED TO RESTRICT ACCESS TO THE AREA.

Figure 18-19. This illustration shows the requirements for a controlled decking zone.

EMPLOYEE
• SHALL BE PROTECTED FROM FALL HAZARDS OF MORE THAN 2 STORIES OR 30 FT., WHICHEVER IS LESS

**CONTROLLED DECKING ZONE (CDZ)
OSHA 1926.760(c)**

For example: Single story, high bay warehouse structures and pre-engineered metal buildings often require decking operations more than 30 ft. above lower levels. The exception would not apply in these situations.

Access to the controlled decking zone is limited exclusively to those employees who are actually engaged in and trained in the hazards involved in leading edge work.

The boundaries of a controlled decking zone shall be designated and clearly marked. The controlled decking zone shall not be more than 90 ft. wide and 90 ft. deep from any leading edge, and control lines, or the equivalent (for example, the perimeter wall), shall be used to restrict access to the area.

The controlled decking zone section requires that the boundaries of the zone be designated and clearly marked and that the access be limited exclusively to those employees engaged in leading edge work. One means of fulfilling this obligation is to erect control lines. While other methods might also be used, control lines are commonly used to restrict access to the unprotected area by creating a highly visible boundary. Their high visibility readily defines the area in which employees will work without conventional fall protection, and visually warns employees that access is limited to authorized personnel. Warning line systems, however, are erected close to the edge of a roof (as close as 6 ft.). They delineate the area where mechanical equipment may be used on roofs, and warn employees when they are approaching a fall hazard.

Safety Tip: Employees working in a controlled decking zone shall complete controlled decking zone training.

Each employee working in a controlled decking zone shall complete the controlled decking zone training, as specified in this subpart. Employees shall be trained to recognize the hazards associated with working in a controlled decking zone, and trained in the establishment, access, safe installation techniques and work practices required by certain sections of this subpart, such as OSHA 1926.754(e) — Decking and OSHA 1926.760(c) — Controlled Decking Zone.

During initial placement, deck panels shall be placed to ensure full support by structural members. This provision addresses the specific hazard that results when full support is absent when placing metal decking.

For example: In steel joist construction, metal deck sheets are typically 20 ft. or longer and may span more than 4 joists (typically spaced 5 ft. apart). A hazard is created if the deck is placed so that only three joists are supporting the sheet and the deck ends are unsupported. A worker not using fall protection and stepping onto the unsupported end of a deck sheet so placed is exposed to a potentially fatal fall hazard.

Unsecured decking in a controlled decking zone shall not exceed 3000 square feet. OSHA 1926.760(c)(5) is intended to limit the area of unsecured decking in which employees work. Because metal decking sheets are typically not uniformly sized and can create alignment problems, it is common practice to install a series of unsecured sheets on the structural member prior to fastening. The 3000 sq. ft. is necessary for the metal decking to be placed and then properly aligned prior to tack welding.

Safety deck attachments shall be performed in the controlled decking zone from the leading edge back to the control line and shall have at least two attachments per panel. This provision was intended to address the hazard in leading edge work that arises when an employee turns his/her back to the leading edge while attaching deck sheets. This provision will help prevent employees from inadvertently stepping off the leading edge. Safety deck attachments are usually accomplished with tack welds but can also be achieved with a mechanical attachment, such as self-drilling screws or pneumatic fasteners.

Final deck attachments and the installation of shear connectors shall not be performed in the controlled decking zone. Activities such as these are not leading edge work, and employees performing this type of work can be readily protected from falls by the use of conventional fall protection.

CRITERIA FOR FALL PROTECTION EQUIPMENT
OSHA 1926.760(d)

Guardrail systems, safety net systems, personal fall arrest systems, positioning device systems and their components shall conform to the criteria in OSHA 1926.502. OSHA 1926.502 does contain requirements for components of personal fall arrest systems, many of which are also used in restraint systems.

Components used in a restraint system in steel erection work shall meet the requirements in OSHA 1926.502 for those components.

In brief, a positioning device enables an employee to work in a position that allows the employee to fall, but only up to 2 ft. A fall restraint system prevents the employee from reaching an open side or edge, thus preventing the employee from falling.

Perimeter safety cables shall comply with the relevant criteria for guardrail systems in OSHA 1926.502.

Fall protection equipment shall be maintained even after steel erectors have completed their work. Usually, perimeter safety cables are initially installed and maintained by the steel erector, but the cables remain on site after steel erection work is completed. With this provision, the fall protection equipment will only be left in place if the controlling contractor (or its authorized representative) has taken responsibility for ensuring that it will be properly maintained. Without this provision, the fall protection could fall into disrepair and become ineffective.

When safety protection provided by the erector is left remaining in an area to be used by other trades after steel erection activity is completed, the owner shall be responsible for accepting and maintaining this protection, assuring that it is adequate for the protection of all other affected trades, assuring that it complies with all applicable safety regulations when being used by other trades, indemnifying the erector from any damages incurred as a result of the safety protection's use by other trades, removing the safety equipment when no longer required and returning it to the erector in the same condition as it was received.

TRAINING
OSHA 1926.761

The OSHA steel erection standard has many new requirements involving more widespread use of personal fall protection equipment and special procedures for making multiple lifts, for decking activities in controlled decking zones and for connecting. OSHA recognizes the need for a separate training section to address these and other requirements. The requirements in OSHA 1926.761 supplement OSHA's general training and education requirements for construction contained in OSHA 1926.21.

Safety Tip: Employees performing steel erection work shall be properly trained.

Since the employer can choose the provider, method and frequency of training that are appropriate for the employees being trained, the employer has flexibility in developing and implementing a training program. The program shall meet the requirements of this section, and each employee shall be provided the training prior to exposure to the hazard. The employer can choose the provider, method and frequency of training that are appropriate for the employees being trained. The provider may be an outside, professional training organization or other qualified entity, or the employer may develop and conduct the training in-house.

The requirement to provide training is met only when the training is effective in providing the knowledge stipulated in these provisions. An effective training program necessarily involves some means of determining whether the instruction is understood by the employee. This can be done in a variety of ways, such as formal oral or written tests, observation or through discussion.

While retraining/refresher training is not specifically addressed, the employer is responsible for making sure that it has programs necessary to comply with the training requirements in OSHA 1926.21(b)(2): "The employer shall instruct each employee in the recognition and avoidance of unsafe conditions and the regulations applicable to his work environment to control or eliminate any hazards or other exposure to illness or injury."

Steel erection involves progressive sequences of erection, so that the work environment on any day may involve entirely different or unique new hazards than the day before and new employees may enter the erection process when it is already underway. In order to apply OSHA 1926.21 during steel erection activities, an employer would have to assess the type of training needed on a continuing basis as the environment and changes in personnel occur. It is the employer's responsibility to determine if an employee needs retraining in order to strengthen skills required to safely perform the assigned job duties, and whenever the work environment changes to include newly recognized or encountered hazards. This is a key element in the employer's accident prevention program.

Where an employer hires a worker, such as a connector, who is already trained and skilled, OSHA anticipates that the employee's high level of knowledge will be readily apparent and easily ascertained by informal discussion and observation.

OSHA agrees that additional training will be required to ensure that the employees are aware of and understand the regulations applicable to their work environment. However, the Agency believes that the new requirements in this rule are needed to make steel erection safer, and the additional training requirements will play a major role in achieving that increased safety. The following provisions supplement the requirements of OSHA 1926.21:

- Training personnel
- Fall hazard training
- Special training programs

TRAINING PERSONNEL
OSHA 1926.761(a)

All training required by this section shall be provided by a qualified person. As discussed earlier, a "qualified person" is defined in OSHA 1926.751 as one who, by possession of a recognized degree, certificate or professional standing, or who by extensive knowledge, training and experience, has successfully demonstrated the ability to solve or resolve problems relating to the subject matter, the work or the project.

FALL HAZARD TRAINING
OSHA 1926.761(b)

Employers shall provide a training program for all employees exposed to fall hazards. The program shall include the following:

- Training and instruction in recognition and identification of fall hazards in the work area

- The use and operation of guardrail systems, personal fall arrest systems, fall restraint systems, safety net systems, controlled decking zones and other protection to be used

- The correct procedures for erecting, maintaining, disassembling and inspecting the fall protection systems to be used

- The procedures to be followed to prevent falls to lower levels and through or into holes and openings in walking/working surfaces and walls

- The fall protection requirements of OSHA 1926.760(b)(5)

SPECIAL TRAINING PROGRAMS
OSHA 1926.761(c)

The employer shall provide specialized training for employees engaged in multiple lift rigging procedures, connecting activities and work in controlled decking zones, due to the hazardous nature of these activities. There were no comments received regarding the provisions in OSHA 1926.761(c)(1), (c)(2) and (c)(3), and they are promulgated without change.

The employer shall provide additional training for employees performing multiple lift rigging in accordance with the provisions in OSHA 1926.753(e). The special training includes, at a minimum, the nature of the hazards associated with multiple lifts and the proper procedures and equipment to perform multiple lifts.

Employers shall ensure that each connector has been provided training in the hazards associated with connecting, and in the establishment, access, proper connecting techniques and work practices required by OSHA 1926.760(b) (fall protection) and OSHA 1926.756(c) (double connections).

Employers shall provide additional training for controlled decking zone employees. The training shall cover the hazards associated with work within a controlled decking zone, and the establishment, access, proper installation techniques and work practices required by OSHA 1926.760(b) (fall protection) and OSHA 1926.754(e) (decking operations).

Steel Erection

Section	Answer	
_____	T	F
_____	T	F
_____	T	F
_____	T	F
_____	T	F
_____	T	F
_____	T	F
_____	T	F
_____	T	F
_____	T	F
_____	T	F
_____	T	F

Steel Erection

1. Site-specific erection plans shall be developed by a qualified person and be available at the worksite.

2. A competent person shall perform a pre-shift visual inspection of the cranes to be used for steel erection.

3. A multiple lift may not involve hoisting more than 7 members during the lift.

4. The multiple lift rigging assembly shall be rigged with the members:

- Attached at their center of gravity and be kept reasonably level
- Be rigged from the top down
- Have a distance of at least 10 ft. between the members.

5. Permanent floors shall be installed as the erection of structural members progresses and that there be not more than eight stories between the erection floor and the upper-most permanent floor, except where the structural integrity is maintained as a result of the design.

6. At no time shall there be more than 5 floors or 60 ft., whichever is less, of unfinished bolting or welding above the foundation or uppermost permanently secured floor, except where the structural integrity is maintained as a result of the design.

7. A fully planked or decked floor or nets shall be maintained within 2 stories or 30 ft., whichever is less, directly under any erection work being performed.

8. When deemed necessary by a competent person, plumbing-up equipment shall be installed in conjunction with the steel erection process to ensure the stability of the structure.

9. When covering roof and floor openings, covers shall be strong enough to withstand the weight of employees, equipment and materials by requiring that covers support 3 times the combined weight.

10. Installed smoke domes and skylight fixtures are considered adequate covers for roof and floor openings.

11. All columns shall be anchored by a minimum of 4 anchor rods/bolts.

12. During the final placing of solid web structural members, the load shall not be released from the hoisting line until the members are secured with at least 4 bolts per connection, of the same size and strength as shown in the construction documents.

Section	Answer
_____	T F

13. Solid web structural members used as diagonal bracing shall be secured by at least one bolt per connection drawn snug tight or secured by an equivalent connection as specified by the project structural engineer of record.

14. When making a double connection, the first member shall remain connected to a supporting member by at least two connection bolts at all times unless a connection seat or equivalent connection device is supplied with the members to secure the first member and prevent the column from being displaced.

15. Each column splice shall be designed to resist a minimum eccentric gravity load of 300 pounds located 18 in. from the extreme outer face of the column in each direction at the top of the column shaft.

16. A vertical stabilizer plate a minimum of 6 in. by 6 in. shall be provided at each column for steel joists and extend at least 3 in. below the bottom chord of the joist with a 13/16 in. hole to provide an attachment point for guying or plumbing cables.

17. Except for steel joists that have been pre-assembled into panels (panelized), connections of individual steel joists to steel structures in bays of 60 ft. or more shall not be made unless they have been fabricated to allow for field bolting during erection.

18. No construction loads are allowed on steel joists until all bridging is installed and anchored and all joist-bearing ends are attached.

19. Each employee engaged in a steel erection activity who is on a walking/working surface with an unprotected side or edge more than 12 ft. above a lower level shall be protected by conventional fall protection.

20. Each connector shall be protected from fall hazards of more than 2 stories or 30 ft. above a lower level, whichever is less.

Demolition

This chapter covers the preparatory operations; engineering survey; utility location; chute openings; removal of material through floor openings; removal of walls, masonry sections and chimneys; manual removal of floors; mechanical removal of floors, walls and materials; storage; removal of steel construction; mechanical demolition; medical services and first aid; and police and fire contact for demolition of buildings.

PREPARATORY OPERATIONS
OSHA 1926.850

Before the start of every demolition job, the demolition contractor should take a number of steps to safeguard the health and safety of workers at the job site. These preparatory operations involve the overall planning of the demolition job, including the methods to be used to bring the structure down, the equipment necessary to do the job and the measures to be taken to perform the work safely. Planning for a demolition job is as important as actually doing the work. Therefore all planning work should be performed by a competent person experienced in all phases of the demolition work to be performed.

The American National Standards Institute (ANSI) in its ANSI A10.6-1983 – Safety Requirements For Demolition Operations states:

"No employee shall be permitted in any area that can be adversely affected when demolition operations are being performed. Only those employees necessary for the performance of the operations shall be permitted in these areas."

ENGINEERING SURVEY

Prior to starting all demolition operations, OSHA Standard 1926.850(a) requires that an engineering survey of the structure shall be conducted by a competent person. The purpose of this survey is to determine the condition of the framing, floors and walls so that measures can be taken, if necessary, to prevent the premature collapse of any portion of the structure. When indicated as advisable, any adjacent structure(s) or improvements should also be similarly checked. The demolition contractor shall maintain a written copy of this survey. Photographing existing damage in neighboring structures is also advisable.

The engineering survey provides the demolition contractor with the opportunity to evaluate the job in its entirety. The contractor should plan for the wrecking of the structure, the equipment to do the work, manpower requirements and the protection of the public. The safety of all workers on the job site should be a prime consideration. During the preparation of the engineering survey, the contractor should plan for potential hazards such as fires, cave-ins and injuries.

OSHA Required Training:
1926.850(a)

If the structure to be demolished has been damaged by fire, flood, explosion or some other cause, appropriate measures, including bracing and shoring of walls and floors, shall be taken to protect workers and any adjacent structures. It shall also be determined if any type of hazardous chemicals, gases, explosives, flammable material or similar dangerous substances have been used or stored on the site. If the nature of a substance cannot be easily determined, samples should be taken and analyzed by a qualified person prior to demolition.

During the planning stage of the job, all safety equipment needs should be determined. The required number and type of respirators, lifelines, warning signs, safety nets, special face and eye protection, hearing protection and other worker protection devices should be determined during the preparation of the engineering survey. A comprehensive plan is necessary for any confined space entry.

UTILITY LOCATION

One of the most important elements of the pre-job planning is the location of all utility services. All electric, gas, water, steam, sewer and other services lines should be shut off, capped or otherwise controlled, at or outside the building before demolition work is started. In each case, any utility company which is involved should be notified in advance, and its approval or services, if necessary, shall be obtained.

If it is necessary to maintain any power, water or other utilities during demolition, such lines shall be temporarily relocated as necessary and/or protected. The location of all overhead power sources should also be determined, as they can prove especially hazardous during any machine demolition. All workers should be informed of the location of any existing or relocated utility service.

Safety Tip: Walls and/or floors shall be shored or braced when employees must work inside a structure that has been damaged.

When employees must work inside a structure that has been damaged by fire, flood, explosion or other means, the walls and/or floors shall be shored or braced.

Worker entrances to multistory structures being demolished shall be protected by sidewalk sheds or canopies that extend out a minimum of 8 ft. from the building. All such canopies shall be 2 ft. wider than the building access (1 ft. to each side) and capable of sustaining a load of 150 pounds per square foot.

Whenever the possibility exists that a worker might fall through a wall opening in a structure being demolished, the opening shall be protected to a height of 42 in.

If power, water or other utilities must be maintained during demolition operations, these lines shall be temporarily relocated and protected as necessary.

Floor openings not used as material drops shall be covered over with material substantial enough to support the weight of any load placed on them, and the coverings shall be secured with nails or bolts. Permanent flooring may also be used over floor openings.

Safety Tip: Floor openings not used as material drops shall be covered.

If the presence of hazardous chemicals, gases, explosives, flammable materials or other dangerous substances in pipes, tanks or other equipment on the property is suspected or apparent, this equipment shall be tested and purged; the hazard shall be eliminated before demolition is started.

Glass that may potentially shatter shall be removed before work is done in the area.

When debris is dropped through holes in the floor (without the use of chutes), the area onto which it is dropped shall be enclosed with barricades at least 42 in. high and at least 6 ft. from the projected edge of the opening above. Signs stating "DANGER-OVERHEAD HAZARD" shall be posted at each level. Debris shall not be removed in lower areas until persons dropping debris from above have ceased operations. **(See Figure 19-1)**

NOTE 1: FLOOR OPENINGS NOT USED AS MATERIAL DROPS SHALL BE COVERED OVER WITH MATERIAL SUBSTANTIAL ENOUGH TO SUPPORT THE WEIGHT OF ANY LOAD PLACED ON THEM, AND THE COVERING SHALL BE SECURED WITH NAILS OR BOLTS.

NOTE 2: ALL MATERIAL CHUTES OR SECTIONS OF CHUTES AT AN ANGLE OF MORE THAN 45° FROM THE HORIZONTAL SHALL BE ENTIRELY ENCLOSED EXCEPT FOR OPENINGS USED FOR INSERTION OF MATERIAL. THESES OPENINGS SHALL BE NO MORE THAN 4 FT. AND SHALL BE CLOSED WHEN NOT IN USE.

Figure 19-1. This illustration shows the requirements for debris that is dropped through holes in floor without the use of chutes.

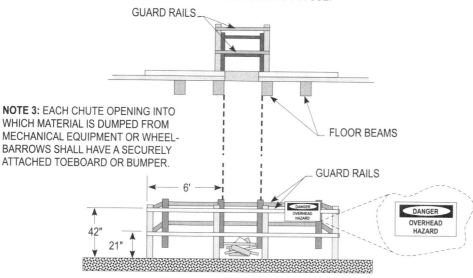

NOTE 3: EACH CHUTE OPENING INTO WHICH MATERIAL IS DUMPED FROM MECHANICAL EQUIPMENT OR WHEEL-BARROWS SHALL HAVE A SECURELY ATTACHED TOEBOARD OR BUMPER.

PREPARATORY OPERATIONS
OSHA 1926.850

CHUTES
OSHA 1926.852

OSHA Required Training:
1926.852(c)

All material chutes or sections of chutes at an angle of more than 45° from the horizontal shall be entirely enclosed except for openings used for insertion of material. These openings shall be no more than 4 ft. high and shall be closed when not in use.

Chute openings into which workmen dump debris shall be protected by guardrails 42 in. above the floor. Any space between the chute and the edge of floor openings through which debris passes shall be covered securely.

No material shall be dropped to points lying outside the exterior walls of a structure unless the area is closed to workers and the public.

The area surrounding the discharge end of a chute shall be securely closed off when operations are not in progress.

Gates shall be installed at or near the discharge end of each chute. A competent person shall supervise operation of the gate and backing-up and loading of trucks.

Each chute opening into which material is dumped from mechanical equipment or wheelbarrows shall have a securely attached toeboard or bumper.

Chutes shall be strong enough to withstand expected loads.

REMOVAL OF MATERIAL THROUGH FLOOR OPENINGS
OSHA 1926.852

Openings cut in a floor for the disposal of material shall be no more than 25 percent of the total floor area. Openings may be larger if the lateral supports of the removed floor remain in place.

Floors weakened or made otherwise unsafe by demolition operations shall be shored.

REMOVAL OF WALLS, MASONRY SECTIONS AND CHIMNEYS
OSHA 1926.854

Masonry walls or sections of these walls shall be permitted to fall upon the floors of the building only in quantities within the safe carrying capacities of the floor.

Walls more than one-story high shall be laterally braced unless they were originally constructed to stand without lateral support and are in a condition to be self-supporting.

Workers shall not work on the top of walls when weather conditions are hazardous.

Structural or load-supporting members on any floor shall be cut or removed only after all stories above such a floor have been demolished and removed, unless otherwise discussed in paragraphs 3 and 5 of this section.

Walkways or ladders shall be provided for workers to safely reach or leave any scaffold or wall.

Retaining walls that support earth or adjoining structures shall not be demolished until such earth has been properly braced or the adjoining structures have been properly underpinned.

Debris shall be piled only against walls capable of supporting it.

Floor openings within 10 ft. of any wall being demolished shall be planked solidly when workers are required to work beneath such openings.

MANUAL REMOVAL OF FLOORS
OSHA 1926.855

Openings cut in floors shall extend the full span of the arch between supports.

Workers who break down floor arches shall stand on planks that are at least 2 in. by 10 in. in cross section. These planks shall be a maximum of 16 in. apart and positioned to provide a safe support for workers, should the arch between the beams collapse.

Walkways not less than 18 in. wide and formed of planks not less than 2 in. thick or metal of equivalent strength shall be used by workers to walk between exposed beams.

Stringers of sufficient strength to support floor planks shall be installed; the ends of such stringers shall be supported by floor beams or girders rather than by floor arches alone. **(See Figure 19-2)**

NOTE 1: WORKERS WHO BREAK DOWN FLOOR ARCHES SHALL STAND ON PLANKS THAT ARE AT LEAST 2 IN. BY 10 IN. IN CROSS SECTION. THESE PLANKS SHALL BE A MAXIMUM OF 16 IN. APART AND POSITIONED TO PROVIDE A SAFE SUPPORT FOR WORKERS, SHOULD THE ARCH BETWEEN THE BEAMS COLLAPSE.

NOTE 2: WALKWAYS NOT LESS THAN 18 IN. WIDE AND FORMED OF PLANKS NOT LESS THAN 2 IN. THICK OR METAL OF EQUIVALENT STRENGTH SHALL BE USED BY WORKERS TO WALK BETWEEN EXPOSED BEAMS.

Figure 19-2. This illustration shows the requirements for stringers to support floor planks.

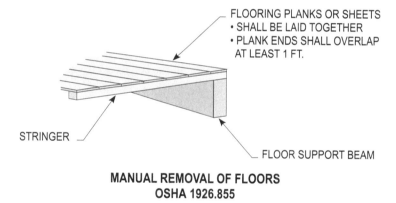

FLOORING PLANKS OR SHEETS
• SHALL BE LAID TOGETHER
• PLANK ENDS SHALL OVERLAP
  AT LEAST 1 FT.

STRINGER

FLOOR SUPPORT BEAM

**MANUAL REMOVAL OF FLOORS
OSHA 1926.855**

Planks shall be laid together over solid floor supports (bearings) with plank ends overlapping at least 1 ft.

When floor arches are being removed, the area directly beneath shall be barricaded to prevent access to it.

Demolition of floor arches shall begin only after debris and other unnecessary materials have been removed from the floor arch and within 20 ft. of it.

MECHANICAL REMOVAL OF WALLS, FLOORS AND MATERIAL
OSHA 1926.856

Mechanical equipment shall be used only on floors or work surfaces that have sufficient strength to support the equipment.

Floor openings shall have curbs (stoplogs) to prevent equipment from running over the edge.

STORAGE
OSHA 1926.857

The storage of waste material and debris on any floor shall be within the allowable floor loads.

Wooden floor boards may be removed from one floor above grade to provide storage space for debris only if falling material does not endanger the stability of the structure.

When wooden floor beams serve to brace interior walls or free-standing exterior walls, the beams shall be left in place until other supports can be installed to replace them.

Floor arches up to 25 ft. above grade may be removed to provide a storage area for debris if such removal does not endanger the stability of the structure. **(See Figure 19-3)**

Figure 19-3. This illustration shows the requirements for storage space of debris.

NOTE 1: WOODEN FLOOR BOARDS MAY BE REMOVED FROM ONE FLOOR ABOVE GRADE TO PROVIDE STORAGE SPACE FOR DEBRIS ONLY IF FALLING MATERIAL DOE NOT ENDANGER THE STABILITY OF THE STRUCTURE.

NOTE 2: WHEN WOODEN FLOOR BEAMS SERVE TO BRACE INTERIOR WALLS OR FREE-STANDING EXTERIOR WALLS, THE BEAMS SHALL BE LEFT IN PLACE UNTIL OTHER SUPPORTS CAN BE INSTALLED TO REPLACE THEM.

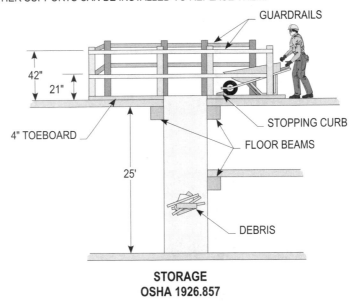

GUARDRAILS

42"

21"

4" TOEBOARD

STOPPING CURB

FLOOR BEAMS

25'

DEBRIS

STORAGE
OSHA 1926.857

Storage space into which material is dumped shall be blocked off and closed except when removing materials.

REMOVAL OF STEEL CONSTRUCTION
OSHA 1926.858

After floor arches have been removed, workers shall stand on at least 2 in. by 10 in. planks while razing the steel frame.

Steel construction shall be dismantled at first column length by column length and then tier by tier (columns may be in two-story lengths).

I-beams or columns being dismembered shall not be overstressed.

MECHANICAL DEMOLITION
OSHA 1926.859

To prevent twisting of the load line, the ball shall be attached to it by a swivel-type connection and by a positive means, that prevents accidental disconnecting.

Workers shall not be allowed in areas where bailing or clamming operations take place.

The weight of the demolition ball shall not exceed 50 percent of the crane's rated load or 25 percent of the nominal breaking strength of the wire rope or cable on which it is suspended, whichever results in the lesser value.

All affected steel members shall be cut free before pulling over walls or portions of walls.

All ornamental stonework, including cornices, shall be removed before the walls are pulled over.

Continuous inspections shall be made by a competent person to detect hazards resulting from weakened or deteriorated floors or walls or from loosened material. Where such hazards to employees exist, operations shall continue only after the hazards are corrected by shoring, bracing or equivalent means.

OSHA Required Training:
1926.852(g)

MEDICAL SERVICES AND FIRST AID

Prior to starting work, provisions should be made for prompt medical attention in case of serious injury. The nearest hospital, infirmary, clinic or physician shall be located as part of the engineering survey. The job supervisor should be provided with instructions for the most direct route to these facilities. Proper equipment for prompt transportation of an injured worker, as well as a communication system to contact any necessary ambulance service, shall be available at the job site. The telephone numbers of the hospitals, physicians or ambulances shall be conspicuously posted.

Safety Tip: In case of serious injury prior to starting work, provisions shall be made for prompt medical attention.

In the absence of an infirmary, clinic, hospital or physician that is reasonably accessible in terms of time and distance to the worksite, a person who has a valid certificate in first aid training from the U.S. Bureau of Mines, the American Red Cross or equivalent training should be available at the worksite to render first aid.

A properly stocked first aid kit, as determined by an occupational physician, shall be available at the job site. The first aid kit should contain approved supplies in a weatherproof container with individual sealed packages for each type of item. It should also include rubber gloves to prevent the transfer of infectious diseases. Provisions should also be made to provide for quick drenching or flushing of the eyes should any person be working around corrosive materials. Eye flushing must be done with water containing no additives. The contents of the kit shall be checked before being sent out on each job and at least weekly to ensure the expended items are replaced.

POLICE AND FIRE CONTACT

The telephone numbers of the local police, ambulance and fire departments should be available at each job site. This information can prove useful to the job supervisor in the event of any traffic problems, such as the movement of equipment to the job, uncontrolled fires or other police/fire matters. The police number may also be used to report any vandalism, unlawful entry to the job site or accidents requiring police assistance.

Demolition

Section	Answer	
_____	T	F
_____	T	F
_____	T	F
_____	T	F
_____	T	F
_____	T	F
_____	T	F
_____	T	F
_____	T	F
_____	T	F
_____	T	F

Demolition

1. Prior to starting all demolition operations, OSHA Standard 1926.850(a) requires that an engineering survey of the structure shall be conducted by a competent person.

2. If the structure to be demolished has been damaged by fire, flood, explosion or some other cause, appropriate measures, including bracing and shoring of walls and floors, shall be taken to protect workers and any adjacent structures.

3. Worker entrances to multistory structures being demolished shall be protected by sidewalk sheds or canopies that extend out a minimum of 6 ft. from the building. All such canopies shall be 2 ft. wider than the building access (1 ft. to each side) and capable of sustaining a load of 150 pounds per square foot.

4. Whenever the possibility exists that a worker might fall through a wall opening in a structure being demolished, the opening shall be protected to a height of 42 in.

5. When debris is dropped through holes in the floor (without the use of chutes), the area onto which it is dropped shall be enclosed with barricades at least 36 in. high and at least 6 ft. from the projected edge of the opening above.

6. All material chutes or sections of chutes at an angle of more than 45° from the horizontal shall be entirely enclosed except for openings used for insertion of material. These openings shall be no more than 4 ft. high and shall be closed when not in use.

7. Chute openings into which workmen dump debris shall be protected by guardrails 24 in. above the floor. Any space between the chute and the edge of floor openings through which debris passes shall be covered securely.

8. Gates shall be installed at or near the discharge end of each chute. A competent person shall supervise operation of the gate and backing-up and loading of trucks.

9. Floor openings within 12 ft. of any wall being demolished shall be planked solidly when workers are required to work beneath such openings.

10. Workers who break down floor arches shall stand on planks that are at least 2 in. by 10 in. in cross section. These planks shall be a maximum of 16 in. apart and positioned to provide a safe support for workers, should the arch between the beams collapse.

11. Demolition of floor arches shall begin only after debris and other unnecessary materials have been removed from the floor arch and within 20 ft. of it.

_____   T   F

_____   T   F

_____   T   F

_____   T   F

12. Floor arches up to 35 ft. above grade may be removed to provide a storage area for debris if such removal does not endanger the stability of the structure.

13. The weight of the demolition ball shall not exceed 60 percent of the crane's rated load or 25 percent of the nominal breaking strength of the wire rope or cable on which it is suspended, whichever results in the lesser value.

14. Continuous inspections shall be made by a competent person to detect hazards resulting from weakened or deteriorated floors or walls or from loosened material.

15. A properly stocked first aid kit, as determined by an occupational physician, shall be available at the job site.

Blasting and the Use of Explosives

This chapter covers the general provisions, blaster qualifications, surface transportation of explosives, underground transportation of explosives, storage of explosives and blasting agents, loading of explosives or blasting agents, initiation of explosive charges – blasting agents, use of safety fuse, use of detonating cord, firing the blast, inspection after blast, misfires, underwater blasting and blasting in excavation work under compressed air.

GENERAL SAFE WORK PRACTICES
OSHA 1926.900

The following shall be considered for general safe work practices:

- Blasting survey and preparation
- Fire precautions

BLASTING SURVEY AND SITE PREPARATION

Prior to the blasting of any structure or portion thereof, a complete written survey shall be made by a qualified person of all adjacent improvements and underground utilities. When there is a possibility of excessive vibration due to blasting operations, seismic or vibration tests should be taken to determine proper safety limits to prevent damage to adjacent or nearby buildings, utilities or other property.

The preparation of a structure for demolition by explosives may require the removal of structural columns, beams or other building components. This work should be directed by a structural engineer or a competent person qualified to direct the removal of these structural elements. Extreme caution must be taken during this preparatory work to prevent the weakening and premature collapse of the structure.

The use of explosives to demolish smokestacks, silos, cooling towers or similar structures should only be permitted if there is a minimum of 90 degrees of open space extended for at least 150 percent of the height of the structure or if the explosives specialist can demonstrate consistent previous performance with tighter constraints at the site.

FIRE PRECAUTIONS

The presence of fire near explosives presents a severe danger. Every effort should be made to ensure that fires or sparks do not occur near explosive materials. Smoking, matches, firearms, open flame lamps and other fires, flame or heat-producing devices shall be prohibited in or near explosive magazines or in areas where explosives are being handled, transported or used. In fact, persons working near explosives should not even carry matches, lighters or other sources of sparks or flame. **(See Figure 20-1)**

Open fires or flames should be prohibited within 100 ft. of any explosive materials. In the event of a fire that is in imminent danger of contact with explosives, all employees shall be removed to a safe area. **(See Figure 20-1)**

Electrical detonators can be inadvertently triggered by stray RF (radio frequency) signals from two-way radios. RF signal sources should be restricted from or near the demolition site, if electrical detonators are used. **(See Figure 20-1)**

Figure 20-1. This illustration shows the requirements to be taken for fire precautions.

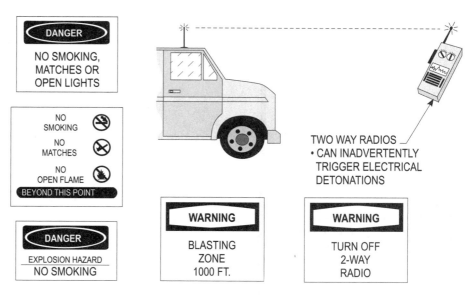

NOTE 1: OPEN FIRES OR FLAMES SHOULD BE PROHIBITED WITHIN 100 FT. OF ANY EXPLOSIVE MATERIALS.
NOTE 2: IN THE EVENT OF A FIRE THAT IS IN IMMINENT DANGER OF CONTACT WITH EXPLOSIVES, ALL EMPLOYEES SHALL BE REMOVED TO SAFE AREA.

FIRE PRECAUTIONS
OSHA 1926.900

PERSONNEL SELECTION
OSHA 1926.901

A blaster is a competent person who uses explosives. A blaster shall be qualified by reason of training, knowledge or experience in the field of transporting, storing, handling and using explosives. In addition, the blaster should have a working knowledge of state and local regulations which pertain to explosives. Training courses are often available from manufacturers of explosives and blasting safety manuals are offered by the Institute of Makers of Explosives (IME) as well as other organizations.

Blasters shall be required to furnish satisfactory evidence of competency in handling explosives and in safely performing the type of blasting required. A competent person should always be in charge of explosives and should be held responsible for enforcing all recommended safety precautions in connection with them.

TRANSPORTATION OF EXPLOSIVES
OSHA 1926.902

Vehicles used for transporting explosives shall be strong enough to carry the load without difficulty and shall be in good mechanical condition. All vehicles used for the transportation of explosives shall have tight floors, and any exposed spark-producing metal on the inside of the body shall be covered with wood or some other non-sparking material. Vehicles or conveyances transporting explosives shall only be driven by, and shall be under the supervision of, a licensed driver familiar with the local, state and Federal regulations governing the transportation of explosives. No passengers should be allowed in any vehicle transporting explosives.

Explosives, blasting agents and blasting supplies shall not be transported with other materials or cargoes. Blasting caps shall not be transported with other materials or cargoes. Blasting caps shall not be transported in the same vehicle with other explosives. If an open-bodied truck is used, the entire load should be completely covered with a fire and water-resistant tarpaulin to protect it from the elements. Vehicles carrying explosives should not be loaded beyond the manufacturer's safe capacity rating, and in no case should the explosives be piled higher than the closed sides and ends of the body.

Every motor vehicle or conveyance used for transporting explosives shall be marked or placarded with warning signs required by OSHA and the DOT. **(See Figure 20-2)**

Each vehicle used for transportation of explosives shall be equipped minimally with at least a 10 pound rated serviceable ABC fire extinguisher. All drivers should be trained in the use of the extinguishers on their vehicle. **(See Figure 20-2)**

In transporting explosives, congested traffic and high density population areas should be avoided, where possible, and no unnecessary stops should be made. Vehicles carrying explosives, blasting agents or blasting supplies shall not be taken inside a garage or shop for repairs or servicing. No motor vehicle transporting explosives shall be left unattended.

STORAGE OF EXPLOSIVES
OSHA 1926.904

The following shall be considered for storage of explosives:

• Inventory handling and safe handling
• Storage conditions

Figure 20-2. This illustration shows the requirements for transportation of explosives.

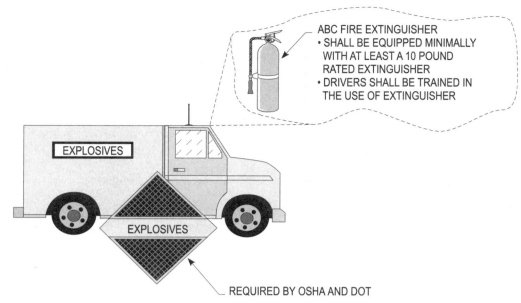

ABC FIRE EXTINGUISHER
• SHALL BE EQUIPPED MINIMALLY WITH AT LEAST A 10 POUND RATED EXTINGUISHER
• DRIVERS SHALL BE TRAINED IN THE USE OF EXTINGUISHER

EXPLOSIVES

EXPLOSIVES

REQUIRED BY OSHA AND DOT

NOTE 1: VEHICLES OR CONVEYANCES TRANSPORTING EXPLOSIVES SHALL ONLY BE DRIVEN BY, AND SHALL BE UNDER THE SUPERVISION OF, A LICENSED DRIVER FAMILIAR WITH THE LOCAL, STATE AND FEDERAL REGULATIONS GOVERNING THE TRANSPORTATION OF EXPLOSIVES.

NOTE 2: IF AN OPEN-BODIED TRUCK IS USED, THE ENTIRE LOAD SHOULD BE COMPLETELY COVERED WITH A FIRE AND WATER-RESISTANT TARPAULIN TO PROTECT IT FROM THE ELEMENTS.

VEHICULAR SAFETY
OSHA 1926.902

INVENTORY HANDLING AND SAFE HANDLING

Safety Tip: Explosives not being used shall be kept in a locked magazine. All explosives shall be accounted for at all times.

All explosives shall be accounted for at all times, and all not being used shall be kept in a locked magazine. A complete detailed inventory of all explosives received and placed in, removed from and returned to the magazine should be maintained at all times. Appropriate authorities shall be notified of any loss, theft or unauthorized entry into a magazine.

Manufacturers' instructions for the safe handling and storage of explosives are ordinarily enclosed in each case of explosives. The specifics of storage and handling are best referred to these instructions and the aforementioned IME manuals. They should be carefully followed. Packages of explosives should not be handled roughly. Sparking metal tools should not be used to open wooden cases. Metallic slitters may be used for opening fiberboard cases, provided the metallic slitter does not come in contact with the metallic fasteners of the case.

The oldest stock should always be used first to minimize the chance of deterioration from long storage. Loose explosives or broken, defective or leaking packages can be hazardous and should be segregated and properly disposed of in accordance with the specific instructions of the manufacturer. If the explosives are in good condition it may be advisable to repack them. In this case, the explosives supplier should be contacted. Explosives cases should not be opened or explosives packed or repacked while in a magazine.

STORAGE CONDITIONS

Providing a dry, well-ventilated place for the storage of explosives is one of the most important and effective safety measures. Exposure to weather damages most kinds of explosives, especially dynamite and caps. Every precaution should be taken to keep them dry and relatively cool. Dampness or excess humidity may be the cause of misfires resulting in injury or loss of life. Explosives should be stored in properly constructed fire and bullet-resistant structures, located according to the IME American Table of Distances and kept locked at all times except when opened for use by an authorized person. Explosives should not be left, kept or stored where children, unauthorized persons or animals have access to them, nor should they be stored in or near a residence.

Detonators should be stored in a separate magazine located according to the IME American Table of Distances.

DETONATORS SHOULD NEVER BE STORED IN THE SAME MAGAZINE WITH ANY OTHER KIND OF EXPLOSIVES

Ideally, arrangements should be made whereby the supplier delivers the explosives to the job site in quantities that will be used up during the work day. An alternative would be for the supplier to return to pick up unused quantities of explosives. If it is necessary for the contractor to store his explosives, he should be familiar with all local requirements for such storage.

PROPER USE OF EXPLOSIVES
OSHA 1926.905

Blasting operations shall be conducted between sunup and sundown, whenever possible. Adequate signs should be sounded to alert to the hazard presented by blasting. Blasting mats or other containment should be used where there is danger of rocks or other debris being thrown into the air or where there are buildings or transportation systems nearby. Care should be taken to make sure mats and other protection does not disturb the connections to electrical blasting caps.

Radio, television and radar transmitters create fields of electrical energy that can, under exceptional circumstances, detonate electric blasting caps. **(See Figure 20-3)**

PRECAUTIONS TAKEN TO PREVENT ACCIDENTAL DISCHARGE OF BLASTING CAPS
• ENSURING THAT MOBILE RADIO TRANSMITTERS ON THE JOB SITE THAT ARE LESS THAN 100 FT. AWAY FROM ELECTRICAL BLASTING CAPS, IN OTHER THAN ORIGINAL CONTAINERS, SHALL BE DEENERGIZED AND EFFECTIVELY LOCKED.
• THE PROMINENT DISPLAY OF ADEQUATE SIGNS, WARNING AGAINST THE USE OF MOBILE RADIO TRANSMITTERS, ON ALL ROADS WITHIN 1000 FT. OF THE BLASTING OPERATIONS.
• MAINTAINING THE MINIMUM DISTANCES RECOMMENDED BY THE IMES BETWEEN THE NEAREST TRANSMITTER AND ELECTRIC BLASTING CAPS.
• THE SUSPENSION OF ALL BLASTING OPERATIONS AND REMOVAL OF PERSONS FROM THE BLASTING AREA DURING THE APPROACH AND PROGRESS OF AN ELECTRIC STORM.

Figure 20-3. This illustration shows that radio, television and radar transmitters can create fields of electrical energy that could detonate electric blasting caps.

**PROPER USE OF EXPLOSIVES
OSHA 1926.905**

Certain precautions shall be taken to prevent accidental discharge of electric blasting caps from current induced by radar, radio transmitters, lightning, adjacent power lines, dust storms or other sources of extraneous or static electricity. These precautions shall include:

- Ensuring that mobile radio transmitters on the job site that are less than 100 ft. away from electric blasting caps, in other than original containers, shall be deenergized and effectively locked;

- The prominent display of adequate signs, warning against the use of mobile radio transmitters, on all roads within 1000 ft. of the blasting operations;

- Maintaining the minimum distances recommended by the IMES between the nearest transmitter and electric blasting caps;

- The suspension of all blasting operations and removal of persons from the blasting area during the approach and progress of an electric storm.

After loading is completed, there should be as little delay as possible before firing. Each blast should be fired under the direct supervision of the blaster, who should inspect all connections before firing and who should personally see that all persons are in the clear before giving the order to fire. Standard signals, that indicate that a blast is about to be fired and a later all clear signal have been adopted. It is important that everyone working in the area be familiar with these signals and that they be strictly obeyed.

Blasters shall keep accurate, up-to-date records of explosives, blasting agents and blasting supplies used in a blast. Blasters shall keep an accurate running inventory of all explosives and blasting agents stored on the construction site.

INITIATION OF EXPLOSIVE CHARGES – ELECTRIC CHARGES OSHA 1926.906

Where extraneous sources of electricity make the use of electric blasting caps dangerous, electric blasting caps shall not be used. Blasting-cap leg wires shall be kept short-circuited (shunted) until they are connected into the circuit for firing. During any individual blasting, all caps shall be of the same style, function and manufacture.

Blasting circuits or power circuits used during electric blasting shall be set up following the recommendations of the electric blasting cap manufacturer, an approved contractor or a designated representative.

Before adopting any system of electrical firing, blasters shall conduct a thorough survey for extraneous electrical currents; all such currents shall be eliminated before the holes are loaded.

When firing a circuit of electric blasting caps, care shall be taken to ensure that adequate current can be delivered according to the manufacturer's recommendation.

Connecting wires and lead wires shall be insulated, solid, single wires with sufficient current-carrying capacity.

Bus wires shall be solid, single wires with sufficient current-carrying capacity.

The insulation on all firing lines shall be adequate and in good condition.
The power circuit used to fire electric blasting caps shall not be grounded.

Blasting machines shall be in good condition and shall be tested periodically to make sure that they can deliver power at the rated capacity. **(See Figure 20-4)**

Figure 20-4. This illustration shows that blasting machines shall be in good condition and shall be tested periodically.

NOTE 1: CONNECTING WIRES AND LEAD WIRES SHALL BE INSULATED, SOLID, SINGLE WIRES WITH SUFFICIENT CURRENT-CARRYING CAPACITY.

NOTE 2: BUS WIRES SHALL BE SOLID, SINGLE WIRES WITH SUFFICIENT CURRENT-CARRYING CAPACITY.

NOTE 3: THE POWER CIRCUIT USED TO FIRE ELECTRIC BLASTING CAPS SHALL NOT BE GROUNDED.

NOTE 4: WHEN TESTING CIRCUITS LEADING TO CHARGED HOLES, BLASTERS SHALL USE BLASTING GALVANOMETERS THAT ARE EQUIPPED WITH A SILVER CHLORIDE CELL ON A BLASTING MULTIMETER.

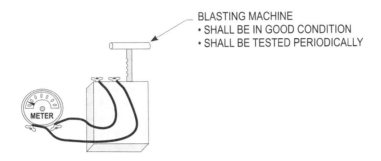

BLASTING MACHINE
• SHALL BE IN GOOD CONDITION
• SHALL BE TESTED PERIODICALLY

METER

INITIATION OF EXPLOSIVE CHARGES - ELECTRIC CHARGES
OSHA 1926.906

The connections made on firing machines shall be made according to the recommendations of the manufacturer of the blasting caps.

When a power circuit is used for firing, the firing switch shall be locked in the "OFF" position before the round is fired. The switch shall be designed so that the firing lines to the cap circuit are automatically short-circuited when the switch is in the "OFF" position. No one other than the blaster shall control the keys to the firing switch.

The number of electric blasting caps connected to a blasting machine shall not exceed the machine's rated capacity. In primary blasting, the number of blasting caps connected in series is limited by the cap manufacturer.

Blasters shall be in charge of the blasting machines. No one other than the blaster in charge shall connect the lead wires to the blasting machine.

When testing circuits leading to charged holes, blasters shall use blasting galvanometers that are equipped with a silver chloride cell or a blasting multimeter.

Whenever the possibility exists that a lead line or a blasting wire might be thrown over a live power line by the force of an explosion, care shall be taken to (1) make sure that the total length of wires is kept to a minimum so that the length is too short to contact the line or (2) make sure that the wires are securely anchored to the ground. If neither precaution can be taken, a nonelectrical system shall be used.

The person connecting the lead wire shall fire the shot. All connections shall be made from the bore hole back to the source of firing current. Until the charge is to be fired, the lead wires shall remain shorted and shall not be connected to the blasting machine or other source of current.

USE OF SAFETY FUSES
OSHA 1926.908

Safety fuses shall be used only when extraneous sources of electricity make the use of electric blasting caps dangerous. Hammered, crushed or damaged fuses shall not be used.

Each blasting fuse shall have a fresh-cut end. To ensure a fresh cut, a short length of fuse shall be cut from the end of the supply reel before the safety fuse is capped. The cut piece should be taken to a safe location and burned.

Only a cap crimper of approved design shall be used to attach blasting caps to safety fuses. Crimpers shall be kept in good repair and accessible for use.

When blasting with safety fuses, consideration shall be given to the burning rate and length of the safety fuse to allow sufficient time for the blaster to reach a safe place before the blast goes off. It is advisable to cut a measured 1 ft. of fuse and then test burn it at a remote safe location to confirm the burn and burn rate of the safety fuse.

Hanging fuses on nails or other projections that may form sharp bends in the fuses is prohibited. **(See Figure 20-5)**

Figure 20-5. This illustration shows that hanging fuses on nails or other projections is prohibited.

NOTE 1: SAFETY FUSES SHALL BE USED ONLY WHEN EXTRANEOUS SOURCES OF ELECTRICITY MAKE THE USE OF ELECTRIC BLASTING CAPS DANGEROUS.

NOTE 2: WHEN BLASTING WITH SAFETY FUSES, CONSIDERATION SHALL BE GIVEN TO THE BURNING RATE AND LENGTH OF THE SAFETY FUSE TO ALLOW SUFFICIENT TIME FOR THE BLASTER TO REACH A SAFE PLACE BEFORE THE BLAST GOES OFF.

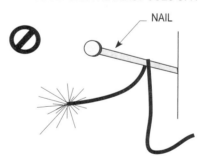

USE OF SAFETY FUSES
OSHA 1926.908

No unused cap or short-capped fuse shall be placed in any hole to be blasted. Unused detonators shall be removed from the work area and destroyed.

Safety Tip: Workers shall not carry detonators or primers in unapproved area.

Fuses shall not be capped and primers shall not be made up in any magazine or near any possible ignition source.

Workers shall not carry detonators or primers when they are in an unapproved area. Workers shall not carry detonators or primers in their pockets, around their necks or on any part of their person. Safety fuses used in blasting shall be at least 30 in. long.

Ignitor fuses are the recommended means of lighting fuse rounds. If ignitor fuses cannot be used and multiple cap and fuse blasting is done using hand-lighting methods, at least two people shall be present.

When the multiple cap and fuse blasting is done by hand-lighting methods, not more than 12 fuses shall be lit by each blaster.

USE OF DETONATING CORD
OSHA 1926.908

Care shall be taken to select detonating cord appropriate for the type and physical condition of the bore hole and the type of explosive used.

Before loading the remainder of a hole or placing additional charges, the line of detonating cord extending from a bore hole or from a charge shall be cut from the supply spool. Detonating cord shall be handled and used with care to avoid damaging or cutting the cord during or after loading and hookup.

Detonating cord connections shall be made following approved, recommended methods. Knot-type or other cord-to-cord connections shall be made only with detonating cord in which the explosive core is dry. Before firing the blast, all detonating cord connections shall be inspected.

All detonating cord trunk lines and branch lines shall be free of loops, sharp kinks or angles that direct the cord back toward the oncoming line of detonation. When using detonating cord with millisecond-delay connectors or short-interval delay electric blasting caps, manufacturer's recommendations shall be strictly followed. Detonators for firing the trunk line shall not be brought to the loading area or attached to the detonating cord until the blast is about to be fired.

FIRING THE BLAST
OSHA 1926.909

OSHA Required Training:
1926.909(a)

A code of blasting signs shall be posted in one or more prominent places at the operation. All workers shall be familiar with the code and shall conform to it. Danger signs shall be placed at appropriate locations. The following blasting signals shall be used:

- **WARNING SIGNAL**
  A 1-MINUTE SERIES OF LONG BLASTS 5 MINUTES BEFORE BLAST SIGNAL

- **BLAST SIGNAL**
  A SERIES OF SHORT BLASTS 1 MINUTE BEFORE THE SHOT

- **ALL-CLEAR SIGNAL**
  A PROLONGED BLAST FOLLOWING THE INSPECTION OF BLAST AREA

Before firing a blast, the blaster in charge shall give a loud warning signal. The blaster shall ensure that all surplus explosives are in a safe place and that all workers, vehicles and equipment are at a safe distance or under sufficient cover.

Note: Be aware of probable close air traffic because it may become necessary to file firing schedules and traffic control plans with the Federal Aviation Administration (FAA).

Flaggers shall be safely stationed on highways that pass through the danger zone so that they can stop traffic during blasting operations. Blasting times shall be arranged and coordinated by the blaster.

PROCEDURES AFTER BLASTING
OSHA 1926.910

Immediately after the blast has been fired, the firing line shall be disconnected from the blasting machine and short-circuited. Where power switches are used, they shall be locked open or in the off position. Sufficient time shall be allowed for dust, smoke and fumes to leave the blasted area before returning to the spot. An inspection of the area and the surrounding rubble shall be made by the blaster to determine if all charges have been exploded before employees are allowed to return to the operation. All wires should be traced and the search for unexploded cartridges made by the blaster.

MISFIRES
OSHA 1926.911

Safety Tip: The blaster must take proper precautions to ensure that all workers are kept from the danger zone if a misfire is found.

If a misfire is found, the blaster shall take proper precautions to ensure that all workers are kept from the danger zone. No more work than what is necessary to remove the hazard of the misfire shall be done, and only those workers necessary to remove the hazard shall remain in the danger zone.

No attempt shall be made to remove explosives from any charged or misfire hole. If an explosive misfires, a new primer shall be put in the hole and the hole shall be reblasted. If refiring the misfire hole presents a hazard, the explosives shall be washed out of the hole with water or, if the misfire is underwater, it shall be blown out with air.

If there are any misfires while using a cap and fuse method, all workers shall stay away from the charge for at least 1 hour. Misfires shall be handled under the direction of the person in charge of the blasting. All wires and fuses shall be carefully traced, and a search shall be made for unexploded charges.

No drilling, digging or picking shall be allowed until

(1) all missed holes have been detonated or
(2) the authorized representative gives approval.

DISPOSAL OF EXPLOSIVES

Explosives, blasting agents and blasting supplies that are obviously deteriorated or damaged should not be used; they should be properly disposed of. Explosives distributors will usually take back old stock. Local fire marshals or representatives of the United States Bureau of Mines may also arrange for its disposal. Under no circumstances should any explosives be abandoned.

Wood, paper, fiber or other materials that have previously contained high explosives should not be used again for any purpose, but should be destroyed by burning. These materials should not be burned in a stove, fireplace or other confined space. Rather, they should be burned at an isolated outdoor location, at a safe distance from thoroughfares, magazines and other structures.

It is important to check that the containers are entirely empty before burning. During burning, the area should be adequately protected from intruders and all persons kept at least 100 ft. from the fire.

Blasting and the Use of Explosives

Section	Answer	
_____	T	F
_____	T	F
_____	T	F
_____	T	F
_____	T	F
_____	T	F
_____	T	F
_____	T	F
_____	T	F
_____	T	F
_____	T	F
_____	T	F
_____	T	F

Blasting and the Use of Explosives

1. Prior to the blasting of any structure or portion thereof, a complete written survey shall be made by a qualified person of all adjacent improvements and underground utilities.

2. The use of explosives to demolish smokestacks, silos, cooling towers or similar structures should only be permitted if there is a minimum of 90 degrees of open space extended for at least 125 percent of the height of the structure or if the explosives specialist can demonstrate consistent previous performance with tighter constraints at the site.

3. Open fires or flames should be prohibited within 100 ft. of any explosive materials. In the event of a fire which is in imminent danger of contact with explosives, all employees shall be removed to a safe area.

4. A blaster shall be qualified by reason of training, knowledge or experience in the field of transporting, storing, handling and using explosives.

5. Each vehicle used for transportation of explosives shall be equipped minimally with at least a 2 pound rated serviceable ABC fire extinguisher. All drivers should be trained in the use of the extinguishers on their vehicle.

6. All explosives shall be accounted for at all times and all not being used shall be kept in a locked magazine.

7. Detonators should be stored in a separate magazine located according to the IME International Table of Distances.

8. Radio, television and radar transmitters create fields of electrical energy that can, under exceptional circumstances, detonate electric blasting caps.

9. Mobile radio transmitters on the job site that are less than 85 ft. away from electric blasting caps, in other than original containers, shall be deenergized and effectively locked.

10. Where extraneous sources of electricity make the use of electric blasting caps dangerous, electric blasting caps shall not be used.

11. Connecting wires and lead wires shall be insulated, stranded, single wires with sufficient current-carrying capacity.

12. Blasting machines shall be in good condition and shall be tested periodically to make sure that they can deliver power at the rated capacity.

13. Blasters shall be in charge of the blasting machines. No one other than the blaster in charge shall connect the lead wires to the blasting machine.

Section	Answer	
_____	T	F

14. The person connecting the lead wire shall fire the shot.

15. No unused cap or short-capped fuse shall be placed in any hole to be blasted. Unused detonators shall be removed from the work area and thrown in the trash.

16. When the multiple cap and fuse blasting is done by hand-lighting methods, not more than 15 fuses shall be lit by each blaster.

17. Before firing a blast, the blaster in charge shall yell a warning signal.

18. If a misfire is found, the blaster shall take proper precautions to ensure that all workers are kept from the danger zone.

19. If there are any misfires while using a cap and fuse method, all workers shall stay away from the charge for at least 1 hour.

20. Wood, paper, fiber or other materials that have previously contained high explosives should not be used again for any purpose, but should be destroyed by burning. During burning, the area should be adequately protected from intruders and all persons kept at least 75 ft. from the fire.

Power Transmission and Distribution

Employees working on and around power transmission and distribution systems face grave dangers from electrocution, falls and other serious hazards. Workers employed in the construction of these systems have a risk of between 17 and 23 deaths per thousand workers over a working lifetime of 45 years. These workers are not covered by the 1994 OSHA electric power generation standard, which applies only to general industry. The construction industry standard is over twenty years old and inconsistent with the newer, more protective general industry standard. To improve worker protection, OSHA is designating power transmission and distribution in construction as a priority for rulemaking to revise the existing standard.

HAZARD DESCRIPTION

More than 110,000 power line workers who construct/repair power transmission and distribution systems face a wide range of serious and potentially fatal injuries, including electrocutions, falls from elevation and injuries from falling objects. Major causes of non-fatal injuries include over-exertion, electrical shock injuries (burns), sprains and strains, cuts and lacerations and contusions.

The National Institute for Occupational Safety and Health's (NIOSH) National Traumatic Occupational Fatalities (NTOF) surveillance system identified power line workers as a high-risk occupation group for work-related deaths. According to NTOF data, the average annual fatality rate for power line workers is 56.3 deaths per 100,000 employees. The Bureau of Labor Statistics' (BLS) Census of Fatal Occupational Injuries (CFOI) identified 42 fatalities among electric power installers and repairers in 1993 (38 deaths per 100,000 workers). These rates correspond to a risk of between 17 and 23 deaths per thousand workers over a working lifetime of 45 years. The risk may actually be higher, however, because available data do not provide specific numbers for construction workers.

CURRENT STATUS

OSHA has separate standards for construction and general industry. OSHA's construction standard is over 20 years old, and is based on technology and practices that reflect its age. The general industry standard, published in January 1994, established new requirements addressing work practices to be used during the operation and maintenance of electric power generation, transmission and distribution facilities. It also revised existing requirements for electrical protective equipment in general industry. Ambiguities and inconsistencies exist between the old construction standard and the new general industry standard.

Consensus standards have been developed by the American National Standards Institute's (ANSI) National Electrical Safety Code (ANSI C2-1993; NESC) (4); the American Society for Testing and Materials (ASTM); and the Institute of Electrical and Electronic Engineering (IEEE).

RATIONALE

The hazards associated with the construction of electric power transmission and distribution systems meet the criteria for designation as an OSHA priority. A moderately large number of workers are at extremely high risk of fatal injury and methods of protection are well known. In addition, construction workers, who are at greater risk, receive inferior protection compared to general industry employees; there has been strong support from both industry and labor groups for action on reducing hazards in this industry; and the action will complete the second phase of rulemaking begun with the general industry standard.

GENERAL REQUIREMENTS
OSHA 1926.950

The following general requirements shall be considered for power transmission and distribution in construction:

- Application
- Initial inspections, tests or determinations
- Clearances
- Deenergizing lines and equipment
- Emergency procedures and first aid
- Night work
- Work near and over water
- Sanitation facilities
- Hydraulic fluids

APPLICATION
OSHA 1926.950(a)

The occupational safety and health standard contained in this Subpart V applies to the construction of electric transmission and distribution lines and equipment. The term "construction" as used here includes the erection of new electric transmission and distribution lines and equipment, and the alteration, conversion and improvement of existing electric transmission and distribution lines and equipment.

Existing electric transmission and distribution lines and electrical equipment need not be modified to conform to the requirements of applicable standards in this subpart, until "construction" work as described above is to be performed on such lines or equipment.

The standards set forth in this Subpart V provide minimum requirements for safety and health. Employers may require adherence to additional standards that are not in conflict with these standards. OSHA has promulgated a general industry standard to protect employees from the hazards arising out of the operation or maintenance of electric power generation, transmission and distribution installations (29 CFR 1910.269). Although this standard does not apply to construction work, the reader is encouraged to refer to it for general information, since it is provides comprehensive safe work practices.

INITIAL INSPECTIONS, TESTS OR DETERMINATIONS
OSHA 1926.950(b)

Existing conditions shall be determined before starting work, by an inspection or a test. Such conditions shall include, but not be limited to, energized lines and equipment, conditions of poles, and the location of circuits and equipment, including power and communication lines, CATV and fire alarm circuits.

Electric equipment and lines shall be considered energized until determined to be deenergized by tests or other appropriate methods or means.

Operating voltage of equipment and lines shall be determined before working on or near energized parts.

CLEARANCES
OSHA 1926.950(c)

No employee shall be permitted to approach or take any conductive object without an approved insulating handle closer to exposed energized parts than shown in **Figure 21-1**, unless:

- The employee is insulated or guarded from the energized part (gloves or gloves with sleeves rated for the voltage involved shall be considered insulation of the employee from the energized part), or

- The energized part is insulated or guarded from him/her and any other conductive object at a different potential, or

- The employee is isolated, insulated or guarded from any other conductive object(s), as during live-line bare-hand work.

The minimum working distance and minimum clear hot stick distances stated in **Figure 21-1** shall not be violated. The minimum clear hot stick distance is that for the use of live-line tools held by linemen when performing live-line work.

Conductor support tools, such as link sticks, strain carriers and insulator cradles, may be used: provided that the clear insulation is at least as long as the insulator string or the minimum distance specified in **Figure 21-1** for the operating voltage.

Figure 21-1. This illustration shows the minimum working and clear hot stick clearances.

ALTERNATING CURRENT - MINIMUM DISTANCES	
VOLTAGE RANGE (PHASE-TO-PHASE) (KILOVOLTS)	MINIMUM WORKING DISTANCE AND CLEAR HOT STICK DISTANCE
2.1 TO 15	2 FT. 0 IN.
15.1 TO 35	2 FT. 4 IN.
35.1 TO 46	2 FT. 6 IN.
46.1 TO 72.5	3 FT. 0 IN.
72.6 TO 121	3 FT. 4 IN.
138 TO 145	3 FT. 6 IN.
161 TO 169	3 FT. 8 IN.
230 TO 242	5 FT. 0 IN.
345 TO 362	7 FT. 0 IN.*
500 TO 552	11 FT. 0 IN.*
700 TO 765	15 FT. 0 IN.*

* **NOTE:** FROM 345 kV - 362 kV, 500 kV - 552 kV and 700 kV - 765 kV, THE MINIMUM WORKING DISTANCE AND THE MINIMUM CLEAR HOT STOCK DISTANCE MAY BE REDUCED PROVIDED SUCH DISTANCES ARE NOT LESS THAN THE SHORTEST DISTANCE BETWEEN THE ENERGIZED PART AND A GROUNDED SURFACE.

CLEARANCES
OSHA 1926.950(c)

OSHA Required Training:
1926.950(d)(1)(ii)(a) thru (c), (vi) and (vii)

DEENERGIZING LINES AND EQUIPMENT
OSHA 1926.950(d)

When deenergizing lines and equipment operated in excess of 600 volts, and the means of disconnecting from electric energy is not visibly open or visibly locked out, the following provisions shall be complied with:

• The particular section of line or equipment to be deenergized shall be clearly identified, and it shall be isolated from all sources of voltage.

• Notification and assurance from the designated employee shall be obtained that:

(a) All switches and disconnectors through which electric energy may be supplied to the particular section of line or equipment to be worked have been deenergized;

(b) All switches and disconnectors are plainly tagged, indicating that employees are at work;

(c) And that where design of such switches and disconnectors permits, they have been rendered inoperable.

• After all designated switches and disconnectors have been opened, rendered inoperable and tagged, visual inspection or tests shall be conducted to ensure that equipment or lines have been deenergized.

• Protective grounds shall be applied on the disconnected lines or equipment to be worked on.

• Guards or barriers shall be erected as necessary to adjacent energized lines.

• When more than one independent crew requires the same line or equipment to be deenergized, a prominent tag for each such independent crew shall be placed on the line or equipment by the designated employee in charge.

• Upon completion of work on deenergized lines or equipment, each designated employee in charge shall determine that all employees in his/her crew are clear, that protective grounds installed by his/her crew have been removed and he/she shall report to the designated authority that all tags protecting his/her crew may be removed.

• When a crew working on a line or equipment can clearly see that the means of disconnecting from electric energy are visibly open or visibly locked-out, the following provisions shall apply:

(a) Guards or barriers shall be erected as necessary to adjacent energized lines.

(b) Upon completion of work on deenergized lines or equipment, each designated employee in charge shall determine that all employees in their crew are clear, that protective grounds installed by their crew have been removed and he/she shall report to the designated authority that all tags protecting their crew may be removed.

OSHA Required Training:
1926950(d)(2)(ii)

EMERGENCY PROCEDURES AND FIRST AID
OSHA 1926.950(e)

The employer shall provide training or require that his/her employees are knowledgeable and proficient in:

OSHA Required Training:
1926.950(e)(1)(i); (ii) and (2)

• Procedures involving emergency situations, and

• First-aid fundamentals including resuscitation.

In lieu of the above requirements, the employer may comply with the provisions of 1926.50(c) regarding first-aid requirements.

NIGHT WORK
OSHA 1926.950(f)

When working at night, spotlights or portable lights for emergency lighting shall be provided as needed to the perform the work safely.

WORK NEAR AND OVER WATER
OSHA 1926.950(g)

When crews are engaged in work over or near water and when danger of drowning exists, suitable protection shall be provided as stated in OSHA 1926.104, 1926.105 or 1926.106.

SANITATION FACILITIES
OSHA 1926.950(h)

The requirements of OSHA 1926.51 of Subpart D of Part 1926, Occupational Health and Environmental Controls, shall be complied with for sanitation facilities.

HYDRAULIC FLUIDS
OSHA 1926.950(i)

All hydraulic fluids used for the insulated sections of derrick trucks, aerial lifts and hydraulic tools that are used on or around energized lines and equipment shall be of the insulating type. The requirements for fire resistant fluids of OSHA 1926.302(d)(1) do not apply to hydraulic tools covered by this paragraph.

TOOLS AND PROTECTIVE EQUIPMENT
OSHA 1926.951

The following shall be considered for tools and protective equipment:

- Protective equipment
- Personal climbing equipment
- Ladders
- Live-line tools
- Measuring tapes or measuring ropes
- Hand tools

PROTECTIVE EQUIPMENT
OSHA 1926.951(a)

Rubber protective equipment shall be in accordance with the provisions of the American National Standards Institute (ANSI). **(See Figure 21-2)**

Figure 21-2. This illustration shows the standards for rubber protective equipment.

RUBBER PROTECTIVE EQUIPMENT STANDARDS	
ITEM	STANDARD
RUBBER INSULATING GLOVES	ANSI/ASTM D120-1984
RUBBER MATTING FOR USE AROUND ELECTRIC APPARATUS	ANSI/ASTM D178-1988
RUBBER INSULATING BLANKETS	ANSI/ASTM D1048-1981
RUBBER INSULATING HOODS	ANSI/ASTM D1049-1983
RUBBER INSULATING LINE HOSE	ANSI/ASTM D1050-1985
RUBBER INSULATING SLEEVES	ANSI/ASTM D1050-1981

PROTECTIVE EQUIPMENT
OSHA 1926.951(a)

Rubber protective equipment shall be visually inspected prior to use. In addition, an "air" test shall be performed for rubber gloves prior to use.

Protective equipment of material other than rubber shall provide equal or better electrical and mechanical protection.

Protective hats shall be in accordance with the provisions of ANSI Z89.2 - 1971, Industrial Protective Helmets for Electrical Workers, Class B, and shall be worn at the jobsite by employees who are exposed to the hazards of falling objects, electric shock or burns. (Also see ANSI Z89.1-1986.)

PERSONAL CLIMBING EQUIPMENT
OSHA 1926.951(b)

Body belts with straps or lanyards shall be worn to protect employees working at elevated locations on poles, towers or other structures except where such use creates a greater hazard to the safety of the employees, in which case other safeguards shall be employed.

Body belts and safety straps shall meet the requirements of OSHA 1926.959. In addition to being used as an employee safeguarding item, body belts with approved tool loops may be used for the purpose of holding tools. Body belts shall be free from additional metal hooks and tool loops other than those permitted in OSHA 1926.959.

Body belts and straps shall be inspected before use and each day to determine that they are in safe working condition. Life lines and lanyards shall comply with the provisions of OSHA 1926.104.

Safety lines are not intended to be subjected to shock loading and are used for emergency rescue such as lowering a person to the ground. Such safety lines shall be a minimum of 2 in. diameter and three or four strand first-grade manila or its equivalent in strength (2650 lb) and durability. Defective ropes shall be replaced.

LADDERS
OSHA 1926.951(c)

Portable metal or conductive ladders shall not be used near energized lines or equipment except as may be necessary in specialized work such as in high voltage substations, where nonconductive ladders might present a greater hazard than conductive ladders. Conductive or metal ladders shall be prominently marked as conductive and all necessary precautions shall be taken when used in specialized work.

Hook or other type ladders used in structures shall be positively secured to prevent the ladder from being accidentally displaced.

LIVE-LINE TOOLS
OSHA 1926.951(d)

Only live-line tool poles having a manufacturer's certification to withstand the following minimum tests shall be used:

- 100,000 volts per foot of length for 5 minutes when the tool is made of fiberglass; or

- 75,000 volts per foot of length for 3 minutes when the tool is made of wood; or

- Other tests equivalent to these tests as appropriate.

All live-line tools shall be visually inspected before use each day. Tools to be used shall be wiped clean and if any hazardous defects are indicated, removed from service.

MEASURING TAPES OR MEASURING ROPES
OSHA 1926.951(e)

Measuring tapes or measuring ropes that are metal or contain conductive strands shall not be used when working on or near energized parts.

HAND TOOLS
OSHA 1926.951(f)

Switches for all powered hand tools shall comply with OSHA 1926.300(d). All portable electric hand tools shall:

- Be equipped with three-wire cord having the ground wire permanently connected to the tool frame and means for grounding the other end; or

- Be of the double insulated type and permanently labeled as "double insulated"; or

- Be connected to the power supply by means of an isolating transformer, or other isolated power supply.

All hydraulic tools that are used on or around energized lines or equipment shall use nonconducting hoses having adequate strength for the normal operating pressures. It should be noted that the provisions of OSHA 1926.302(d)(2) shall also apply.

All pneumatic tools that are used on or around energized lines or equipment shall:

- Have nonconducting hoses having adequate strength for the normal operating pressures and

- Have an accumulator on the compressor to collect moisture.

MECHANICAL EQUIPMENT
OSHA 1926.952

The following shall be considered for mechanical equipment:

- General
- Aerial lifts
- Derrick trucks, cranes and other lifting equipment

GENERAL
OSHA 1926.952(a)

Visual inspections of the equipment shall be made to determine that it is in good condition each day the equipment is to be used.

Tests shall be made at the beginning of each shift during which the equipment is to be used to determine that the brakes and operating systems are in proper working condition.

No employer shall use any motor vehicle equipment having an obstructed view to the rear unless:

- The vehicle had a reverse signal alarm audible above the surrounding noise level or

- The vehicle is backed up only when an observer signals that it is safe to do so.

AERIAL LIFTS
OSHA 1926.952(b)

The provisions of OSHA 1926.556 shall apply to the utilization of aerial lifts.

When working near energized lines or equipment, aerial lift trucks shall be grounded or barricaded and considered as energized equipment, or the aerial lift truck shall be insulated for the work being performed.

Equipment or material shall not be passed between a pole or structure and an aerial lift while an employee working from the basket is within reaching distance of energized conductors or equipment that are not covered with insulating protective equipment.

DERRICK TRUCKS, CRANES AND OTHER LIFTING EQUIPMENT
OSHA 1926.952(c)

All derrick trucks, cranes and other lifting equipment shall comply with Subparts N and O of Part 1926 except:

- As stated in OSHA 1926.550(a)(15)(i) and (ii) relating to clearance (for clearances in this chapter see **Figure 21-1**), and

- Derrick truck (electric line trucks) shall not be required to comply with OSHA 1926.550(a)(7)(vi), (a)(17), (b)(2) and (e).

With the exception of equipment certified for work on the proper voltage, mechanical equipment shall not be operated closer to any energized line or equipment than the clearances set forth in **Figure 21-1** unless:

- An insulated barrier is installed between the energized part and the mechanical equipment, or

- The mechanical equipment is grounded, or

- The mechanical equipment is insulated or

- The mechanical equipment is considered as energized.

MATERIAL HANDLING
OSHA 1926.953

The following shall be considered for material handling:

- Unloading
- Pole hauling
- Storage
- Tag line
- Oil filled equipment
- Framing
- Attaching the load

UNLOADING
OSHA 1926.953(a)

Prior to unloading steel, pole cross arms and similar material, the load shall be thoroughly examined to ascertain if the load has shifted, binders or stakes have broken or the load is otherwise hazardous to employees.

POLE HAULING
OSHA 1926.953(b)

During pole hauling operations, all loads shall be secured to prevent displacement and a red flag shall be displayed at the trailing end of the longest pole.

Precautions shall be exercised to prevent blocking of roadways or endangering other traffic.

When hauling poles during the hours of darkness, illuminated warning devices shall be attached to the trailing end of the longest pole.

STORAGE
OSHA 1926.953(c)

No materials or equipment shall be stored under energized bus, energized lines or near energized equipment, if it is practical to store them elsewhere.

When materials or equipment are stored under energized lines or near energized equipment, applicable clearances shall be maintained as stated in Table V-1; and extraordinary caution shall be exercised when moving materials near such energized equipment.

TAG LINE
OSHA 1926.953(d)

Where hazards to employees exist, tag lines or other suitable devices shall be used to control loads being handled by hoisting equipment.

OIL FILLED EQUIPMENT
OSHA 1926.953(e)

During construction or repair of oil filled equipment, the oil may be stored in temporary containers other than those required in OSHA 1926.152, such as pillow tanks.

FRAMING
OSHA 1926.953(f)

During framing operations, employees shall not work under a pole or a structure suspended by a crane, A-frame or similar equipment unless the pole or structure is adequately supported.

ATTACHING THE LOAD
OSHA 1926.953(f)

The hoist rope shall not be wrapped around the load. This provision shall not apply to electric construction crews when setting or removing poles.

GROUNDING FOR PROTECTION OF EMPLOYEES
OSHA 1926.954

The following shall be considered for grounding for protection of employees:

- General
- New construction
- Communications conductors
- Voltage testing
- Attaching grounds
- Grounds placement
- Testing without grounds
- Grounding electrode
- Grounding to tower
- Ground lead

GENERAL
OSHA 1926.954(a)

All conductors and equipment shall be treated as energized until tested or otherwise determined to be deenergized or until grounded.

NEW CONSTRUCTION
OSHA 1926.954(b)

New lines or equipment may be considered deenergized and worked on as such where:

- The lines or equipment are grounded or

- The hazard of induced voltages is not present, and adequate clearances or other means are implemented to prevent contact with energized lines or equipment and the new lines or equipment.

COMMUNICATIONS CONDUCTORS
OSHA 1926.954(c)

Bare wire communications conductors on power poles or structures shall be treated as energized lines unless protected by insulating materials.

VOLTAGE TESTING
OSHA 1926.954(d)

Deenergized conductors and equipment that are to be grounded shall be tested for voltage. The results of this test shall determine the subsequent procedures as required in OSHA 1926.950(d).

ATTACHING GROUNDS
OSHA 1926.954(e)

When attaching grounds, the ground end shall be attached first, and the other end shall be attached and removed by means of insulated tools or other suitable devices.

When removing grounds, the grounding device shall first be removed from the line or equipment using insulating tools or other suitable devices.

GROUNDS PLACEMENT
OSHA 1926.954(f)

Grounds shall be placed between the work location and all sources of energy and as close as practicable to the work location, or grounds shall be placed at the work location. If work is to be performed at more than one location in a line section, the line section shall be grounded and short circuited at one location in the line section and the conductor to be worked on shall be grounded at each work location. The minimum distance shown in Table V-1 shall be maintained from ungrounded (phase) conductors at the work location. Where the making of a ground is impracticable, or the conditions resulting there from would be more hazardous than working on the lines or equipment without grounding, the grounds shall be permitted to be omitted and the line or equipment worked as energized.

TESTING WITHOUT GROUNDS
OSHA 1926.954(g)

Grounds shall be permitted to be temporarily removed only when necessary for test purpose,s and extreme caution shall be exercised during the test procedures.

GROUNDING ELECTRODE
OSHA 1926.954(h)

When grounding electrodes are utilized, such electrodes shall have a resistance to ground low enough to remove the danger of harm to personnel or permit prompt operation of protective devices.

GROUNDING TO TOWER
OSHA 1926.954(i)

Grounding to tower shall be made with a tower clamp capable of conducting the anticipated fault current.

GROUND LEAD
OSHA 1926.954(j)

A ground lead, to be attached to either a tower ground or driven ground, shall be capable of conducting the anticipated fault current and have a minimum conductance of 2 AWG copper.

OVERHEAD LINES
OSHA 1926.955

The following shall be considered for overhead lines:

- Overhead lines
- Metal tower construction
- Stringing or removing deenergized conductors
- Stringing next to energized lines
- Live-line, bare-hand work

OVERHEAD LINES
OSHA 1926.955(a)

When working on or with overhead lines, the following provisions shall be complied with:

- Prior to climbing poles, ladders, scaffolds or other elevated structures, an inspection shall be made to determine that the structures are capable of sustaining the additional or unbalanced stresses to which they will be subjected.

- Where poles or structures may be unsafe for climbing, they shall not be climbed until made safe by guying, bracing or other adequate means.

- Before installing or removing wire or cable, strains to which poles and structures will be subjected shall be considered and necessary action taken to prevent failure of supporting structures.

- When setting, moving or removing poles using cranes, derricks, gin poles, A-frames or other mechanized equipment near energized lines or equipment, precautions shall be taken to avoid contact with energized lines or equipment, except in bare-hand live-line work, or where barriers or protective devices are used.

- Equipment and machinery operating adjacent to energized lines or equipment shall comply with OSHA 1926.952(c)(2).

- Unless using suitable protective equipment for the voltage involved, employees standing on the ground shall avoid contacting equipment or machinery working adjacent to energized lines or equipment.

- Lifting equipment shall be bonded to an effective ground, or it shall be considered energized and barricaded when utilized near energized equipment or lines.

- Pole holes shall not be left unattended or unguarded in areas where employees are currently working.

- Tag lines shall be of a nonconductive type when used near energized lines.

METAL TOWER CONSTRUCTION
OSHA 1926.955(b)

When excavating or augering in unstable material, pad- or pile-type footings more than 5 ft. deep shall be sloped to the angle of repose or shored if worker entry is required. Ladders shall be provided when footings in excavations are more than 4 ft. deep. See Chapter 16 "Excavations," of this guide.

OSHA Required Training:
1926.955(b)(3)(i)
1926.955(b)(8) and (d)(7)

When towers are erected near energized lines, the lines shall be deenergized or appropriate clearances listed in **Figure 21-1** shall be maintained. During lifts, a spotter shall determine the required clearances.

No one shall be permitted to remain in a footing while equipment is being spotted or moved for placement.

STRINGING OR REMOVING DEENERGIZED CONDUCTORS
OSHA 1926.955(c)

Before stringing operations begin, all workers shall receive a pre-job briefing to include a review of work assignments, equipment required and precautions to be taken for the operation.

When there is a possibility of a conductor accidentally contacting an energized circuit, the conductor being installed or removed shall be grounded; or workers shall be insulated or isolated. If the existing line is deenergized, proper clearance authorization shall be secured and the line shall be grounded on both sides of the crossover; or the strung line shall be considered and worked as if it were energized.

When crossing over energized conductors of more than 600 volts, rope necks or guard structures shall be installed unless provisions are made to insulate or isolate the live conductor or worker. Where practical, the automatic reclosing feature of the circuit-interrupting device shall be made inoperative. In addition, the line being strung shall be grounded either side of the crossover or worked as energized.

Conductors being strung or removed shall be controlled with tension reels, guard structures, tielines or other equivalent means to prevent contact with energized lines.

Conductor grips shall not be used on wire rope, unless the conductor grips are designed for this application.

Clipping crews shall have a minimum of two structures clipped in between the crew and the conductor being sagged. When working on bare conductors, crews shall work between grounds at all times; and grounds shall remain intact until work is completed.

Reliable communications between the reel tender and the pulling-rig operator shall be provided. Each pull shall be snubbed or dead-ended at both ends before subsequent pulls.

STRINGING NEXT TO ENERGIZED LINES
OSHA 1926.955(d)

Before stringing parallel to an existing live line, it shall be determined if dangerous induced-voltage buildups will occur, especially during switching and ground fault conditions. When there is a possibility of dangerous induced voltages, the requirements in this section shall be followed unless the lines are worked as energized.

All pulling and tension equipment shall be isolated, insulated or effectively grounded. A ground shall be installed between the tensioning reel and the first structure in order to ground each bare conductor, sub-conductor or overhead ground during stringing operations.

During stringing operations, each of the conductors in the paragraph above shall be grounded at the first tower adjacent to both the tension and pulling setup and in increments so that no point is more than 2 miles from the ground. Grounds shall be left in place until work is completed, removed as a last phase of aerial cleanup or removed with a hot stick. Such conductors shall be grounded at all dead-end or catch-off points.

A ground shall be located on each side and within 10 ft. from the working area where conductors, sub-conductors, or overhead ground conductors are being spliced. The two ends of the conductors to be spliced together shall be bonded together. It is recommended that splicing be carried out on an insulated platform or metallic grounding mat that is bonded to both grounds. All conductors, sub-conductors and ground conductors shall be bonded to the tower at any isolated tower when necessary to complete work on the transmission line. When deadend towers are worked, all deenergized lines shall be grounded.

When performing work from structures, all workers on conductors shall be protected by individual grounds at every work location.

LIVE-LINE, BARE-HAND WORK
OSHA 1926.955(e)

OSHA Required Training:
1926.955(e)(1) and (4)

Workers shall be instructed and trained in live-line, bare-hand techniques and safety requirements before beginning such work.

Handlines shall not be used between buckets, booms or ground, and there shall be no conductive objects over 36 in. long in a bucket (except for appropriate length jumpers, armor rods and tools).

Workers shall know the voltage rating of the circuit, clearances to ground and other energized parts and the voltage limitations of the aerial-lift equipment before they begin live-line, bare-hand work.

Only equipment and tools designed, tested and intended for live-line, bare-hand work shall be used. Tools and equipment shall be maintained clean and dry. All work shall be personally supervised by a trained and qualified worker in live-line, bare-hand work.

When practical, the automatic reclosing feature of an interrupting device shall be made inoperative before working on any energized line or equipment.

A conductive bucket liner or other conductive device shall be used for bonding the insulated aerial device to the live line or equipment. Workers shall be connected to the bucket liner with conductive shoes, leg clips or other equivalent means. When necessary for electrostatic protections, appropriate electrostatic shielding or conductive clothing shall be provided.

Before the boom is elevated, the outriggers on the aerial truck shall be extended and adjusted to stabilize the truck, and the body of the truck shall be bonded to an effective ground or barricaded and considered as energized equipment. The controls of the aerial lift shall be inspected and tested (ground level and bucket) before it is moved into the work position.

Arm current tests shall be made before starting work each day and any time a new higher voltage is to be worked.

Aerial lifts shall have upper and lower controls; lower controls shall have override capabilities. Bucket controls shall be within easy reach of workers; lower controls shall be located near the base of the boom. Controls shall be overriding, and the lower control shall not be used without approval of the worker in the bucket except in the event of an emergency.

The minimum clearance distances for live-line, bare-hand work are specified in **Figure 21-3.** These distances shall be maintained from all grounded objects; from lines and equipment, including the grounded frame of the lift truck; and from lines of a different potential to those bonded to the insulated aerial device, unless these objects are covered by insulated guards. The clearance distances shall be maintained when live circuits are approached, or left, or when bonded to live circuits.

Before workers contact the energized parts to be worked, the conductive bucket liner shall be bonded to the energized conductor and remain so until work is completed.

The minimum clearances as stated in **Figure 21-2** shall be printed on durable, nonconductive material and posted inside the bucket where they may easily be seen by workers inside the bucket.

Figure 21-3. This illustration shows the minimum clearances for live-line bare-hand work.

MINIMUM CLEARANCE DISTANCES FOR LIVE-LINE BARE HAND WORK (ALTERNATING CURRENT)		
VOLTAGE RANGE (PHASE-TO-PHASE) KILOVOLTS (kV)	DISTANCE IN FT. AND IN. FOR MAXIMUM VOLTAGE	
	PHASE-TO-GROUND	PHASE-TO-PHASE
2.1 TO 15	2 FT. 0 IN.	2 FT. 0 IN.
15.1 TO 35	2 FT. 4 IN.	2 FT. 4 IN.
35.1 TO 46	2 FT. 6 IN.	2 FT. 6 IN.
46.1 TO 72.5	3 FT. 0 IN.	3 FT. 0 IN.
72.6 TO 121	3 FT. 4 IN.	4 FT. 6 IN.
138 TO 145	3 FT. 6 IN.	5 FT. 0 IN.
161 TO 169	3 FT. 8 IN.	5 FT. 6 IN.
230 TO 242	5 FT. 0 IN.	8 FT. 4 IN.
345 TO 362*	7 FT. 0 IN.	13 FT. 4 IN.
500 TO 552*	11 FT. 0 IN.	20 FT. 0 IN.
700 TO 765*	15 FT. 0 IN.	31 FT. 0 IN.

*FOR NOTED kV's, THE MINIMUM DISTANCE MAY BE REDUCED, PROVIDED THE DISTANCES ARE NOT MADE LESS THAN THE SHORTEST DISTANCE BETWEEN THE ENERGIZED PART AND A GROUNDED SURFACE.

LIVE-LINE, BARE-HAND WORK
OSHA 1926.955(e)

UNDERGROUND LINES
OSHA 1926.956

The following shall be considered for underground lines:

- Guarding street opening used for access to underground lines or equipment
- Work in manholes
- Trenching and excavating

GUARDING STREET OPENING USED FOR ACCESS TO UNDERGROUND LINES OR EQUIPMENT
OSHA 1926.956(a)

Appropriate warning signs shall be promptly placed when covers of manholes, handholes or vaults are removed. The nature and location of the hazards involved determine what is an appropriate warning sign.

Before an employee enters a street opening, such as a manhole or an unvented vault, it shall be promptly protected with a barrier, temporary cover or other suitable guard.

When work is to be performed in a manhole or unvented vault:

• No entry shall be permitted unless forced ventilation is provided or the atmosphere is found to be safe by testing for oxygen deficiency and the presence of explosive gases or fumes;

• Where unsafe conditions are detected, by testing or other means, the work area shall be ventilated and otherwise made safe before entry;

• Provisions shall be made for an adequate continuous supply of air.

WORK IN MANHOLES
OSHA 1926.956(b)

OSHA Required Training:
1926.956(b)(1)

While work is being performed in manholes, an employee shall be available in the immediate vicinity to render emergency assistance as may be required. This shall not preclude the employee in the immediate vicinity from occasionally entering a manhole to provide assistance, other than an emergency. This requirement does not preclude a qualified employee working alone from entering for brief periods of time a manhole where energized cables or equipment are in service, for the purpose of inspection, housekeeping, taking readings or similar work if such work can be performed safely.

When open flames must be used or smoking is permitted in manholes, extra precautions shall be taken to provide adequate ventilation.

Before using open flames in a manhole or excavation in an area where combustible gases or liquids may be present, such as near a gasoline service station, the atmosphere of the manhole or excavation shall be tested and found safe or cleared of the combustible gases or liquids.

TRENCHING AND EXCAVATING
OSHA 1926.956(c)

During excavation or trenching, in order to prevent the exposure of employees to the hazards created by damage to dangerous underground facilities, efforts shall be made to determine the location of such facilities and work conducted in a manner designed to avoid damage.

Trenching and excavation operations shall comply with OSHA 1926.651 and 1926.652. See Chapter 16.

When underground facilities are exposed (electric, gas, water, telephone, etc.) they shall be protected as necessary to avoid damage.

Where multiple cables exist in an excavation, cables other than the one being worked on shall be protected as necessary.

When multiple cables exist in an excavation, the cable to be worked on shall be identified by electrical means unless its identity is obvious by reason of distinctive appearance.

Before cutting into a cable or opening a splice, the cable shall be identified and verified to be the proper cable.

When working on buried cable or on cable in manholes, metallic sheath continuity shall be maintained by bonding across the opening or by equivalent means.

CONSTRUCTION IN ENERGIZED SUBSTATIONS
OSHA 1926.957

The following shall be considered for construction in energized substations:

- Work near energized equipment facilities
- Deenergized lines or equipment
- Barricades and barriers
- Control panels
- Mechanical equipment
- Storage
- Substation fences
- Footing excavation
- Lineman's body belts

OSHA Required Training:
1926.957(a)(1)

WORK NEAR ENERGIZED EQUIPMENT FACILITIES
OSHA 1926.957(a)

When construction work is performed in an energized substation, authorization shall be obtained from the designated, authorized person before work is started, and the following shall be determined:

- What facilities are energized and

- What protective equipment and precautions are necessary for the safety of personnel.

Extraordinary caution shall be exercised in the handling of busbars, tower steel, materials and equipment in the vicinity of energized facilities. The requirements set forth in OSHA 1926.950(c) shall be complied with.

DEENERGIZED LINES OR EQUIPMENT
OSHA 1926.957(b)

When it is necessary to deenergize equipment or lines for protection of employees, the requirements of OSHA 1926.950(d) shall be complied with.

BARRICADES AND BARRIERS
OSHA 1926.957(c)

Barricades or barriers shall be installed to prevent accidental contact with energized lines or equipment.

Where appropriate, signs indicating the hazard shall be posted near the barricade or barrier. These signs shall comply with OSHA 1926.200. See Chapter 7.

OSHA Required Training:
1926.957(d)(1)

CONTROL PANELS
OSHA 1926.957(d)

Work on or adjacent to energized control panels shall be performed by designated employees. Precaution shall be taken to prevent accidental operation of relays or other protective devices due to jarring, vibration or improper wiring.

MECHANICAL EQUIPMENT
OSHA 1926.957(e)

Use of vehicles, gin poles, cranes and other equipment in restricted or hazardous areas shall at all times be controlled by designated employees.

All mobile cranes and derricks shall be effectively grounded when being moved or operated in close proximity to energized lines or equipment, or the equipment shall be considered energized.

Fenders shall not be required for lowboys used for transporting large electrical equipment, transformers, or breakers.

STORAGE
OSHA 1926.957(f)

The storage requirements of OSHA 1926.953(c) shall be complied with.

SUBSTATION FENCES
OSHA 1926.957(g)

When a substation fence must be expanded or removed for construction purposes, a temporary fence affording similar protection when the site is unattended shall be provided. Adequate interconnection with ground shall be maintained between temporary fence and permanent fence. All gates to all unattended substations shall be locked except when work is in progress.

FOOTING EXCAVATION
OSHA 1926.957(h)

Excavation for auger, pad and piling type footings for structures and towers shall require the same precautions as for metal tower construction [see OSHA 1926.955(b)(1)].

No employee shall be permitted to enter an unsupported auger-type excavation in unstable material for any purpose. Necessary clean-out in such cases shall be accomplished without entry.

LINEMAN'S BODY BELTS
OSHA 1926.957(i)

Lineman's belts, safety straps and lanyards shall meet the requirements of the American Society of Testing Materials (ASTM) Standard B117-64. PPE (body belts with straps and lanyards) shall be worn by those working at elevated locations, except in operations in which use of these items may create a greater hazard; then other safeguards shall be used.

Before each use, body belts and straps shall be inspected for a deformed buckle; cracked or broken "D" ring; failure of the snap hook; parted, torn or cracked fabric or leather; and other damaged items. Safety lines are not intended for shockloading; they are used to lower workers during an emergency rescue. Defective lines shall be replaced. The cushion support of a body belt shall contain no exposed rivets on the inside.

Power Transmission and Distribution

Section	Answer	
_____	T	F
_____	T	F
_____	T	F
_____	T	F
_____	T	F
_____	T	F
_____	T	F
_____	T	F
_____	T	F
_____	T	F
_____	T	F
_____	T	F
_____	T	F

Power Transmission and Distribution

1. Electric equipment and lines shall be considered energized until determined to be deenergized by tests or other appropriate methods or means.

2. The minimum working distance and minimum clear hot stick distances for 7.2 kV is 2 ft. 4 in.

3. Hydraulic fluids used for the insulated sections of derrick trucks, aerial lifts and hydraulic tools that are used on or around energized lines and equipment shall not be required to be of the insulating type.

4. Rubber protective equipment shall be visually inspected at least weekly. In addition, an "air" test shall be performed for rubber gloves prior to use.

5. Body belts and straps shall be inspected before use and each day to determine that they are in safe working condition.

6. Safety lines are not intended to be subjected to shock loading and are used for emergency rescue such as lowering a person to the ground. Such safety lines shall be a minimum of 1/4 in. diameter and three or four strand first-grade manila or its equivalent in strength (2650 lb.) and durability.

7. All live-line tools shall be visually inspected semi-annually.

8. When working near energized lines or equipment, aerial lift trucks shall be grounded or barricaded and considered as energized equipment, or the aerial lift truck shall be insulated for the work being performed.

9. When attaching grounds, the ground end shall be attached last.

10. Grounding to tower shall be made with a tower clamp capable of conducting the anticipated voltage.

11. A ground lead, to be attached to either a tower ground or driven ground, shall be capable of conducting the anticipated fault current and have a minimum conductance of 4 AWG copper.

12. Prior to climbing poles, ladders, scaffolds or other elevated structures, an inspection shall be made to determine that the structures are capable of sustaining the additional or unbalanced stresses to which they will be subjected.

13. When excavating or augering in unstable material, pad- or pile-type footings more than 6 ft. deep shall be sloped to the angle of repose or shored if worker entry is required.

T   F

14. Before stringing operations begin, all workers shall receive a weekly pre-job briefing to include a review of work assignments, equipment required and precautions to be taken for the operation.

T   F

15. If the existing line is deenergized, proper clearance authorization shall be secured and the line shall be grounded on both sides of the crossover; or the strung line shall be considered and worked as if it were energized.

T   F

16. When crossing over energized conductors of more than 750 volts, rope necks or guard structures shall be installed unless provisions are made to insulate or isolate the live conductor or worker.

T   F

17. Clipping crews shall have a minimum of two structures clipped in between the crew and the conductor being sagged. When working on bare conductors, crews shall work between grounds at all times; and grounds shall remain intact until work is completed.

T   F

18. During stringing operations, conductors shall be grounded at the first tower adjacent to both the tension and pulling setup and in increments so that no point is more than 4 miles from the ground.

T   F

19. A ground shall be located on each side and within 10 ft. from the working area where conductors, subconductors or overhead ground conductors are being spliced.

T   F

20. While work is being performed in electrical vaults and manholes, an employee shall be available within 20 minutes to render emergency assistance as may be required.

Stairways and Ladders

Stairways and ladders are a major source of injuries and fatalities among construction workers.

OSHA estimates that there are 24,882 injuries and as many as 36 fatalities per year due to falls from stairways and ladders used in construction. Nearly half of these injuries are serious enough to require time off the job - 11,570 lost workday injuries and 13,312 non-lost workday injuries occur annually due to falls from stairways and ladders used in construction. This data demonstrates that work on and around ladders and stairways is hazardous. More importantly, they show that compliance with OSHA's requirements for the safe use of ladders and stairways could have prevented many of these injuries.

This discussion serves as a quick and easy reference for use on job sites. The requirements of OSHA safety regulations for the safe use of ladders and stairs (Subpart X, Title 29 Code of Federal Regulations, Part 1926.1050 through 1926.1060) are explained in this discussion.

SCOPE AND APPLICATION
OSHA 1926.1050(a)

The OSHA rules apply to all stairways and ladders used in construction, alteration, repair (including painting and decorating) and demolition of work sites covered by OSHA's construction safety and health standards. They also specify when stairways and ladders must be provided. They do not apply to ladders that are specifically manufactured for scaffold access and egress, but do apply to job-made and manufactured portable ladders intended for general purpose use and then used for scaffold access and egress.

GENERAL REQUIREMENTS
OSHA 1926.1051

Safety Tip: At all worker points of access where there is a break in elevation of 19 in. or more, a stairway or ladder shall be provided.

A stairway or ladder shall be provided at all worker points of access where there is a break in elevation of 19 in. (48 cm) or more and no ramp, runway, embankment or personnel hoist is provided.

When there is only one point of access between levels, it shall be kept clear to permit free passage by workers. If free passage becomes restricted, a second point of access shall be provided and used.

When there are more than two points of access between levels, at least one point of access shall be kept clear.

All stairway and ladder fall protection systems required by these rules shall be installed and all duties required by the stairway and ladder rules shall be performed before employees begin work that requires them to use stairways or ladders and their respective fall protection systems.

STAIRWAYS
OSHA 1926.1052

The following general requirements apply to all stairways used during the process of construction, as indicated:

- Stairways that will not be a permanent part of the structure on which construction work is performed shall have landings at least 30 in. deep and 22 in. wide (76 x 56 cm) at every 12 ft. (3.7 m) or less of vertical rise.

- Stairways shall be installed at least 30 degrees, and no more than 50 degrees, from the horizontal.

- Variations in riser height or stair tread depth shall not exceed 1/4 in. in any stairway system, including any foundation structure used as one or more treads of the stairs.

- Where doors or gates open directly onto a stairway, a platform shall be provided that is at least 20 in. (51 cm) in width beyond the swing of the door.

- Metal pan landings and metal pan treads shall be secured in place before filling.

- All stairway parts shall be free of dangerous projections such as protruding nails.

• Slippery conditions on stairways shall be corrected.

• Spiral stairways that will not be a permanent part of the structure may not be used by workers.

The following requirements apply to stairs in temporary service during construction:

• Except during construction of the actual stairway, stairways with metal pan landings and treads shall not be used where the treads and/or landings have not been filled in with concrete or other material, unless the pans of the stairs and/or landings are temporarily filled in with wood or other material. All treads and landings shall be replaced when worn below the top edge of the pan.

• Except during construction of the actual stairway, skeleton metal frame structures and steps shall not be used (where treads and/or landings are to be installed at a later date) unless the stairs are fitted with secured temporary treads and landings.

• Temporary treads shall be made of wood or other solid material and installed the full width and depth of the stair.

STAIRRAILS AND HANDRAILS
OSHA 1926.1052(c)

The following general requirements apply to all stairrails and handrails:

• Stairways having four or more risers, or rising more than 30 in. (76 cm) in height, whichever is less, shall have at least one handrail. A stairrail also shall be installed along each unprotected side or edge. When the top edge of a stairrail system also serves as a handrail, the height of the top edge shall not be more than 37 in. (94 cm) or less than 36 in. (91.5 cm) from the upper surface of the stairrail to the surface of the tread.

• Winding or spiral stairways shall be equipped with a handrail to prevent using areas where the tread width is less than 6 in. (15 cm).

• Stairrails installed after March 15, 1991, shall not be less than 36 in. (91.5 cm) in height.

• Midrails, screens, mesh, intermediate vertical members or equivalent intermediate structural members shall be provided between the top rail and stairway steps of the stairrail system.

• Midrails, when used, shall be located midway between the top of the stairrail system and the stairway steps.

• Screens or mesh, when used, shall extend from the top rail to the stairway step, and along the opening between top rail supports.

• Intermediate vertical members, such as balusters, when used, shall not be more than 19 in. (48 cm) apart.

• Other intermediate structural members, when used, shall be installed so that there are no openings of more than 19 in. (48 cm) wide.

• Handrails and the top rails of the stairrail systems shall be capable of withstanding, without failure, at least 200 pounds (890 N) of weight applied within 2 in. (5 cm) of the top edge in any downward or outward direction, at any point along the top edge.

Safety Tip: At least on handrail must be provided for stairways having four or more risers or rising more than 30 in.

- The height of handrails shall not be more than 37 in. (94 cm) or less than 30 in. (76 cm) from the upper surface of the handrail to the surface of the tread.

- The height of the top edge of a stairrail system used as a handrail shall not be more than 37 in. (94 cm) or less than 36 in. (91.5 cm) from the upper surface of the stairrail system to the surface of the tread.

- Stairrail systems and handrails shall be surfaced to prevent injuries such as punctures or lacerations and to keep clothing from snagging.

- Handrails shall provide an adequate handhold for employees to grasp to prevent falls.

- The ends of stairrail systems and handrails shall be constructed to prevent dangerous projections such as rails protruding beyond the end posts of the system.

- Temporary handrails shall have a minimum clearance of 3 in. (8 cm) between the handrail and walls, stairrail systems and other objects.

- Unprotected sides and edges of stairway landings shall be provided with standard 42 in. (1.1 m) guardrail systems.

LADDERS
OSHA 1926.1053

The following general requirements apply to all ladders, including job-made ladders:

- A double-cleated ladder or two or more ladders shall be provided when ladders are the only way to enter or exit a work area having 25 or more employees, or when a ladder serves simultaneous two-way traffic.

- Ladder rungs, cleats and steps shall be parallel, level and uniformly spaced when the ladder is in position for use.

- Rungs, cleats and steps of portable and fixed ladders (except as provided below) shall not be spaced less than 10 in. (25 cm) apart, nor more than 14 in. (36 cm) apart, along the ladder's side rails. **(See Figure 22-1)**

Figure 22-1. This illustration shows the requirements for spacing of ladder rungs.

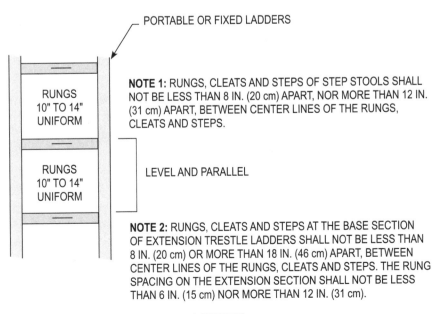

PORTABLE OR FIXED LADDERS

RUNGS 10" TO 14" UNIFORM

RUNGS 10" TO 14" UNIFORM

LEVEL AND PARALLEL

NOTE 1: RUNGS, CLEATS AND STEPS OF STEP STOOLS SHALL NOT BE LESS THAN 8 IN. (20 cm) APART, NOR MORE THAN 12 IN. (31 cm) APART, BETWEEN CENTER LINES OF THE RUNGS, CLEATS AND STEPS.

NOTE 2: RUNGS, CLEATS AND STEPS AT THE BASE SECTION OF EXTENSION TRESTLE LADDERS SHALL NOT BE LESS THAN 8 IN. (20 cm) OR MORE THAN 18 IN. (46 cm) APART, BETWEEN CENTER LINES OF THE RUNGS, CLEATS AND STEPS. THE RUNG SPACING ON THE EXTENSION SECTION SHALL NOT BE LESS THAN 6 IN. (15 cm) NOR MORE THAN 12 IN. (31 cm).

LADDERS
OSHA 1926.1053

- Rungs, cleats and steps of step stools shall not be less than 8 in. (20 cm) apart, nor more than 12 in. (31 cm) apart, between center lines of the rungs, cleats and steps.

- Rungs, cleats and steps at the base section of extension trestle ladders shall not be less than 8 in. (20 cm) nor more than 18 in. (46 cm) apart, between center lines of the rungs, cleats and steps. The rung spacing on the extension section shall not be less than 6 in. (15 cm) or more than 12 in. (31 cm).

- Ladders shall not be tied or fastened together to create longer sections unless they are specifically designed for such use.

- A metal spreader or locking device shall be provided on each stepladder to hold the front and back sections in an open position when the ladder is being used.

- When splicing side rails, the resulting side rail shall be equivalent in strength to a one-piece side rail made of the same material.

- Two or more separate ladders used to reach an elevated work area shall be offset with a platform or landing between the ladders, except when portable ladders are used to gain access to fixed ladders.

- Ladder components shall be surfaced to prevent injury from punctures or lacerations, and prevent snagging of clothing.

- Wood ladders shall not be coated with any opaque covering, except for identification or warning labels, which may be placed only on one face of a side rail.

PORTABLE LADDERS

Non-self-supporting and self-supporting portable ladders shall support at least four times the maximum intended load; extra heavy-duty type 1A metal or plastic ladders shall sustain 3.3 times the maximum intended load. The ability of a self-supporting ladder to sustain loads shall be determined by applying the load to the ladder in a downward vertical direction. The ability of a non-self-supporting ladder to sustain loads shall be determined by applying the load in a downward vertical direction when the ladder is placed at a horizontal angle of 75.5 degrees.

The minimum clear distance between side rails for all portable ladders shall be 11.5 in. (29 cm).

The rungs and steps of portable metal ladders shall be corrugated, knurled, dimpled, coated with skid-resistant material or treated to minimize slipping.

FIXED LADDERS

A fixed ladder shall be capable of supporting at least two loads of 250 pounds (114 kg) each, concentrated between any two consecutive attachments. Fixed ladders also shall support added anticipated loads caused by ice buildup, winds, rigging and impact loads resulting from the use of ladder safety devices.

Individual rung/step ladders shall extend at least 42 in. (1.1 m) above an access level or landing platform either by the continuation of the rung spacings as horizontal grab bars or by providing vertical grab bars that shall have the same lateral spacing as the vertical legs of the ladder rails.

Each step or rung of a fixed ladder shall be capable of supporting a load of at least 250 pounds (114 kg) applied in the middle of the step or rung.

The minimum clear distance between the sides of individual rung/step ladders and between the side rails of other fixed ladders shall be 16 in. (41 cm).

The rungs of individual rung/step ladders shall be shaped to prevent slipping off the end of the rungs.

The rungs and steps of fixed metal ladders manufactured after January 14, 1991, shall be corrugated, knurled, dimpled, coated with skid-resistant material or treated to minimize slipping.

The minimum perpendicular clearance between fixed ladder rungs, cleats and steps, and any obstruction behind the ladder shall be 7 in. (18 cm), except that the clearance for an elevator pit ladder shall be 4.5 in. (11 cm).

The minimum perpendicular clearance between the centerline of fixed ladder rungs, cleats and steps, and any obstruction on the climbing side of the ladder shall be 30 in. (76 cm). If obstructions are unavoidable, clearance shall be permitted to be reduced to 24 in. (61 cm), provided a deflection device is installed to guide workers around the obstruction. **(See Figure 22-2)**

Figure 22-2. This illustration shows the minimum perpendicular clearance between the centerline of fixed ladder rungs, cleats and steps.

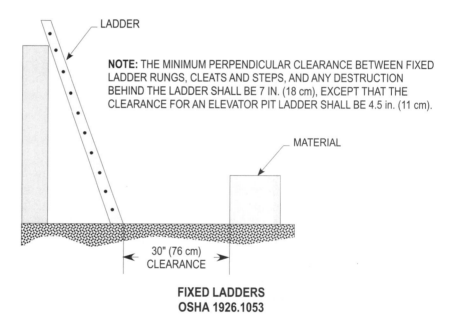

LADDER

NOTE: THE MINIMUM PERPENDICULAR CLEARANCE BETWEEN FIXED LADDER RUNGS, CLEATS AND STEPS, AND ANY DESTRUCTION BEHIND THE LADDER SHALL BE 7 IN. (18 cm), EXCEPT THAT THE CLEARANCE FOR AN ELEVATOR PIT LADDER SHALL BE 4.5 in. (11 cm).

MATERIAL

30" (76 cm) CLEARANCE

FIXED LADDERS
OSHA 1926.1053

The step-across distance between the center of the steps or rungs of fixed ladders and the nearest edge of a landing area shall be no less than 7 in. (18 cm) and no more than 12 in. (30 cm). A landing platform shall be provided if the step-across distance exceeds 12 in. (30 cm).

Fixed ladders without cages or wells shall have at least a 15 in. (38 cm) clear width to the nearest permanent object on each side of the centerline of the ladder.

Fixed ladders shall be provided with cages, wells, ladder safety devices or self-retracting lifelines where the length of climb is less than 24 ft. (7.3 m) but the top of the ladder is at a distance greater than 24 ft. (7.3 m) above lower levels.

If the total length of a climb on a fixed ladder equals or exceeds 24 ft. (7.3 m), the following requirements shall be met: fixed ladders shall be equipped with either (a) ladder safety devices; (b) self-retracting lifelines and rest platforms at intervals not to exceed 150 ft. (45.7 m); or (c) a cage or well, and multiple ladder sections, each ladder section not to exceed 50 ft. (15.2 m) in length. These ladder sections shall be offset from adjacent sections, and landing platforms shall be provided at maximum intervals of 50 ft. (15.2 m). **(See Figure 22-3)**

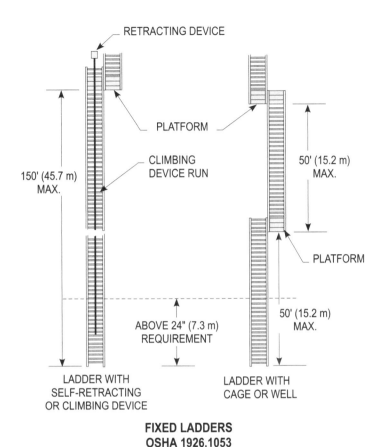

FIXED LADDERS
OSHA 1926.1053

Figure 22-3. This illustration shows the requirements for the total length of a climb of a fixed ladder that equals or exceeds 24 ft. (7.3 m).

The side rails of through or side-step fixed ladders shall extend 42 in. (1.1 m) above the top level or landing platform served by the ladder. For a parapet ladder, the access level shall be at the roof if the parapet is cut to permit passage through it; if the parapet is continuous, the access level is the top of the parapet.

Steps or rungs for through-fixed-ladder extensions shall be omitted from the extension; and the extension of side rails shall be flared to provide between 24 in. (61 cm) and 30 in. (76 cm) clearance between side rails.

When safety devices are provided, the maximum clearance between side rail extensions shall not exceed 36 in. (91 cm).

CAGES FOR FIXED LADDERS

Horizontal bands shall be fastened to the side rails of rail ladders, or directly to the structure, building or equipment for individual-rung ladders.

Vertical bars shall be on the inside of the horizontal bands and shall be fastened to them.

Cages shall not extend less than 27 in. (68 cm), or more than 30 in. (76 cm) from the centerline of the step or rung, and shall not be less than 27 in. (68 cm) wide.

The inside of the cage shall be clear of projections.

Horizontal bands shall be spaced at intervals not more than 4 ft. (1.2 m) apart measured from centerline to centerline.

Vertical bars shall be spaced at intervals not more than 9.5 in. (24 cm) apart measured from centerline to centerline.

The bottom of the cage shall be between 7 ft. (2.1 m) and 8 ft. (2.4 m) above the point of access to the bottom of the ladder. The bottom of the cage shall be flared not less than 4 in. (10 cm) between the bottom horizontal band and the next higher band.

The top of the cage shall be a minimum of 42 in. (1.1 m) above the top of the platform, or the point of access at the top of the ladder. Provisions shall be made for access to the platform or other point of access.

WELLS FOR FIXED LADDERS

Wells shall completely encircle the ladder. Wells shall be free of projections.

The inside face of the well on the climbing side of the ladder shall extend between 27 in. (68 cm) and 30 in. (76 cm) from the centerline of the step or rung. The inside width of the well shall be at least 30 in. (76 cm). The bottom of the well above the point of access to the bottom of the ladder shall be between 7 ft. (2.1 m) and 8 ft. (2.4 m).

LADDER SAFETY DEVICES AND RELATED SUPPORT SYSTEMS FOR FIXED LADDERS

All safety devices shall be capable of withstanding, without failure, a drop test consisting of a 500-pound weight (226 kg) dropping 18 in. (41 cm).

All safety devices shall permit the worker to ascend or descend without continually having to hold, push or pull any part of the device, leaving both hands free for climbing.

All safety devices shall be activated within 2 ft. (.61 m) after a fall occurs, and limit the descending velocity of an employee to 7 ft./sec. (2.1 m/sec.) or less.

The connection between the carrier or lifeline and the point of attachment to the body belt or harness shall not exceed 9 in. (23 cm) in length.

MOUNTING LADDER SAFETY DEVICES FOR FIXED LADDERS

Mountings for rigid carriers shall be attached at each end of the carrier, with intermediate mountings, spaced along the entire length of the carrier, to provide the necessary strength to stop workers' falls.

Mountings for flexible carriers shall be attached at each end of the carrier. Cable guides for flexible carriers shall be installed with a spacing between 25 ft. (76 m) and 40 ft. (12.2 m) along the entire length of the carrier, to prevent wind damage to the system.

The design and installation of mountings and cable guides shall not reduce the strength of the ladder.

Side rails and steps or rungs for side-step fixed ladders shall be continuous in extension.

USE OF ALL LADDERS (INCLUDING JOB-MADE LADDERS)

When portable ladders are used for access to an upper landing surface, the side rails shall extend at least 3 ft. (.9 m) above the upper landing surface. When such an extension is not possible, the ladder shall be secured, and a grasping device such as a grab rail shall be provided to assist workers in mounting and dismounting the ladder. A ladder extension shall not deflect under a load that would cause the ladder to slip off its support.

Ladders shall be maintained free of oil, grease and other slipping hazards.

Ladders shall not be loaded beyond the maximum intended load for which they were built or beyond their manufacturer's rated capacity.

Ladders shall be used only for the purpose for which they were designed.

Non-self-supporting ladders shall be used at an angle where the horizontal distance from the top support to the foot of the ladder is approximately one-quarter of the working length of the ladder. Wood job-made ladders with spliced side rails shall be used at an angle where the horizontal distance is one-eighth the working length of the ladder. **(See Figure 22-4)**

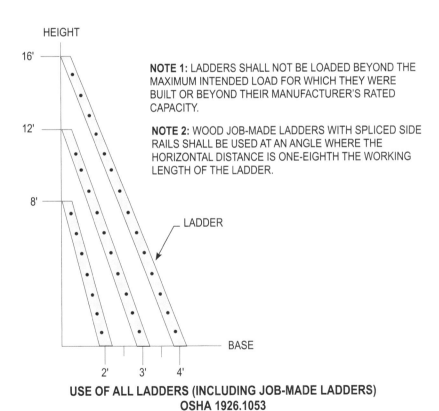

NOTE 1: LADDERS SHALL NOT BE LOADED BEYOND THE MAXIMUM INTENDED LOAD FOR WHICH THEY WERE BUILT OR BEYOND THEIR MANUFACTURER'S RATED CAPACITY.

NOTE 2: WOOD JOB-MADE LADDERS WITH SPLICED SIDE RAILS SHALL BE USED AT AN ANGLE WHERE THE HORIZONTAL DISTANCE IS ONE-EIGHTH THE WORKING LENGTH OF THE LADDER.

USE OF ALL LADDERS (INCLUDING JOB-MADE LADDERS)
OSHA 1926.1053

Figure 22-4. This illustration shows that non-self-supporting ladders shall be used at an angle where the horizontal distance from the top support to the foot of the ladder is approximately one-quarter of the working length of the ladder.

Fixed ladders shall be used at a pitch no greater than 90 degrees from the horizontal, measured from the back side of the ladder.

Ladders shall be used only on stable and level surfaces unless secured to prevent accidental movement.

Ladders shall not be used on slippery surfaces unless secured or provided with slip-resistant feet to prevent accidental movement. Slip-resistant feet shall not be used as a substitute for the care in placing, lashing or holding a ladder upon slippery surfaces.

Ladders placed in areas such as passageways, doorways or driveways, or where they can be displaced by workplace activities or traffic shall be secured to prevent accidental movement, or a barricade shall be used to keep traffic or activities away from the ladder.

The area around the top and bottom of the ladders shall be kept clear.

The top of a non-self-supporting ladder shall be placed with two rails supported equally unless it is equipped with a single support attachment.

Ladders shall not be moved, shifted or extended while in use.

Ladders shall have nonconductive siderails if they are used where the worker or the ladder could contact exposed energized electrical equipment.

The top or top step of a stepladder shall not be used as a step. **(See Figure 22-5)**

Figure 22-5. The illustration shows that the top or top step of stepladder shall not be used as a step.

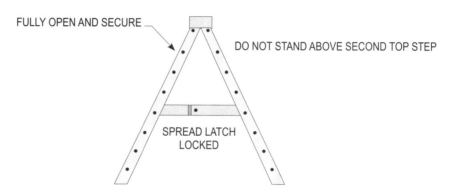

FULLY OPEN AND SECURE

DO NOT STAND ABOVE SECOND TOP STEP

SPREAD LATCH LOCKED

NOTE 1: LADDERS SHALL HAVE NONCONDUCTIVE SIDERAILS IF THEY ARE USED WHERE THE WORKER OR THE LADDER COULD CONTACT EXPOSED ENERGIZED ELECTRICAL EQUIPMENT.

NOTE 2: A WORKER ON A LADDER SHALL NOT CARRY ANY OBJECT OR LOAD THAT COULD CAUSE THE WORKER TO LOSE BALANCE AND FALL.

**USE OF ALL LADDERS (INCLUDING JOB-MADE LADDER)
OSHA 1926.1053**

Cross-bracing on the rear section of stepladders shall not be used for climbing unless the ladders are designed and provided with steps for climbing on both front and rear sections.

Ladders shall be inspected by a competent person for visible defects on a periodic basis and after any incident that could affect their safe use.

Single-rail ladders shall not be used.

When ascending or descending a ladder, the worker shall face the ladder.

Each worker shall use at least one hand to grasp the ladder when moving up or down the ladder.

A worker on a ladder shall not carry any object or load that could cause the worker to lose balance and fall.

STRUCTURAL DEFECTS

Portable ladders with structural defects - such as broken or missing rungs, cleats or steps, broken or split rails, corroded components or other faulty or defective components - shall immediately be marked defective, or tagged with "Do Not Use" or similar language and withdrawn from service until repaired.

Fixed ladders with structural defects - such as broken or missing rungs, cleats or steps, broken or split rails, or corroded components - shall be withdrawn from service until repaired.

Defective fixed ladders are considered withdrawn from use when they are (a) immediately tagged with "Do Not Use" or similar language; (b) marked in a manner that identifies them as defective; or (c) blocked (such as with a plywood attachment that spans several rungs).

Ladder repairs shall restore the ladder to a condition meeting its original design criteria, before the ladder is returned to use.

TRAINING REQUIREMENTS

Under the provisions of the standard, employers shall provide a training program for each employee using ladders and stairways. The program shall enable each employee to recognize hazards related to ladders and stairways and to use proper procedures to minimize these hazards. For example, employers shall ensure that each employee is trained by a competent person in the following areas, as applicable:

- The nature of fall hazards in the work area;

- The correct procedures for erecting, maintaining and disassembling the fall protection systems to be used;

- The proper construction, use, placement and care in handling of all stairways and ladders; and

- The maximum intended load-carrying capacities of ladders used. In addition, retraining shall be provided for each employee, as necessary, so that the employee maintains the understanding and knowledge acquired through compliance with the standard.

Stairways and ladders

Section	Answer	
_____	T	F
_____	T	F
_____	T	F
_____	T	F
_____	T	F
_____	T	F
_____	T	F
_____	T	F
_____	T	F
_____	T	F
_____	T	F
_____	T	F
_____	T	F

Stairways and ladders

1. Stairways and ladders are a major source of injuries and fatalities among construction workers.

2. A stairway or ladder shall be provided at all worker points of access where there is a break in elevation of 15 or more **and** no ramp, runway, embankment or personnel hoist is provided.

3. When there are more than two points of access between levels, at least one point of access shall be kept clear.

4. Stairways that will not be a permanent part of the structure on which construction work is performed shall have landings at least 30 in. deep and 22 in. wide at every 12 ft. or less of vertical rise.

5. Variations in riser height or stair tread depth shall not exceed 1/4 in. in any stairway system, including any foundation structure used as one or more treads of the stairs.

6. Stairways having four or more risers, or rising more than 24 in. in height, whichever is less, shall have at least one handrail.

7. Handrails and the top rails of the stairrail systems shall be capable of withstanding, without failure, at least 150 pounds of weight applied within 2 in. of the top edge in any downward or outward direction, at any point along the top edge.

8. The height of the top edge of a stairrail system used as a handrail shall not be more than 37 in. or less than 36 in. from the upper surface of the stairrail system to the surface of the tread.

9. A double-cleated ladder or two or more ladders shall be provided when ladders are the only way to enter or exit a work area having 25 or more employees, or when a ladder serves simultaneous two-way traffic.

10. Rungs, cleats, and steps of step stools shall not be less than 6 in. apart, nor more than 12 in. apart, between center lines of the rungs, cleats and steps.

11. Ladders shall not be tied or fastened together to create longer sections unless they are specifically designed for such use.

12. Non-self-supporting and self-supporting portable ladders shall support at least four times the maximum intended load; extra heavy-duty type 1A metal or plastic ladders shall sustain 3.3 times the maximum intended load.

13. A fixed ladder shall be capable of supporting at least two loads of 350 pounds each, concentrated between any two consecutive attachments.

_____   T   F

14. The minimum clear distance between the sides of individual rung/step ladders and between the side rails of other fixed ladders shall be 16 in.

_____   T   F

15. The minimum perpendicular clearance between the centerline of fixed ladder rungs, cleats and steps, and any obstruction on the climbing side of the ladder shall be 36 in.

_____   T   F

16. Fixed ladders shall be provided with cages, wells, ladder safety devices or self-retracting lifelines where the length of climb is less than 24 ft. but the top of the ladder is at a distance greater than 24 ft. above lower levels.

_____   T   F

17. Steps or rungs for through-fixed-ladder extensions shall be omitted from the extension; and the extension of side rails shall be flared to provide between 24 in. and 30 in. clearance between side rails.

_____   T   F

18. Cages for fixed ladders shall not extend less than 27 in., or more than 30 in. from the centerline of the step or rung, and shall not be less than 27 in. wide.

_____   T   F

19. Wells for fixed ladders shall completely encircle the ladder.

_____   T   F

20. When portable ladders are used for access to an upper landing surface, the side rails shall extend at least 2 ft. above the upper landing surface.

Abbreviations

A

(A) amps
(AFG) above finished grade
(AC) alternating current
(A/C) air conditioner
(AEGCP) assured equipment grounding conductor program
(AHJ) authority having jurisdiction
(Alu.) aluminum
(ASCC) available short-circuit current

B

(BC) branch-circuit
(BCSC) branch-circuit selection current
(BJ) bonding jumper

C

(°C) Celsius
(CB) circuit breaker
(CEE) concrete encased electrode
(CL) code letter
(CM) circular mils
(CMP) code making panel
(Comp.) compressor
(Cond.) condenser
(Cont.) continuous
(cu.) copper
(cu. in.) cubic inches

D

(dia.) diameter
(DC) direct current
(DPCB) double pole circuit breaker

E

(EBJ) equipment bonding jumper

(Eff.) efficiency
(EGB) equipment grounding bar
(EGC) equipment grounding conductor
(EMT) electrical metallic tubing
(ENT or ENMT) electrical nonmetallic tubing
(Ex.) exception

F

(°F) Fahrenheit
(FC) feeder-circuit
(FLA) full-load amperage
(FLC) full-load current
(FMC) flexible metal conduit
(FPN) fine print note
(ft.) foot

G

(G) ground
(GE) grounding electrode
(GEC) grounding electrode conductor
(GES) grounding electrode system
(GFL) ground-fault limiter
(GFCI) ground-fault circuit interrupter
(GFPE) ground-fault protection of equipment
(GSC) grounded service conductor

H

(H) hot conductor
(HACR) heating, air conditioning, cooling, and refrigeration
(HP) horsepower
(Htg.) heating
(Hz) hertz

I

(I) amperage or current
(IG) isolated ground
(in.) inches
(IRA) inrush amps

K

(KCMIL) kilo circular mils
(kFT) 1,000 ft.
(kV) kilo volts
(kVA) kilo volt-amps
(kW) kilo watts
(kWh) kilo watt-hour

L

(L) length of conductor
(Ld.) load
(LRC) locked-rotor current
(LTR) long time rated

M

(mA) milli-amps
(MBJ) main bonding jumper
(max.) maximum
(MEL) maximum energy level
(MF) microfarads
(MGFA) maximum ground-fault available
(min.) minimum
(min.) minute
(MR) momentary rated
(Mt.) motor

N

(N) neutral
(NACB) non automatic circuit breaker
(NB) neutral bar
(NEC) National Electrical Code

(NEMA) National Electrical Manufacturers Association
(NFD) nonfused disconnect
(NFPA) National Fire Protection Association
(NLTFMC) nonmetallic liquidtight flexible metal conduit
(NTD) nontime delay
(NTDF) nontime-delay fuse

O

(OCP) overcurrent protection
(OCPD) overcurrent protection device
(OL) overload
(OLP) overload protection

P

(Ph.) phases (Hots)
(pri.) primary
(PSA) power supply assembly
(PF) power factor
(PC) pull chain

R

(R) ohms or resistance
(RMC) rigid metal conduit

S

(SBS) structural building steel
(SIA) seal-in amps
(S/P) single phase
(SP) single pole
(SCC) short-circuit current
(sec.) secondary
(SF) service factor
(SPCB) single-pole circuit breaker
(sq. ft.) square foot (feet)
(STR) short time rated
(sq. in.) square inches
(SWD) switched disconnect

T

(TDF) time delay fuse
(TP) thermal protector
(TR) temperature rise
(trans.) transformer
(TS) trip setting
(TV) tough voltage

V

(V) volts
(VA) volt-amps
(VD) voltage drop

W

(W) watts

II

Topic Index

C

F

G

J

K

L

M

N

O

P

R

S

T

U

V

W

X

Appendix

This appendix contains the forms that are contained within a Company Safety Program.

JOBSITE SAFETY AUDIT

Company Name	Company Address	
Project Location	**Project Number**	**Inspection Date**

Note: All negative answers (x) require explanation and corrective action.

Check (√) if satisfactory	(X) if needs attention	(NA) not applicable

General Conditions

A. OSHA sign posted
B. OSHA form 200 up-to-date
C. Supervisor aware of OSHA inspection rights
D. Weekly safety meeting held
E. Appropriate safety reference material on jobsite
F. Emergency numbers posted
G. Basic safety rules posted
H. New employee orientation procedures in place
I. Permits signed and in place
J. Approved and stocked first aid kits on jobsite
K. All safety activities documented and on file

L. Safety signs prominently displayed
M. Subcontractors aware of company and government responsiblities
N. Client safety procedures disseminated to all employees
O. Accidents/incidents investigated in timely manner - documentation on file
P. Precautions taken to protect customers, property, land and employees
Q. Proper clothing worn by employees
R. First-report-of-injury files up-to-date and first aid log current
S. Job hazard analysis in use
T. MSHA-required and task training documented

Health and Environment

A. Toilets adequate and well maintained
B. Drinking water marked, cups supplied
C. Procedures in place for hazardous waste removal
D. Illumination adequate for worksite and trailers
E. Oxygen deficiency, toxicity and flammability tests performed, where required
F. Ventilation adequate, where needed

G. Certified first aid/CPR person on jobsite
H. Hand washing facilities available
I. Chemical showers and eyewash stations available
J. Fuels and flammable material storage area bermed with plastic liner
K. Confined space program in place
L. Noisy work areas posted
M. Bloodborne pathogen procedures in place

Hazard Communication Program

A. Written program on site
B. All personnel trained - attendance documented
C. MSDS data sheets available - location known

D. Labeling procedure in place
E. Job-specific chemicals and MSDS's discussed - acids, caustics, etc.
F. Chemical inventory on site

Personal Protective Equipment

A. Approved Z89.1 hard hats required - no bump caps
B. ANSI Z87.1 eye protection worn
C. Double eye protection required for special work tasks
D. Hearing protection available and used near loud equipment and where posted
E. Proper gloves worn
F. Safety nets required

G. Approved respirators available - employees medically qualified and fit tested
H. Safety harness and lanyards in good condition and used properly
I. Proper footwear in place
J. Additional PPE required for work with chemicals or in toxic material areas

JOBSITE SAFETY AUDIT

Check (√) if satisfactory	(X) if needs attention	(NA) not applicable

Material Handling Equipment - Rigging

A. Canopy guards or roll-over protection on all heavy equipment [_____]
B. Back-up alarms and/or flashing lights provided [_____]
C. No equipment allowed closer than 10 ft. to power lines [_____]
D. Fire extinguishers in cab of all moving equipment [_____]
E. All equipment operators trained and certified [_____]
F. Seat belts mandatory for operators [_____]
G. Outriggers in place, when needed [_____]
H. Maximum use made of available material handling equipment by employees [_____]
I. Monthly maintenance records maintained [_____]
J. Monthly crane inspections on file [_____]
K. Horns used during lifts [_____]

L. Slings proper size and in safe condition [_____]
M. All hooks moused except pelican hooks [_____]
N. Cable clips approved type, proper size and properly installed [_____]
O. Wire rope inspected - monthly inspections on file [_____]
P. Capacity charts at operator's station [_____]
Q. Spreader bars marked with capacity [_____]
R. Hand signal charts posted and followed [_____]
S. Tail-swing of crane barricaded [_____]
T. Signal person designated and used [_____]
U. Lifting lugs welded by qualified person and checked by supervisor [_____]
V. Tag lines used [_____]
W. Heavy loads weighed and marked, when possible [_____]

Housekeeping and Material Storage

A. Construction materials neatly and safely stored at lay down yards and jobsite [_____]
B. Housekeeping performed on daily basis [_____]
C. Nails clenched or removed from scrap lumber [_____]
D. Adequate trash bins/barrels provided [_____]
E. Aisles kept clear of debris, power cords, cables and other tripping hazards [_____]
F. Office trailer clean and orderly [_____]

G. Metal banding straps removed quickly [_____]
H. Air hoses and electrical cords covered to prevent damage from moving equipment [_____]
I. Trash removed from jobsite on regular basis [_____]
J. Tool trailers neat and orderly [_____]
K. Mushrooms used or rebar, where needed [_____]
L. Racks, bins or buckets used for small parts storage [_____]

Ladders and Scaffolds

A. Ladders extend 3 ft. above landings and tied off [_____]
B. Scaffolds properly erected under qualified supervision [_____]
C. No riders on moving platforms [_____]
D. Wheels locked on rolling scaffolds [_____]
E. Scaffolding guyed or tied-off to structure [_____]
F. Fiberglass ladders in use by electricians [_____]
G. Handrails in place [_____]
H. Handrails, midrails and toe boards required [_____]
I. Ladders in safe condition and defective ladders removed from service [_____]
J. Monthly ladder inspections on file [_____]

K. Weekly scaffold inspections on file [_____]
L. Proper scaffolding tags used [_____]
M. All stairways have handrails [_____]
N. Scaffold boards-banded, tested, lapped and cleated [_____]
O. Brackets plumb, arms secured and welded by certified welders [_____]
P. Folding ladders not used as straight ladders [_____]
Q. Folding step ladders placed on solid footings, brackets fully extended, no work off top two sides [_____]
R. Ladder base-to-height requirements acceptable [_____]
S. Tool pails or ropes used for material hoisting [_____]

JOBSITE SAFETY AUDIT

Check (√) if satisfactory	(X) if needs attention	(NA) not applicable

Tools and Equipment

A. Air for cleaning properly used [____]
B. Air hose connections wired together [____]
C. Pressure reducing valves in flow lines (larger than 1/2 in.) [____]
D. Guards in place on power tools [____]
E. Abrasive wheels properly handled and stored [____]
F. Power activated tool operators properly trained [____]

G. Air receivers certified and equipped with pressure gauge and pop-off valve [____]
H. Defective hand tools repaired or removed from jobsite [____]
I. Radioactive testing equipment and procedures in compliance with governmental standards [____]
J. Constant pressure switches on hand-held power tools [____]

Fire Prevention

A. Fuels and flammables properly stored [____]
B. No gas cans allowed in trailers
C. Safety cans used for gasoline storage - equipped with screens and properly labeled [____]
D. Extinguishers check on monthly basis and properly tagged [____]
E. Fire extinguisher available for each work area and conspicuously located [____]
F. "No Smoking" or "Smoking Permitted" locations designated with signs
G. Gasoline never used for cleaning

H. Paints and other flammables with broken seals stored in metal cabinets or separate outside storage bins [____]
I. Adequate precautions taken during spray painting operations [____]
J. Gasoline equipment refueled outside and shut off [____]
K. Flammable fuel containers grounded and bonded [____]
L. Oxygen and gas cylinders stored upright, secured and separated by 20 ft. or fire-resistant barrier [____]
M. Pipelines blanked off or disconnected [____]
N. Employees trained in fire extinguishment procedures [____]
O. FRC's required

Comments on corrective action taken or recommended. Indicate date item abated - use extra sheet, if necessary.

_____

_____

_____

Unsafe conditions or acts noticed (not on audit form). Itemize and note corrections - indicate date completed.

_____

_____

_____

Project Manager's Approval	**Date**

SAFETY COMMITTEE OR FIELD SAFETY INSPECTOR'S SIGNATURE

FORM 2

OSHA DATA FORM

Job Location _____   **Date of Inspection** _____

Job Number _____   **Submitted By**

Pre-Inspection

A. Who assigned to accompany the OSHA inspector?
Name: _____
Position: _____

B. Whom did the inspector first contact at the jobsite?
Name: _____
Position: _____

C. Did the inspector show his credentials?
Yes _____
No _____

D. Name of inspector: _____
Federal: _____
State: _____

E. Time inspector arrived: _____

Opening Conference

A. Who was present? (Name and Firm)
1. _____
2. _____
3. _____
4. _____
5. _____
6. _____

B. What was the purpose of the visit as explained by the Inspector?
_____
_____
_____
_____
_____

FORM 3

Company Name
_____

Emergency Phone Number
_____

Physicians: _____

Hospital: _____

Ambulance: _____

Carrier: _____

Safety Department: _____

_____

POST IN A CONSPICUOUS LOCATION
CLOSE TO ALL TELEPHONES, BULLETIN BOARDS, ETC.

FORM 4

Company Name

Return to Work Form

This form must be completed by the Supervisor and taken to the physician by the employee for each visit. This form must then be completed by the attending physician and returned to the employee. The employee must present this form to his or her Supervisor immediately after returning from each visit to the physician.

Please render medical services to _____, who is employed by [COMPANY NAME] and is presumed to have sustained an injury or illness in the course of employment. If hospitalization and/or medical treatment is required in excess of the limit provided by the Workers' Compensation Law of the state in which the injury occurred, the Company will not be liable for the excess unless written authorization for such treatment is first obtained from [COMPANY NAME] or its designated representative.

[COMPANY NAME] requires all employees with doctor-treatable work-related injuries or illnesses to undergo drug and alcohol screening. Call the following number listed below for details.

Date: _____   [COMPANY NAME]: _____

Phone Number: _____   By: _____

Title: _____

Physician's Release

_____ , whom I have treated for an injury or illness, is hereby released to:

_____ return to work with attention given to not aggravate the injury

_____ Other _____

_____ no additional treatment is required

_____ a follow-up appointment is scheduled on _____
(Date)

_____
Physician Name (Print)

_____
Physician Signature

_____
Date

FORM 5

Waiver of Hepatitis B Vaccination

I, _____, understand that due to my potential occupational exposure to blood or other potentially infectious materials I may be at risk of acquiring the hepatitis B virus (HBV) infection. I have been given the opportunity to be vaccinated with the hepatitis B vaccine at no charge to myself. However, I decline hepatitis B vaccination at this time. I understand that by declining this vaccine, I continue to be at risk of acquiring hepatitis B, a serious disease. If in the future I continue to have occupational exposure to blood or other potentially infectious materials and I want to be vaccinated with the hepatitis B vaccine, I can receive the vaccination series at no charge to me.

_____
Date

_____
Employee Signature

_____
Date

_____
Witness

FORM 6

Employee Declination Statement

The following is Appendix A to Section CFR 1910.1030 (OSHA) - Hepatitis B Vaccine. Declination is mandatory if you **DO NOT** take the shots.

I understand that due to my occupational exposure to blood or other potentially infectious materials I may be at risk of acquiring hepatitis B virus (HBV) infection. I have been given the opportunity to be vaccinated with hepatitis B vaccine, at no charge to myself. However, I decline hepatitis B vaccination at this time. I understand that by declining this vaccine, I continue to be at risk of acquiring hepatitis B, a serious disease. If in the future I continue to have occupational exposure to blood or other potentially infectious materials and I want to be vaccinated with hepatitis B vaccine, I can receive the vaccination series at no charge to me.

_____

Signature

_____

Date

FORM 7

 First Aid Safety Procedures

Before attempting to give first aid for bleeding or other potential exposure to bloodborne pathogens, follow these rules.

A. Wear rubber gloves. Do not reuse rubber gloves. Wash your hands with soap and water after removing gloves.
B. Wear safety goggles if there is potential for contaminants to splash into the eyes.
C. Wear a mask if there is potential for contaminants to splash into the mouth or nose.
D. Wear additional protective clothing if your skin is not covered.
E. If you become exposed to a bloodborne pathogen, wash the area immediately and report it to management so professional medical attention can be provided, including administration of the Hepatitis B vaccine, if prescribed by a physician.
F. Ensure that regulated waste is properly bagged, labeled and disposed of, according to Infection Control Procedures.

Recommended Medical Facility: _____

Telephone: _____

Paramedics: _____

Hospital: _____

FORM 8

 Employee Incident Report
(Bloodborne Pathogen Exposure)

To be completed by employee administering first aid.

Name	Job Classification
Jobsite Location	Job Number
Age	Social Security Number	Date Employed
Witness (if any)	Signature	Date
Injured Employee Signature	Date

Accident Data

Date and time you were notified of incident:

Where exactly did it occur?

Describe exactly what occurred:

What has been done to prevent this type incident from occurring again?

Cause of incident (check all that apply):

_____ Cut	_____ Splash	_____ Improper work procedure
_____ Abrasion	_____ Unsafe act of worker	_____ Lack of PPE
_____ Bite	_____ Contaminated waste	_____ Other

Signature of Supervisor (Safety or Jobsite)   **Name**   **Date**

FORM 9

Bloodborne Pathogen Training Program

I affirm that I have been trained regarding the following elements of the Bloodborne Pathogen Control Program:

- Hepatitis B virus vaccinations
- Use of personnel protective equipment
- Clean-up measures
- Waste disposal
- Record-keeping requirements

Employee Signature: _____

Employee Name: _____

Employee Social Security Number: _____

Regular Work Assignment: _____

Date: _____

Instructor: _____

FORM 10

Company Name
Supervisor's First Report of Injury

Project Location	Project Number	Date of Report	Client

Must be completed and submitted within 24 hours. Purpose - to discover and prevent recurrence of on-the-job injuries. Injured employee's supervisor is responsible for the accuracy and completion of all forms. All serious injuries must be reported immediately by telephone to corporate safety. Jobsite safety supervisor will assist in completion of this form and all incident documentation's.

FORM 11.1

Company Name

▼! **Pre-Project Safety Review and Safety Meeting** ▼!

Job Location _____ **Customer** _____

Date _____ **Job Number** _____ **Job Task** _____

Supervisor: Complete a walk-through of the jobsite at the beginning of each day or shift. Use this form for guidance in identifying existing and potential hazards at the jobsite. Circle Y for yes or N or no, as appropriate. Review the complete form with the entire crew before beginning work for the day or shift. Safety concerns expressed by any of the crew must be resolved before beginning the work shift. Have all crew members sign this form.

Required Check No/Yes N/A If Not Applicable

A.	Work permit required	N Y	Type:	_____
B.	Plant alarm sounds and actions	N Y	Describe:	_____
C.	Plant emergency numbers	N Y	Numbers:	_____
D.	First aid station locations	N Y	Locations:	_____
E.	Emergency evacuation routes	N Y	Routes:	_____
F.	Evacuation "safe-refuge" areas	N Y	Locations:	_____
G.	Slip/trip/fall hazard	N Y	Describe:	_____
H.	Pinch/cut/shear/crush hazard	N Y	Describe:	_____
I.	Excessive noise hazard	N Y	Protection:	_____
J.	Eye injury hazard	N Y	Type and protection:	_____
K.	Pressure release hazard	N Y	Source:	_____
L.	Dust/mist/fume/gas/vapor hazard	N Y	Type and protection:	_____
M.	Temperature stress hazard	N Y	Describe:	_____
N.	Fire hazard	N Y	Describe:	_____
O.	Fire extinguisher	N Y	Location:	_____
P.	Electrical hazard	N Y	Describe:	_____
Q.	Overhead hazard	N Y	Describe:	_____
R.	Traffic/moving vehicle hazard	N Y	Describe:	_____

FORM 11.2

Supervisor: Complete a walk-through of the jobsite at the beginning of each day or shift. Use this form for guidance in identifying existing and potential hazards at the jobsite. Circle Y for yes or N or no, as appropriate. Review the complete form with the entire crew before beginning work for the day or shift. Safety concerns expressed by any of the crew must be resolved before beginning the work shift. Have all crew members sign this form.

Required Check No/Yes N/A If Not Applicable

S.	Static grounding needed	N Y	Location:	_____
T.	Barricaded/signs required	N Y	Type and location:	_____
U.	Lockout/tagout required	N Y	Hazard and location:	_____
	• Procedure available	N Y	Attach procedure:	_____
V.	Chemical hazard	N Y		
	• Chemical identified	N Y	Identity:	_____
	• MSDS reviewed	N Y	Location:	_____
	• Irritant or corrosive to eyes and skin	N Y	Specify:	_____
	• Dust/mist/fume/gas/vapor/oxygen hazard	N Y	Specify:	_____
	• Fire/explosive/reactive/oxidizer hazard	N Y	Specify:	_____
	• Safety shower/eye wash available	N Y	Location:	_____
W.	Confined space entry			
	• Permit required	N Y	Source:	_____
	• Procedure available	N Y	Attach procedure:	_____
X.	Job procedure			
	• Verbal/written job procedures required	N Y	Specify:	_____
	• Safety procedures reviewed	N Y	Describe:	_____
	• Special equipment required	N Y	Comment:	_____
	• Tools/equipment inspected	N Y	Specify:	_____
Y.	Protective equipment required	N Y	Describe:	_____
Z.	Other contractors/workers in area	N Y		

Job Leader/Supervisor:_____ **Crew Member:**_____

1. In the blank space to the right of each "N/Y," comment on the reason for each circled yes (Y) on the sheet. If more space is needed, use the back of this sheet.
2. Retain this sheet at the jobsite until the job is complete. Include this sheet in the job package.

FORM 12

Company Name

Employee Training Record

Employee Name _____   **Employee Number** _____

	Training Date	Employee Initials	Instructor Initials
A. Safety orientation	_____	_____	_____
B. Hazard communication	_____	_____	_____
C. Respiratory protection*	_____	_____	_____
D. Personal protective equipment	_____	_____	_____
E. Hearing protection/conservation*	_____	_____	_____
F. Lockout/tagout	_____	_____	_____
G. Confined spaces procedures	_____	_____	_____
H. First responder (awareness)*	_____	_____	_____
I. Back safety	_____	_____	_____
J. Fire extinguishers	_____	_____	_____
K. Company substance abuse policy	_____	_____	_____
L. Forklift operation	_____	_____	_____
M. Trenching/excavation safety	_____	_____	_____
N. Scaffold safety	_____	_____	_____
O. Hoisting/lift safety	_____	_____	_____
P. Hazardous waste site operations*	_____	_____	_____
Q. Other	_____	_____	_____
R. Job skill or task and safety	_____	_____	_____
Skill/task	_____	_____	_____
Skill/task	_____	_____	_____
Skill/task	_____	_____	_____

Date and initial each block after completion of the training.

* Requires annual refresher training

FORM 13

Company Name

Customer Safety Training Record

Customer: _____

Employees are required to complete specific training before working in locations containing highly hazardous chemicals.

To ensure compliance, a review of the applicable customer safety procedures has been completed. Based on the review, customer-specific safety training has been completed by the following employees who will be working within the customer location that contains highly hazardous chemicals:

_____   _____
_____   _____
_____   _____
_____   _____

Completed Customer Safety Training   **Date Completed**

A. Customer safety orientation   _____
B. Customer hazard communication program   _____
C. Customer permit procedures   _____
D. Customer emergency procedures   _____
E. Customer evacuation procedures   _____
F. Customer benzene safety procedures   _____
G. Customer lockout/tagout procedures   _____
H. Customer confined space procedures   _____
I. Customer heavy metal safety procedures   _____
J. Other applicable safety procedures   _____

_____   _____
Date   **Representative**

FORM 14

Company Name

Master Isolation List

Job Identified: _____   Page: _____

Device Type • Lock = L • Tag = T • Blind = B	Location
1.	
2.	
3.	
4.	
5.	
6.	
7.	

FORM 15.1

Job Number: _____

▼ Job-Specific Lockout/Tagout Survey ▼

This lockout/tagout plan is specific to the following project: _____

Location of job: _____

Customer: _____

Date plan prepared/modified: _____

Plan prepared by: _____   Safety coordinator: _____

Plan approved by: _____   Safety coordinator and site supervisor: _____

Plan supervised by: _____   Site supervisor: _____

A copy of this plan is to be maintained at the job site. The implementation of this plan is the responsibility of the site supervisor.

This plan addresses the use of lockout and tagout procedures at this job site and identifies required activities.

This plan is designed to enable supervisors and employees to recognize the hazards on this job and to establish the procedures that are to be followed to isolate a potentially hazardous energy source. Each employee will be trained in these procedures and must strictly adhere to them, except when doing so would expose the employee to a greater hazard. If, in the operation of the employee, this is the case, the employee is to notify his or her supervisor at the jobsite. The concerns of the employee must be addressed before proceeding.

Lockout/Tagout to be used on this Project

The following lockout and tagout devices will be used on this job:

_____

_____

_____

FORM 15.2

Job Number: _____

⚠ Job-Specific Lockout/Tagout Survey ⚠

Typical Lockout Device

Lockout devices will be used for the following tasks:

_____
_____
_____

Typical Tagout Device

Tagout devcies will be used for the following purposes:

_____
_____
_____

Typical Multiple Lockout HASP

Multiple lockout HASP will be used for the following purposes:

_____
_____
_____

Identification of Hazards at this Jobsite

_____
_____
_____
_____
_____
_____
_____
_____
_____
_____
_____

FORM 16

Company Name

Employee Training Record

This is to certify that I have received Basic Personal Protective Equipment Training.

Name: _____

Signature: _____

Employee #: _____

Date: _____

Training Conducted By:

_____   _____

Job Location   **Customer**

FORM 17.1

Company Name

▼ Confined Space Entry Permit ▼

Permit only authorized for one shift.

Specific Location: _____
Confined Space Description: _____
Purpose of Entry: _____

Hazards of Space: Identify which are applicable

A. Contains or has the potential to contain a hazardous atmosphere
B. Contains a material that has the potential to engulf an entrant
C. Has internal configuration that could trap or asphyxiate entrant by inwardly converging walls
D. Other

_____
_____

I. Atmosphere Tests: (Perform in the following order)

Test Item	Initial Results	Allowable Limits	Re-entry	
A. Oxygen		19.5% - 23.5%		
B. Flammability		<10% LEL		
C. Hydrogen Sulfide*		<10 ppm		
D. Benzene*		<1 ppm		
E. Norm*		<2 mrems/hr		
F. Other				

*(If Applicable)

Monitoring Equipment:

Location of atmospheric gas test(s): _____
Time of initial test(s): _____
Person conducting initial tests: _____

Make _____ **Serial #** _____

FORM 17.2

▼ Confined Space Entry Permit ▼

II. Continuous Monitoring Required	Yes	No
Following Items Completed		
Contents removed/purged		
Energy isolation		
Blinding/disconnecting		
Ventilation provided		
Rescue plan complete (Sec. VI)		
Rescue equipment on site		
Adequate/safe lighting		
Calibration check of monitor prior to use		
Tailgate safety meeting		
Hot work permit required		
Other		

Explain other _____

Explain (N/A) answers _____

Authorized entrants/or methods used

Name	Time in	Time out
_____		
Name	Time in	Time out
_____		
Name	Time in	Time out

(Additional signature lines on back)

III. Entry is Only Approved with Respiratory Protective Equipment

Protective and Rescue Equipment Required	Yes	N/A
Goggles		
Chemical gloves		
Chemical boots		
Face shield		
Slicker suit		
Acid suit		
ALR/SCBA		
FRC		
Full body harness/life line		
Other		

Explain other _____

Explain (N/A) answers _____

Communication procedure _____

Additional permits issued _____

Foreman verbally notified _____

Signature of entry supervisor _____

Sign when permit cancelled _____

I have witnessed that the conditions listed above are as indicated and understand they may change.

Contractor signature _____

Contractor signature _____

FORM 17.3

▼! Confined Space Entry Permit ▼!

IV. Attendant(s)
V. Special precautions required for entry _____
VI. Rescue plan _____
_____
_____

A. Location of safe briefing area(s) _____
B. Emergency numbers (medical/rescue) _____

(Fire)   (Supervisor)

Problems encountered/comments _____

_____

This permit is void when acceptable entry conditions are exceeded.

Authorized Entrants

Name	Time in	Time out
Name	Time in	Time out
Name	Time in	Time out
Name	Time in	Time out
Name	Time in	Time out
Name	Time in	Time out
Name	Time in	Time out
Name	Time in	Time out
Name	Time in	Time out

FORM 18.1

⚠ Personnel Hoist Procedures ⚠

Project Location	Project Number	Lift Number	Date	Client

It is Company policy to provide employees with safe hoisting equipment and procedures when personnel are required to work at higher levels using crane-suspended personnel platforms. These personnel lifts will be done only when no other method is available for the work to be performed.

Before a personnel lift is made, the job superintendent shall determine that the lift is necessary and can be safely completed.

This checklist must be completed by superintendent for each personnel lift. After completion of the checklist, all employees involved shall acknowledge compliance with all applicable procedures by signing at the end of the checklist.

Prelift Checklist

OSHA Standard 1926.550(G)		Yes	No
A.	By supervisor		
B.	By crane operator		
Equipment			
A.	Personnel platform is of a proper design		
B.	Personnel platform is posted with its intended maximum load capacity		
C.	Crane is equipped with a positive acting anti-two-block system		
Rigging			
A.	Rigging is capable of supporting five times its intended load		
B.	Hooks are a type that can be closed and locked		
C.	All eyes in wire rope slings have thimbles		
D.	The load is evenly divided to prevent tilting		
E.	Personnel basket rigging is not used for any other purpose		
Safety Equipment			
A.	Hard hats and safety glasses are worn		
B.	Safety harnesses are worn and lanyards properly occurs during, not before lift used		
Crane Operation			
A.	Crane is in proper location and within 1 percent of level grade		
B.	Crane outriggers are fully deployed and placed on crane mats or firm footing		
C.	Total weight of loaded personnel platform does not exceed 50 percent of the rated capacity for the radius and configuration of the crane		
D.	Load line is capable of supporting a minimum of seven times the intended maximum load		
E.	No other load lines are being used while personnel are suspended on a platform		
F.	Cranes having live booms are prohibited		

FORM 18.2

Prelift Checklist		
Trial Lift, Inspection and Proof Testing	**Yes**	**No**
A. Trial lift made with an evenly distributed deadman weight of 125 percent of maximum platform capacity attached to the hoist for a period of 5 minutes at each working location has succeeded		
B. Visual reinspection of all equipment and rigging by competent person immediately after trial lift shows equipment is stable		
C. Platform is lifted a few inches from the ground just before use to ensure proper balance		
D. Trial lift is repeated before lifting personnel whenever the crane is moved and set up in new location or returned to a previously used location		
E. Hoist line is free of kinks and not twisted around multiple lines		
F. Primary attachment must be centered over the platform		
Work Practices		
A. Employees know to keep all body parts inside the personnel platform at it is raised, lowered and positioned		
B. Materials and tools for use during a personnel lift are secured to prevent displacement		
C. The personnel platform is not being used to lift only material or tools		
D. Workers know to stand firmly on the floor of the platform and do not sit or climb on the rail of frame of the hoist. No planks, ladders or other devices are being used for a work position		
E. Hoisting will be discontinued immediately upon weather or other impending conditions		
F. Crane operator knows to remain at controls at all times when the platform is occupied		
G. Operator or signal person inside the platform knows to remain in continuous sight and communication with employees being hoisted		
H. Employees are fastened by a harness system to the load block or other structural member of the platform		

Prelift Meeting

All persons involved in the lift have reviewed all aspects and requirements of the lift.

Jobsite Superintendent: _____

Safety supervisor: _____

Lift operator: _____

Signalman: _____

Employee: _____

Employee: _____

Employee: _____

FORM 19.1

Monthly Crane and Side Boom Inspection Report

Date: _____ Equipment Number: _____ Serial Number: _____

Equipment Type: _____ Make/Model: _____ Hours: _____

Boom Length: _____ Jib Length: _____

Wire Rope: _____ Size: _____ Classification: _____

Other Comments: _____
_____

Inspection (Check if OK; detail exceptions on line 4)

A. General

____ Capacity chart
____ Controls marked
____ Operator's manual
____ Proximity signs
____ Signal charts
____ Signal horn
____ Backup alarms
____ Tailswing protection
____ Fire extinguisher, type _____
____ Boom angle indicator
____ Load movement device
____ First aid kit

____ Telescoping length indicator
____ Load indicator
____ Cab
____ Safety glass
____ Ladder or handholds
____ Levels
____ Exhaust pipes
____ Machinery guards
____ Fuel filler (location)
____ Appearance/housekeeping
____ Instrument check

B. Machinery

____ Controls
____ Brakes and clutches
____ Drum guards
____ Drum rotation indicator
____ Power boom hoist
____ Boom hoist pawl
____ Boom hoist kickout
____ Power load lowering
____ Brake locks
____ Safety brakes
____ Check valves

____ Swing mechanism/circle
____ Swing brakes
____ Travel mechanism/chains
____ Travel brakes
____ Hydraulic brakes
____ Air leaks
____ Pressure settings
____ Car body and carrier
____ Removing frame
____ Gantry
____ Turntable mounting

FORM 19.2

Inspection (Check if OK; detail exceptions on line D)

C. Boom

____ Boom
____ Boom stops
____ Point sheaves
____ Jib
____ Jib stops
____ Hook and block
____ Jib hook
____ Counterweight

____ Reeving
____ Wire ropes
____ Rope sockets
____ Cable clamps
____ Pendants
____ Outriggers and controls
____ Tires and tracks

D. Exceptions: _____
_____
_____

E. Date of last inspection was conducted: _____ Inspector: _____

F. I certify that the manufacturer's recommended daily and monthly checks and inspections have been performed.

_____ (Initials)

Inspection conducted by: _____ Date: _____

FORM 20

▼! Hot Work Permit ▼!

Work location _____

Specific location _____ **Date** _____

Contact for hot work _____

Time limits	
Issued	AM/PM
Expires	AM/PM
Hot work permits limited	

Person performing hot work _____

Specifically describe hot work to be done and object being worked on: _____

_____

_____

Calibration data _____ **Model** _____

Survey results: _____ % **LEL (10% maximum allowable):** _____ **Initials or person conducting survey:** _____

The representative must initial the following items, indicating that precautions where taken to prevent accidental ignition if acceptable for hot work and note any additional conditions in the comments section.

Items completed before hot work tailgate safety training flammable/combustible material:	Completed	N/A
A. Surveyed the area including floor cracks and openings, walls, partitions, ceilings and roofs for the presence of combustible material and for a flammable/combustible atmosphere		
B. Shut down conveyors and air ducts that could convey sparks		
C. Taken precautions for heat transmission		
D. Relocated or protected combustible materials		
E. Deactivated fire-eye monitors and notified control room		
F. Assigned fire watch		
G. Appropriate fire extinguisher on site		
H. Energy isolation		

Items completed after expiration of hot work:

	Completed	N/A
A. Surveyed area for hazards		
B. Activated fire-eye monitors and notified control room		

Continuous monitoring should be provided in areas where changing conditions are likely or in high risk areas such as in tanks or in the process areas of plants.

If continuous monitoring is not performed, the work area should be re-surveyed following breaks in the job.

Comments: _____

_____ **Date:** _____ **Time:** _____ **AM/PM:** _____

Name of area foreman verbally notified

The signature below indicate that the above information has been reviewed and that all conditions set will be complied with. It is understood that this permit is void if the conditions change.

_____ **Date:** _____ **Time:** _____ **AM/PM:** _____

Signature of employee approving hot work

_____ **Date:** _____ **Time:** _____ **AM/PM:** _____

Signature of person performing hot work

FORM 21

Company Name

Use of Company Vehicles

The following items define Company policy for the use of Company vehicles:

A. The motor vehicle record (MVR) of all employee who operate or who may operate a company vehicle will be checked.

B. Company employees who have a driving while intoxicated (DWI) or driving under the influence (DUI) citation listed on their respective MVR's will be denied use of company vehicles.

C. Company employees issued company vehicles are expected to report all traffic violations, whether received in a company or personal vehicle, to the Company home office.

D. Employees with two or more accidents or traffic violations in a three-year period must take a defensive driving course before being allowed to use a company vehicle.

E. The Company forbids the drinking of alcoholic beverages or intake of drugs before or while driving company vehicles. Operating a company vehicle while under the effects of alcohol or drugs can result in termination of employment.

F. Company vehicles are to be used for Company business only.

G. Company employees must provide valid drivers licenses before being issued Company vehicles.

H. Seal belts are to be worn by all occupants in a Company vehicle.

I. The maintenance of the vehicle is the responsibility of the person to whom the vehicle is assigned; necessary expenses will be reimbursed by the Company.

J. Damage to vehicles must be reported immediately to the Corporate Safety Department followed up by a written report of the incident to the Risk Management Department.

Please sign the following statement and return this memorandum to _____ at the home office.

I, _____, have read and understood the Company policy statement regarding use of Company vehicles.

Signed: _____ Dated: _____

FORM 22

Company Name

Competent Person Assignment

Project Number: _____ **Project Location:** _____

Client Date: _____

Fill in designated person's name or N/A if not applicable

The following personnel are assigned as the competent persons for the activities listed below on this specific jobsite:

Activity Description	Personnel Assigned
Accident prevention responsibility and training	
Medical services and first aid	
Excavations, trenching and shoring	
Scaffold erection	
Ladders	
Lockout/tagout procedure	
Assured grounding program	
Cranes and lifting equipment	
Assigning qualified operators	
Rigging equipment inspection	
Fire protection and prevention	
Personnel hoists and manlifts	
Power hand tools	
Respiratory protection	
Fall protection	
Welding and cutting	
Confined space	
Industrial hygiene and atmospheric testing	
Demolition exposure	
Other	
Other	
Other	

Project manager: _____

Project safety supervisor: _____

FORM 23

Job Hazard Analysis

List task	List hazards associated with task	List safe procedures and/or safeguard(s)

FORM 24

▼! Spill Notification Requirements ▼!

United States Coast Guard National Response Center
National Response Center   800-424-8802

Notify immediately for a spill of Federal Reportable Quantity

Environmental Protection Agency
Regional Response Center   214-665-6444
Dallas, Texas

Notify immediately for a spill of Federal Reportable Quantity

State Emergency Response Centers
General Land Office (GLO)   800-832-8224

Report any spill or release of oil into the environment in coastal areas

Texas Natural Resources Conservation Commission (TNRCC)
Regional Office (Amarillo)   806-353-9251

Report any spill or release of oil into the environment in noncoastal areas
Report spills and accidental discharges of hazardous substances, wastes or other substances

Texas Emergency Response Center (TERC)   512-463-7727
24 hours   512-239-2507

Report any spill or release of oil into the environment in noncoastal areas
Report spills and accidental discharges of hazardous substances, wastes or other substances

Texas Railroad Commission (TRC)
Regional Office (Pampa)   806-665-1653

Report fires, leaks or breaks in tanks or pipelines where oil or gas are escaping or have escaped

Texas Department of Public Safety (DPS)   512-465-2850

Report all transportation incidents involving releases of reportable quantities of hazardous materials on public roads or railroads

FORM 25

Excavation Report

Date: _____ Job Number: _____

Project Name: _____

County: _____ State: _____

		A	B	C
A.	What types of soils were encountered?	A	B	C
B.	All open trenches were inspected?		Y	N
C.	All surcharge was located a proper distance from top of trench?		Y	N
D.	Were any tension cracks observed along top of trench?		Y	N
E.	Was any water seepage noted in the trench wall or trench bottom?		Y	N
F.	Was there any evidence of caving or sloughing of soil since the last inspection?		Y	N
G.	Was there any evidence of shrinkage cracks in trench walls?		Y	N
H.	Was there any zones of unusually weak soils or materials not anticipated?		Y	N
I.	Were all short-term trenches covered within 24 hours?		Y	N
J.	Was traffic in area adequately away from trenching operations and barricades?		Y	N
K.	Were trees, boulders or other hazards in area?		Y	N
L.	Were vibrations from equipment or traffic too close to trenching operations?		Y	N
M.	List heavy equipment near operation?			

_____

_____

N. What type of protective system was used?
- Trench box _____
- Manhole box _____
- Side sloping _____
- Bracing _____
- Shoring _____
- Other _____

O. Weather condition: _____ Rainfall: _____

P. Observations:

_____

_____

Signature of competent person approving excavation: _____

FORM 26

Fall Protection Plan

A fall protection plan can only be used for operations involving leading edge work, precast concrete construction work, or residential construction work, where it can be proven that using conventional fall protection systems is infeasible or creates a greater hazard. Each fall protection plan must be site specific.

This fall protection plan is specific for the following project:

_____

Location of job

Erecting company

_____

Date plan prepared or modified

Plan prepared by

It is the responsibility of _____ to carry out this fall protection plan. _____ is responsible for daily inspection of the operation covered by this plan. Supervisors are also responsible for inspecting and correcting any unsafe acts or conditions immediately. Each employee is responsible for understanding and following the procedures of this plan and for following the instructions of a supervisor. Employees must also report any unsafe or hazardous conditions or acts. Any changes to this fall protection plan must be approved in writing by _____.

Reasons for Fall Protection Plan

The following is a written explanation of why the use of conventional fall protection systems are infeasible or create a greater hazard.

A. Guardrail systems are not being used because:

_____
_____
_____

B. Personal fall arrest systems are not being used because:

_____
_____
_____

C. Safety net systems are not being used because:

_____
_____
_____

FORM 26.2

Reasons for Fall Protection Plan

The following is a list of other fall protection measures considered for the project and an explanation of their limitations. If any employee sees an area that could be erected more safely using conventional fall protection measures, the supervisor must be notified.

A. Scaffolds are not being used:

_____
_____
_____

B. Ladders are not being used:

_____
_____
_____

C. Vehicle-mounted platforms are not being used because:

_____
_____
_____

D. Crane-suspended personnel platforms are not being used because:

_____
_____
_____

Specific Areas Covered by this Fall Protection Plan

This fall protection plan addresses the use of other than conventional fall protection at specific areas on the project. This section identifies specific activities and areas that require other means of fall protection. Each of these areas is identified as a Controlled Access Zone (CAZ). These areas include:

A. Connecting activity (point of erection) at:

_____
_____
_____

B. Leading edge work at:

_____
_____
_____

C. Unprotected side or edges at:

_____
_____
_____

FORM 26.3

Fall Protection Systems to be Used on this Project

A. CAZ system
Each CAZ identified in Section III must comply with the following provisions:

- The CAZ will be defined by a control line or another means that restricts access.
- The control line will extend along the entire length of the unprotected or leading edge and must be approximately parallel to the unprotected or leading edge.
- The control line will be connected on each side to a guardrail system or wall.
- Control lines will consist of ropes, wires or other materials with a breaking strength of at least 200 lb and be rigged and supported on stanchions so the line is between 39 and 45 in. above the walking/working surface.
- Each control line will be flagged or otherwise clearly marked with high-visibility at intervals of 6 ft. or less.

Where CAZ's are identified, only those employees necessary to safely accomplish the job will be assigned. The maximum number of workers allowed inside any one CAZ is _____. The following trained employees are allowed to enter the CAZ's and work without the use of conventional fall protection.

_____   _____   _____

_____   _____   _____

The CAZ's workers will be identified by: _____

B. Safety Monitoring System

Where not other fall protection system can be used, work will be performed using a safety monitoring system. Only those employees necessary to safely accomplish the job will be assigned. The maximum number of workers to be monitored by one safety monitor is _____.

The safety monitoring system assigns a competent person who is responsible for recognizing and warning employees of fall hazards. The Safety Monitor must be on the same walking/working surface and within visual contact with the monitored employees. He or she must also be close enough to speak with the employees. The Safety Monitor must not be assigned other duties that could limit his or her ability to monitor the work area. The duties of the Safety Monitor include:

- Warning employees when they are approaching the open edge unsafely
- Warning employees if there is a dangerous situation developing that they cannot readily see.
- Making the CAZ workers aware they are in a dangerous area.
- Warning employees if they appear to be unaware of a fall hazard or are acting unsafely.
- Stopping the work process if communication with the CAZ workers is disrupted.

Safety Monitor:_____

FORM 26.4

The Safety Monitor will be identified by: _____

CAZ workers will be constantly under the control of the Safety Monitor for fall protection and are directed to stay a minimum of 6 ft. from the edge.

The safety monitoring system will not be used when the wind is strong enough to cause loads with large surface areas to swing out of radius, or to cause loss of control of the load, or when whether conditions cause the walking/working surfaces to become icy or slippery.

Employees exposed to falls to 6 ft. or more to lower levels, who are not actively engaged in leading edge work or connecting activity, such as welding, bolting, cutting, bracing, guying, patching, painting or other operations, and who are working less than 6 ft. from an unprotected edge will be tied off at all times, or guardrails will be installed. Employees engaged in these activities who are more than 6 ft. from an unprotected edge as defined by the control lines do not require fall protection, but a warning line or control lines must be erected to remind employees they are approaching an area where fall protection is required.

Training

Only individuals with the appropriate experience, skills and training will be identified as CAZ workers. All employees who will be working as CAZ workers under the CAZ or safety monitoring system will have been trained by a competent person and instructed in the following areas:

• Recognition of the fall hazards in the work area.
• Avoidance of fall hazards using work practices established for employees.
• Recognition of unsafe practices or working conditions that could lead to a fall.
• The function and use of Controlled Access Zones and safety monitoring systems.
• The intended construction sequence or erection plan.
• The specific requirements of this fall protection plan and any approved changes.

Safety Meetings

A daily safety meeting will take place before starting work covered by this Fall Protection Plan. This safety meeting will be conducted by _____ and supervisors in charge of this phase of work and must be attended by all CAZ workers, members of the erection crew and crane crew, and the supervisors of other affected employers. During the preshift safety meeting, erection procedures and sequences pertinent to the day's work will be discussed thoroughly and safety practices being used throughout the day will be specified. All personnel will be informed that the CAZ's are off limits to all personnel other than those CAZ worker specifically trained to work in that area.

Accident Investigations

In the event that an employee fall or a related, serious incident occurs, this plan must be reviewed to determine whether additional practices, procedures or training needs to be implemented to prevent similar types of falls or incidents from occurring.

Changes to Plan

Any changes to this plan will be approved in writing by _____. This plan will be reviewed by on a frequent basis as the job progresses to decide if additional practices, procedures or training needs to be implemented by the competent person to improve or provide additional fall protection. A copy of this plan and a copy of all approved changes must be maintained on the jobsite at all times.

FORM 27

Rotorcraft Operations Test

Name (print)	**SSN**	**Date**

Name (sign)	**Project**

20 questions total. Each question has only one answer.

1.	The crew must conduct a daily job task analysis before the start of operations each day and if the scope of the work changes during the day.	T	F
2.	Communication procedures must be established before the beginning of any rotorcraft operation.	T	F
3.	Unauthorized persons may stay at least 20 ft. from a rotorcraft.	T	F
4.	Employees shall receive instruction on the safest route for approach to and departure from the aircraft.	T	F
5.	Personnel must receive permission from the pilot before approaching or exiting the rotorcraft.	T	F
6.	Only MSA approved headgear with a chin strap attachment may be worn by employees working around and on an operating rotorcraft.	T	F
7.	Objects to be located on an operating rotorcraft may be carried at shoulder level or below.	T	F
8.	At not time may any objects be thrown in the vicinity of an operating rotorcraft.	T	F
9.	Special precaution must be taken when approaching an operating rotorcraft that is positioned on uneven or sloped terrain.	T	F
10.	The static charge on a suspended load must be dissipated by allowing the load to touch the ground or structure or by use of a shunt before employees touch the load.	T	F
11.	Employees need only to wear a body belt with a positioning strap attached to the rotorcraft when riding on the skid of the rotorcraft.	T	F
12.	Each employee must inspect climbing equipment, fall protection equipment, and PPE before each use.	T	F
13.	Employees must transfer to and from the rotorcraft as smoothly as possible and only with the permission of the pilot.	T	F
14.	Employees do not need to attach a shunt when transferring from the rotorcraft to a steel tower.	T	F
15.	Before the start of each rotorcraft operation, the crew must develop an emergency action plan to determine emergency phone number, location of emergency services, and a procedure for caring for injured employees.	T	F
16.	At least two employees per crew must hold a current first aid and CPR card.	T	F
17.	The words "This is an Emergency" may be used for any emergency situation including life threatening injuries and situations.	T	F
18.	The pilot does not need to make a radio check with each affected employee before the start of operations.	T	F
19.	All employees working on the ground during rotorcraft operations must wear safety glasses with side shields.	T	F
20.	Safety of life must outweigh all other considerations	T	F

FORM 28

Rotorcraft Operations
Skills Proficiency Sheet

Name (print)	SSN	Date	Project Number

Name (sign)	Instructor Name (print)	Instrucrtor Name (sign)

Skills - General Knowledge	Passed
Demonstrate understanding of rotorcraft operator's safety requirements	
Demonstrate knowledge of noncompliance disciplinary action procedures	
Demonstrate knowledge of procedure to report an unsafe act or condition	
Demonstrate safe approach to a rotorcraft (skid width)	
Demonstrate safe positioning while awaiting pickup by rotorcraft	
Demonstrate understanding and use of verbal communication	
Demonstrate correct hand signals listed on Figure 1 Nonverbal Communication Hand Signal Chart	
Demonstrate proper Go and No-Go head signals while keeping eye contact with pilot	
Demonstrate knowledge and safe and restricted access zones	
Demonstrate understanding of housekeeping requirements at landing zone	
Demonstrate understanding of requirements for equipment located at the landing zone	
Demonstrate understanding of required PPE	
Demonstrate inspection of PPE	
Demonstrate knowledge of the location of emergency equipment (first aid kit, fire extinguisher)	
Demonstrate knowledge of location of fuel shut down for rotorcraft	
Demonstrate knowledge of location of main power shutoff switch for rotorcraft	
Demonstrate knowledge of location of Emergency Locator Device (ELD) in rotorcraft	
Demonstrate knowledge of terrain as it affects ingress and egress to the rotorcraft	
Demonstrate knowledge of life saving and rescue techniques	

Skills - Site Specific	Passed
Demonstrate safe techniques for attaching and removing suspended loads	
Demonstrate knowledge of when to use shunts	
Demonstrate proper use of shunts	
Demonstrate knowledge of terrain is it affects ingress and egress to the rotorcraft	
Demonstrate inspection of fall protection/climbing equipment before use	
Demonstrate safe positioning while awaiting pickup by rotorcraft	
Demonstrate understanding of fall protection attachment point ratings	
Demonstrate proper attachment of fall protection equipment to rotorcraft	
Demonstrate smooth transfer from rotorcraft to structure	
Demonstrate smooth transfer from structure to structure	
Demonstrate safe approach to rotorcraft with material	
Demonstrate method to load, unload and secure material on and off rotorcraft	
Demonstrate knowledge of site specific PPE requirements	